Advances in the Biosciences 8

Advances in the Biosciences

The Schering Symposia and Workshop Conferences are conducted and sponsored by Schering AG, 1 Berlin 65, Müllerstraße 170

Advances in the Biosciences 8

Workshop on Mechanisms and Prospects of Genetic Exchange
Berlin, December 11 to 13, 1971

Editor: Gerhard Raspé

Associate Editor: S. Bernhard

Editorial Board: Peter Hans Hofschneider
Hilary Koprowski

Pergamon Press · Vieweg

Oxford · Edinburgh · New York · Toronto · Sidney · Braunschweig

Pergamon Press Ltd., Headington Hill Hall, Oxford
Pergamon Press (Scotland) Ltd., 2 & 3 Teviot Place, Edinburgh 1
Pergamon Press Inc., Maxwell House, Fairview Park, Elmsford, New York 10523
Pergamon of Canada Ltd., 207 Queen's Quay West, Toronto 1
Pergamon Press (Aust.) Pty. Ltd., 19a Boundary Street, Rushcutters Bay,
N. S. W. 2011, Australia
Vieweg + Sohn GmbH., Burgplatz 1, Braunschweig

Editorial Assistance: *Gerd Grünewald*

SBN 0 08 0**17290** 3 (Pergamon)

ISBN 3 528 0**7690** 9 (Vieweg)

1972

Library of Congress Catalog Card No. 74–76593

Set by Friedr. Vieweg + Sohn GmbH, Braunschweig
Printed by E. Hunold, Braunschweig
Bookbinder: W. Langelüddecke, Braunschweig
Cover design: Herbert W. Kapitzki, Frankfurt
Printed in Germany-West

Table of Contents

II. Genetic Exchange at the Cellular and Organelle Level

III. Present and Future Application to Medical Practice

Opening Address

Heinz Gibian

Forschungsleitung Pharma der Schering AG, Berlin, Germany

Dr. *Raspé,* who has unfortunately fallen ill, has asked me to extend a very warm welcome to you on behalf of Schering AG in the opening of this eighth workshop.

When the very first conference in this series was held in 1967 on the initiative of Dr. *Raspé*, no one could have foreseen that we would meet today in a workshop on "Mechanisms and Prospects of Genetic Exchange". The reason for this becomes immediately obvious when I mention the subject of our first meeting "Symposium on Endocrinology". That was, of course, a subject which was very closely connected with our own research; the same was true for the following workshops, at which we had our own contributions to make. This time, however, we are merely listeners.

Despite this, the seeds had already been sown for the selection of today's subject. There are areas of science in the process of an exciting upsurge, which keep us extremely alert intellectually, but which, nevertheless, have no place to call their own since, in the choice of methods, they are interdisciplinary. Perhaps, to quote *Günter Stent*, it is more a case of the representatives of such areas who see themselves in a "romantic phase", to which institutionalization would put an end.

However that may be, from time to time it is necessary for the development of such areas that the *right* people from the various disciplines come together at the *right* place. In the New World this requirement is met admirably by the Cold Spring Harbor meetings, Gordon conferences, and similar meetings. Here, in the Old World, much is left to be desired in this direction; the Schering Workshops are intended to be a contribution to this problem.

New developments should be encouraged in the discussion between the disciplines. Outsiders should be informed about them for it is very possible that they, in particular, will have unexpected possibilities for future research. We attach the greatest importance to the presence of young colleagues here. Every invited speaker was given the opportunity of naming, as a participant, a younger scientist from whatever country he wished. We are trying to realize this principle for each workshop. As with everything else concerned with this workshop, there are no hard and fast rules; this also applied to the selection of topics.

The fascinating subject of this conference is the result of a discussion between *Hilary Koprowski* and *Peter Hans Hofschneider.* We agreed to their brilliant proposal very quickly. I should think that even the initiators of this workshop were somewhat surprised – and a bit worried – about this speed as it meant a special responsibility for them in view of the short time available for preparations. The same is true for *Silke Bernhard*, whom I should like to thank very much for her magnificent efforts in organizing this workshop.

My first wish for the next few days, during which molecularbiologists, geneticists, cell biologists, biophysicists, and immunologists will be communicating with each other, is that new bridges will be built during the official discussions, over a glass of wine, or on a visit to one of the charming Berlin restaurants. My second wish will then become true all by itself: We will obtain a competent intermediate balance within a field which is of interest not only to scientists, but also to the public.

Since the memorable conference "Man and His Future" in London, speculations, fears, and utopia about a genetic dictatorship of mankind through science have not ceased. It would be in everyone's interest if the basis to the facts could be presented. With this in mind, my colleagues and I are very happy to be your hosts.

I should like to thank all of you very cordially for having come here; my special gratitude goes to the lecturers without whom this workshop would not have been possible.

Introduction to Session I

Robert L. Sinsheimer

California Institute of Technology, Division of Biology, Pasadena, California

When one starts to think seriously about the possibilities for the development of genetic therapies or genetic change, one quickly confronts a set of problems that has to do with the insertion or the deletion or the transposition of pieces of genetic material – DNA pieces which might range from a few nucleotides in length to several genes – into an existing genome. Now in a formal sense, models for the solution of such questions clearly exist in nature in the processes of genetic recombination, of provirus and episomal insertion, of genetic inversion, in heterochromatic condensation, and in the postulated gene expansion and contraction. As yet, of course, we know very little of the molecular events involved in such action.

At least two general mechanisms of genetic recombination appear to have been developed. One seems to require at least a fair degree of genetic homology between the recombining DNAs; it seems to be favoured by the presence of nicks in DNA, and to be mediated in bacteria at least by the so-called rec genes. The second mechanism for which provirus insertion is a good model does not appear to require any extensive homology between the insertant and the insertee, and appears to be mediated by enzymes specifically evolved with recognition capacities for that purpose.

Variants of these mechanisms can certainly be imagined and most likely exist. And, indeed, wholly distinct processes of recombination may await discovery. I am not at all sure that the events that lead to recombination during meiosis – the formation of the synaptinemal complex and so on – bear any simple relationship to those events that I have already mentioned.

At the molecular level, these events clearly involve processes of molecular recognition of undefined extent, processes of chain scission, chain extension, and chain ligation for which we have putative model catalysts. They also very likely involve processes of chain initiation and chain termination which are still obscure. The papers that will be presented today will be concerned with our understanding of the molecular and genetic events underlying processes of gene interaction and gene exchange.

Enzymology of Genetic Recombination

Charles M. Radding and Era Cassuto

Departments of Medicine and Molecular Biophysics and Biochemistry, Yale University School of Medicine, New Haven, Conn. 06510, USA

Summary: Studies of genetic recombination in prokaryotes have shown (1) that recombination occurs by breakage and reunion of DNA, sometimes, but not always, associated with new DNA synthesis, and (2) that the parental contributions to a recombinant molecule are commonly joined by a short heteroduplex or hybrid region. In the past few years, some of the enzymes involved in recombination in prokaryotes have been identified, such as the exonuclease made by bacteriophage λ. Recent studies of λ exonuclease make it possible to rationalize most of the properties of the enzyme in terms of its role in producing a perfect heteroduplex joint between homologous molecules of DNA. λ exonuclease cleaves 5′ mononucleotides from the 5′ end of native DNA in a processive fashion, extensively degrading any molecule of DNA before detaching and attacking another molecule of substrate. The latter property suggests that some control prevents the enzyme from playing an exclusively degradative role. 5′ phosphoryl termini located at gaps in one strand of duplex DNA are resistant to the enzyme. Although 5′ phosphoryl termini at nicks are even more resistant, the enzyme appears to bind weakly at such sites. The significance of these properties may be seen in the enzyme's action at the site of a redundant single stranded branch, such as one might expect to find at a joint between two fragments of DNA. A redundant strand is assimilated into the helix, behind λ exonuclease, as the enzyme processively degrades the homologous helical strand. The enzyme recognizes the presence of the redundant strand both for initiation and termination of hydrolysis. Removal of the redundant single strand by the prior action of exonuclease I blocks the action of λ exonuclease on the helical strand. Moreover when a redundant strand has been completely assimilated through the action of λ exonuclease, the enzyme stops at the precise point which permits the interrupted polynucleotide strand to be sealed by polynucleotide ligase. The sequential action of λ exonuclease and polynucleotide ligase on redundant joint molecules of λ DNA produces intact polynucleotide strands that are biologically active. Several models have been suggested to relate the assimilation of single strands to the genetic recombination of λ and possibly to recombination in other systems as well. Molecules of DNA with double-stranded branches have also been synthesized to test one of the

---

Manuscript received: 13 December 1971

models. The models suggest that λ exonuclease may catalyze a concerted reaction that (1) exposes complementary nucleotide sequences, (2) forms or extends the heteroduplex region, and (3) eliminates redundant branches, precisely restoring a duplex structure that can be sealed covalently by polynucleotide ligase. The λ enzyme, and similar exonucleases, might drive otherwise reversible interactions of a single strand with a recipient duplex, including certain kinds of interactions between two molecules of double-stranded DNA.

Introduction

Genetic recombination is the set of processes that leads to new linkage relationships. By greatly accelerating genetic permutations, recombination has probably played an important role in evolution. For the present, recombination is of interest as an aspect of DNA metabolism in which complicated and poorly understood relationships exist among replication, repair, and recombination. In the future, our understanding of genetic recombination may influence our ability to deal with such medically important phenomena as resistance transfer factors, carcinogenesis, and gene therapy.

There are three possible objects to study in recombination: the progeny, the intermediates, or the enzymes. In the past few years, explorations of recombination deficient mutants in prokaryotes have strongly implicated certain enzymes in recombination, thus making it possible to approach the enzymology with greater assurance [9, 33]. One such enzyme, the exonuclease made by bacteriophage λ, is the product of a gene that is essential for recombination of the phage [25, 31, 34]. Recent studies of λ exonuclease make it possible to rationalize most of the properties of the enzyme in terms of its role in perfectly splicing homologous portions of DNA. These experiments will be summarized after a brief digression on the features of recombination in prokaryotes that are particularly relevant to an enzymic analysis.

Types

Most recombination can occur more or less anywhere along the length of two homologous genophores, and is called *general recombination.* A special type of recombination that occurs only at a restricted number of specific sites is called *site specific recombination* [14]. Still a third type, sometimes called unusual or *illegitimate recombination* [10], occurs between genophores which may have localized regions of homology, but which are carrying distinctly different genetic messages. In this paper we shall deal only with general recombination.

Reciprocity

An individual act of recombination between distant markers *(AM* × *am)* may give rise to reciprocal products (both *Am* and *aM*) or nonreciprocal products (either *Am* or *aM).* Different recombination systems appear to be either largely reciprocal or largely nonreciprocal [4, 23, 33].

Material exchange and DNA synthesis

Studies of the progeny of recombination show that the parents usually contribute most of the material of which the recombinant molecule is made. Newly synthesized DNA is detectable in some recombinant molecules but not in others [22, 23, 36, 37, 45]. Three possible relationships between DNA synthesis and recombination may be imagined: (1) New DNA synthesis repairs gaps in certain intermediate structures. (2) Replication, by producing interruptions and branches in DNA, provides the substrate for recombination which then proceeds by mechanisms that are independent of replication. (3) New DNA synthesis is an intrinsic part of the mechanism of recombination (see Discussion).

Precision

Nucleotides are rarely gained or lost in recombination. In a cross represented by *ABCD* × *abcd,* all four letters are represented in recombinant progeny, for example *ABcd.* Repetitions or deletions such as *ABBcd* or *Acd* are rare. This precision is accomplished in general recombination through the pairing of homologous bases, but the details of that process are obscure. Does recognition of the sequence homology of two molecules of DNA occur before or after the breakage of one or both strands of each parent? The two general possibilities might be represented as *break and join* vs. *join and break.* The work of *Alberts* et al. has provided new insights into the biological mechanisms for making and breaking hydrogen bonds [1].

Heterozygosis

In prokaryotes, the parental contributions of DNA are usually joined by a short region in which each parent contributes one strand of this duplex DNA. If mutations are present in the heteroduplex region, heterozygosity may be observed [23]. Heterozygosity based on a heteroduplex junction between the two parental arms of a recombinant molecule is intimately related to the basic mechanism of recombination and to its precision. Studies of intermediates in recombination have revealed a stage at which the parental contributions are united only by hydrogen bonds which are presumably located in the heteroduplex region [17, 39]. Such intermediates are called joint molecules [39]. An unanticipated way in which heteroduplex regions may be generated is revealed by the studies on λ exonuclease (see below).

Multiple exchanges

Genetic exchanges tend to be clustered; the site of an exchange is frequently the site of nearby exchanges [33]. This may turn out to be the most difficult property of recombination to analyze by an enzymologic approach, unless clustering is shown to result from the excision and repair of mismatched bases in the heteroduplex junction [30, 40, 45].

Properties of λ exonuclease

To find out what λ exonuclease does in genetic recombination, we began to reevaluate the properties of the enzyme with particular attention to its possible action at internal sites in DNA such as nicks, gaps, and branches. Some of the elementary properties of λ exonuclease are outlined in Fig. 1. The observation that the enzyme acts processively, extensively degrading any molecule of DNA on which it starts, led to the inference that some control exists which prevents the enzyme from playing an exclusively degradative role [6, 26]. The specificity of the enzyme for native DNA suggested that the enzyme did not act by degrading redundant single-stranded branches. In spite of the specificity of λ exonuclease for native DNA, binding of the enzyme to denatured DNA has also been observed [28]. Examination

1. (a) highly specific for native DNA [20],
   (b) but binds to denatured DNA [28]
2. processively cleaves 5′ mononucleotides from the 5′ phosphoryl end of native DNA [6, 20]
3. (a) does not initiate digestion at a 5′ phosphoryl terminus located at a nick [6, 21],
   (b) but binds to nicks [26]
4. does not readily initiate digestion at a 5′ phosphoryl terminus located at a gap [6, 12, 21, 27]

Fig. 1. Properties of λ exonuclease.

of the action of λ exonuclease at internal sites, first nicks and then gaps, gave negative results [6], but another apparent paradox was noted. Although the enzyme shows no tendency to act at nicks, it appears to bind at such sites [26]. On the basis of these properties, we made the hypothesis shown in Fig. 2 [8]. According to this notion, λ exonuclease degrades native DNA at the site of a single-stranded branch (called a redundant joint) making way for assimilation of the branch into the helix. When the redundant strand has been completely assimilated, further digestion by the enzyme stops, leaving a nick that can be sealed by polynucleotide ligase. The hypothesis rationalizes the properties subsumed under 1–3 in Fig. 1.

Single-strand assimilation

Two different substrates have been synthesized to test the action of λ exonuclease on redundant joints (Figs. 3, 4). In each case, the normal ends of λ DNA were protected and the only 5′ terminus that was potentially available to the enzyme was the one located at a joint in the middle of the DNA molecule. Protection of the ends was achieved in one of two ways: (1) The complementary ends of λ DNA were annealed to form either circles or polymers. (2) The 5′ single-stranded termini of λ DNA were dephosphorylated by *E. coli* alkaline phosphatase. This treatment reduced

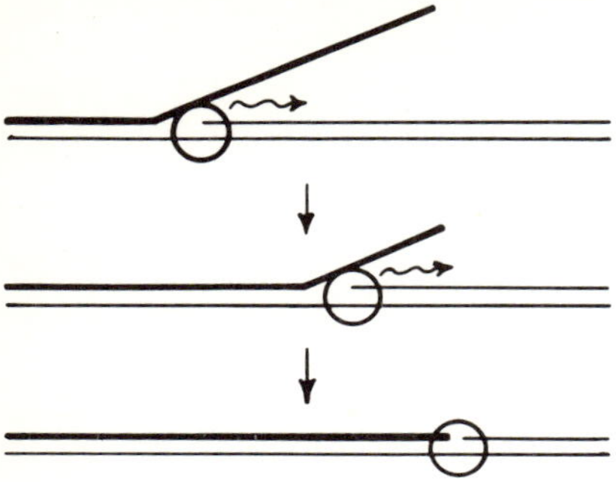

Fig. 2
Single-strand assimilation.
A homologous single strand at a redundant joint is assimilated into the helix by λ exonuclease (represented by circles) which digests the 5′ terminated helical strand until assimilation is complete. The interruption left in one strand of the helix can be closed by polynucleotide ligase [7, 8].

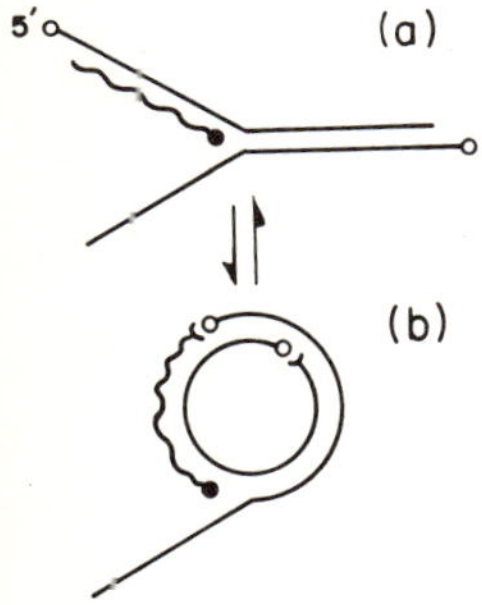

Fig. 3
Diagram of redundant joint molecules of λ DNA.
(a) Linear form. The original 5′ termini (○) were dephosphorylated to protect them from attack by λ exonuclease (see below).
(b) Circular form with homologous single-stranded ends annealed in a Hershey joint which is resistant to attack by λ exonuclease. The two structures are interconvertible by annealing or melting.
(——) ^{3}H labeled DNA; (∿∿) ^{32}P labeled DNA. The dot (•) indicates the only 5′ end available to λ exonuclease in either substrate.

Dephosphorylation of the original 5′ termini of form (a) above was done prior to preparation of the redundant joint substrate (cf. [8]). The reaction mixture contained, in a total volume of 0.35 ml, 0.2 μ moles of DNA, 30 μ moles of Tris-HCl, pH 8, 2.5 μ moles of $MgCl_2$, 13 μ moles of NaCl and 1 unit of *E. coli* alkaline phosphatase *(Worthington)*, purified according to *Weiss* et al. [44]. After incubation for 30 min at 37 °C, the phosphatase was eliminated by phenol extraction, and intact λ DNA strands were separated from broken strands by sedimentation in alkaline sucrose gradients. The susceptibility of λ DNA to digestion by an excess of λ exonuclease was reduced at least tenfold by dephosphorylation (e.g., Fig. 5 B 1).

the action of λ exonuclease to a few percent of its normal activity on λ DNA (e.g., Fig. 5, line B 1) because the enzyme acts specifically on a 5′ phosphoryl terminus of native DNA [20].

λ exonuclease specifically degrades the 5′ terminated helical strand at a redundant joint and yields a product that is no longer susceptible to partial digestion by exonuclease I. (Fig. 5, Part A). In a complementary fashion, exonuclease I specifically degrades the 3′ terminated redundant single strand, and yields a product that is no longer susceptible to partial digestion by λ exonuclease (Fig. 5, Part B). These experiments show directly that the redundant single strand plays a role in the initiation of hydrolysis by λ exonuclease at a joint, and that the redundant strand is assimilated after hydrolysis of the helical strand.

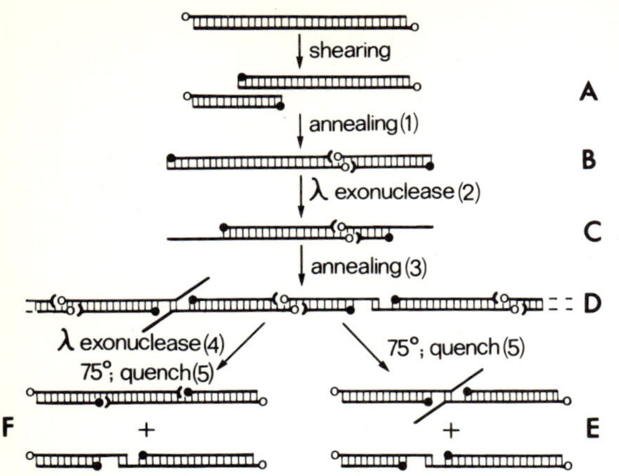

Fig. 4

Diagram of the preparation of redundant joint molecules of λ DNA with redundancies of nonuniform length [7]. The normal 5′ termini are indicated by open circles, the 5′ termini exposed by shearing are indicated by closed circles. The ball and socket symbols signify complete joints: —(○— Hershey joint; —(●— internal joint. In preparation D, the normal ends of λ DNA were protected by polymerization (cf. Fig. 3b). The diagrams of preparations E and F illustrate the types of joint molecules: (a) *redundant,* with extra DNA, (b) *incomplete,* with nucleotides missing on either side of the heteroduplex region, and (c) *complete,* with no missing nucleotides. Preparation F yielded intact polynucleotide chains when treated with polynucleotide ligase; preparation E did not.

The scheme for the synthesis of a second kind of redundant joint molecule is shown in Fig. 4. Joint molecules produced in this way should include some that have redundant single strands of variable lengths. The sequential action of λ exonuclease and polynucleotide ligase produced intact polynucleotide chains as determined by sedimentation in alkaline sucrose. Assimilation of redundant single strands was further reflected in a decrease in the number of DNA molecules which had some single-stranded character as measured by chromatography on BN cellulose. Covalent closure after assimilation was further reflected in the biological activity of molecules that survived heating under conditions which destroy the activity of joint molecules ([7], see also legend to Fig. 4). These experiments represent the best evidence that λ exonuclease stops precisely when the redundant strand is completely assimilated, leaving an interruption that can be closed by polynucleotide ligase.

Expt.	Treatment	% Acid soluble DNA	
		^{3}H	^{32}P
A	1) exo I	20	0
	2) λ exo	5	79
	3) λ exo, then exo I	7	72
B	1) λ exo	3	75
	2) exo I	19	0
	3) exo I, then λ exo	25	0

Fig. 5. Sequential treatment of redundant joints with λ exonuclease and exonuclease I

All incubations were at 37 °C

A) Incubation mixture No. 1 containing in 0.15 ml: 5.7 nmoles of redundant joint substrate, 10 μmoles of glycine buffer (pH 9.6), 1 μmole of $MgCl_2$, 0.2 μmole of 2-mercaptoethanol, and

3 units of exonuclease I [19], was incubated for 60 minutes. Incubation mixture No. 2 contained in 0.3 ml: 11 nmoles of substrate, 20 μmoles of glycine buffer, 1.5 μmoles of $MgCl_2$, 3 nmoles of tRNA and 5 units of λ exonuclease [24]. After incubation for 60 minutes, an aliquot of 0.15 ml was withdrawn and acid soluble radioactivity was measured. The remaining 0.15 ml (incubation mixture No. 3) was heated at 75 °C for 10 minutes to inactivate λ exonuclease, and 0.7 μmole of $MgCl_2$, 0.2 μmole of 2-mercaptoethanol, and 3 units of exonuclease I were added. The mixture was incubated for 60 minutes. Control experiments showed that the presence of λ exonuclease did not inhibit the action of exonuclease I on denatured DNA, and that denatured DNA heated in the presence of λ exonuclease was unchanged in its susceptibility to exonuclease I. The redundant joint substrate used in these experiments was of type b shown in Fig. 3. The above data are reproduced from *Cassuto* and *Radding* [8] for comparison with part B.

B) Incubation mixture No. 1 containing in 0.15 ml: 4 nmoles of substrate, 2 μmoles of Tris buffer (pH 7.5), 0.1 μmole of $MgCl_2$, 3 nmoles of tRNA, and 3 units of λ exonuclease was incubated for 60 minutes. Incubation mixture No. 2 contained in 0.3 ml: 7.5 nmoles of redundant joint substrate, 20 μmoles of glycine buffer (pH 9.6), 2 μmoles of $MgCl_2$, 0.5 μmole of 2-mercaptoethanol, and 15 units of exonuclease I. After incubation for 30 minutes, an aliquot of 0.15 ml was withdrawn and acid soluble radioactivity measured. To the remaining 0.15 ml (incubation mixture No. 3), 0.37 μmole of EDTA (pH 7) was added to bring the free magnesium concentration to the optimum value for λ exonuclease, and the mixture was incubated for 60 minutes in the presence of 3 units of λ exonuclease. Control experiments showed that adjustment of the magnesium concentration with EDTA allowed full activity for λ exonuclease on native DNA from phage P22, and that exonuclease I did not inhibit the action of λ exonuclease on native DNA. The redundant joint substrate used in these experiments was of the type diagramed in Fig. 3a.

Models

The following models suggest different ways in which λ exonuclease might participate in genetic recombination and also illustrate the potentially broader significance of single-strand assimilation.

Figure 6a depicts perhaps the simplest way in which λ exonuclease might act. This model indicates the several roles that λ exonuclease may play in a concerted fashion: (1) The exposure of homologous sequences, (2) the formation and extension of the heteroduplex region, and (3) the precise elimination of redundant single strands by assimilation. The synthesis of redundant joint molecules, then complete joint molecules, and finally closure of the latter by polynucleotide ligase (cf. Fig. 4) represents a simulation in vitro of this model.

Figure 6b illustrates how a single-stranded fragment might be inserted into double-stranded DNA. This reaction, which produces an insertion heteroduplex, might be one source of multiple exchanges (see Introduction). A similar process might occur in bacterial transformation in which insertion heteroduplexes result from the recombination of donor and recipient DNA [15].

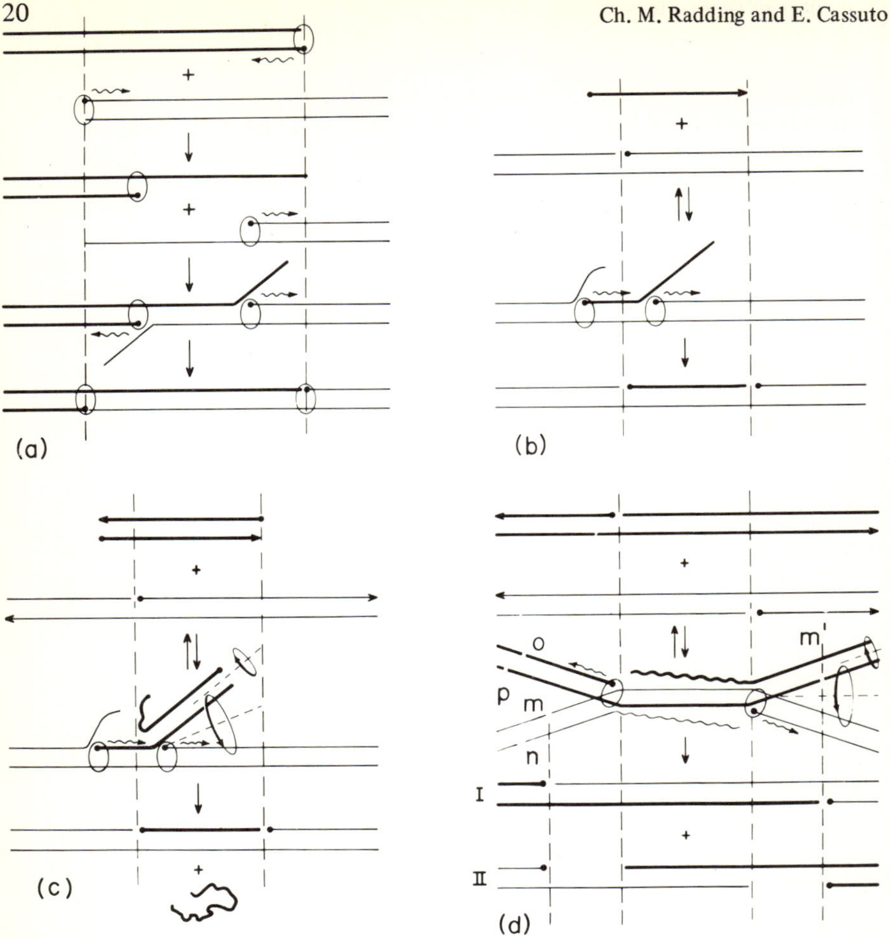

Fig. 6. Hypothetical variants of single-strand assimilation in genetic recombination.

(a) Reunion of homologous double-stranded fragments. (b) Interaction of a single-strand fragment with duplex DNA. (c) and (d) Interaction of two duplexes.

Ellipses in the diagrams represent enzymes. In (d) distal nicks at *m* an *m'* would terminate single-strand assimilation (see [5]) leaving a complete joint molecule (I), and possibly an incomplete joint molecule (II). The outcome, as pictured, is the reciprocal recombination of distant markers, and the nonreciprocal recombination of close markers. Nicks at *n* or *o* would probably have no effect on the primary recombinant (I), but would destroy (II). A nick at *p* would destroy (II) and probably produce an insertion heteroduplex by switching the arm that is conserved in the complete joint recombinant. In the sequence pictured, the complex was initiated by nicks in strands of opposite polarity. The relative orientation of the nicks affects the orientation of 3′ and 5′ ends that might initially be presented to an enzyme like λ exonuclease that acts only from the 5′ ends. It is likely that branch migration [18] over short distances makes 3′ and 5′ terminated structures interchangeable with respect to enzyme specificity. A complex initiated by nicks in strands of the same polarity would be expected to yield an insertion heteroduplex in a fashion very similar to that shown in part (c).

By extension of diagram b, Fig. 6c shows that a single strand might be extracted from a duplex donor and simultaneously assimilated into a duplex recipient. The required rotations of helical arms around their own axes and around a common axis are indicated in c and d. The energy requirement of the ATP-dependent exonucleases might be related to such rotational motions.

The model in part d is an extension of c and was stimulated by the model of *Broker* and *Lehman* [5]. Nicks in DNA might be the basis for trial strand exchanges catalyzed perhaps by enzymes like the gene 32 protein of phage T4 [1]. The formation of a biparental complex should be reversible (see also parts b and c), but the intervention of an exonuclease as pictured would drive the reaction irreversibly toward recombination. In the particular sequence shown (Fig. 6d), new DNA synthesis would accompany the formation of one of the products. The genetic outcome would be reciprocal recombination of distant markers and nonreciprocal recombination of markers close to the region of crossover. (The regions of new DNA synthesis would be associated with 3 : 1 segregation ratios.) Other genetic possibilities are indicated in the legend to Fig. 6.

The synthesis of DNA molecules with double-stranded branches

An effort to construct defined substrates with double-stranded branches was stimulated by the models shown in Fig. 6c and d. Our aim was to produce the structure shown in Fig. 7a. Such a substrate would test whether λ exonuclease can operate on a double-stranded branch and would provide an internal control since

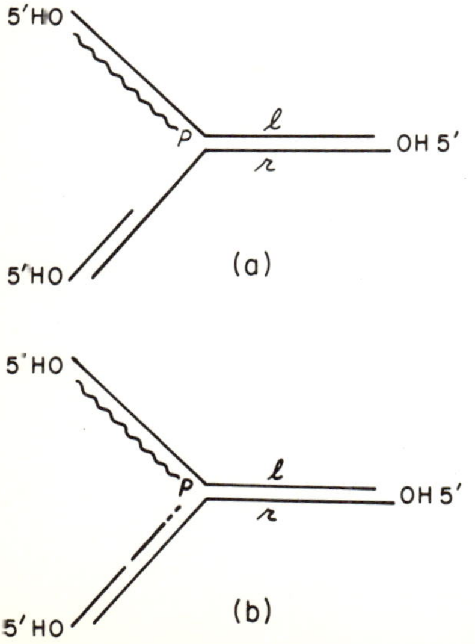

Fig. 7

Diagrams of double-stranded branched molecules of λ DNA.

(a) Substrate designed to test the action of λ exonuclease on double-stranded branches.

(b) Proposed structure of branched molecule actually synthesized.

The isolated l_L piece (see Fig. 8) had approximately 1 break per molecule. The steps in the synthesis of this material are described in the text. (∿∿) ^{32}P labeled DNA; (——) ^{3}H labeled DNA.

the minimum expectation is that λ exonuclease would digest 50 % of the specifically labeled ^{32}P strand (compare Figs. 3a and 7a). For technical reasons, we have so far succeeded in making a less defined substrate, the probable structure of which is illustrated in Fig. 7b. The scheme for the synthesis of double-stranded branches is

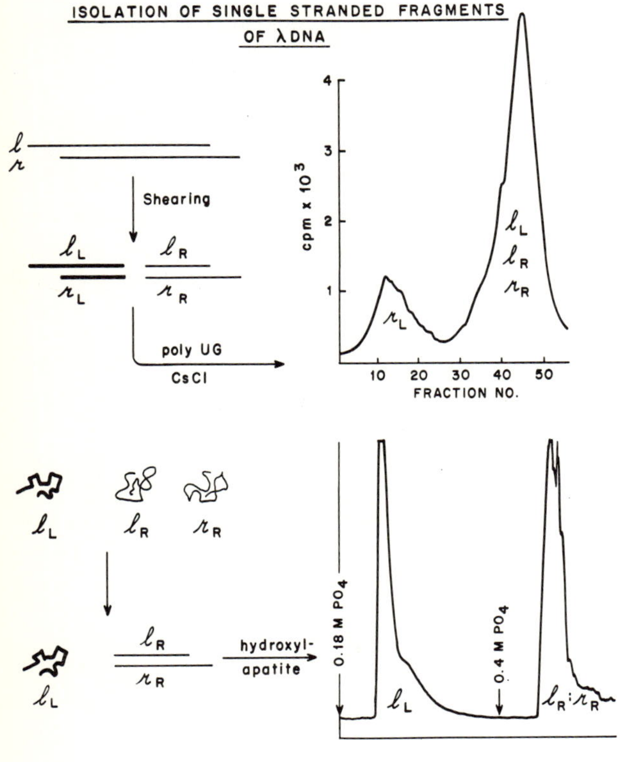

Fig. 8. Isolation of specific fragments of λ DNA.

The methods for shearing and isolation of the r_L fragment have been described previously [8, 16]. For the isolation of the l_L fragment, the fractions from the CsCl gradient containing l_L, l_R, and r_R were pooled and dialyzed against SSC (0.15 M NaCl, 0.015 M sodium citrate). The poly UG was eliminated by treatment with 10 units/ml of T_1 RNase (Calbiochem) [38] at 65 °C and pH 8 for 10 minutes. The l_R and r_R fragments were annealed by heating at 75 °C for 2 minutes and cooling slowly in 0.6 M NaCl, 0.05 M phosphate buffer, pH 7. The single-stranded l_L fragment was then separated from the native $l_R : r_R$ fragment by chromatography on hydroxylapatite; the l_L fragment was eluted with 0.18 M phosphate buffer, pH 7, the native fragment with 0.4 M phosphate buffer, pH 7. The elution diagram is an automatic recording of percent transmission of UV light and is not directly proportional to the amount of nucleic acid. The abscissa is an arbitrary time scale.

a simple extention of that for making the redundant joint molecules shown in Fig. 3a [8]. It involves the isolation of the left half of the *l* strand (l_L) of λ DNA, which was accomplished as indicated in Fig. 8. The left half of the *r* strand (r_L) was isolated from a ^{32}P labeled preparation of λ DNA as described before [8, 16]. $^{32}P\ r_L$ was annealed with ^3Hl strand, ^3Hl_L was annealed with ^3Hr strand, and the products of these reactions were then annealed to each other. Examination of the final product by sedimentation in neutral sucrose revealed a major species of DNA, the molecular weight of which was approximately 1.5 times that of intact λ DNA (Fig. 9b). The molar ratios of ^3H labeled and ^{32}P labeled DNA in the peak of a neutral sucrose gradient (Fig. 9a) were 4.5, 4.0, 4.5, 6.5, and 4.5. Consistent with the structure

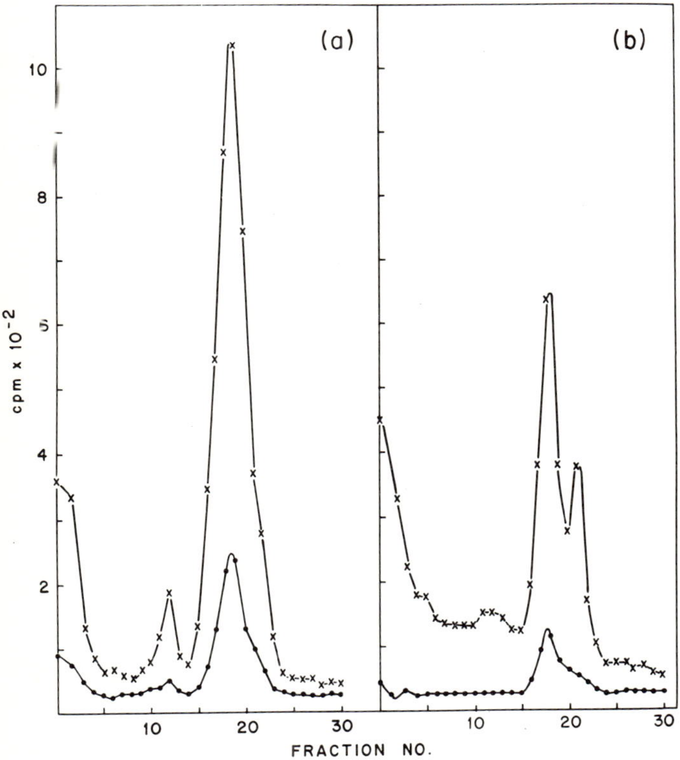

Fig. 9. Characterization of double-stranded branched molecules of λ DNA.

(a) Sedimentation in a 5 to 20 % neutral sucrose gradient (see Fig. 7b and details in the text). Centrifugation was performed in a Spinco SW50. 1 rotor at 50.000 rpm for 2 hours at 4 °C, and 45 fractions were collected.

(b) Same as (a), but 0.3 n mole of ^3H λ DNA was added as a marker.
(– x – x –) ^3H labeled DNA; (–•–•–) ^{32}P labeled DNA.

proposed in Fig. 7b, neither ^{32}P nor ^{3}H labeled DNA was attacked by an amount of exonuclease I that digested 98 % of both labels in a heat denatured sample of the substrate. (Cf. action of exonuclease I on the redundant joint substrate, Fig. 5.)

λ exonuclease did not digest the ^{32}P labeled arm of this substrate under any conditions that we have tried so far. These conditions included, (1) pH 7.5 or pH 9.6, (2) with or without added ATP, and (3) with or without β protein [25, 29]. The possible implications of these negative results are discussed below.

The proposed structure in Fig. 7 may be viewed as a simplified version of a single-strand crossover between two helices.

Discussion

Among interior sites of a DNA molecule, i.e., all sites except simple double-stranded ends, λ exonuclease acts specifically at a branch point. The interaction of the enzyme with the branch is manifested in several ways. If the single-stranded branch is removed by the prior action of exonuclease I, the λ enzyme is unable to initiate digestion (Fig. 5B). When the single stranded-branch is completely assimilated, λ exonuclease ceases further digestion [7]. The experiments that we have described here on a substrate with a double-stranded branch may mean that λ exonuclease specifically recognizes a single-stranded branch. Binding of the enzyme to denatured DNA might be another reflection of this specificity. Recognition of the single-stranded branch permits λ exonuclease to discriminate among interior sites in DNA; but occupation of a recognition site on the enzyme by single-stranded DNA is not essential under all circumstances since λ exonuclease rapidly initiates digestion at the double-stranded ends of native DNA. We have no explanation for this paradox.

A class of ATP-dependent exonucleases has been described and implicated in genetic recombination in *E. coli* and other bacteria [2, 3, 11, 13, 42, 46]. Like λ exonuclease, these enzymes appear to act processively [2, 11]; the enzyme from *E. coli* fails to act at nicks [46], the enzyme from *H. influenza* acts at neither nicks nor gaps [35]. It is attractive to think that these shared properties reflect the action of recombination enzymes on replicating DNA, in which nicks and gaps stimulate strand exchanges. Accordingly, the resistance of nicks and gaps may be seen as protection from premature exonucleolytic attack.

Much more needs to be known before we can describe the pathway of λ recombination. One other phage product, the β protein [25], is an essential component of the λ recombination system, but its function remains completely unknown. Another phage gene, γ, plays a smaller role in recombination [47] through an interaction with the host system [41]. As indicated in the introduction, the nature – even the order – of the steps leading to breakage and recognition of specific sequence homology is unknown. In agreement with the view stressed by *Clark* [9], even a complete description of λ recombination would not suffice for understanding all systems of general recombination. On the other hand, the strand assimilation catalyzed by λ exonuclease is likely to be important in λ recombination, and provides a unifying view of heteroduplex formation that may signify the general importance of this mechanism.

Single-strand assimilation could account for recombinants that have no newly synthesized DNA, if such exist. The sensitivity of available methods tells us only that some recombinants have little or no newly synthesized DNA [22, 23, 37, 45]. Consistent with the largely nonreciprocal nature of general λ recombination [43], single-strand assimilation is basically nonreciprocal. Yet, as illustrated in Fig. 6d, the exonucleolytic mechanism can be reconciled with processes that might be partially or even wholly reciprocal. Single-strand assimilation is a conservative mechanism that does not require new DNA synthesis, but it can be reconciled with all of the general ways in which DNA synthesis might be involved in recombination. Involvement of new DNA synthesis in repair is illustrated in Fig. 6d. The stimulation of strand exchanges by interruptions in replicating DNA could obviously produce substrates for the action of λ exonuclease. *Sigal* and *Alberts* [32] have recently proposed a model in which DNA polymerase might facilitate strand assimilation by λ exonuclease in a "push-pull" mechanism of strand exchange between helices. Their model is closely related to that in Fig. 6d.

With respect to mechanisms in recombination, the diagrams in Fig. 6 illustrate the potentially concerted, driven, and yet controlled nature of single-strand assimilation. Exposure of complementary sequences is coupled to the formation of new base pairs; single- or double-strand branches are simultaneously eliminated; and strand exchanges are irreversibly driven by an exonucleolytic action which is limited to precisely what is necessary. The possible generality of single-strand assimilation in genetic recombination is also suggested by Fig. 6 which shows how this one mechanism might rationalize, on the one hand, interactions between single-stranded and double-stranded DNA (Fig. 6b) and, on the other hand, interactions between double-stranded molecules (Fig. 6a, c, and d).

For the reasons just discussed, the kind of model exemplified by Fig. 6d retains its attractiveness in spite of the failure of λ exonuclease to attack a possibly relevant substrate (Fig. 7b). This negative result might reflect the enzyme's true specificity. It might also mean that we don't understand the structure of the substrate or haven't found the right conditions for the action of λ exonuclease. Among such conditions might be a requirement for the simultaneous action of other enzymes (see reference above to the "push-pull" mechanism). Our facility in making various kinds of branched substrates is improving, and we hope to return to this question with the more defined substrate illustrated in Fig. 7a, as well as with a truer analogue of a single-strand crossover between two helices. It seems likely that we have only begun to realize the usefulness of complicated but well-defined branched substrates in the enzymologic study of recombination. With these more sophisticated substrates, we may also be able to define the role of other enzymes or proteins in strand displacement and unwinding (see Fig. 6d), operations that are central to general genetic recombination.

Acknowledgements: The research described in this paper was sponsored by grants from the National Institutes of Health (5 Ro1 AI 08 160) and the Jane Coffin Childs Memorial Fund for Cancer Research (268).

The authors gratefully acknowledge the interest and critical judgement contributed to this work by our colleagues *Terry Lash* and *K. S. Sriprakash.*

References

[1] *Alberts, B. M., and Frey, L.* 1970: T4 bacteriophage gene 32: A structural protein in the replication and recombination of DNA. Nature **227:** 1313–18.

[2] *Anai, M., Hirahashi, T., and Takagi, Y.* 1970: A deoxyribonuclease which requires nucleoside triphosphate from *Micrococcus lysodeikticus.* I. Purification and characterization of the deoxyribonuclease activity. J. Biol. Chem. **245:** 767–74.

[3] *Barbour, S. D., and Clark, A. J.* 1970. Biochemical and genetic studies of recombination proficiency in *E. coli.* I. Enzymatic activity associated with *rec* B^+ and *rec* C^+ genes. Proc. Nat. Acad. Sci., U. S. **65:** 955–61.

[4] *Berg, D. E., and Gallant, J. A.* 1971. Tests of reciprocality in crossing over in partially diploid F′ strains of *E. coli.* Genetics **68:** 457–72.

[5] *Broker, T. R., and Lehman, I. R.* 1971. Branched DNA molecules: Intermediates in T4 recombination. J. Mol. Biol. **60:** 131–49.

[6] *Carter, D. M., and Radding, C. M.* 1971. The role of exonuclease and β protein of phage λ in genetic recombination. II. Substrate specificity and the mode of action of λ exonuclease. J. Biol. Chem. **246:** 2502–10.

[7] *Cassuto, E., Lash, T., Sriprakash, K. S., and Radding, C. M.* 1971. Role of exonuclease and β protein of phage λ in genetic recombination. V. Recombination of λ DNA in vitro. Proc. Nat. Acad. Sci., U. S. **68:** 1639–43.

[8] *Cassuto, E., and Radding, C. M.* 1971. Mechanism for the action of λ exonuclease in genetic recombination. Nature New Biology **229:** 13–16, **230:** 128.

[9] *Clark, A. J.* 1971. Towards a metabolic interpretation of genetic recombination of *E. coli* and its phages. Ann. Rev. Microbiol. **25:** 437–464.

[10] *Franklin, N. C.* 1971. Illegitimate recombination. In: *Hershey* (ed.): The Bacteriophage λ. Cold Spring Harbor Laboratory, New York.

[11] *Friedman, E., and Smith, H. O.* 1971. An ATP-dependent DNAse from *Hemophilus influenza.* Fed. Proc. **30,** 1155 Abs.

[12] *Fuke, M., Dudnick, P., and Thomas, C. A., Jr.* personal communication.

[13] *Goldmark, P. J.,* and *Linn, S.* 1970. An endonuclease activity from *E. coli* absent from certain *rec*⁻ strains. Proc. Nat. Acad. Sci., U. S. **67:** 434–41.

[14] *Gottesman, M. E., and Weisberg, R. A.* 1971. Prophage insertion and excision. In: *Hershey* (ed.): The Bacteriophage λ. Cold Spring Harbor Laboratory, New York.

[15] *Gurney, T., Jr., and Fox, M. S.* 1968. Physical and genetic hybrids formed in bacterial transformation. J. Mol. Biol. **32:** 83–100.

[16] *Hradecna, Z., and Szybalski, W.* 1967. Fractionation of the complementary strands of coliphage λ DNA based on the asymmetric distribution of the poly I, G-binding sites. Virol. **32:** 633–43.

[17] *Kozinski, A. W., and Kozinski, P. B.* 1964. Replicative fragmentation in T4 bacteriophage DNA. II. Biparental molecular recombination. Proc. Nat. Acad. Sci., U. S. **52:** 211–18.

[18] *Lee, C. S., Davis, R. W., and Davidson, N.* 1970. A physical study by electron microscopy of the terminally repetitious, circularly permuted DNA from the coliphage particles of *E. coli.* J. Mol. Biol. **48:** 1–22.

[19] *Lehmann, I. R., and Nussbaum, A. L.* 1964. The deoxyribonucleases of *E. coli.* V. On the specificity of exonuclease I (phosphodiesterase). J. Biol. Chem. **239:** 2628–36.

[20] *Little, J. W.* 1967. An exonuclease induced by bacteriophage λ. II. Nature of the enzymatic reaction. J. Biol. Chem. **242:** 679–86.

[21] *Masamune, Y., Fleischman, R. A., and Richardson, C. C.* 1971. Enzymatic removal and replacement of nucleotides at single strand breaks in DNA. J. Biol. Chem. **246:** 2680–91.

[22] *McMilin, K. D., and Russo, V. E. A.* Maturation and recombination of phage λ DNA molecules in the absence of DNA duplication. Submitted for publication.

[23] *Meselson, M.* 1967. The molecular basis of genetic recombination. In: *Brink* (ed.): Heritage from Mandel. The University of Wisconsin Press, Madison.

[24] *Radding, C. M.* 1966. Regulation of λ exonuclease. I. Properties of λ exonuclease purified from lysogens of λ_{T11} and wild type. J. Mol. Biol. **18:** 235–50.

[25] *Radding, C. M.* 1970. The role of exonuclease and β protein of bacteriophage λ in genetic recombination. I. Effects of *red* mutants on protein structure. J. Mol. Biol. **52:** 491–99.

[26] *Radding, C. M., and Carter, D. M.* 1971. The role of exonuclease and β protein of phage λ in genetic recombination. III. Binding to DNA. J. Biol. Chem. **246:** 2513–18.

[27] *Radding, C. M., and Her, M. O.* unpublished observations.

[28] *Radding, C. M., Her, M. O., Rosenzweig, J., and Cassuto, E.* unpublished observations.

[29] *Radding, C. M., Rosenzweig, J., Richards, F., and Cassuto, E.* 1971. Separation and characterization of exonuclease, β protein, and a complex of both. J. Biol. Chem. **246:** 2510–12.

[30] *Russo, V. E. A., Stahl, M. M., and Sthal, F. W.* 1970. On the transfer of information from old to new chains of DNA duplexes in phage λ: Destruction of heterozygotes. Proc. Nat. Acad. Sci.,U. S. **65:** 363–67.

[31] *Shulman, M. J., Hallick, L. M., Echols, H., and Signer, E. R.* 1970. Properties of recombination deficient mutants of phage λ. J. Mol. Biol. **52:** 501–20.

[32] *Sigal, N., and Alberts, B.* personal communication.

[33] *Signer, E.* 1971. General Recombination. In: *Hershey* (ed.): The Bacteriophage λ. Cold Spring Harbor Laboratory, New York.

[34] *Signer, E. R., Echols, H., Weil, J., Radding, C. M., Shulman, M., Moore, L., and Manly, K.* 1968. The general recombination system of bacteriophage λ. Cold Spring Harb. Symp. Quant. Biol. **33:** 711–14.

[35] *Smith, H. O.* personal communication.

[36] *Stahl, M. M., and Stahl, F. W.* 1971. DNA synthesis associated with recombination. I. Recombination in a DNA-negative host. In: *Hershey* (ed.): The Bacteriophage λ. Cold Spring Harbor Laboratory, New York.

[37] *Stahl, F. W., and Stahl, M. M.* 1971. DNA synthesis associated with recombination. II. Recombination between repressed chromosomes. In: *Hershey* (ed.): The Bacteriophage λ. Cold Spring Harbor Laboratory, New York.

[38] *Takahashi, K.* 1961. The structure and function of ribonuclease T. I. Chromatographic purification and properties of ribonuclease T_1. J. Biochem. (Tokyo) **49:** 1–8.

[39] *Tomizawa, J.* 1967. Molecular mechanisms of genetic recombination in bacteriophage: Joint molecules and their conversion to recombinant molecules. J. Cell. Physiol. **70:** Sup. 1: 201–13.

[40] *Trautner, T. A.* et al. 1972. Exchange between complementary strands of DNA? See this volume p. 79 ff.

[41] *Unger, R. C., and Clark, A. J.* In press. Inhibition of the *rec*B$^+$–*rec*C$^+$ gene product following infection of *E. coli* by bacteriophage λ.

[42] *Vovis, G. F., and Buttin, G.* 1970. An ATP-dependent deoxyribonuclease from *Diplococcus pneumonia.* II. Evidence for its involvement in bacterial recombination. Biochim. Biophys. Acta **224:** 42–54.

[43] *Weil, J.* 1969. Reciprocal and non-reciprocal recombination in bacteriophage λ. J. Mol. Biol. **43:** 351–55.

[44] *Weiss, B., Live, T. R., and Richardson, C. C.* 1968. Enzymatic breakage and joining of DNA. V. End group labeling and analyses of DNA containing single strand breaks. J. Biol. Chem. **243:** 4530–42.

[45] *Whitehouse, H. L. K.* 1970. The mechanism of genetic recombination. Biol. Rev. **45:** 265–315.

[46] *Wright, M., Buttin, G., and Hurwitz, J.* 1971. The isolation and characterization from *E. coli* of an adenosine triphosphate-dependent deoxyribonuclease directed by *rec* B, C genes. J. Biol. Chem: **246,** 6543–6555.

[47] *Zissler, J., Signer, E., and Schaefer, F.* 1971. The role of recombination in growth of bacteriophage λ. I. The gamma gene. In: *Hershey* (ed.): The Bacteriophage λ. Cold Spring Harbor Laboratory. New York.

Discussion

Russo: I simply would like to make a small comment on the physical structure of recombinant DNA. You spoke about the break and reunion model, and then you said that duplication plays a good role in recombination. Results in my lab with λ show that you can have normal recombination without duplication. (I observed less than a 4 % duplication.) Furthermore, my results also show that recombination is probably due to a partial-single-strand hybridization starting from one end of the DNA molecule. On the other hand, I don't think there is any proof of the break and reunion model.

Radding: You are saying that strand insertion also occurs?

Russo: Yes, if started from one end of the DNA molecule. One of the beauties of this model is that you don't need to recognize homologous stretch of stranded DNA for pairing and recombination.

Sinsheimer: I was wondering how the enzyme knows where it is and what to do. If you have double-stranded DNA with a nick, and if you were to add to this some extraneous single-stranded DNA, would that stimulate the enzyme to start work?

Radding: We haven't done that experiment yet.

Audience: I am confused by your differentiation between a gap, a redundant joint which has a gap, and the different mode of activity. You say that the enzyme works poorly at gaps, but it does work at gaps to a limited extent. *Richardson* has also shown that, I think.

Radding: That is right. By gap, I mean a place where nucleotides are missing. At a redundant joint, there are no nucleotides missing although there is an interruption in the DNA. The enzyme appears to recognize such a branch point.

Audience: Your redundant joint has single strands. Is the enzyme binding at that single strand which is sticking out?

Radding: Well, I suggested that indirectly. Binding to denatured DNA might be related to recognition of the redundant joint. There is a paradox however: the enzyme does not always need the redundant strand. It can start at the end of double-stranded DNA and proceed down the molecule perfectly happily with no strand behind it. But, it appears that if the enzyme is to start at an interruption somewhere in the middle of DNA, then it specifically needs that single strand of DNA to recognize that it is going to start there.

Klingmüller: I am alluding to yeasts and fungi where the occurrence of nonreciprocal recombination or "gene conversion" has been demonstrated. Present interpretations of this phenomenon invoke the involvement of specific enzymes, but those enzymes have not yet been found in these organisms. Would you dare to extrapolate on gene conversion in fungi from your findings in phage lambda? Would you think one could invoke your enzyme or a similar one in effecting gene conversion?

Radding: I was instructed a little bit in this area by *Frank Stahl,* who tells me that the specifics of gene conversion in higher organisms cannot be explained by the model shown in Fig. 6d. You don't get the right segregation ratio. I put it in a more general way when I spoke about that model, namely that it could explain *some* nonreciprocal phenomena for closely linked markers.

Bodmer: Have you demonstrated genetically that you have a recombinant strand? It should be possible after all, if you take your half-molecules from genetically different λ, coupled with density markers, to construct molecules that are molecularly and genetically recombinant. This would be the ultimate test of having made a recombinant in vitro.

Radding: So far we have not put in markers. Half-molecules have no biological activity in the spheroplast assay. We showed augmented biological activity in covalently sealed (i. e., heat stable) molecules, so we think we have put half-molecules together. Moreover, I think that the ligase experiments with the alkaline sucrose gradients demonstrate quite convincingly that we put half-molecules together. The use of genetic markers as you suggest would be equally convincing.

I should emphasize, however, that our principle aim has not been to simulate recombination in vitro but rather it has been to study the mechanism of action of an enzyme that is essential in one recombination pathway. The principal result so far is a new view of how single-strand exchanges may actually occur.

Alternative Modes of Prophage Insertion and Excision

Michael B. Yarmolinsky

Institut de Biologie Moléculaire,
Faculté des Sciences de l'Université de Paris, Paris 5ème, France

Summary: This review surveys prokaryotic processes that may have their counterparts in eukaryotic cells. Material for drawing analogies is provided along with cautionary examples of the pitfalls into which such reasoning may lead.

Alternative responses to infection by a temperate bacteriophage include cell death with release of progeny phage, and cell survival with maintenance of the viral DNA (prophage) integrated or not within the bacterial chromosome. Prophage integration can be reversed exactly or inexactly, with or without cell death. The critical regulatory genes involved in deciding among these alternatives are few.

Normal prophage insertion is mediated by the product of the phage *int* gene; no other gene products, bacterial or viral, are known to be required. The *int* protein promotes site-specific exchanges between a ring form of the viral DNA and the bacterial chromosome. An additional site-specific phage protein, determined by the *xis* gene, is required for prophage excision. Thus insertion and excision are not symmetrical, a conclusion also reached from studies of various permutations of pairwise recombinations. Paradoxes revealed in these studies have been resolved by the recognition of kinetic variables that influence site-specificity.

Int-mediated recombination occasionally inserts prophage into abnormal sites and also causes certain bacterial and viral deletions. Other abnormal recombinations that do not appear to involve homology generate various kinds of transducing particles, deletions, duplications, and insertions and are not *int*-mediated. In most such cases the relevant functions are unknown.

The versatility of viral DNA appears to depend upon few viral functions. To believe that small animal viruses cannot "afford" determinants of site-specificity or regulated integration is to underestimate the power of modest means.

Manuscript received: 17 January 1972

"Life is the secret of DNA" – *Martin Quigley* [139]

Introduction

Implicit in the organization of this workshop is the assumption that knowledge about prophage integration in enteric bacteria is potentially relevant to medical practice. This assumption stems from an intuition about the "unity of life" and from experimental evidence that viral DNA is found integrated in the DNA of animal cells transformed by certain oncogenic viruses, just as bacteriophage λ DNA is found integrated into the *E. coli* chromosome. In addition, there is a possibility that some kind of integration process may be used to fabricate hybrids of the form *Homo sapiens-lambda.* Whether the species designation for such hybrids would remain appropriate is a matter of more than taxonomic interest. It is also a matter of serious concern whether such acts of fabrication lie within the legitimate domain of medical practice. We shall be hearing more about these possibilities and the questions they may raise.

In accepting your invitation to discuss prophage insertion and excision, I must warn you against facile extrapolations from temperate phages to their oncogentic distant cousins. Shorter jumps have proven dangerous. Let me illustrate with two examples from the lore of bacteriophage λ. The realization that λ prophage DNA integrates within the bacterial chromosome by a recombination event led to the assumption, by analogy with general recombination, that the sites at which such recombinations occur are within regions of extensive homology. The realization that transducing particles could originate by anomalous prophage excision led to the additional assumption, by another extrapolation, that such excision corresponds to a malfunction of the normal excision mechanism. Both these assumptions have proven to be incorrect. In what follows, I will expand upon the risk of assuming uniformity in the modes of the genetic exchange, even in situations where the recombination modes would appear to be extremely limited.

The varieties of integration and excision in which different phages participate is quite extraordinary. If we are to attempt successful extrapolation from the behavior of bacteriophages to the behavior of animal viruses, then we also must be aware that the "integration problem", as it has been called, has a variety of solutions. Accordingly, I will concentrate here on the comparative aspects of the integration problem and, in so doing, hope to provide a brief review that will complement the recent, thorough coverage of λ prophage insertion and excision in the Cold Spring Harbor Laboratory volume, "The Bacteriophage Lambda."

The lytic and lysogenic alternatives

Whatever mode of phage development follows infection, circularization of the DNA is an essential first step, whether ensured by host or phage enzymes [19, 68, 130, 179]. Regulatory phage functions are expressed early after infection, subject to metabolic influences by the host [9, 31, 77, 85, 100, 119, 142]. These functions affect the establishment of repression, determining whether the phage takes the lytic or lysogenic pathway. In the lytic response, the further expression of phage genes occurs in an orderly sequence that normally results in cell lysis and the liberation of many progeny phages. In the lysogenic response, the expression of the lethal phage functions is reversibly restrained by at least one phage-determined, phage-specific repressor that also confers immunity to superinfecting phages of the same type. In addition, the perpetuation of the repressed viral DNA, now called prophage, is ensured by integration within the host chromosome or by a more autonomous replication as a plasmid. Normally, the prophage of λ is integrated within the bacterial chromosome and passively replicated like any other continuous segment of the bacterial genetic material.

Prior to integration, the infecting λ DNA probably replicates transiently in the cytoplasm, since λ mutants unable to replicate are relatively inefficient at lysogenizing singly infected cells [24, 101; see also 56]. Moreover, cells singly infected with *P2* can give rise to lysogens carrying more than one copy of prophage [11]. In certain unusual (mutant or hybrid) bacteria, wild type λ DNA may continue to replicate cytoplasmically, establishing itself as a plasmid [52, 63, 69]. In these and in certain possibly related hosts there is a specific decrease in the effectiveness of the product of phage gene *N* [6, 70, 70a, 95, 104, 137]. A corresponding decrease also occurs by direct mutation of this gene. The impaired function of the regulatory gene *N* prevents repressor synthesis, prophage integration, and phage-induced lethality, and it reduces expression of phage replication genes to a level that is still adequate to maintain the prophage as a plasmid [46, 110, 111, 157, 158]. Thus, alteration in the expression of a single viral gene can completely change the way in which prophage replication is accomplished.

Our current view of how λ DNA enters (and leaves) the bacterial chromosome originated from a curious discrepancy between the genetic maps of λ determined by different procedures. A map based on recombination between λ prophages provided a permuted viral gene order relative to that established on the basis of recombination between coinfecting phages [30]. *Campbell's* demonstration that the permutation is indeed cyclic led him to propose explicitly that λ integration occurs by a site-specific reciprocal recombination between a yet-to-be discovered ring form of λ DNA and the bacterial chromosome [35]. The recombination (attachment) site and the site of ring closure (end-join) have proved to be widely separated. Thus, the initial observation did not depend upon a very limited choice of markers (see Fig. 1).

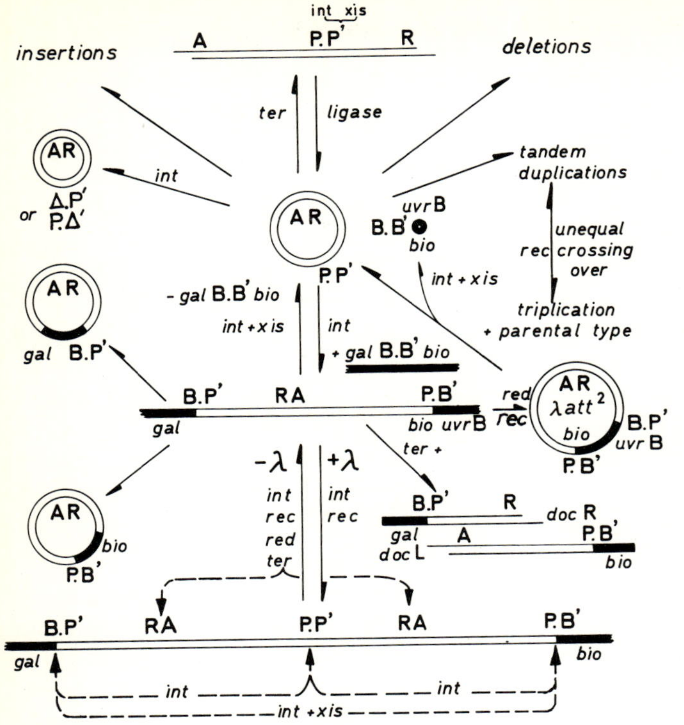

Fig. 1. Rearrangements of λ DNA.

The infecting linear form of λ DNA is shown at the top. In the middle and bottom of the figure are shown a single prophage and a tandem pair of prophages respectively. Functions involved in the reactions are indicated beside the arrows; in some cases several alternatives are listed. Bacterial DNA is indicated by heavy lines, phage DNA by thin lines.

To account for the fact that inactivation of the viral repressor allows prophage excision to occur, *Campbell* proposed that derepression causes an exact reversal of the integration process, restoring the bacterial chromosome intact. Normally this restoration cannot be observed because the lytic response follows derepression. However, if repression is promptly reestablished, the excised prophage is prevented from replication or expression of lethal functions, and cells "cured" of prophage are produced in subsequent cell divisions [187].

Campbell further proposed that anomalous excision events result in the occasional inclusion within phage particles of some bacterial genes originally adjacent to the prophage, roughly compensated by the omission of some prophage genes distal to the addition. The origin and genetic structure of the transducing particles produced upon induction are thus explained (Fig. 1). A large number of experiments on λ as well as several other temperate phages substantiate this interpretation of the facts.

Integration and general recombination compared

Site specificity

The most striking difference between general recombination and the recombination that normally accomplishes prophage integration relates to site specificity. The specificity of prophage integration can be quite strict and depends upon functions of the phage. Lambda normally integrates by a crossover between a particular site on the phage DNA and a particular chromosomal site. Phage *P2* integrates in any of some ten or more chromosomal sites with varying degrees of preference, and the prophage may face "clockwise" or "counterclockwise" on the *E. coli* chromosome, according to its location [17, 32, 163]. The attachment site within the *P2* DNA appears unique [32]. Unlike λ or *P2*, phage μ–1 [177] appears totally indiscriminate in its selection of a chromosomal site. It exhibits no obvious site preference in non-selective lysogenizations [50], and some 76 integration sites selected to lie within the bacterial *z* gene of the *lac* operon appear scattered at random within that gene [27]. However, evidence on μ–1 prophage structure suggests that μ–1 DNA may itself bear a unique or preferred attachment site [86].

Proteins involved

Integration is accomplished by such phages as λ, *P2,* or *P22* with the aid of a phage-determined protein, the product of an *int* gene conveniently located adjacent to *att,* the prophage attachment site [39, 73, 74, 108, 113, 168, 197]. The *int* proteins are site-specific with regard to both the viral and bacterial *att* sites. An int^+ phage can permit the integration of an int^- phage of the same attachment-specificity without itself necessarily becoming integrated, i. e., the *int* protein is diffusible.

Although it would appear likely that certain enzymes should play a role in both generalized and site-specific recombination, there is no evidence to support this contention. Phage-determined generalized recombination functions (*red* in λ and *erf* in P22) are not essential for the integration process [19, 160]. Of several host mutants altered in DNA repair (including *uvr*A, *uvr*B, *uvr*C, *uvr*D, *uvr*E, *uvr*F, *lex (exr*A), *pol*A, and *lon)* or recombination (including *rec*A, B, C) none are integration-deficient when tested with λ [41, 121, 129, 172, 172a, 176, 190]. Recombination deficiency in the host does not interfere with *int*-promoted recombination in *P2* [17, 113]. Neither is *rec*A function required for integration of *P22* or μ–1 [192, 18, 51]. However, to convert terminally redundant, linear *P22* DNA into a ring capable of integration requires a generalized recombination function, either *rec* or *erf* [19]. No such requirement exists for excision.

Independence of DNA synthesis

Interference with phage DNA synthesis by a *rep* gene bacterial mutation in the case of *P2* [34] and by *O* or *P* phage gene mutations in λ does not prevent integration [24, 25, 101, 158]. Furthermore, unreplicated λ DNA can be shown, by density labeling techniques, to participate in various kinds of *int*-promoted reactions in vivo [53, 105, 127, 138, 169].

Although these experiments clearly show that phage DNA need not be extensively synthesized for either insertion or excision to occur, a complete independence of DNA synthesis has not been demonstrated. The question is a difficult one to resolve, but it bears importantly upon the models of site-specific recombination that can be adopted; certain models of recombination involve a limited DNA synthesis [156]. As it is now possible to block DNA synthesis in several independent ways and to measure site-specific recombination with considerable sensitivity and accuracy, the question may well deserve reexamination in a more rigorous quantitative way.

Under certain conditions, DNA replication may be needed for λ integration. A λ phage unable to replicate its DNA can lysogenize only on multiple infection [24, 101]. It has been suggested that each infecting phage assumes one of two alternative transcriptional states, and that products of each state are needed for rapid establishment of repression [101, 102]. Thus, the stable insertion of λ upon single infection *does* appear to require phage DNA synthesis, although apparently for reasons not connected with the recombination process itself.

Phage *P22* presents other complications. Certain phage functions are required for phage DNA to be replicated (even once [117]) and to be integrated [109]. The thermosensitive mutants by which these functions (or function) are recognized, *P22 ts18* and *P22 ts12,* complement each other for DNA synthesis and phage production. Paradoxically, they do not complement for lysogenization in mixed infection, and each interferes with lysogenization by wild type P22 [109]. Thus, the inability of either mutant to lysogenize at the nonpermissive temperature is not due to a failure of phage DNA synthesis. The precise function of these genes remains unknown; cartographically and functionally they would appear to be analogs of lambda genes *0* and *P,* known to be involved specifically and cooperatively in the initiation of phage DNA synthesis [182].

Relative efficiency and high negative interference

The efficiency of *int*-promoted recombination can be measured in crosses between two phages. This is so because *int* can promote recombination not only between the *att* sites of host and phage, but also between the *att* sites of two phages. An observed distortion of phage genetic maps can be attributed to the relative efficiency of site-specific *int*-promoted recombination. This distortion is particularly striking in phage *P2*. *P2*, for reasons possibly related to phage DNA conformation in the vegetative state, exhibits a drastically reduced level of generalized recombination. No such reduction is seen in recombination between two *P2* prophages [191].

Recombination (during lytic growth) between any two markers of *P2* that bracket the phage *att* site is largely *int*-promoted. Therefore, a map of *P2* that represents genetic distances based upon crossover frequencies is highly distorted, with 80 % of its length containing only the *att* site [113]. Similarly distorted maps of λ are generated if general recombination is eliminated by host and phage mutations [48].

The efficiency of *int*-promoted recombination relative to generalized recombination can be altered by UV-irradiation of the input phages. *Int*-promoted recombination in λ appears to be decreased by UV [160]; generalized recombination is stimulated several fold [90, 156]. The latter effect of UV is particularly striking in phage *P2* where, by UV irradiation, generalized recombination may be augmented selectively by two orders of magnitude [112].

Among phage particles selected for having undergone recombination in one DNA segment, the frequency of a crossover event in the second segment may be increased relative to what it would be in the unselected population. A large increase, known as "high negative interference," is usually seen only when the chosen segments are adjacent and short. The regions of multiple crossover that result in high localized negative interference are believed to correspond to regions of heterozygosity [4, 98]. In phage *P2*, negative interference may amount to several thousandfold even when the markers under consideration are widely separated [112]. This result suggests that there exists a special obstacle to *P2* recombination that, once overcome, affects the entire phage chromosome. The question arises whether, in the act of *int*-promoted recombination, the same obstacle is overcome and the *P2* chromosome rendered more susceptible to generalized recombination. Crosses that involve markers on either side of the *P2* attachment site show less negative interference than other crosses [113]. This observation has been taken to suggest that *int*-promoted recombination may not remove the barrier to generalized recombination, or that the removal is too transient to be observed [17]. The question deserves further study.

Reciprocity

"Ring-into-ring-insertion" with conservation of genetic material implies a double-strand break in each parental chromosome and a crosswise rejoining that is exactly reciprocal. Total conservation of genetic material in *int*-promoted insertion and excision is suggested by two observations. First, *E. coli* can be subjected to repeated cycles of λ lysogenization and curing without apparent damage to the *att* site [155, 161]. Secondly, *int*-promoted curing of λ from a site of mutagenic insertion restores normal gene function [152]. These results, taken together with evidence for efficient ring-into-ring insertion in prophage integration, imply that *int*-promoted recombination is generally reciprocal. The same conclusion has been reached in a study of the distribution of reciprocal recombinant types in the yields of phages released from individual cells coinfected with two marked phages [186].

This study further suggests that whereas reciprocity characterizes *int*-promoted recombination, *red*-promoted recombination is nonreciprocal. More recent evidence suggests that red- and rec- promoted recombination also can be reciprocal, although the relative frequencies of reciprocal and nonreciprocal events is not evident [99].

Transcriptional effects on recombination

Transcription across the substrate of recombination has been observed to have a modest influence upon recombination rates [81, 83]. Recently, a dramatic transcriptional stimulation of transduction by phage *T1* has been reported. Derepression of an operon in the donor DNA may increase the number of (odd) crossovers in the actively transcribed region by more than one hundredfold at the time of fragment integration [45]. Evidence both for and against transcriptional activation of *int*-promoted excision has been adduced [106, 171, 189]. The evidence in favor of transcriptional activation of prophage excision is indirect, relying on a comparison of the yields of certain prophage types in the lysates of variously superinfected induced lysogens. A more direct measure of prophage excision (such as visualization of prophage loss from an F′ episome [67] may be required to provide a picture of the molecular events in recombination sufficiently clear to resolve this controversy.

Recombination that affects transcription

Recombination that integrates one DNA within another can be mutagenic if integration occurs within functional genes. *Mu–1,* for example, is broadly mutagenic by virtue of its nonspecific integration mechanism [18, 177]. Accordingly, this phage has been used to inactivate a variety of genes, including those of prophage λ [183]. "Mu-tations" are not reversible, either by direct selection procedures or by attempts (unsuccessful to date) at curing μ–1 lysogens of their prophage [177]. Like μ–1, λ becomes mutagenic when its range of integration sites is enlarged. When the normal bacterial attachment site for λ prophage (between *gal* and *bio)* is deleted, inefficient attachment of λ occurs at abnormal sites, sometimes causing loss of gene function [152]. Abnormal integration of λ appears to be exactly reversed by supplying transiently the proteins normally required for λ prophage excision, i. e., curing of prophage reverses the lambda-induced mutations.

Rearrangement of the prophage itself, by virtue of the integration act, may play a significant role in the control of *P2* integration. The *int* gene of *P2* is inactive in the prophage perhaps largely because it becomes disjoined from the strong promoter that permitted its repressor-insensitive transcription prior to insertion [15]. It has been suggested that for lack of such disjunction, tandem *P2* prophages do not stably coexist unless one of them is *int* [16]. However, the argument is weakened by the finding of exceptional cases of *int*$^+$ tandem *P2* prophages [16], and by the finding that a minor *att* site alteration, called *saf* (see "Site affinity alterations"), may also allow tandem *P2* prophages to coexist stably [164]. If the integrated *P2* prophage is stabilized by a separation of the *int* gene from the high-level promoter, it must be recognized that a second (inefficient) promoter for the *int* gene must exist to account for the occasional spontaneous *int*-dependent release of phage particles by lysogenic strains. Examples of more than one promoter controlling a single gene in different ways is reminiscent of the complex regulation of the λ repressor [142] or the *trp* operon [7, 124].

Whereas, in *P2,* insertion may normally turn off the viral *int* function, insertion may also serve to turn on phage genes. Of some 60 independent insertions of μ–1 prophage within *lac,* 22 allow *lac* induction to turn on the functions of μ–1 needed to produce an appreciable phage yield [178]. This arbitrary mode of prophage induction apparently involves "read-through," i. e., transcription across normal operon limits, here from *lac* into μ–1. Transcription originating in an induced prophage and passing some distance into bacterial DNA also occurs. Some of the increased *gal* and *bio* expression seen following induction of λ prophage or increased *ara* and *trp* expression following induction of the corresponding transducing phages, and known as "escape synthesis," is due to "read-through" [29, 60, 103, 145]. Thus, the novel juxtapositions of DNA sequences that occur upon integration can alter transcription even beyond the proximal operons that may be broken or fused in the process.

Consequences of replicon fusion

Replication units, like transcription units, can also make new alliances as a result of integration. Integrations that fuse two replicons can bring one of them, passive for physiological or genetic reasons, under control of the other. The principal advantage to a virus of possessing an *int* gene would appear to be to provide for this kind of coddling. Under appropriate conditions, such a dependency relationship can be reversed. Suitable mutants of phage *P2* [114], perhaps also of λ [20] and *P1* [38] and mutants of the fertility factor F [128], when approximately integrated within the DNA of bacterial mutants defective in the initiation of DNA synthesis, can suppress these genetic defects. The integrated episome may furnish to the defective host both the diffusible initiation functions and the site(s) on which they act. The suppression can be good enough to permit a mutant host to divide and form colonies. Genetic elements less disparate in size also enter into dependency relationships in which one element becomes passively replicated, e. g., F can be driven by λ [67], one prophage by another [75, 180], or a unit harboring drug-resistance genes (r-determinant) by a transfer factor *(RTF)* [143].

Integration can also result in replicon misalliances. The slowly replicating plasmid λ*Nc*I, and the bacterial chromosome can coexist without interference until prophage integration fuses these two replicons. Fusion is normally followed by cell death, although in certain hosts (where replication of the plasmid may be more sluggish) the cells respond by filamentation and slower growth [49, 64, 111].

Insertion and excision compared

Site specificity and the asymmetry of insertion and excision

If, as was originally proposed, the DNA of phage and host were to become paired and recombined at regions of homology in the act of lysogenization, such regions would remain unaltered by the recombinations within them. For example, assigning to both viral and bacterial *att* sites the genetic structure AB would imply that the structure of an integrated prophage is AB-prophage-AB, the two new *att* sites being equivalent to one another and to the original *att* sites. The error of this simple formulation became clear as various examples of *int*-promoted recombination were examined and found to differ widely in frequency. Consider the case of λ b2, a plaque-forming deletion mutant with a structural defect in *att.* This phage (although it can furnish *int* function to a coinfecting *int*$^-$ phage) fails to lysogenize a sensitive host, except at very low frequency [75, 195]. It lysogenizes a heteroimmune lysogen (or in homoimmune coinfection) very efficiently, the λ b2 phage always inserting to the right of the wild type prophage [37, 55, 195]. These results reveal that the *att* site to the right of a prophage differs from both the left-hand site and from an unoccupied site. The inverse insertion pattern of the transducing phage λ*gal* reveals that the *att* site to the left of a prophage differs from both the right-hand site and from an unoccupied site [34a, 47, 189]. Thus, it would appear that the recognition elements in the *att* sites are precisely the *nonhomologous* features of the regions bracketing the locus or loci of crossing-over [80, 161, 188].

This insight is formally expressed in the notation for the bacterial and phage attachment sites as composite structures B.B′ and P.P′, the dot representing the crossover locus. When B.B′ and P.P′ recombine during integration, two new attachment sites are generated – B.P′ and P.B′, found at the prophage ends. The excision of λ*gal* or λ*bio* transducing phages will give genomes containing B.P′ respectively. A cross at "dot" between the transducing phages *gal* B.P′ and P.B′ *bio* generates a λ that is *gal* B.B′ *bio.* Therefore it is possible to study in phage crosses the recombinations between all pairwise combinations of the attachment sites P.P′, B.B′, B.P′, P.B′, and even P.Δ′ and Δ.P′ (deletion phages). From the appropriate crosses, performed in otherwise recombination-deficient backgrounds, a table of efficiencies of site-specific recombinations may be constructed [73]. It can be seen from such a table that the *int*-promoted recombination of λb2 x λb2 (Δ.P′) is only one-twentieth as frequent as *int*-promoted recombination of λb2 (Δ.P′) x λ*bio* (P.B′). That is, non-homology near the crossover site can actually enhance *int*-promoted recombination.

The requirement for a separate function in excision

Perhaps the principal value of the above analysis has been to reveal that integration and excision need not be considered symmetrical events and accordingly might exhibit different requirements. Excision of prophages λ and *P22 does* require an additional site-specific phage protein, determined by a gene called *xis* [78, 93, 167]. Moreover, phage *P2* has a function, *cox,* that is required specially for excision [115]. Of the various possible recombinations among the four normal attachment sites, the *xis* protein is known to be required only for the recombination B.P′ x P.B′, and then in conjunction with the *int* protein [78, 93]. The genes for the functions that cooperate in excision are adjacent in λ and are separated in *P2* by the gene for the immunity repressor. It has been observed that mutations in *cox* strikingly increase recombination across *att,* probably by relieving an inhibition of *int* function [115].

Studies of the relatedness of lambdoid phages by DNA heteroduplex electron microscopy suggests that λ, 434, and 82 have identical or very similar attachment sites and *int* proteins, but have different *xis* proteins [162]. No attempts have yet been made, to my knowledge, to determine whether these differences in *xis* structure are reflected in a functional incompatibility of *int* and *xis* of different lambdoid phages.

Phage *P2* excision is normally inefficient. This phage is noninducible in two senses. First, the immunity repressor is refractory to treatments that induce the lambda-related phages (UV irradiation, thymine deprivation, etc.). Secondly, even mutants with thermosensitive immunity are not thermally induced to yield phage because excision is defective [14]. Mutants that excise prophage more readily after thermal induction have been selected. In these *nip*⁻ mutants, the *int* gene may be expressed efficiently even in the repressed integrated prophage. That a *nip*⁻ prophage is nonetheless relatively stable until induced suggests that an additional function required for excision may remain under immunity control [33]. This function may be *cox.*

Both insertion and excision of λ prophage are said to be "efficient" processes. Infected cells can each give rise to a stable lysogenic clone, and transiently induced cells can give rise to a large proportion of cured cells. It is reasonable to suppose that in both reactions *int* acts as a catalyst, as *int* is required not only in these but in many other site-specific recombinations involving various combinations of B, B′, P, P′. However, *int* alone does not suffice to catalyze excision. This result is surprising because the principle of microscopic reversibility requires that if *int* catalyzes X → Y, it also should catalyze Y → X.

One explanation offered is that the forward and reverse reactions are completely different and each irreversible, like the synthesis and breakdown of glycogen or DNA. In the two pathways, *int* would assume different catalytic forms. A simpler hypothesis is that *int* and *xis* have roles as participants, as well as catalysts, in these reactions (42). It was recognized many years ago that enzymes, when present in stoichiometric amounts, can differentially influence rate constants by differential binding to substrates or products [72]. The idea that *xis* protein may influence the binding of *int* protein to a particular substrate or pair of substrates is particularly attractive because of analogy with the *red* recombination system. The binding of λ-exonuclease (*red* α product) to DNA appears to be increased by an accessory protein, the product of *red* β [140].

The efficiency of insertion and excision as measured by recombination frequency in phage crosses or by lysogenization frequency may reflect changes in concentration of the participating proteins, including the enzymes. If one or the other should decay (perhaps even at a rate which exceeds the rate of adjustment of the catalyzed equilibrium), then the final stable state can reflect a transient condition not readily reproducible. Evidence concerning the importance of such kinetic considerations follows.

Relative stability of *int* and *xis* proteins

The concentration of *int* and *xis* gene products is governed by their rates of formation and decay. The two genes are under coordinate control in λ, as the viral recombination functions belong to a single operon, with *int* distal to the promoter. Coordination at the level of gene expression need not be reflected in coordination at the level of the functional proteins. When the capacity for site-specific recombination decays in an infected cell, following establishment of repression, it is probably the *xis* protein that is the first to become rate-limiting [189]. A greater lability of the *xis* protein could account for, or contribute to, the greater requirement for *xis* expression than *int* expression in prophage excision [93].

Evidence concerning the decay of *int* and *xis* functions comes from experiments in which is measured the internal recombination in a specially constructed transducing phage, λatt^2, bearing both *att* B.P′ and *att* P.B′ and defective in either *xis* or *int* [154, 189]. Internal recombination between *att* B.P′ and *att* P.B′, dependent upon both *int* and *xis,* excises the carried bacterial DNA (see Fig. 1). The original transducing phage λ*att*^{2} may be distinguished from the nontransducing product of internal recombination by its greater DNA content and therefore greater sensitivity to agents that chelate magnesium ions [134]. The decay of *int* and *xis* activity following transient induction of an *int*$^+$*xis*$^+$ normal prophage is measured as the loss of capacity for complementing an *int* or *xis* defect of λ*att*^{2} superinfecting at various times thereafter. The direct enumeration of those phages that derive from λ*att*^{2} by the action of the complementing *int* and *xis* proteins is accomplished by phage titration in the presence of a chelating agent. *Xis* activity decays to less than five percent the initial value within a half hour; *int* activity decays only fifty percent in one hour [189]. The decay rates, during the roughly logarithmic phase, differ by a factor of at least six.

Two paradoxes are explained by this differential stability of *int* and *xis.* The first is the efficiency of both integration and excision. The time at which integration becomes irreversible is late; excision normally occurs early, i. e., before, or not long after, replication begins [67, 75, 118, 168, 189]. Insertion is stabilized by the decay of the *xis* protein. The effects of excision are rendered irreversible by phage (or bacterial) replication before *xis* decay has shifted the equilibrium back toward insertion.

The second (or Guerrini) paradox is really a special case of the first [80]. It is the apparently absolute preference of λ*gal (att* B.P′) for integration at the homologous left-hand (B.P′ site) of λ prophage, although integration at the right-hand P.B′ site would correspond to normal λ excision. In fact λ *gal* does integrate at both *att* B.P and *att* P.B′, but integration at the right-hand site is transient [47, 188]. With the decay of *xis* activity, insertion within the P.B′ site ceases, but excision from this site requires no *xis* protein and so persists. When the λ*gal* finally becomes stably integrated, it is into the left-hand site. Stable *xis*-requiring integration of λ*gcl* into the P.B′ site present in λ*bio* prophage does occur efficiently. This is because the λ*bio*-λ*gal* arrangement, though subject to *int*-promoted loss of either prophage, will not readily lose the remaining prophage [47, 189].

Without going through the preceding arguments in detail, it should be clear that they explain certain fundamental aspects of the biology of λ integration: how both insertion and excision can be favored and how certain prophage arrangements, theoretically possible, are nevertheless proscribed.

Site affinity alterations

Up to this point we have been considering as identical the phage DNA which is inserted and the phage DNA which is excised, but even this degree of symmetry does not always obtain.

Whereas phages released upon induction are normally indistinguishable from phages produced on lytic infection, in some cases prior residence within the host chromosome genetically alters them. Gross alterations, such as seen in transducing particles, occur very rarely and their generation involves separate modes of excision, considered in the section "abnormal insertion and excision." Minor alterations have been observed to occur at an appreciable frequency in phage *P2*. These are recognized by an increased affinity for a particular chromosomal attachment site, conferred as a result of interaction with it. The new hereditary property (called *saf* for site affinity) is apparently the result of acquiring genetic information from the site of prior attachment [17, 163]. This could be done as a result of cutting and rejoining, but only if cutting may occur at more than one point within a given *att* site. The existence of *P2 saf* phages (presumably generated by *int*-promoted recombination) make it less attractive to assume that the crossover region is, as normally written, a point. A less prejudicial notation has been suggested, honoring the region with the Greek letter, χ [162]. We may be obliged to acknowledge that homology can play a role in favoring attachment site interactions even if this role normally is a minor one.

The recognition of *saf* mutants in *P2* depends on the existence of more than one frequently used but different bacterial attachment site for this phage. Now that novel bacterial attachment sites for λ have been discovered [152], alterations corresponding to λ*saf* can be sought. So far, no evidence of *saf* mutations in λ have been found.

Reciprocity again

Some doubts have been raised about whether excision is necessarily a single reciprocal event resulting in free rings of phage DNA and cured host DNA. When λ is excised from an inefficiently used site [75, 152] or under certain conditions of heteroimmune superinfection [106], the rate-limiting step in phage production appears to be the low level of prophage excision. A corresponding reduction in the appearance of empty λ attachment sites (curing) is expected, but not always found [152, 171]. Moreover, λ mutants have recently been obtained that appear deficient in prophage liberation, but not in curing [171]. As the measures of excision are indirect, these situations are still poorly understood. However, there is good reason to believe that *int* + *xis*-mediated recombination *can* be reciprocal, as when λ*gal* integrates into a λ*bio* prophage, thus fusing two DNA rings (see the previous section, "Relative stability of *int* and *xis* proteins"). Furthermore, the quantitative recovery of covalent rings of λ and F′ DNA, following curing of F′(λ) by transient derepression in the absence of DNA synthesis [67], is highly suggestive that excision is normally reciprocal.

The possibility that some excision events might be nonreciprocal could influence our interpretation of the basis of certain site-specificities and, more importantly, could influence how we think about site-specific recombination in molecular detail.

Towards understanding the mechanism of *int*-mediated recombination

A model for site-specific recombination in λ has been suggested in which the non-homologous regions recognized by proteins flank a short core of homology where DNA–DNA recognition can occur [155]. Single-strand breaks are postulated to occur in the homology region on the opposite strands of the DNA helix, much as they are presumed to occur in the generation of the complementary single-stranded ends of phage DNA [94], or as they are known to occur by the action of certain ATP-dependent nucleases associated with DNA restriction [122]. If the postulated homology region is to be commensurate with the cohesive DNA ends of λ or *186* (12 and 19 nucleotides respectively [193, 132]), then the region must be some 10 to 20 nucleotide pairs in length.

This model represents a compromise between two extremes: the preliminary proposal of *Campbell* [35] that homology is the basis for *att* site recognition and a proposal by *Dove* [42] that no common core of homology exists in the *att* sequences and, in each form of *att*, exchanges occur at one particular internucleotide position.

The dissection of *att*-site structure either by electron microscopy of DNA heteroduplexes or by genetic analysis has provided a resolution either too coarse or too fine to test the validity of the compromise model. However, upper-size limits for the crossover region and the attachment determinants flanking it have been defined by electron microscopy of heteroduplexes between DNA derived from variously substituted and deleted λ phages. Heteroduplexes constructed from pairs of λ-related phages have displayed the degree of relatedness of different attachment sites and *int* and *xis* genes in λ, *434* and *82,* phages which integrate between *gal* and *bio* in the *E. coli* chromosome. The sequences about the integration sites, including those that specify the *int* protein, appear to be identical in these three phages [162]. The crossover locus appears fixed within a region not larger than 100 nucleotides long [40, 87]. By a rather indirect argument, it may be concluded that the B portion of the bacterial attachment site does not exceed 20 nucleotide pairs in length. Finally, homology between *att* B.B′ and *att* P.P′ cannot be seen in heteroduplexes, although homology extending over as few as a dozen nucleotide pairs could be detected in principle [40]. This finding is consistent with the observation that the degree of DNA homology at the phage and bacterial attachment sites appears less than required to allow preferential λ integration at the normal site by way of a generalized recombination pathway when *int* is defective [75].

If a very small number of base pairs make up the *att* sites, then one might expect that point mutations in *att* should affect site-specific recombination. Mutations in *att,* induced by hydroxylamine and by nitrosoguanidine, have been isolated recently [153, 154]. They are selected on the basis of the stability they confer on λatt^2 when this phage is grown through several cycles on an induced lysogenic host that furnishes the *int* and *xis* proteins. There is no evidence that the putative punctual mutations in *att* revert to wild type.

One *att* mutation is of particular interest because it is variably associated with *att* P.P′ or *att* B.B′, suggesting that it may lie within the core homology sequence postulated to be common to both [153]. The λatt^2 bearing this mutation breaks down (inefficiently) to give two types of λ phages, those that appear mutant in *att* P.P′, and those that appear normal. The bacterial *att* site and associated bio^+ genes released in the breakdown process as a DNA ring may be recovered by generalized recombination with the chromosome of a host deleted for *att* B.B′ and part of *bio.* The resultant bio^+ transductants are likewise of two types, those that appear mutant in *att* B.B′, and those that appear normal. This situation is reminiscent, in some respects, of the behavior of *saf* mutants in *P2* (see previous section, "Site affinity alterations"). One can visualize the origin of the *att* site heterogeneity as due to variability in the site of recombination relative to the fixed site of mutation. Alternatively, staggered single-strand breaks at fixed sites followed by recombination could render the mutation heterozygous. Replication or nonselective repair would yield both types of homozygotes.

Determinations of nucleotide sequences at the *att* sites are clearly required, and perhaps already feasible by chemical means. To this end it has been suggested [175] that advantage might be taken of the observation that λ (as well as μ–1) can integrate into known structural genes [152]. Should it be possible to select lysogens bearing λ prophage integrated within a gene that codes for a protein of known sequence, an analysis of the hybrid polypeptide, or its messenger RNA, might reveal the *att* structure in detail.

Another approach to understanding the mechanism of recombination is the identification of intermediates in the process, such as the unsealed "joint molecules" that appear in the course of generalized recombination in phage *T4* [23a, 181]. Various simplified systems are being used to follow the course of site-specific recombination. These include λ and F′ [67], λ and λ [79, 127, 169], and λatt^2 [154]. A study of site-specific recombination between density-labeled, replication-defective λ phages support the view that *int*-promoted recombination occurs by a breakage and reunion mechanism in the absence of substantial semiconservative DNA replication [79, 127, 169]. These in vivo studies, however, have not resulted in the isolation of identifiable intermediates in the breakage and reunion reactions.

A kinetic study has been undertaken of the appearance of single-strand breaks in rings of F′ DNA in the course of λ lysogenization or in F′ (λ) DNA following thermal induction. The DNA is radiochemically labeled, and lysates made at various times after infection or induction are centrifuged through alkaline sucrose to separate the intact from the nicked DNA molecules [57, 67]. The effect of mutations in DNA replication (*O* and *P*) and prophage excision (*int* and *xis*) suggest that there is a background level of nicking associated with DNA replication that precedes nicking associated with prophage excision [65, 67]. *Int* function appears responsible for nicks in F′ rings during lysogenization. Moreover, the *int* protein apparently requires an attachment site pair to promote, or stabilize, the single-strand breaks that precede insertion, since such nicks are not detected upon infection with *att*-defective int^{+} phage λb2 [57]. The nicking activity of *int* in excision is completely dependent upon the presence of *xis.* A nicking activity observed in the apparent absence of *int* and *P* function is attributed to *xis.* These results suggested that both the *int* and *xis* proteins have endonuclease activity or control such activity [67].

Endonucleolytic activity of the *int* protein of λ has been postulated to explain an accessory activity – the generation of viral deletions with one terminus in *att* [40, 133, 134]. These deletions might originate by DNA degradation from an *int*-generated scission, followed by resealing of the molecular ends. Among such *int*-dependent deletions is one, b2, that arises repeatedly and might originate in a different way, viz., by recombination of an internal *att*-like site with the normal phage *att* so as to excise the $b2^{+}$ region. A significant feature of this gratuitous activity of *int* is that the crossover region (χ) is not lost in the deletions. This suggests that the *int*-promoted nicks are precisely localized, rather than being anywhere in a short region of homology [40].

A promising approach to an understanding of site-specific recombination is the isolation and characterization of the *int* and *xis* proteins. The protein monomers have been identified by gel-electrophoresis and have molecular weights of 42,000 and 32,000, respectively [82]. The *int* protein, purified from heavily irradiated λ-infected bacteria by differential labeling, binds to λ DNA but exhibits only slightly greater affinity for wild type λ DNA relative to attachment site deletion DNA [5, 126]. An isolation procedure that omits the irradiation step has been developed and should permit large-scale isolation of the recombination proteins, facilitating their study [126].

Thus far, no in vitro enzymatic activity of *int* protein, endonucleolytic or otherwise, has been noted, possibly because proper conditions for bringing enzyme and substrates together have not been realized. On the other hand, no evidence available excludes the possibility that the function of *int* protein is solely one of juxtaposing the *att* regions, and that *int,* on its own, has no catalytic powers. Whether or not this is the case, it should be realized that recombination is probably too complex a process to be catalyzed by a single protein. Information about the participation of bacterial functions is conspicuously lacking. To my knowledge, no systematic attempt has been made to find bacterial mutants that are defective (conditionally or not) in the process of insertion or excision.

Abnormal insertion and excision

Methods

We shall consider here those modes of prophage insertion and excision that are relatively rare or that are seen only when certain modes of recombination are not available. Recombinational modes may be blocked by various means, and insertion and excision events, even if rare, may be scored by special techniques.

To block general recombination in a phage-infected cell may require that both the host and the phage be mutant (e. g., *rec*A and *red* in the case of *E. coli* – λ). Interference with site-specific recombination is accomplished by mutations affecting the relevant proteins or *att* sites or even by the simple expedient of thermally inactivating the *int*-protein, which, in the case of λ and *P22* is relatively heat labile [160, 168, 195]. Normal excision is prevented in xis^- or int^- lysogens. The latter are constructed with the aid of complementing int^+ phage. If the int^+ helper phage differs in immunity from the phage that is helped, the desired int^- lysogens may be readily selected.

Rare lysogenization events are easily scored because the lysogens, if stable, resist continual challenge by a repressor-sensitive phage. On a plate appropriately seeded with phage, lysogens (and lysogens almost exclusively) can form a colony. In principle, the challenging phage should be neutral vis-à-vis the phage tested. It should influence neither of the component processes in lysogenization, neither repression nor integration. Accordingly, repressor-negative int^- phage is commonly used. However, it has proved inadequate to the task of measuring the frequency with which λN^- mutants stably lysogenize [157, 158]. The *N* function supplied by the challeng-phage can induce an unintegrated λN^- phage to produce both *int* and repressor, allowing lysogenization to occur in time to prevent the lethal effect of the first (and subsequent) challenges.

The detection of rare excision events can be accomplished with limitless sensitivity, provided a phage particle is produced. The markers present in genetically defective phage particles can be assayed in complementation or recombination tests. If the particles themselves are defective, the phage DNA they contain can be assayed, in some cases, by transfection (i.e., measurement of the biological activity of infecting free DNA). Separation on the basis of buoyant density allows various classes of particles to be distinguished. The reciprocal products of excision, cured cells and "eductants" (i. e., cells that have lost some host genetic material along with prophage), can also be scored readily. When the curing or eduction process eliminates a thermoinducible prophage, the resultant cells alone survive an inducing temperature.

Consequences of interference with normal prophage excision

Given the existence in phages which integrate of a system specialized to accomplish excision, one might expect that interference with this specialized system would completely eliminate the prophage release. This expectation is not fulfilled. Moreover, the interference can reveal and even accentuate modes of excision that were previously masked.

Int$^-$ lysogens fall into two classes: those that yield < 0.1 phage per cell upon induction, and those that give a normal burst size. In the class of "high yielders" are lysogens harboring tandem λ*int*$^-$ prophages [55, 75]. A single λ*int*$^-$ prophage bracketed by homologous segments of bacterial DNA also gives an appreciable yield [61]. For this result, it is required that the lysogenic host be *rec*$^+$ or the phage be *red*$^+$. Apparently, if regions of homology bracket a prophage, they allow *red*- or *rec*-promoted recombination to substitute for the normal site-specific mode of prophage excision.

Induction of tandem λ prophages in a completely recombination-deficient situation (*int*$^-$ *red*$^-$ *rec*$^-$) also releases several phages per cell. This surprising result is attributed to the following sequence of events. First, viral replication occurs in situ (probably in both directions from a fixed origin [149, 170]). Second, λ chromosomes are cut out by *"ter,"* a function normally involved in generating the cohesive termini of phage DNA [125, 194]. The genetic structure of the phage chromosome that is excised from a tandem pair of prophages in the absence of *int, red,* and *rec* functions is exactly as expected if two *ter*-mediated double-strand scissions excise one recombinant phage from within two tandem prophages (see Fig. 1). The consequences for prophage excision of also eliminating *ter* function are not easily studied. This is because end-cutting plays a crucial role both in packaging the λ DNA into infectious particles and in releasing the naked DNA in a form that is infectious in the normal assay system [43, 92].

The few λ phage particles released upon induction of *int*$^-$, *xis*$^-$, or *att*-defective singly lysogenic bacteria are not of uniform buoyant density, owing to a variation in their DNA content, the range extending from about one-fourth less to one-tenth more than normal [10, 184]. Because of packaging requirements, this DNA contains at least one cohesive end. It may contain bacterial DNA as well, either that which is normally to one side, to the other side, or even to both sides of the prophage.

The spectrum of transducing particles present in a lysate is greatly influenced by mutations affecting the recombination functions and, in the case of λ, even by deletions of apparently "neutral" phage DNA. The topic is sufficiently complex that we will devote to it a separate section.

Origin of transducing particles

Bacterial markers can be transferred from one cell to another by exceptional phage particles in either of two ways, which are called generalized and specialized transduction. Generalized transduction is the transfer of nonspecific segments of the bacterial chromosome within phage capsids. Prior residence of the phage DNA in the host chromosome appears not to be required and for some phages capable of generalized transduction, such as *P1* [88, 107] or *T1* [44], may occur rarely or not at all [89, 150]. In each case studied, the exceptional particles are formed on lytic infection and contain little if any viral DNA [89, 146, 148]. Of the temperate phages considered here, *P22* [146, 148, 196] and *μ*–1 [86] form generalized transducing particles; a promising analysis of the *P22* phage genes involved in their formation has been begun [147].

More pertinent to our discussion here is specialized transduction, the transfer in phage particles of specific segments of the bacterial chromosome as part of the DNA molecule of the phage (e. g., λ, *P22*, *μ*–1). Prior residence of the phage DNA in the host chromosome is essential; specialized transducing particles are not formed on lytic infection [131]. Because only a limited amount of DNA may be accommodated in a phage particle, the addition of DNA from the host usually entails a compensatory loss from the phage chromosome. One region of the λ chromosome may not be deleted – the end-join, a region that gives rise to the cohesive ends. For reasons presently obscure, at least one cohesive end is needed for λ DNA to be packaged [116]. No restrictions as to the region deleted are apparent in the case of phage *P22* [166]. The clustering at the ends of λ prophage of genes nonessential for phage development is such that the compensating deletion of viral genes involved in the formation of λ*bio* need not render the phage defective [159]. On the other hand, λ*gal* is invariably defective unless the *gal* genes have been brought closer to the prophage by deletion of intervening bacterial DNA [144].

Defective or not, λ transducing particles supplied with *int* function lysogenize only rarely unless a helper phage (or prophage) provides a suitable attachment site. This is because the hybrid *att* sites of transducing λ recombine poorly with *att* B.B′. In addition to the class of transducing phages able to lysogenize with helper is a class of transducing phages unable to lysogenize under any circumstances. The lysogenizers usually transduce by addition, making unstable partial diploids; the nonlysogenizers transduce by substitution, making stable haploids [36]. The DNA of the lysogenizing particles can circularize, whereas the DNA of the nonlysogenizing particles cannot [71, 75, 116]. In the lysates of an induced wild type λ lysogen, there is only about one transducing particle per 10^6 phages. Of these, about one-third are of the nonlysogenizing class [36].

When normal prophage excision is prevented by an *int*⁻ mutation, the number of *gal*-transducing particles in the lysate increases as much as fiftyfold. Moreover, a *red*⁻ mutation increases *gal*-transducing particles in the lysate as much as 200-fold, independent of the *rec* alleles of the host. Finally, in the *int*⁻ *red*⁻ *rec*⁻ case, where all known recombination systems are defective, the number of transducing particles is augmented slightly further, and may even exceed the number of plaque formers in the lysate by orders of magnitude. The over-produced particles are of the nonlysogenizing class [71, 75, 116]. Those of the lysogenizing class are actually reduced in number in lysates of λ*red*⁻ [97]. In λ*int*⁻ lysates, there is also a striking increase of particles containing DNA capable of transforming *bio*⁻ cells to *bio*⁺ by substitution. The DNA that is capable of transforming to *bio*⁺ must first be extracted from particles that are themselves noninfectious [116].

The existence of a special kind of transducing DNA in lysates of induced λ*int*⁻ lysogens had been postulated once it was appreciated that *ter* function might cleave a single unexcised prophage just as it cleaves a prophage pair [75]. This cleavage, if accompanied by cuts in the adjacent bacterial DNA, could give rise to λ*gal* DNA bearing a right cohesive end and to λ*bio* DNA bearing a left cohesive end (Fig. 1). It has now been demonstrated unequivocally that the extracted DNAs of these infectious *gal*- and noninfectious *bio*- transducing defective particles bear only one cohesive end, right and left, respectively [116]. These *gal* and *bio* particles are therefore called *doc*R and *doc*L.

It has been estimated that an appreciable proportion of induced λ*int*⁻ *red*⁻ lysogens gives rise to at least one λ *doc* particle per cell [116]. Thus "*doc* excision" may be within an order of magnitude as efficient as *int*-promoted excision. The potential for this alternative mode of excision is normally masked by the functioning of *int, xis,* and *red* genes prior to replication and prior to the appearance of *ter* function. Furthermore, relative to complete λ prophage, *doc* DNA released from the chromosome is at a disadvantage, as it fails to replicate [116; see also 19].

The fact that λ*doc* particles include bacterial genes adjacent to the prophage is a consequence of DNA size requirements in packaging and comes about as the result of various scissions within bacterial DNA. These scissions differ from those responsible for generating lysogenizing transducing phage in that the latter are coincident with, or followed by, union of the excised ends to form DNA rings. Either the *ter*-cut ends of *doc* molecules cannot fuse with the other cut end or such fusions block subsequent release of cohesive ends, and the resultant DNA rings are simply not packaged. It has been suggested that *red*$^+$ function contributes to the fusion of nonhomologous ends [96]. This is consistent with the opposing effects of *red*$^-$ mutations on the genesis of lysogenizing and nonlysogenizing transducing particles, but fails to explain the disproportionately large increase in *doc* particles that occur upon introduction of certain *red*$^-$ alleles.

The extraordinary potential for excisions that do not fuse the broken ends raises the question of whether the cuts in bacterial DNA required for the generation of λ*doc* are formed upon induction or whether they preexist. If we admit that such cuts are irreversible and distributed more or less randomly along the bacterial chromosome, then their preexistence is excluded; otherwise each bacterial chromosome would be severed prior to induction. The argument that cuts in bacterial DNA arise largely as a consequence of induction, rather than beforehand, is supported by more direct evidence. Firstly, it has been demonstrated that a normal yield of transducing particles is produced in zygotic induction, i. e., in the course of conjugal transfer of integrated prophage to a nonlysogenic recipient [1]. Secondly, transient induction of λ lysogens under suitable conditions produces *gal* deletions at high frequency (about one percent of survivors) [135]. Some of the same preferred regions of scission are revealed in the frequency distribution of end-points within *gal* of independent deletions [135] and of independent transducing particles [136]. This finding is consistent with the hypothesis that at least some of the surviving bacteria in which *gal* deletions have been induced represent the reciprocal products of transducing particle genesis. To what extent these reciprocal products are generated in single reciprocal events is not known. If we admit that λ induction causes cuts in bacterial DNA, then we must consider how it is that transducing particles are not produced on heteroimmune superinfection of a λ lysogen [1]. What *cis*-specific consequence of induction might result in scissions outside the prophage borders? The most obvious candidates would appear to be prophage-initiated replication and DNA packaging in situ.

Although most known functions affecting DNA repair and recombination, including *int* itself, have been found nonessential for transducing particle generation [59], functions having to do with DNA replication and packaging have not been tants. It should be much simpler to determine whether mutations that affect DNA replication (or packaging) influence the formation of induced *gal* deletions. Unfortunately, both cutting and joining of nonhomologous ends are required to recover a viable cell, and we cannot discriminate between the two processes. Cutting without fusion can occur frequently, as in the formation of λ*doc,* but we do not know the parameters that determine whether nonhomologous ends fuse.

The mode of transforming particle formation seems clear in only one exceptional case, that of λ*att*^{2}, the phage bearing both *att* B.P′ and *att* P.B′ and the *bio* and *uvrB* operons between them [75, 154]. Formation of this phage depends upon a prior deletion of dispensable phage genes sufficiently large to permit packaging of the entire prophage as well as host DNA on either side. In addition, an excision defect is required, protecting λatt^2 from destruction by internal recombination between the two *att* sites. If these two requirements are met, the vast majority of plaque-formers in an induced lysate will be λatt^2. A distinctive feature of this phage, its well defined buoyant density, suggests that it is formed differently from other transducing phages. Either *red* or *rec* function seems to be required for its formation; hence its origin may be due to generalized recombination between short regions of homology bracketing *att* B.P′, and *uvr*B. Lambda att^2 is thus a very special phage in its genesis as well as in its properties.

Use of abnormal *att* sites

Although no known variety of transducing particle appears generated by erroneous *int*-mediated excision, *int can* make errors in the sense that it can act inefficiently at sites other than those normally used. In hosts deleted for the normal bacterial *att* site, *int*-promoted insertions of λ or *P22* into "odd" sites can occur [141, 152]. Curing of λ from odd sites is dependent upon *int* and *xis* and normally causes an exact reversal of the integration event since mutations produced by the original insertions are reversed upon excision [152]. For this reason, the inefficient use of secondary sites by *int* cannot be considered sloppy.

The requirement for *xis* to excise from all secondary λ sites so far tested suggests that these sites resemble *att* B.B′ more closely than *att* P.P′, *att* B.P′, or *att* P.B′. On the other hand, there appear to be one or more secondary sites for *P2* prophage in *E. coli* K12 that may be more like *att* B.P′ than *att* B.B′. These sites, perhaps vestiges of a cryptic *P2* prophage, are utilized in a reaction called "eduction" (defined as the loss of host genetic material along with the loss of an integrated episome). The prophage is lost from a site near *his,* along with a series of bacterial genes nearby and "clockwise" to it [173, 174]. If the sites were sufficiently close to *P2* prophage, this *int*-promoted reaction could generate, in principle, a class of transducing particles.

The range of bacterial markers that can be transduced by λ integrated at a given *att* site is severely limited. It has been much amplified by taking advantage of infrequently occupied *att* sites [152], or by bringing bacterial markers nearer to a normal *att* site by insertion of an F′ factor and by selection of appropriate deletions or inversions [8, 76, 144, 165]. A large number of bacterial markers have been fused to λ DNA in these ways. The resultant transducing DNAs may or may not be defective for phage production depending on those viral genes, if any, that have been replaced.

Deletions, duplications, and insertions

So far we have considered insertion and excision reactions in which an entire phage chromosome or a phage equivalent of DNA interacts with host DNA. Rearrangements of a similar kind can occur within the phage chromosome itself, resulting in deletions, duplications, and insertions. These are generally very rare events, but there exist potent selection methods to enrich for mutants that have experienced one or another of these changes. A variety of methods are given below.

In selecting for phage mutants with altered DNA content, advantage is taken of the effects of such alterations on particle buoyant density [10, 185] and viral stability to heat and chelating agents [13, 134]. Deletions crossing the prophage-host boundary are found among bacteria that must lose a pair of functions, e. g. both a phage- and a host-determined function to survive the selective conditions [2, 62, 135, 174]. Of course, the lost segment of host DNA must contain no indispensable genes. A large deletion of dispensable phage genes can result in strong selection pressure for the acquisition of additional DNA. For example, a deletion mutant of a hybrid (ϕ 80–λ) phage making small plaques and missing about one-fourth of its genome, produces lysates containing several percent "add" mutants, i. e., phages of increased density and plaque-size owing to DNA additions [10, 58]. Changed DNA content, insertions or deletions, are also found in phages that have been selected for certain classes of functional defects that do not readily arise by point mutation. Insertions of short segments of genetic material with effects on operon expression have been found in *gal* [84a, 91, 151] and *lac* [120] operons, in λ [21] and *P2* [12, 114] phages. In addition, certain acquired viral functions in λ, *Q*-bypass and *fec*-bypass, are attributed to substitutions of genetic material, although with certainty only in the first case [54]. At least one *N*-bypass mutation is known to be a small deletion [54]. That the effort to select bypass mutations should be rewarded with success attests to the extent of latent genetic capacities. Perhaps the forceful term "saltation" deserves revival to denote these extraordinary jumps.

Some additions of genetic material in phage are known to be additions of homologous DNA. The "add" mutants, for example, arise spontaneously by tandem duplication. The segment duplicated may come from any part of the phage genome. Special properties ensue from internal homology as a result of the opportunities afforded for inter- and intra-molecular crossing-over [10]. Provided the duplication is small enough, and generalized recombination functions are available, the phage DNA can grow further by "unequal crosssing-over", i. e., intermolecular recombination of homologous segments in nonaligned molecules. This recombination generates tandem triplications and, subsequently, even higher oligomers. Formation of the initial tandem duplications, in contrast to subsequent DNA growth by unequal crossing-over, is independent of generalized recombination functions. Intra- as well as inter- molecular recombination can occur between tandem duplications.

The former yields products of diminished length; the latter can result in products of increased or diminished length. Intermolecular recombination appears to be favored over intramolecular.

The importance of tandem duplications for evolution has long been appreciated [22, 26]. The persistence of a capacity for generating multiple gene copies may have an evolutionary value even in an organism that possesses relatively few repeated sequences that have remained protected from divergent evolution. Multiple gene copies appear more common in eukaryotes [23], where this mode of amplifying gene expression may convey some special advantages.

How tandem duplications are formed is not known. It is tempting to connect the formation of tandem duplications and deletions via a common initial step, a reciprocal recombination that produces both a deletion and a small ring of excised DNA. The latter could recombine within a homologous region of an intact sister chromosome to form a tandem duplication. The temptation must be resisted, since "add" mutants arise in the absence of the generalized recombination functions (*recA* and *red)* expected to be required for the second step.

Alternative models for deletion and tandem duplication may be entertained. It is not necessary, for example, to assume that breakage and reunion are involved at all. Aberrant events could result from errors in DNA replication. Skipping of a looped-out sequence could produce a deletion; "back-sliding" of DNA polymerase, leading to recopying of the same template, could produce a tandem duplication. The initial products would be heterozygous [59].

Except in the special case of *int*-promoted deletions (see previous section, "Towards understanding the mechanism . . ."), no functions have been directly implicated in the formation of deletions, duplications, or insertions. The recent finding of high frequency deletions of bacterial genes upon prophage curing makes another mechanism for obtaining deletions accessible to experimental analysis [135]. Again, a critical function is carried by λ. There are other means of increasing the frequency of deletions in λ and *E. coli;* certain bacterial mutations or mutagenic agents that induce or perpetuate DNA lesions have this effect [59]. It would be interesting to determine how these various deletion-forming factors interact. A beginning in this direction has been made in the case of *int*-promoted and UV-promoted deletion in λ [40, 133, 134]. Certain functions, notably those involved in DNA synthesis, have not been tested for their effect on these events.

We may group together the excisions and fusions that result in transducing particles and the equally obscure events that lead to deletions, duplications, and insertions. In each case a "novel joint," or new juxtaposition of genetic sequences is produced [84]. Processes that give rise to a "novel joint" are subsumed under the term "illegitimate recombination" [59]. "Novelty" here implies insufficient homology with the parental DNA to promote heteroduplex formation. this definition may appear somewhat loose or arbitrary, but it is probably adequate to set apart a class of recombinations that differ in mechanism from generalized recombination, since they prove to be independent of *recA* function. Illegitimate recombinations are not necessarily a homogeneous class, and it seems probable that the response of different reactions belonging to this class to UV-irradiation, or to other treatments, will reveal various subclasses.

It is not hard to imagine how a particular illegitimate recombination might become favored in the course of evolution. A possible stepping stone between a process that is rare and haphazard and one that is efficient and precise may be exemplified by a particular insertion that renders phage *P2* immunity insensitive. This small added segment of DNA appears to be lost in a precise fashion, simultaneously restoring sensitivity to immunity [12].

Those illegitimate recombinations responsible for generating transducing particles were originally considered to be cases of faulty site-specific recombination [35]. We may consider an alternative relationship – the one of progenitor.

Conclusions

Prophage insertions and excisions require few protein components in addition to those normally present in the uninfected cell (and in some circumstances none). A very limited number of elements, interacting according to certain thermodynamic and kinetic prescriptions, suffices to account for the varieties of site preference in site-specific recombinations. These reactions efficiently provide for the establishment of integrated prophage and ensure its maintenance until signs of the unreliability of the host provoke the efficient release of the parasite. Any of several simple genetic or physiological changes may alter the elaborate strategies by which host and parasite coexist, without necessarily disrupting the stability of the system.

The essential feature of recombinations that add or subtract DNA sequences, rather than simply commute equivalent segments, could depend upon topology alone. In fact, viral recombination functions have evolved that are specialized for the addition and subtraction of one or another prophage and display various degrees of site preference. There are, furthermore, the means to add or subtract DNA sequences that are not specialized. Generalized recombinations (that involve pairing of homologs), as well as illegitimate recombinations (that have no such requirement within the present limits of detection), or even the cutting of phage DNA prerequisite for its encapsulation may serve these purposes.

In the processes of insertion and excision, alliances between units of transcription and replication are made and broken. These changes may influence the expression of *int* (among other genes), thereby providing for a regulation of integration (among other functions). This regulation is so simple as to be suitable for even a small animal virus of modest means.

Abnormal excisions that produce transducing particles, in which some host DNA remains joined to viral DNA, are also variously achieved. Attempts to interfere with the genesis of transducing particles by elimination of particular recombination functions have stimulated the production of unusual forms. The protean creativity in this small compass is a feature of hereditary material; it increases the capacity for adaptive change.

The aphorism quoted at the outset of this article, "life is the secret of DNA," expresses succinctly an appreciation of a creativity that counteracts causal determinism. In more general terms, this idea has been stated by *Mario Bunge* [28]: "As *Hegel* said, 'whatever has life does not allow the cause to reach the effect, that is, cancels it as cause'; what *Hegel* could not suspect in his time is that the same transmutation of external causes by inner conditions, the same specificity of response, is to be found everywhere in reality, not only in the domains of life and spirit."

Acknowledgements: Several persons have generously provided me with accounts of their unpublished work, and with good council. For this aid I thank *S. Adhya, R. L. Baldwin, G. Bertani, A. Bukhari, R. Calendar, W. F. Dove, H. Drexler, N. C. Franklin, M. E. Gottesman, P. Kourilsky, M. M. Howe, S. Kumar, M. Levine, G. Lindahl, D. Pfeifer, M. Shulman, H. B. Strack, M. Sunshine, R. Thomas, R. A. Weisberg,* and *D. Zipser. Naomi Franklin* deserves special thanks for her persistent, perceptive criticism of the manuscript during its successive, often illegitimate, recombinations.

References

N. B. "TBL" (in several references, e.g., [5]) stands for:

Hershey, A. D., ed., 1971. The Bacteriophage Lambda. Cold Spring Habor Laboratory, New York.

[1] *Adhya, S.* Personal communication.

[2] *Adhya, S., Cleary, P., and Campbell, A.* 1968. A deletion analysis of prophage lambda and adjacent genetic regions. Proc. Nat. Acad. Sci. **61:** 956–62.

[3] *Adhya, S., and Court, D.* Unpublished results.

[4] *Amati, P., and Meselson, M.* 1964. Localized negative interference in bacteriophage λ. Genetics **51:** 369–79.

[5] *Ausubel, F., Voynow, P., Signer, E., and Mistry, J.* Purification of protein determined by two nonessential genes in lambda. "TBL," 395–405.

[6] *Baron, L. S., Penido, E., Ryman, I. R., and Falkow, S.* 1970. Behavior of coliphage lambda in hybrids between Escherichia coli and Salmonella. J. Bacteriol. **102:** 221–33.

[7] *Bauerle, R. H., and Margolin, P.* 1967. Evidence for two sites for initiation of gene expression in the tryptophan operon of Salmonella typhimurium. J. Mol. Biol. **26:** 423–36.

[8] *Beckwith, J. R., and Signer, E. R.* 1966. Transposition of the *lac* region of Escherichia coli. I. Inversion of the *lac* operon and transduction of *lac* by ϕ 80. J. Mol. Biol. **19:** 254–65.

[9] *Belfort, M., and Wulff, D. L.* A mutant of Escherichia coli that is lysogenized with high frequency. "TBL," 739–42.

[10] *Bellett, A. J. D., Busse, H. G., and Baldwin, R. L.* Tandem genetic duplications in a derivative of phage lambda. "TBL." 501–13.

[11] *Bertani, G.* 1962. Multiple lysogeny from single infection. Virology **18:** 131–39.

[12] *Bertani, G., and Bertani L. E.* Personal communication.

[13] *Bertani, G., and Walker, D.* Unpublished observations cited in [17].

[14] *Bertani, L. E.* 1968. Abortive induction of bacteriophage P2. Virology **36:** 87–103.

[15] *Bertani, L. E.* 1970. Split-operon control of a prophage gene. Proc. Nat. Acad. Sci. **65:** 331–36.

[16] *Bertani, L. E.* 1971. Stabilization of P2 tandem double lysogens by *int* mutations in the prophage. Virology **46:** 426–36.

[17] *Bertani, L. E., and Bertani, G.* 1971. Genetics of P2 and related phages. Advan. Genetics, M. Demerec memorial volume, **16:** 199–237.

[18] *Boram, W., and Abelson, J.* 1971. Bacteriophage Mu integration: On the mechanism of Mu-induced mutations. J. Mol. Biol. **62:** 171–78.

[19] *Botstein, D., and Matz, M.* 1970. A recombination function essential to the growth of bacteriophage P22. J. Mol. Biol. **54:** 417–40.

[20] *Brachet, P.* Personal communication.

[21] *Brachet, P., Eisen, H., and Rambach, A.* 1970. Mutations of coliphage λ affecting the expression of replicative functions O and P. Mol. Gen. Genetics **108:** 266–76.

[22] *Bridges, C. B.* 1935. Salivary chromosome maps. J. Hered. **26:** 60–64.

[23] *Britten, R. J., and Kohne, D. E.* 1968. Repeated sequences in DNA. Science **161:** 529–40.

[23 a] *Broker, T. R., and Lehman, I. R.* 1971. Branched DNA molecules: Intermediates in T4 recombination. J. Mol. Biol. **60:** 131–49.

[24] *Brooks, K.* 1965. Studies in the physiological genetics of some supressor-sensitive mutants of bacteriophage λ. Virology **26:** 489–99.

[25] *Brown, A., and Arber, W.* 1964. Temperature-sensitive mutants of coliphage lambda. Virology **24:** 237–39.

[26] *Bryson, V., and Vogel, H. J.*, eds. 1965. Evolving Genes and Proteins. New York: Academic Press.

[27] *Bukhari, A. I., and Zipser, D.* 1972. Random insertion of Mu–1 DNA within a single gene. Nature New Biology **236:** 240–43.

[28] *Bunge, M.* 1959. Causality, the Place of the Causal Principle in Modern Science, pp. 179–80. Cambridge, Mass: Harvard University Press.

[29] *Buttin, G., Dechamps, M., Eisen, M., and Pereira da Silva, L.* Submitted for publication. Regulation of galactose enzyme biosynthesis in Escherichia coli IV. A dual mechanism for "escape synthesis" of host *gal* enzymes following induction of λ prophages.

[30] *Calef, E., and Liciardello, G.* 1960. Recombination experiments on prophage host relationships. Virology **12:** 81–103.

[31] *Calendar, R.* 1970. The regulation of phage development. Ann. Rev. Microbiol. **24:** 241–96.

[32] *Calendar, R., and Lindahl, G.* 1969. Attachment of prophage P2: gene order at different host chromosomal sites. Virology **39:** 867–81.

[33] *Calendar, R., Lindahl, G., Marsh, M., and Sunshine, M.* 1972. Temperature-inducible mutants of P2 phage. Virology **47:** 68–75.

[34] *Calendar, R., Lindqvist, B., Sironi, G., and Clark, A. J.* 1970. Characterization of REP⁻ mutants and their interaction with P2 phage. Virology **40:** 72–83.

[34a] *Campbell, A.* 1957. Transduction and segregation in Escherichia coli K12. Virology **4:** 366–84.

[35] *Campbell, A.* 1962. Episomes. Advances in Genetics **11:** 101–45.

[36] *Campbell, A.* 1964. Transduction. In The Bacteria, eds. *I. C. Gunsalus,* and *R. Y. Stanier,* Vol. 5, pp. 49–85. New York: Academic Press.

[37] *Campbell, A.* 1965. The steric effect in lysogenization by bacteriophage lambda. II. Chromosomal attachment of the *b2* mutant. Virology **27:** 340–45.

[38] *Caro, L.* Personal communication.

[39] *Choe, B. K.* 1969. Integration defective mutants of bacteriophage P2. Mol. Gen. Genetics **105:** 275–84.

[40] *Davis, R. W., and Parkinson, J. S.* 1971. Deletion mutants of bacteriophage lambda. III. Physical structure of *att.* J. Mol. Biol. **56:** 403–23.

[41] *Donch, J., Greenberg, J., and Green, M. H.* 1970. Repression of induction by U. V. of λ phage by *exrA* mutations in Escherichia coli. Genetic. Res. Camb. **15:** 87–97.

[42] *Dove, W.* 1970. An energy level hypothesis for λ prophage insertion and excision. J. Mol. Biol. **47:** 585–89.

[43] *Dove, W. F., and Weigle, J. J.* 1965. Intracellular state of the chromosome of bacteriophage lambda. I. The eclipse of infectivity of the bacteriophage DNA. J. Mol. Biol. **12:** 620–29.

[44] *Drexler, H.* 1970. Transduction by bacteriophage T1. Proc. Nat. Acad. Sci. **66:** 1083–88.

[45] *Drexler, H.* 1972. The transduction of *gal*⁺ by coliphage T1: II. The role of λ transcription control on the efficiency of transduction. J. Virol. **9:** 280–85.

[46] *Echols, H.* Regulation of lytic development. "TBL," 247–70.

[47] *Echols, H., and Court, D.* The role of helper phage in *gal* transduction. "TBL," 701–10.

[48] *Echols, H., Gingery, R., and Moore, L.* 1968. The integrative recombination function of bacteriophage λ. Evidence for a site-specific recombination enzyme. J. Mol. Biol. **34:** 251–60.

[49] *Eisen, H. A., Siminovitch, L., and Mohide, P. T.* 1968. Excision of lambda prophage: Effects on host survival. Virology **34:** 97–103.

[50] *Faelen, M.* Personal communication from *R. Thomas.*

[51] *Faelen, M., Toussaint, A., and Couturier, M.* 1971. Mu–1 promoted integration of a λ-*gal* phage in the chromosome of E. coli. Mol. Gen. Genetics **113:** 367–70.

[52] *Falkow, S., and Baron, L. S.* 1970. Plasmid formation after lambda bacteriophage infection of *Escherichia coli-Salmonella typhosa* hybrids. J. Bacteriol. **102:** 288–90.

[53] *Fangman, W. L., and Feiss, M.* 1969. The fate of λ DNA in a bacterial host defective in DNA synthesis. J. Mol. Biol. **44:** 103–16.

[54] *Fiandt, M., Hradecna, Z., Lozeron, H., and Szybalski, W.* Electron micrographic mapping of deletions, insertions, inversions, and homologies in the DNAs of coliphages lambda and phi 80. "TBL," 329–54.

[55] *Fischer-Fantuzzi, L.* 1967. Integration of λ and λb2 genomes in nonimmune host bacteria carrying a λ cryptic prophage. Virology **32:** 18–32.

[56] *Folkmanis, A.* In preparation.

[57] *Folkmanis, A., and Freifelder, D.* 1972. Studies on lysogeny in *Escherichia coli* with bacteriophage λ. Physical observation of the insertion process. J. Mol. Biol. **65:** 63–73.

[58] *Franklin, N. C.* 1967. Deletions and functions of the center of the φ 80-λ phage genome. Evidence for a phage function promoting genetic recombination. Genetics **57:** 301–18.

[59] *Franklin, N. C.* Illegitimate recombination. "TBL," 175–94.

[60] *Franklin, N. C.* The *N* operon of lambda: Extent and regulation as observed in fusions to the tryptophan operon of *Escherichia coli.* "TBL," 621–38.

[61] *Franklin, N. C.* Personal communication.

[62] *Franklin, N. C., Dove, W. F., and Yanofsky, C.* 1965. The linear insertion of a prophage into the chromosome of *E. coli* shown by deletion mapping. Biochem. Biophys. Res. Comm. **18:** 910–23.

[63] *Friedman, D. I.* A bacterial mutant affecting lambda development. "TBL," 733–38.

[64] *Friedman, D. I., and Yarmolinsky, M. B.* In press. Prevention of the lethality of induced λ prophage by an isogenic λ plasmid. Virology.

[65] *Freifelder, D.*, and *Kirschner, I.* 1971. A phage λ endonuclease controlled by genes *O* and *P.* Virology **44:** 223–25.

[67] *Freifelder, D., Kirschner, I., and Goldstein, R.* In press. Physical studies of prophage curing and excision. J. Mol. Bio.

[68] *Gellert, M.* 1967. Formation of covalent circles of lambda DNA by *E. coli* extracts. Proc. Nat. Acad. Sci. **57:** 148–55.

[69] *Georgopoulos, C. P.* A bacterial mutation affecting N function. "TBL," 639–45.

[70] *Georgopoulos, C. P.* 1971. Bacterial mutants in which the gene N function of bacteriophage lambda is blocked have an altered RNA polymerase. Proc. Nat. Acad. Sci. **68:** 2977–81.

[70a] *Ghysen, A. and Pironio M.* 1972. Relationship between the *N* function of bacteriophage λ and host RNA polymerase. J. Mol. Biol. **65:** 259–272.

[71] *Gingery, R., and Echols, H.* 1968. Integration, excision and transducing particle genesis by bacteriophage λ. Cold Spring Harbor Symp. Quant. Biol. **33:** 721–27.

[72] *Goldstein, A.* 1944. The mechanism of enzyme-inhibitor-substrate reactions. Illustrated by the cholinesterase-physostigmine-acetylcholine system. J. Gen. Physiol. **27:** 529–80.

[73] *Gottesman, M. E., and Weisberg, R. A.* Prophage insertion and excision. "TBL," 113–38.

[74] *Gottesman, M. E., and Yarmolinsky, M. B.* 1968. Integration-negative mutants of bacteriophage lambda. J. Mol. Biol. **31:** 487–505.

[75] *Gottesman, M. E., and Yarmolinsky, M. B.* 1968. The integration and excision of the bacteriophage lambda genome. Cold Spring Harbor Symp. Quant. Biol. **33:** 735–47.

[76] *Gottesman, S., and Beckwith, J. R.* 1969. Direct transposition of the arabinose operon. A technique for isolation of specialized transducing bacteriophages for any *Escherichia coli* gene. J. Mol. Biol. **44:** 117–27.

[77] *Grodzicker, T., Arditti, R. R., and Eisen, H.* 1972. Establishment of repression by lambdoid phage in catabolite activator protein and adenylate cyclase mutants of *Escherichia coli.* Proc. Nat. Acad. Sci. **69:** 336–70.

[78] *Guaneros, G., and Echols, H.* 1970. New mutants of bacteriophage λ with a specific defect in excision from the host chromosome. J. Mol. Biol. **47:** 565–74.

[79] *Guarneros, G., and Echols, H.* Personal communication.

[80] *Guerrini, F.* 1969. On the asymmetry of λ integration sites. J. Mol. Biol. **46:** 523–42.

[81] *Helling, R. B.* 1967. The effect of arabionose-specific enzyme synthesis on recombination in the arabinose genes of *E. coli.* Genetics **57:** 665–75.

[82] *Hendrix, R. W.* Identification of proteins coded in phage lambda. "TBL," 355–70.

[83] *Hermann, R. K.* 1968. Effect of gene induction on frequency of intragenic recombination of chromosome and F merogenote in *E. coli* K 12. Genetics **58:** 55–67.

[84] *Hershey, A. D.* 1969. Genes and hereditary characteristics. Carnegie Inst. Wash. Yearbook **68:** 655.

[84a] *Hirsch, H. J., Saedler, H.,* and *Starlinger, P.* 1972. Insertion mutations in the control region of the galactose operon of *E. coli* II. Physical characterization of the mutations. Mol. Gen. Genetics **115:** 266–276.

[85] *Hong, J. S., Smith, G. R., and Ames, B. N.* 1971. Adenosine-3′:5′-cyclic monophosphate concentration in the bacterial host regulates the viral decision between lysogeny and lysis. Proc. Nat. Acad. Sci. **68:** 2258–62.

[86] *Howe, M. M.* Personal communication.

[87] *Hradecna, Z., and Szybalski, W.* 1969. Electron micrographic maps of deletions and substitutions in the genomes of transducing coliphages λ*dg* and λ*bio.* Virology **38:** 473–77.

[88] *Ikeda, H., and Tomizawa, J.* 1965. Transducing fragments in generalized transduction by phage P1. I. Molecular origin of the fragments. J. Mol. Biol. **14:** 85–109.

[89] *Ikeda, H., and Tomizawa, J.* 1968. Prophage P1, an extra-chromosomal replication unit. Cold Spring Harbor Symp. Quant. Biol. **33:** 791–98.

[90] *Jacob, F., and Wollman, E. L.* 1955. Etude génétique d'un bactériophage tempéré d' *Escherichia coli.* III. Effet du rayonnement ultraviolet sur la recombinaison génétique. Ann. Inst. Pasteur **88:** 724–49.

[91] *Jordan, E., Saedler, H., and Starlinger, P.* 1968. 0° and strong-polar mutations in the *gal* operon are insertions. Mol. Gen. Genetics **102:** 353–63.

[92] *Kaiser, A. D., and Inman, R. B.* 1965. Cohesion and the biological activity of bacteriophage lambda DNA. J. Mol. Biol. **13:** 78–91.

[93] *Kaiser, A. D., and Masuda, T.* 1970. Evidence for a prophage excision gene in lambda. J. Mol. Biol. **47:** 557–64.

[94] *Kaiser, A. D., and Wu, R.* 1968. Structure and function of DNA cohesive ends. Cold Spring Harbor Symp. Quant. Biol. **33:** 729–34.

[95] *Kayajanian, G.* 1970. Plating of λ derivatives on an *Escherichia coli-Salmonella typhosa* hybrid. Virology **40:** 763–67.

[96] *Kayajanian, G.* 1970. *Gal* transduction by phage λ: on the origin and nature of LFT tranducing genomes. Mol. Gen. Genetics **108:** 338–48.

[97] *Kayajanian, G.* Personal communication cited in [59].

[98] *Kellenberger, G., Zichichi, M. L., and Epstein, H. T.* 1962. Heterozygosis and recombination of bacteriophage λ. Virology **17:** 44–55.

[99] *Kellenberger-Gujer, G.* Recombination in bacteriophage lambda. II. The mechanism of general recombination. "TBL," 417–30.

[100] *Kirby, E. P., Jacob, F., and Goldthwait, D. A.* 1967. Prophage induction and filament formation in a mutant strain of *Escherichia coli.* Proc. Nat. Acad. Sci. **58:** 1903–10.

[101] *Kourilsky, P.* 1971. Lysogenization by bacteriophage lambda and the regulation of lambda repressor synthesis. Virology **45:** 853–57.

[102] *Kourilsky, P.* Personal communication.

[103] *Krell, K., Gottesman, M. E., and Parks, J. S.* In press. Escape synthesis of the biotin operon in induced λb2 lysogens. J. Mol. Biol.

[104] *Kumar, S., and Fuerst, C. R.* isolated a class of thermosensitive *E. coli* mutants called *traL* in which λ^+ behaves like λN^- (personal communication from *S. Kumar).*

[105] *Lash, T. and Rubinstein, I.* Personal communication cited in [73].

[106] *Davies, R. W., Dove, W. F., Inokuchi, H., Lehman, J. F.* and *Roehrdanz, R. L.* In press. Regulation of λ prophage excision by the transcriptional state of DNA. Nature New Biology.

[107] *Lennox, E. S.* 1955. Transduction of linked genetic characters of the host by bacteriophage P1. Virology **1:** 190–206.

[108] *Levine, M.* In press. Replication and lysogeny with phage P22 in *Salmonella typhimurium.* In Current Topics in Microbiology and Immunology.

[109] *Levine, M., and Schott, C.* 1971. Mutations of phage P22 affecting phage DNA synthesis and lysogenization. J. Mol. Biol. **62:** 53–64.

[110] *Lieb, M.* 1970. λ mutants that persist as plasmids. J. Virol. **6:** 218–25.

[111] *Lieb, M.* Lysogenization by lambda lacking repressor and N function. "TBL," 679–90.

[112] *Lindahl, G.* 1969. Genetic map of bacteriophage P2. Virology **39:** 839–60.

[113] *Lindahl, G.* 1969. Multiple recombination mechanisms in bacteriophage P2. Virology **39:** 861–66.

[114] *Lindahl, G., Hirota, Y., and Jacob, F.* 1971. On the process of cellular division in *Escherichia coli:* replication of the bacterial chromosome under control of prophage P2. Proc. Nat. Acad. Sci. **68:** 2407–11.

[115] *Lindahl, G. and Sunshine, M.* Submitted for publication. Excision-deficient mutants of bacteriophage P2.

[116] *Little, J. W., and Gottesman, M.* Defective lambda particles whose DNA carries only a single cohesive end. "TBL," 371–94.

[117] *Loveday, and Botstein, D.* Personal communication cited in [109].

[118] *Luria, S. E., Fraser, D. K., Adams, J. N., and Burrous, J. W.* 1958. Lysogenization, transduction, and genetic recombination in bacteria. Cold Spring Harbor Symp. Quant. Biol. **23:** 71–82.

[119] *Lwoff, A.* 1953. Lysogeny. Bacteriol. Rev. **17:** 269–337.

[120] *Malamy, M.* 1970. Some properties of insertion mutations in the *lac* operson. In The Lactose Operon, eds. *J. Beckwith,* and *D. Zipser,* pp. 359–73. Cold Spring Harbor Laboratory.

[121] *Mattern, I. E., Van Winden, M. P., and Rörsch, A.* 1965. The range of action of genes controlling radiation sensitivity of *Escherichia coli.* Mutation Res. **2:** 111–31.

[122] *Meselson, M., and Yuan, R.* 1968. DNA restriction enzyme from *E. coli.* Nature **217:** 1110–14.

[123] *Monk, M., Peacey, M., and Gross, J. D.* 1971. Repair of damage induced by ultraviolet light in DNA polymerase-defective *Escherichia coli* cells. J. Mol. Biol. **58:** 623–30.

[124] *Morse, D. E., and Yanofsky, C.* 1969. A transcription-initiating mutation within a structural gene of the tryptophan operon. J. Mol. Biol. **41:** 317–28.

[125] *Mousset, S., and Thomas, R.* 1969. Ter, a function which generates the ends of the mature λ chromosome. Nature **221:** 242–44.

[126] *Nash, H. A.* Personal communication.

[127] *Nash, H. A., and Robertson, C. A.* 1971. On the mechanism of *int*-promoted recombination. Virology **44:** 446–49.

[128] *Nishimura, Y., Caro, L., Berg, C. M., Hirota, Y.* 1971. Chromosome replication in *Escherichia coli.* IV. Control of chromosome replication and cell dividion. J. Mol. Biol. **55:** 441–56.

[129] *Ogawa, H., Shimada, K., and Tomizawa, J. I.* 1968. Studies on radiation-sensitive mutants of *E. coli.* I. Mutants defective in the repair synthesis. Molec. Gen. Genetics **101:** 227–44.

[130] *Ortin, J., Viñuela, E., Salas, M., and Vasquez, C.* 1971. DNA protein complex in circular DNA from phage ϕ 29. Nature New Biology **234:** 275–77.

[131] *Ozeki, H., and Ikeda, H.* 1968. Transduction mechanisms. Ann. Rev. Genetics **2:** 245–78.

[132] *Padmanabhan, R., and Wu, R.* 1972. Nucleotide sequence analysis of DNA IV. Complete sequence of the left-hand cohesive and of bacteriophage 186 DNA. J. Mol. Bol. **65:** 447–67.

[133] *Parkinson, J. S.* 1971. Deletion mutants of bacteriophage lambda. II. Genetic properties of *att*-defective mutants. J. Mol. Biol. **56:** 385–401.

[134] *Parkinson, J. S., and Huskey, R. J.* 1971. Deletion mutants of bacteriophage lambda. I. Isolation and initial characterization. J. Mol. Biol. **56:** 369–84.

[135] *Pfeifer, D.* Manuscript in preparation.

[136] *Pfeifer, D., and Royspal, St.* 1969. Deletion mapping of galactose mutants with transducing λ phages. Mol. Gen. Genetics **104:** 116–23.

[137] *Pironio, M., and Ghysen, A.* 1970. A Bacterial mutation which affects recognition of the N gene product of bacteriophage λ. Mol. Gen Genetics **108:** 374–75.

[138] *Ptashne, M.* 1965. The detachment and maturation of conserved lambda prophage DNA. J. Mol. Biol. **11:** 90–96.

[139] *Quigley, M.* 1963. Quoted in Science and Survival, *B. Commoner,* p. 43. New York: Viking Press.

[140] *Radding, C. M., and Carter D. M.* 1971. The role of β protein of phage λ in genetic recombination. J. Biol. Chem. **246:** 2513–18.

[141] *Rao, H. N., and Smith, H. O.* 1968. Phage P22 lysogens of a *Salmonella typhimurium* mutant deleted at the normal prophage attachment site. Virology **36:** 328–30.

[142] *Reichardt, L., and Kaiser, A. D.* 1971. Control of λ repressor synthesis Proc. Nat. Acad. Sci. **68:** 2185–89.

[143] *Rownd, R., and Mickel, S.* 1971. Dissociation and reassociation of RTF and r-determinants of the R-factor NR1 in *Proteus mirabilis.* Nature New Biology **234:** 40–43.

[144] *Sato, K., and Campbell, A.* 1970. Specialized transduction of galactose by lambda phage from a deletion lysogen. Virology **41:** 474–87.

[145] *Schleif, R., Greenblatt, J., and Davis, R. W.* 1971. Dual control of arabinose genes on transducing phage λ *dara.* J. Mol. Biol. **59:** 127–50.

[146] *Schmieger, H.* 1971. The fate of the bacterial chromosome in P22-infected cells of *Salmonella typhimurium.* Mol. Gen. Genetics **110:** 238–44.

[147] *Schmieger, H.* 1971. A method for detection of phage mutants with altered transducing ability. Mol. Gen. Genetics **110:** 378–81.

[148] *Schmieger, H.* 1970. The molecular structure of the transducing particles of *Salmonella*-phage *P22*. II. Density gradient analysis of DNA. Mol. Gen. Genetics **109:** 323–37.

[149] *Schnös, M., and Inman, R.* 1970. Position of branch points in replicating λ DNA. J. Mol. Biol. **51:** 61–73.

[150] *Scott, J. R.* 1970. A defective P1 prophage with a chromosomal location. Virol. **40:** 144–51.

[151] *Shapiro, J.* 1969. Mutations caused by the insertion of genetic material into the galactose operon of *Escherichia coli.* J. Mol. Biol. **40:** 93–105.

[152] *Shimada, K., Weisberg, R. A., and Gottesman, M. E.* 1972. Prophage lambda at unusual chromosomal locations I. Location of the secondary attachment sites and the properties of the lysogens. J. Mol. Biol. **63:** 483–503.

[153] *Shulman, M.* Personal communication.

[154] *Shulman, M., and Gottesman, M.* Lambda att^2: a transducing phage capable of intramolecular *int-xis* promoted recombination. "TBL," 477–87.

[155] *Signer, E.* 1968. Lysogeny: The integration problem. Ann. Rev. Microbiol. **22:** 451–88.

[156] *Signer, E.* General recombination, "TBL," 139–74.

[157] *Signer, E. R.* 1969. Plasmid formation: a new mode of lysogeny by phage λ Nature **223:** 158–60.

[158] *Signer, E. R.* 1970. On the control of lysogeny in phage λ. Virology **40:** 624–33.

[159] *Signer, E. R., Manley, K. F., and Brunstetter, M. A.* 1969. Deletion mapping of the *c*III–*N* region of bacteriophage λ. Virology **39:** 137–41.

[160] *Signer, E. R., and Weil, J.* 1968. Site-specific recombination of bacteriophage λ. Cold Spring Harbor Symp. Quant. Biol. **33:** 715–19.

[161] *Signer, E., Weil, J., and Kimball, P.* 1969. Recombination in bacteriophage λ III. Studies on the nature of the prophage attachment region. J. Mol. Biol. **46:** 543–63.

[162] *Simon, M. N., Davis, R. W., and Davidson, N.* Heteroduplexes of DNA molecules of lambdoid phages: physical mapping of their base sequence relationships by electron microscopy. "TBL," 313–28.

[163] *Six, E.* 1966. Specificity of P2 for prophage site I on the chromosome of *Escherichia coli* C. Virology **29:** 105–25.

[164] *Six, E.* Personal communication cited in [17].

[165] *Smith, G. R.* 1971. Specialized transduction of the *Salmonella hut* operons by coliphage λ: deletion analysis of the *hut* operons employing λ p*hut*. Virology **45**, 206–23.

[166] *Smith, H. O.* 1968. Defective phage formation by lysogens of integration deficient phage P22 mutants. Virology **34:** 203–23.

[167] *Smith, H. O.* Personal communication cited in [108].

[168] *Smith, H. O., and Levine, M.* 1967. A phage P22 gene controlling integration of prophage. Virology **31:** 207–16.

[169] *Stahl, F. W., and Stahl, M. M.* DNA synthesis associated with recombination II. Recombination between repressed chromosomes. "TBL," 443–53 (Fig. 8).

[170] *Stevens, W. F., Adhya, S., and Szybalski, W.* Origin and bidirectional orientation of DNA replication in coliphage lambda. "TBL," 515–34.

[171] *Strack, H. B., and Merkle, E.* In preparation. Also personal communication. Prophage excision in multiple lysogens: Activation of *int* function by heteroimmune superinfection; disparities between curing and prophage liberation.

[172] *Storm, P. K.* 1971. Recombination deficient mutants of Escherichia coli K12. Thesis, University of Utrecht.

[172a] *Storm, P. K.*, and *Zaunbrecher, W. M.* 1972. A new radiation sensitive mutant of *Escherichia coli* K–12. Mol. Gen. Genetics **115:** 89–92.

[173] *Sunshine, M. G.* 1972. Dependence of eduction of P2 *int* product. Virology **47:** 61–67.

[174] *Sunshine, M. G.*, and *Kelly, B.* 1971. Extent of host deletions associated with bacteriophage P2-mediated eduction. J. Bacteriol. **108:** 695–704.

[175] *Szybalski, W.* Personal communication.

[176] *Takano, T.* 1971. Bacterial mutans defective in plasmid formation: requirement for the ion^+ allele. Proc. Nat. Acad. Sci. **68:** 1469–73.

[177] *Taylor, A. L.* 1963. Bacteriophage-induced mutation in *Escherichia coli.* Proc. Nat. Acad. Sci. **50:** 1043–51.

[178] *Taylor, A. L., and Trotter, C. D.* Personal communication.

[179] *Thomas, C. A., Jr, and MacHattie, L. A.* 1967. The anatomy of viral DNA molecules. Ann. Rev. Biochem. **36,** part 2: 485–518.

[180] *Thomas, R., and Mousset, S.* 1970. Passive replication of heteroimmune bacteriophage in a λ-driven replicon. J. Mol. Biol. **41:** 179–91.

[181] *Tomizawa, J-I.* 1967. Molecular mechanisms of genetic recombination in bacteriophage: joint molecules and their conversion to recombinant molecules. J. Cell. Physiol. **70,** suppl. I: 201–14.

[182] *Tomizawa, J-I.* Functional cooperation of genes *O* and *P.* "TBL," 549–52.

[183] *Toussaint, A.* 1969. Insertion of phage Mu I within prophage λ: a new approach for studying the control of the late function in bacteriophage λ. Mol. Gen. Genetics **106:** 89–92.

[184] *Weigle, J.* 1961. Densities of transducing lambda bacteriophage. J. Mol. Biol. **3:** 393–98.

[185] *Weigle, J., Meselson, M., and Paigen, K.* 1959. Density alterations associated with transducing ability in bacteriophage lambda. J. Mol. Biol. **1,** 379–86.

[186] *Weil, J.* 1969. Reciprocal and non-reciprocal recombination in phage λ. J. Mol. Biol. **43:** 351–55.

[187] *Weisberg, R., and Gallant, J.* 1967. Dual function of the λ prophage repressor. J. Mol. Biol. **25:** 537–44.

[188] *Weisberg, R. A., and Gottesman, M.* 1969. The integration and excision defect of bacteriophage λ*dg*. J. Mol. Biol. **46:** 565–80.

[189] *Weisberg, R. A., and Gottesman, M. E.* The stability of *int* and *xis* functions. "TBL," 489–500.

[190] *Willetts, N. S., and Clark, A. J.* 1969. Characteristics of some multiply recombination-deficient strains of *Escherichia coli.* J. Bacteriol. **100:** 231–39.

[191] *Wiman, M., Bertani, G., Kelly, B., and Sasaki, I.* 1970. Genetic map of *Escherichia coli* strain *C.* Mol. Gen. Genetics **107:** 1–31.

[192] *Wing, J. P.* 1968. Integration and induction of phage P22 in a recombination deficient mutant of *Salmonella typhimurium.* J. Virol. **2:** 702–9.

[193] *Wu, R., and Taylor, E.* 1971. Nucleotide sequence analysis of DNA. II. Complete nucleotide sequence of the cohesive ends of bacteriophage λ DNA. J. Mol. Biol. **57:** 451–511.

[194] *Yarmolinsky, M. B.* Making and joining DNA ends. "TBL," 97–111.

[195] *Zichichi, M., and Kellenberger, G.* 1963. Two distinct functions in the lysogenization process: the repression of phage multiplication and the incorporation of the prophage in the bacterial genome. Virology **19:** 450–60.

[196] *Zinder, N. D., and Lederberg, J.* 1952. Genetic exchange in *Salmonella.* J. Bacteriol. **64:** 679–99.

[197] *Zissler, J.* 1967. Integration negative *(int)* mutants of phage λ. Virology **31:** 189.

Discussion

Sinsheimer: I was curious – when you insert into an abnormal site, it requires the int function?

Yarmolinsky: Yes.

Sinsheimer: Then what is the ratio of such an event to the normal site?

Yarmolinsky: It is extremely low; it is four or five orders of magnitude lower.

Hilz: You said that a catalyst can change the direction of a reaction. In terms of chemistry, this is not possible. A catalyst cannot change the direction of an equilibrium.

Yarmolinsky: I think we agree that a catalyst cannot change the direction of an equilibrium, but when you have a substrate which is present in the amount of one or two molecules per cell, and an enzyme which is present in much higher amounts, the enzyme may bind to all of the substrate and then is no longer only a catalyst but also a participant.

Hilz: The presence of an additional enzyme could also indicate a detour reaction. The enzyme as a true catalyst would then make available, for instance, the energy of an ATP molecule, thus pulling the reaction in another direction.

Yarmolinsky: Yes, one alternative that has been suggested is that the insertion of prophage and the excision of that prophage are really in fact very different reactions, just as the formation and breakdown of DNA are different reactions. But since the *int* function is involved in all of the known recombination events between attachment sites, it seems very likely that *int* protein, at least, is a common catalyst. It is hard to imagine that it has different catalystic functions depending upon the reaction, but even that is possible.

Trautner. Maybe I missed a point here. Did you say that when you have deletions which are of the order of a loss of 24 %, that you revert them to semiwild-type?

Yarmolinsky: That is correct. The deletions we are talking about do not impair any function required for phage growth, but the plaques are small.

Trautner: And do you have any idea, first, where this DNA is from, and, second, does it revert to any particular kind of wild-type or do you get a random population of different genotypes?

Yarmolinsky: The phages do not acquire any new functions, only DNA which is needed to give better "packaging". Phage proteins can package the larger piece of DNA more efficiently than the smaller piece. The added DNA is coming from the phage itself and forms tandem duplications. From these, you can get a whole sequence of different densities of phage as unequal crossing-over occurs between them. The duplicated regions can arise in various parts of the phage chromosome.

Trautner: So you do not recover any new function.

Yarmolinsky: No, you do not, although there is the case of a mutant with a deletion including the *red* gene being grown in a *rec*-host and acquiring a *rec*- or *red*-like function. I have not seen anything about it except the preliminary report by Zissler, Signer, and Schaefer in "The Bacteriophage Lambda".

Klingmüller: I think you mentioned, very casually though, that during integration of the prophage there is no DNA synthesis. Is that right?

Yarmolinsky: As far as we know, you do not need much DNA synthesis.

Klingmüller: This seems important because if there were none, the integration mechanism you are studying would be different from the recombination mechanism Radding described this morning. My question is, how carefully have you looked for the involvement of DNA synthesis?

Yarmolinsky: This has been done, for example, with Fangman's mutant FA 22, which is thermolabile for a *dna*B protein required for DNA synthesis. A lysogen of this mutant can be induced, and it will yield a few particles of prophage. I think there is a great deal of difficulty, when one has a mutation that affects DNA synthesis, in concluding that there is no DNA synthesis whatsoever at the level of these little attachment sites. All I can say is that a number of different ways of inhibiting DNA synthesis have been used and none impair site-specific recombination. The *rep* mutants, for example, that prevent P2 replication, allow P2 to integrate perfectly normally. And also, by density labelling, excised prophage DNA appears to be fully conserved, but whether a tiny bit of DNA synthesis occurs, I do not know. So I am not prepared to say that a different mechanism is involved here.

Random Nonhomologous Recombination

David Zipser, Ahmad I. Bukhari, and Jerome B. Zeldis*)

Cold Spring Harbor Laboratory, Cold Spring Harbor, New York, USA

Summary: The totally random insertion of phage Mu-1 into the *Escherichia coli* chromosome is described. Its significance for tumor viruses and genetic engineering is discussed.

Problem

The worst thing that can happen to a species is that its DNA recombines randomly at high frequency. This would reverse evolution in one generation. In fact, cells try very hard not to do this, although they do sneak in these random events at very low frequencies. *E. coli* will do it at between 10^{-7} and 10^{-9}, and these events are recognizable as deletions and insertions. While low, this frequency is nontrivial for evolution. It is important to be able to have an occasional random glueing-together of DNA as a last resort in the hope of getting out of some otherwise hopeless evolutionary problem. It is not yet known whether this low level of random recombination is due to a special enzyme system that is present in very low concentration, to "leakiness" of other enzymes, or to a spontaneous chemical reaction.

Aside from its function in evolution, there would be nothing better for the genetic engineer than to be able to glue pieces of DNA together nonhomologously at high frequencies. It is also possible that systems of this type play a role in higher-organism virus integration systems. In this paper, we discuss the random DNA glueing properties of a phage called Mu-1.

Little is known about Mu-1, and almost everything we know about it is dull in the sense that it is similar to other phages such as lambda, P22, etc. It has a head and a tail and a repressor. We don't know the details, but every time we find a new property of Mu-1 it's like this phage or that one. Mu-1 does have one very clear-cut difference – its mechanism of integration.

*) Brown University, Providence, Rhode Island, USA

Manuscript received: 31 January 1972

Mu-1 was shown by *Taylor* [2] to be a phage that physically inserts into the DNA of *coli* anywhere on the map, causing mutations if it inserts in a gene. He looked at a large number of Mu-1 induced mutants, and they were all at different sites.

When Mu-1 inserts into a gene, it spreads the two ends of this gene apart by its length (10–11 microns), which has a devastating effect on the gene, absolutely obliterating its functions. There is no reversion of these Mu-induced mutants at any reasonable frequency. The very simple question arises: how many different places can Mu-1 insert within a single gene? *A. I. Bukhari* obtained 76 independent Mu-1 insertions in the *z* gene of the *lac* operon. Their sites within the gene were mapped by standard genetic techniques [1].

Results

Initially the mapping was done against four deletions which start from some point to the left of the *lac* promoter and end at different points in the *z* gene. These deletions define five segments of the *z* gene. Segment V, not covered by any of the above deletions, was further divided into six subsegments by five more deletions. The segments, subsegments, and deletions are depicted in Fig. 1.

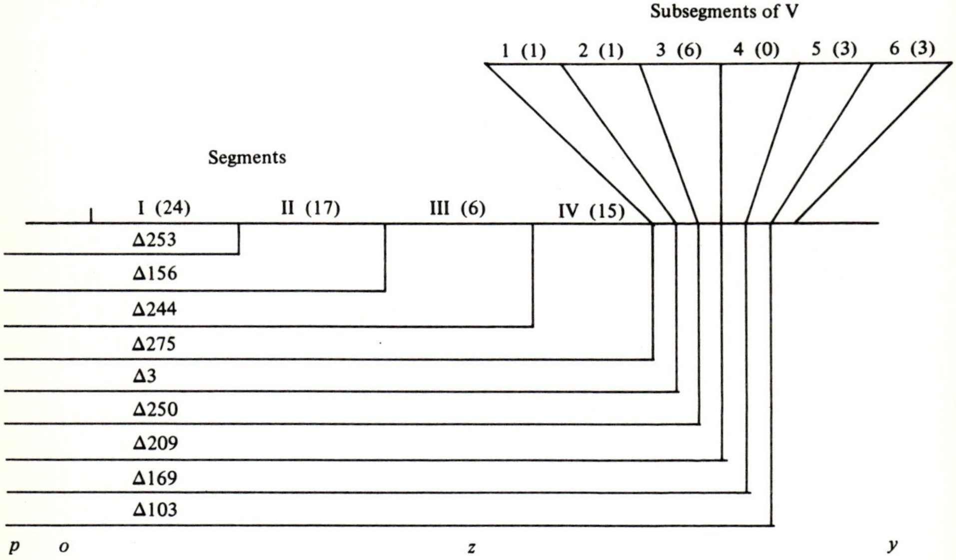

Fig. 1. Deletion mapping of Mu-1 mutants in the *z* gene.

p = promoter,
o = operator,
z = β-galactosidase structural gene,
y = permease structural gene,
Δ = deletions.

Roman numerals are segments. Arabic numerals are subsegments. Arabic numerals in parentheses are the number of mutations in the respective segment.

The distribution of the mutations in the segments and subsegments is also given in Fig. 1. As shown there, the mutations were found to fall in all five segments. Furthermore, the dissection of segment V revealed that mutations in that segment are not localized at one place but are distributed in five of the six subsegments. It is thus clear that Mu-1 can cause *lac* mutations by insertion in essentially every region of the *z* gene.

In order to ascertain whether the mutations in a subsegment are at distinct sites or whether they group together at one place, all possible combinations of crosses between mutations falling in a subsegment were done. The three mutations in subsegment 5 recombined with each other to give lac^{+} recombinants. The three mutations in subsegment 6 gave similar results. Therefore, all six mutations in subsegments 5 and 6 are nonallelic and are at different sites. On the other hand, when the six mutations belonging to subsegment 3 were crossed with each other there were several pairs of mutations that did not give any detectable recombination. Absence of recombination could mean that the two mutations in question are located at exactly the same site. But it was possible to show that pairs of mutations which fail to recombine with each other are in fact at distinct sites by their pattern of recombination with other mutations. For example, if mutation A does not recombine with B, A and B could be located at the identical place. However, if B does not recombine with C, but A does, it follows that A, B, and C are not the same alleles. Furthermore, two mutations unable to recombine with each other can be assumed to be closer to each other than to a third mutation which recombines with both of them. Therefore, the recombination pattern of these mutations would allow their ordering with respect to each other. The results of this analysis of the six subsegment-3 mutations are presented in Fig. 2. It can be seen from the matrix that five mutations display different recombination patterns. A sixth mutation in subsegment 3 (AT2484) gives a recombination pattern like 8118 and therefore is indistinguishable from it. As the matrix shows, the recombination data are consitent with the order

$$8187\text{–}8114\text{–}835\text{–}8157\text{–}\genfrac{}{}{0pt}{}{\text{AT2484}}{8118}\;.$$

		F⁻				
		8187	8114	835	8157	8118
Hfr	8187	–	–	+	+	+
	8114	–	–	–	+	+
	835	+	–	–	–	+
	8157	+	+	–	–	–
	8118	+	+	+	–	–

Fig. 2

Mu-Mu recombination.

The numbers refer to Mu-1 mutations in subsegment 6 of Fig. 1. A + sign in the matrix indicates that the pair of mutants recombine; a – sign, that they do not recombine. The data are redundant in that each cross was done twice, with the Hfr-F reversed.

The point here is that almost every Mu we were looking at went into the *z* gene in a different way. This suggests that there are no hot-spots, no particular entry sequences.

In addition to this mapping, we crossed several point mutants with the nearest Mu's, the idea being to see if the Mu would exclude recombination with the point mutants or whether it had produced a small deletion. The result of this test was very clear-cut: every single point mutant recombined with every single Mu, even within the same deletion segment. There was no detectable removal of DNA by Mu insertion. So the failure of two Mu's to recombine is obviously a consequence of giant pieces of DNA being extremely close and just not being able to carry out whatever the overlap event is for normal recombination.

From data like these, we see that Mu simply goes into the *z* gene at random.

We don't know anything much about the map of Mu itself. But whatever map it has, the genes must have an order. This means that Mu could be pointing in either one of two directions when inserted in the DNA of the *coli* chromosome. To test this, *J. Zeldis* [3] did the following experiment.

An F′*lac* strain with two different Mu's was constructed. One Mu was inserted in the galactose operon on the chromosome and the other in the lactose operon on the F′*lac* episome. The chromosome contained a complete deletion of the *lac* region. In this strain, the only homology between episome and chromosome was in the Mu DNA. A series of these strains were made that differed from one another only in the Mu carried on the episome. These strains were mated to a *leu*$^-$*ura*$^-$ female. Depending on the orientation of the Mu on the episome, either *leu*$^+$ or *ura*$^+$ would be transferred early to the F$^-$. This experiment is diagrammed in Fig. 3. The results were clear cut; each Mu in *lac* on the episome transferred only *leu* or only *ura* first, thus showing that there are two orientations for Mu.

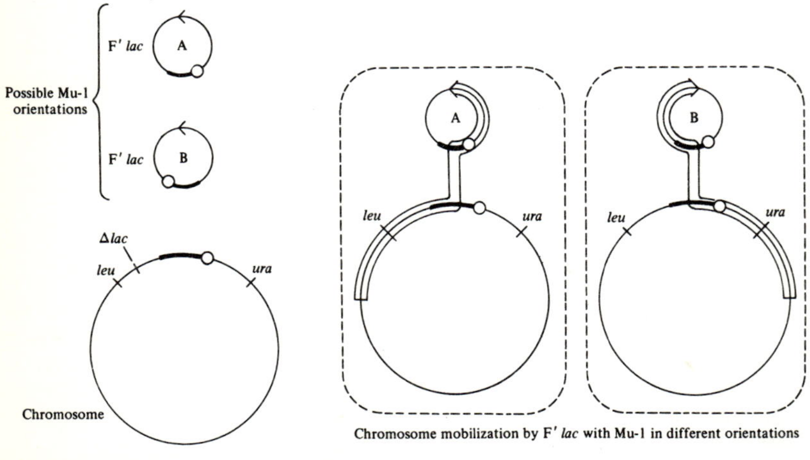

Fig. 3. Orientation of prophage Mu-1.

This figure shows the structure of the male parent in the cross. The female parent was *leu*$^-$*ura*$^-$ *(pyrC*$^-$*)* and naladixic acid resistant. Naladixic acid was used to kill the F′ parent.

Discussion and Conclusions

Mu will enter the chromosome at random in all cells which become lysogenic from Mu, which is about 2 % of the original population infected with Mu. This suggests very strongly that Mu has an enzyme system, operational for a short time during infection, which either glues pieces of DNA together completely at random or glues a specific site on Mu to a random piece of the bacterial chromosome. We know nothing about how this works at the moment, but are quite confident we are going to know exactly how it works very soon. The experiments have essentially been planned for us by previous work with other phages; we just have to do them.

We think Mu is of practical interest for two reasons. First, isolating this enzyme system will be of some interest because it will be a handy thing to have. However, somewhat more important, we think, is the fact that a certain amount of circumstantial evidence suggests that animal viruses of the SV40-polyoma type may work this way, that is, they integrate by some sort of random joining-up to the DNA of the host. If this is true, the enzyme system for Mu is a model for this other enzyme system. Thus, Mu may offer an alternate mechanism, or alternate model, for people studying animal virus integration.

References

[1] *Bukhari, A. I., and Zipser, D.* 1972. Random insertion of Mu-1 DNA within a single gene. Nature, New Biol. **236:** 240–243.

[2] *Taylor, A. L.* 1963. Bacteriophage-induced mutation in *Escherichia coli.* Proc. Nat. Acad. Sci. U.S. **50:** 1043–51.

[3] *Zeldis, J. B.* Manuscript in preparation. The orientation of prophage Mu-1.

Discussion

Audience: I don't understand how you did the last experiment.

Zipser: The experiment can be done in two ways. First, one can ignore the zygotic induction and allow them to induce. There are always many survivors. One simply then asks whether the survivors are leucine$^+$ or uracil$^+$. The answer is that in one class the experiment gives all one kind and in the other class gives all the other kind. Alternatively, one can make the F^- lysogenic and get the same result.

Knippers: I would like to know whether Mu can be inserted into Mu?

Zipser: Yes, but games like this lead to results which are impossible to analyze. The one thing I think should be stressed is that if you have λ inserted in a gene, or P2 inserted in a gene by some unknown method, when λ or P2 comes out the gene gets sewn back together and can refunction. This apparently does not happen with Mu. When one has induced Mu, the gene from which it came is not reverted to its original state. Just what happens to that gene and in which state it now exists is not known, but it is not back to its original condition.

Radding: I would like to add the comment that in evolution it is possible that what Dr. *Zipser* called "random nonhomologous recombination" or "unusual recombination" may have been *more* important or at least *as* important as general recombination. *Bill Dove* has reviewed these arguments [Chapter 15 in *The Bacteriophage* λ, 1971, ed., A. D. Hershey, Cold Spring Harbor].

Epiviruses

Charles C. Brinton

Stanford University Medical Center, Stanford University School of Medicine, Stanford, Calif. 94305, USA

One of the ultimate goals of "genetic engineering" is the efficient introduction of exogenous DNA into somatic cell genomes for the correction of specific genetic defects. This would require the introduction of a specific piece of DNA into each cell of a large population without killing or grossly altering the properties of any of them. The homogeneous population of DNA molecules might be presented to the cells whose defect was to be corrected as free DNA, encapsulated in a virion or pseudovirion, or as part of the genome of another population of cells. Ideally, one would like each cell in the defective population to acquire a piece of the new DNA as a stably integrated and replicating part of its genome. However, it is easy to see that this ideal may be quite difficult to achieve in practice.

After injection into the organism, the DNA must first survive. Free DNA is subject to nuclease attack and virions are subject to serum neutralization. The surviving DNA must then enter all cells without killing them. Free DNA may enter nonspecifically by phagocytosis; virions may enter the same way or by more specific routes involving receptor sites; cellular DNA may enter by cell fusion. Known virions often result in cell death or a carcinogenic transformation, and the infectious process of mammalian viruses is notoriously inefficient. Cell fusion usually occurs with low efficiency and, furthermore, introduces a large excess of unknown and unwanted DNA. The DNA which succeeds in entering must survive intracellularly and become integrated into the cell genome. The molecular mechanisms of integration of exogenously introduced DNA are not completely understood but do not appear to be highly efficient in mammalian cells.

At the moment, the difficulties of correcting somatic genetic defects in differentiated organisms by the introduction of exogenous genetic material appear insurmountable. However, I think it is pertinent to point out that a highly efficient natural mechanism

Manuscript received: 15 February 1972

for correcting specific genetic defects in cell populations by the introduction of exogenous DNA is well known, although only in bacteria. I refer to the much studied system of gene transfer by "conjugation" of bacterial cells. Recent advances in our understanding of the molecular nature of this process reveal some of its novel features which are responsible for its high efficiency and long-term stability. A consideration of these properties may help to stimulate mammalian cell biologists to search for or construct an analogous system.

I will describe the components of this system without supporting evidence, which has been given elsewhere [1]. The essential features are shown schematically in Fig. 1. The central component is an extrachromosomal piece of double-stranded DNA which can be isolated as a covalently closed super-coiled circle or as a more relaxed form with a nick in one strand. This molecule, or plasmid, is capable of autonomous semiconservative replication which is precisely regulated so that usually there are only a few plasmids per cell, and each daughter cell receives at least one plasmid. This replication, which maintains the plasmid within a cell population, is *not* diagrammed in Fig. 1. Presence of the plasmid does no great harm to the cell, although plasmid infected cells may grow more slowly than noninfected cells and are somewhat more susceptible to agents which disrupt membrane function.

The plasmid may contain only genes for its own maintenance, and in this case its presence is only detectable as "satellite" nonchromosomal DNA. However, many plasmids have the property of recombining with host DNA and thereby acquiring a varying number of genes which can change the properties of the host cell and there-

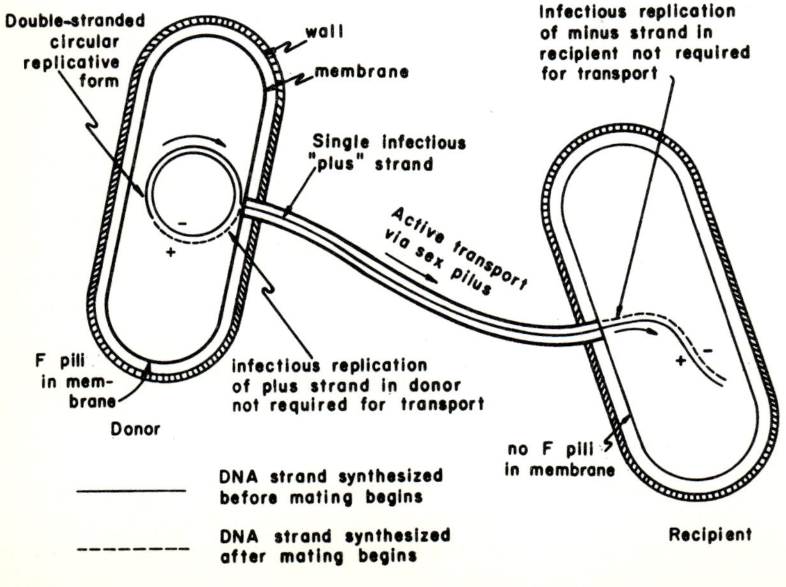

fore reveal the presence of the plasmid. If the chromosome lacks a not absolutely essential but still desirable gene, such as a drug resistant one, the presence of a plasmid having such a gene can "correct" this "genetic deficiency".

Many plasmids are noninfectious, that is they have no way of transferring themselves from one cell to another. However, many infectious plasmids exist which can transfer themselves by means of the mechanism shown in Fig. 1. A certain starting point for transfer exists where one particular DNA strand (the infectious or "plus" strand) is removed from the plasmid and actively transported, via a long thin surface structure called an F pilus, into the recipient cell. The plus strand is not removed until the F pilus contacts the recipient cell's surface. The nature of the signal which triggers plus strand removal is unknown, but it resembles, in many respects, the initial stages of a virus infection. The donor cell loses nothing by virtue of having donated a plasmid plus strand since the damage is immediately repaired by replication. The recipient cell rapidly becomes a donor cell because the entering plus strand is very quickly replicated into a double-stranded linear form which then circularizes to produce another fully functioning and infectious plasmid.

By the repetitive spread of this process, an entire culture of bacteria can very rapidly have its genetic defects corrected. None of the cells involved are killed, and none of their other properties need be seriously affected. Built into the system are two methods for maintaining the added genetic information: replication of the plasmid within the cell and replication of the plasmid by infectious transfer into those cells of the population which may have lost the plasmid by failure of the normal replicating system.

It can be seen that the infectious plasmid system is very viruslike. It has a double-stranded replicative form, a single-stranded infectious form, and an extracellular (but *not* cell-free) infectious particle. I have suggested that such systems be defined as "*epiviruses*", that is, those whose infectious form is obligatorily cell attached. Epiviruses have a distinct advantage over virion viruses as "genetic engineers" since the amount of DNA they can introduce into a cell is not limited by the size of the virion "package".

No epiviruses resembling those in bacteria have yet been found for animal or plant systems which might be exploited for the correction of genetic defects. It should be pointed out, however, that they would probably be quite difficult to detect. Since they produce no cell-free particle and no marked cytopathogenic effect or other alteration of the cells they infect, the usual techniques of virology would be useless. One might have expected that the search for gene transfer systems among mammalian cells in culture would have uncovered any highly efficient epiviral transfer system.

However, all such systems found so far appear to depend upon cellfusion. This does not mean that mammalian epiviruses do not exist. Their potential utility in genetic defect correction at least requires that a search be made. Knowledge of bacterial epiviruses should help in devising suitable techniques in mammalian systems.

Reference

[1] *Brinton, C.*, 1971. The properties of sex pili, the viral nature of "conjugal" genetic transfer systems, and some possible approaches to the control of bacterial drug resistance. Critical Reviews of Microbiology 1: 105–160.

Exchange between Complementary Strands of DNA?

Thomas A. Trautner, Hans-Christof Spatz, Bernhard Behrens, Brigitte Pawlek, Margit Behncke

Max-Planck-Institut für Molekulare Genetik, Abt. Trautner, 1 Berlin 33, Germany

Summary: Heteroduplex molecules of SPP1 phage-DNA involving a large number of point- and deletion mutants were constructed. Transfection of B. subtilis with such molecules followed by single burst analysis of phage progeny revealed a potent heteroduplex correction mechanism (="conversion"). Conversion for ***point mutations*** is specific in the sense that the "heavy" DNA strand determines the direction of conversion in two regions of the map, whereas in another region between these, conversion occurs towards the genotype of the complementary "light" DNA strand. In most heteroduplexes involving ***deletions***, conversion towards only mutant genotype is observed, irrespective of the orientation of the wild type and mutant allele with regard to the complementary strands. These data are discussed in the light of a classical conversion mechanism, which postulates heteroduplex correction as a consequence of the ***recognition*** of a heteroduplex region.

Problem

Various aberrant phenomena of recombinational genetics of eukaryotes like gene conversion, high negative interference, and map expansion have attracted researchers because their interpretation should give clues to an understanding of the molecular mechanisms underlying genetic recombination. Several models have been proposed to account for the observed phenomena (*Whitehouse* [14], *Whitehouse* and *Hastings* [15], *Holliday* [6], *Stahl* [13], *Boon* and *Zinder* [1], *Paszewski* [9]). The Whitehouse-Holliday types of models assume that in recombination single strands of recombining double strands of DNA are first cut, followed by base pairing between different parental complementary strands. If the pairing strands carry different genetic information, a "heteroduplex" DNA region may be formed in the course

Manuscript received: 17 January 1972

of this first step of genetic recombination. The heteroduplex region could be converted by a repair-like mechanism to a "homoduplex" structure. Thus genetic information from one strand would be transferred to its complementary strand. This heteroduplex repair could account for the phenomena mentioned above.

A system of infectious phage DNA ("Transfection") and the possibility to separate and anneal genetically different complementary DNA strands allowed the construction of heteroduplex DNA in *E. coli* phage λ (*Doerfler* and *Hogness* [3]) and B. subtilis phage SPP1 (*Spatz* and *Trautner* [11]), and made possible a direct investigation of heteroduplex repair.

An analysis of phage progeny from cells transfected with such DNA indeed revealed the existence of a potent heteroduplex correction mechanism. We realize that it remains an open question whether this correction is really a consequence of heteroduplex *recognition,* or is an effect occurring *pari passu* with transfection, irrespective of the presence of heteroduplex regions. A summary of our data on SPP1 from the last two years pertinent to heteroduplex correction will be presented.

Materials and methods

Only materials and methods additional to those reported previously [11] are described.

Bacteria:

A sus^+ derivative of MCB [12] was used for the growth of SPP1 amber mutants (*Behncke,* unpublished). This strain was obtained by transformation of competent MCB sus^- with DNA from strain su^+3, originally described by *C. P. Georgopoulos* [4] and kindly provided by *C. Anagnostopoulos.*

Phage:

Amber (am) mutants of SPP1 were obtained from hydroxylamine mutagenesis of SPP1 wild-type phage, and were identified by their failure to form plaques on plates seeded with MCB sus^- after replica plating from MCB sus^+ plates. Amber mutant plaques can be distinguished from wild type when plated on a mixed indicator consisting of sus^+ and sus^- cells. *Deletion Mutants,* distinguishable in their plaque morphology from wild type, were enriched from UV-treated SPP1 lysates by consecutive cycles of inactivation by 0.1 M Na_3 citrate and growth as described by *Parkinson* et al. [8]. The amount of deleted DNA of such mutants was calculated from the difference in buoyant density between such mutants and wild type established in preparative CsCl density gradient centrifugation, using the formula described by *Davis* and *Davidson* [2]. All deletion mutants mapped in the vicinity of plaque type mutant 161. Details of the isolation and characterization of such deletions will be published elsewhere.

Results

All experiments to be reported are of one general type [11]: Complementary H- and L-strands of both SPP1 wild type and various SPP1 mutant DNAs are separated in CsCl density gradients (*Riva* et al. [10]). Complementary strands of wild type and of a respective mutant DNA are annealed to give two types of heteroduplices for each mutant: one in which the wild allele is carried on the H-strand and the mutant on the L-strand ("wild/mutant") and the reciprocal combination ("mutant/wild"). Competent cells of B. subtilis are transfected with such heteroduplex DNA, and the phage progeny of individual cells is analyzed by single-burst technique.

Type of mutant	Genotype of strands		Frequency of bursts with			Number of bursts investigated	Preferred strand	Conversion pattern
	H	L	wild type and mutant	only wild type	only mutant			
Point Mutations plaque type	+	161	0.23	0.69	0.08	876	H	H/H
	161	+	0.15	0.05	0.80	805	H	
plaque type	+	111	0.26	0.20	0.54	110	L	L/L
	111	+	0.32	0.41	0.27	97	L	
amber	+	am 62	0.16	0.80	0.04	55	H	H/H
	am 62	+	0.13	0.15	0.72	46	H	
amber	+	am 117	0.20	0.17	0.63	128	L	L/H
	am 117	+	0.23	0.20	0.57	124	H	
Deletions −2.4 %	+	Z	0.14	0.83	0.03	71	H	H/H
	Z	+	0.21	0.21	0.58	57	H	
−11 %	+	A	0.006	0.027	0.963	2109	L	L/H
	A	+	0.008	0.003	0.989	5901	H	
< 1 %	+	K	0.10		0.90	626	L	L/H
	K	+	0.11		0.89	301	H	

Fig. 1

Specificity of heteroduplex repair

A total of 21 heteroduplex pairs between wild type and plaque type point mutants, 11 amber mutants, and 6 deletions represent the material investigated so far. Some representative results from mutants of each group are given in Fig. 1. A condensed summary of all conversion data is presented in Fig. 2. The following conclusions can be drawn from these data:

1. The three possible kinds of bursts: mixed, pure wild type, and pure mutant are observed in all heteroduplex combinations. The occurrence of pure bursts is indicative of heteroduplex repair.

2. In a given heteroduplex, there is a striking asymmetry in the frequency of the pure bursts. This asymmetry defines as the "preferred" strand the one whose allele occurs with a high frequency in a pure burst. Strand preference imposing a "direction of conversion" is an indication of specificity in heteroduplex repair.
3. Strand preference for a given mutant is in the majority of cases the same for both members of a given heteroduplex pair (HH or LL preference in fig. 1).
4. The direction of asymmetry (H or L) in such pairs is marker dependent.
5. In a few cases, deviations from this specificity pattern are observed. Some heteroduplex pairs appear to show preferential conversion to wild-type genotype in both combinations (H/L of Fig. 2). We had previously attributed this pattern to a secondary effect – a selective growth advantage of wild type over the mutant. However, the majority of heteroduplex molecules with deletions and some with amber mutants show preferential conversion to *mutant* genotype, irrespective of the orientation of the two alleles with regard to the DNA strands, yielding a L/H pattern of strand preference, which could not be attributed to overgrowth of wild type by the mutant.
6. When the distribution of conversion patterns (Fig. 2) is analyzed, it is evident that conversion occurs most frequently towards the genotype of the H-strand.
7. Strand preference for mapped markers is not random with regard to their localization on the SPP1 chromosome (Fig. 3) but appears to be clustered into three adjacent regions, with markers in the central region giving conversion with L-strand preference, and in the bracketing regions with H-strand preference. No clustering of the amber mutants with an L/H conversion pattern could be recognized.

Conversion pattern / Type of mutant	H/H	L/L	H/L	L/H	Sum
plaque type	10	6	5	0	21
amber	8**	0	0	3	11
deletion	2	0	0	4	6
sum	20	6	5	7	38

Fig. 2. Summary of conversion patterns*.

* We are always investigating strand preferences of a given heteroduplex pair in the sequence (1) H wild type/L mutant, and (2) H mutant/L wild type. Correction to wild type in combination (1) and to mutant in combination (2) would yield a H/H conversion pattern.

** Assignment of strand preference to two amber mutants is tentative.

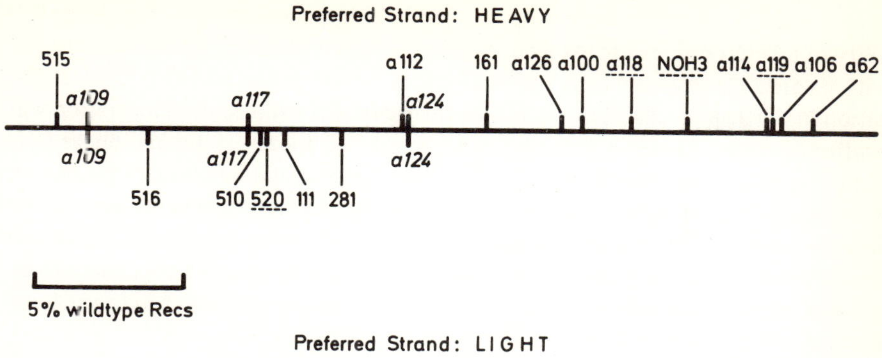

Fig. 3. The distribution of conversion patterns for various mapped amber (a . . .) and plaque-type mutants. All mutants which give a H/H conversion type are entered above the lines, those with an L/L conversion type below the line. Markers entered in italics and appearing below and above the line are of the L/H type. Conversion assignments underlined (- - - - - -) are tentative: in the case of plaque-type mutants 520 and NOH 3 because of selective disadvantage of mutant plaques in comparison to wild type [11], in the case of amber mutants a 119 and a 100 because only one combination (+/am) was analyzed so far.

Discussion

The mechanisms of heteroduplex repair

Although no general mechanism can be formulated at this stage to account for the observed data, some specifications for such mechanisms can be given.

Clearly DNA replication alone would not yield the collection of conversion data unless very implausible specifications of DNA replication were introduced. Direct evidence for the involvement of heteroduplex *correction* rather than DNA replication came from transfections with heteroduplices for *two* markers (111 and 161), in which the conversion behaviour of each marker was independent of the other [11]. Furthermore, destruction of heterozygosity can occur before extensive DNA replication. This was derived on the one hand from experiments in which transfection with heteroduplex molecules was performed with cells whose DNA synthesis was FUdR-inhibited. Under such conditions, the degree of conversion was augmented as the frequency of pure bursts increased at the expense of mixed bursts, which arise predominantly from cells in which heteroduplex replication preceded correction. Other evidence for the independence of heteroduplex repair from replication, indeed the proof that these mechanisms are competing processes, came from an analysis of the frequency of mixed bursts in the various heteroduplex pairs investigated. When the frequency of mixed bursts was plotted against the map positions of the respective markers, a gradient of decreasing frequency of mixed bursts extended from the left to the right end of the map [11]. With the assumption that the probability of a heteroduplex escaping correction through replication would decrease with its distance from the origin of replication, we had postulated that DNA replication in SPP1 proceeds from the left to the right of the molecule. This notion was confirmed from an entirely different experiment: *G. Klotz* [7] isolated phage DNA at various times after infection from SPP1 infected cells and determined frequencies of markers along the SPP1 map in a *Yoshikawa-Sueoka* [16] type experiment. These data gave clear evidence of a unique direction of DNA replication in SPP1 proceeding from the left to the right and of the molecule. It can thus again be concluded that conversion can occur before DNA replication and that replication and conversion are competing processes.

Another specification of heteroduplex correction concerns the heteroduplex region itself. Previous experiments, in which the nature of the mismatch in the heteroduplex region could be described through the use of hydroxylamine as a mutagen and mutagenesis of individual DNA strands, indicated that not the bases of the heterozygous base pair alone determine the direction of conversion [11]. We suggested that bases outside the heteroduplex region exert an influence on the specificity. This interplay of heteroduplex region and outside bases must also account for the marker effects reflected in the amber-heteroduplices yielding in all combinations preferentially mutant genotype. The clustering of identical conversion patterns into contiguous sections of the SPPI map might suggest that other processes showing segmental strand specificity (like transcription) play a role in heteroduplex conversion.

The almost exclusive conversion of most heteroduplexes with deletion mutants to mutant genotype and the fact that here the frequency of mixed bursts is much lower than anticipated from the map position of these mutants suggests that a correction mechanism different from that observed with point mutation heteroduplices is operative. A heteroduplex between wild type and a deletion mutant will contain a segment of single-stranded DNA. It is conceivable that such material is rapidly degraded by endonuclease activity. The loops produced in those delections, which do not follow this scheme, might differ in their secondary structure, and therefore conversion here might follow point mutant conversion. The size of a given deletion is not the factor which determines what conversion pattern is realized.

Is heteroduplex repair a consequence of heteroduplex recognition?

Two mechanisms, of unspecific limited DNA replication, might produce effects which would be recognized as conversion after heteroduplex transfection:

1) Transfecting DNA, in contrast to infecting DNA, may undergo limited single-strand degradation, followed by resynthesis *irrespective* of the presence of mismatched regions. Such degradation must be rather extensive and highly specific

to account for the conversion frequencies which we observe. To establish the contribution of unspecific degradation with subsequent resynthesis of degraded portions, transfection experiments with differential label to distinguish transfecting and newly synthesized DNA are in progress.

2) Previous work [12] has shown that genetic recombination has to precede DNA replication in SPP1 transfection. If such recombination required limited DNA synthesis – as suggested, e. g., by the Boon-Zinder recombination model – one might again observe "conversion" after heteroduplex transfection, which would not be heteroduplex mediated. To control this possibility, conversion experiments of the type described should be performed under conditions where transfection can proceed in the absence of recombination.

Until we have experimental evidence pertinent to these possibilities of unspecific repair, and until direct biochemical evidence proves that the heteroduplex region itself mediates heteroduplex repair, we cannot answer the above question in the affirmative. However, the finding of conversion patterns with some of the amber heteroduplices, which are reminiscent of marker effects found in conversion with fungi [5], and the high specificity of heteroduplex repair represent strong circumstantial evidence for conversion as a consequence of heteroduplex recognition.

References

[1] *Boon, T., and Zinder, N. D.* 1969. A mechanism for genetic recombination generating one parent and one recombinant. Proc. Natl. Acad. Sci. Wash. **64**: 573–77.

[2] *Davis, R. W., and Davidson, N.* 1968. Electron-microscopic visualization of deletion mutations. Proc. Natl. Acad. Sci. Wash. **60**: 243–50.

[3] *Doerfler, W., and Hogness, D. S.* 1968. Gene orientation in bacteriophage lambda as determined from the genetic activities of heteroduplex DNA Formed in vitro. J. Molec. Biol. **33**: 661–78.

[4] *Georgopoulos, C. P.* 1969. Suppressor System in B. subtilis 168. J. Bact. **97**: 1397–1402.

[5] *Gutz, H.* 1971. Site specific induction of gene conversion in schizosaccharomyces pombe. Genetics **69,** 317–337.

[6] *Holliday, R.* 1968. Genetic recombination in fungi. *In* Replication and recombination of genetic material, eds. W. J. Peacock, and R. D. Brock, pp. 157–74. Canberra: Australian Academy of Science.

[7] *Klotz, G.* 1971. Proceedings, European phage meeting. Berlin.

[8] *Parkinson, J. S., and Huskey, R. J.* 1971. Deletion mutants of bacteriophage lambda. I. isolation and initial characterization. J. Mol. Biol. **56**: 369–84.

[9] *Paszewski, A.* 1970. Gene conversion: observations on the DNA hybrid models Genet. Res. **15**: 55–64.

[10] *Riva, S., Posinelli, M., and Falaschi, A.* 1970. A new phage of bacillus subtilis with infectious DNA having separable strands. J. Molec. Biol. **35**: 347–56.

[11] *Spatz, H. Ch., and Trautner, T. A.* 1970. One way to do experiments on gene conversion! Molec. Gen. Genetics **109**: 84–106.

[12] *Spatz, H. Ch., and Trautner, T. A.* 1971. The role of recombination in transfection of B. subtilis. Molec. Gen. Genetics **113**: 174–90.

[13] *Stahl, F. W.* 1969. One way to think about gene conversion. Genetics **61**, Suppl: 1–13.

[14] *Whitehouse, H. L. K.* 1963. A theory of crossing-over by means of hybrid deoxyribonucleic acid. Nature **199**: 1034–40.

[15] *Whitehouse, H. L. K., and Hastings, P. J.* 1965. The analysis of genetic recombination on the polaron hybrid DNA model. Genetic Res. Com. **6**: 27–92.

[16] *Yoshikawa, H., and Sueoka, N.* 1963. Sequential replication of bacillus subtilis chromosome. I. comparison of marker frequencies in exponential and stationary growth phases. Proc. Natl. Acad. Sci. Wash. **49**: 559–66.

Discussion

Starlinger: I have a question concerning the preference of repair to wild-type regardless of the DNA strand which carries the wild-type base sequence. This means that on one strand you have to eliminate the "mutant" purine, and on the other strand the "mutant" pyrimidine. Do you think that this can be explained by the assumption that the mutation has arisen in a symmetrical region, e.g., carrying AAA on the 5′-side of the "heterogentic" base pair, and TTT on the 3′-side? This in connection with specific selection rules for the base to be excised might lead to the observed behaviour.

Trautner: I see, yes, but this is not the case in these amber mutants. If one diagrams the possible combinations between a wild-type and a mutant, one cannot construct a symmetrical situation.

Hurwitz: One thing that should be important here, would be to take a look at the degree of degradation in each of the strands with regard to the selection of the marker. In fact, that is really critical here because you can explain that a lot of what Peter was going to refer to was specifically in terms of a purine or pyrimidine.

Trautner: One point that I failed to make is that all the point mutation experiments which we have done were done with hydroxylamine mutants so that we know we were converting cytosine to thymine. Furthermore, we have mutagenized individual strands. One can first separate the strands, then mutagenize them individually and then rehybridize them with the wild-type complement. By doing this one can establish the orientation of purine and pryrimidine with regard to the strand. In such a case when the heavy strand of one of these hybrids has the pyrimidine and the light strand has the purine, one can ask if this strand orientation is correlated with the direction of correction. The answer is that it is *not* correlated.

Yarmolinsky: I would like to ask whether there is any correlation between this effective clustering and transcription. One reason for asking this is that *Henry Drexler* has found very strong transcriptional effects on recombination in transduction by T1 phage.

Trautner: When *Esche* from our laboratory looked at transcription in vivo, he found that, while there was a small amount of transcription in the light strand, most of the transcription occurred along the heavy strand. Perhaps this is an indication of a correlation, but it may be fortuitous. It certainly is a point which should be investigated.

Sinsheimer: You said that you have looked at double mutations. Were these double ambers?

Trautner: No, they were double point mutants, one oriented to conversion to heavy, the other oriented to conversion to light. They are completely independent in their conversion mechanisms.

Hausmann: Have you looked at two markers so close to each other that they might be converted in one single event, and are they then always on the same strand?

Trautner: Yes, but, since markers which were far apart could also convert in the same direction, one must be careful in interpreting these results.

Klingmüller: Earlier this morning I mentioned fungi and yeasts, and since this is a workshop. I feel free to come back to these nice organisms. The reason is your demonstration of a preference or bias in the direction of correction if DNA heteroduplices derived from wild-type and deletion mutant single strands are used for transduction. A similar bias can be demonstrated in fungi in studies on gene conversion, i.e., in our own recent work on Neurospora, cross wild-type x mutant asco, conversion was predominantly in the direction of the mutant allele (Fig. D1). Don't you find this a very suggestive parallel, pointing to the relevance of your work on phages, for present theories on the molecular mechanism of gene conversion in eukaryotes?

Trautner: Formally, I think there are analogies, but whether the same mechanisms produce the two effects in the phage system and the fungal system is not at all clear.

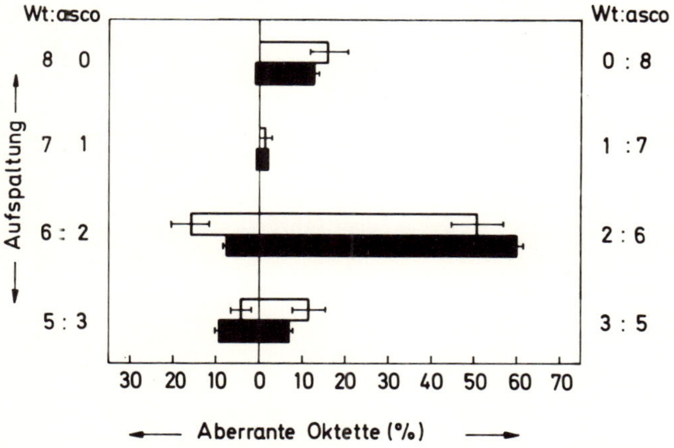

Fig. D.1. Relative frequency of the 8 possible types of aberrant octads, found without (white bars) or with (black bars) UV-irradiation. Averages from 5 independent experiments. Standard errors are indicated.

The Synthesis of Transfer RNA Genes

H. Gobind Khorana

Departments of Biology and Chemistry, Massachusetts Institute of Technology, Cambridge, Massachusetts 02139, USA

Introduction

I would like to review the methods which we are developing for the step-by-step and unambiguous synthesis of double-stranded DNA, and these are best illustrated by describing the synthesis of the DNA duplex, seventy-seven nucleotide units long, which corresponds to the primary sequence of the yeast alanine tRNA (Fig. 1) whose

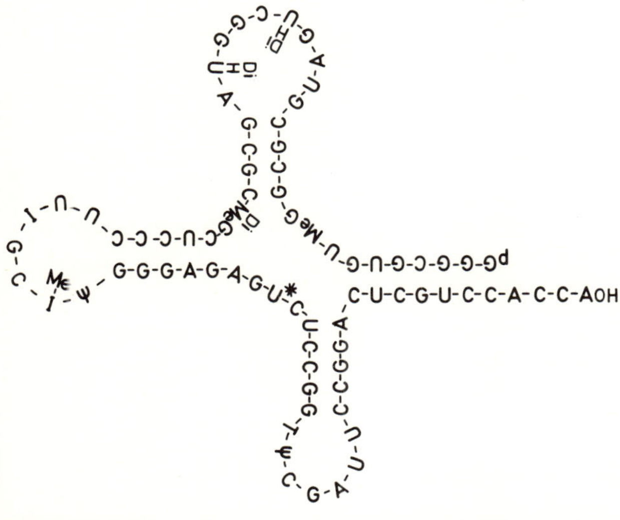

Fig. 1. Structure of the major yeast alanine transfer RNA as deduced by *Holley* and coworkers. The structure is shown in the familiar cloverleaf model for the secondary structure of tRNAs. ψ stands for pseudouridine, T for ribothymidine, I for inosine, Me-I for methylinosine, and DiH-U for 5, 6-dihydrouridine.

Manuscript received: 24 January 1972

sequence was determined by *Holley* and coworkers [2]. Time will not permit any details nor can I afford to go into other topics, such as the synthesis of *E. coli* tyrosine tRNA suppressor DNA, on which work is in progress in my laboratory.

The basic methodology consists of three parts and these are: (1) chemical synthesis of deoxypolynucleotide segments of chain length in the range of 8 to 12 units with 3′ and 5′-hydroxyl end groups free; the segments would represent the entire two strands of the intended DNA, and those belonging to the complementary strands would have an overlap of four to five nucleotides; (2) the phosphorylation of the 5′-hydroxyl group with ATP carrying a suitable label in the γ-phosphoryl group using the T4 polynucleotide kinase; and (3) the head-to-tail joining of the appropriate segments when they are aligned to form bihelical complexes using the T4 polynucleotide ligase.

In early work, actually completed before the discovery of the DNA ligases, we had made two partly complementary icosanucleotides corresponding to the middle part of the intended DNA (Fig. 2) and these had in fact been used in the initial work with the ligases. With the icosanucleotides and the corresponding shorter segments, which were the intermediates in chemical synthesis in hand, we formulated the plan for the total synthesis of the DNA which is shown in Fig. 3 [1]. It would be expected that in the future the segmentation of bihelices for synthetic purposes would be done more rationally, since in the present work the starting point was determined by the availability of the two icosanucleotides (see below).

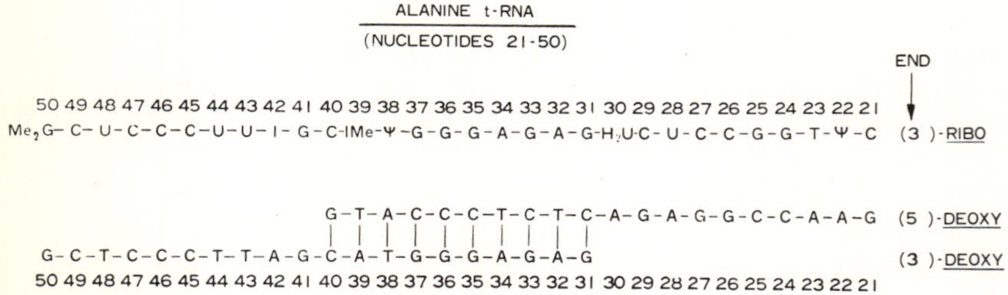

Fig. 2. Chemucally synthesized icosa-deoxyribonucleotides corresponding to sequences 21–50 of yeast alanine t-RNA. I stands for inosine, ψ for pseudouridine.

Total plan for the synthesis of the DNA

The plan shown in Fig. 3 contains a total of 15 segments, indicated by brackets with the serial number being inserted into the brackets. Additional considerations which led to the derivation of the plan were as follows. Starting with the B part (bottom strand) and proceeding to the right, segment 8 consisted of the hexadecanucleotide

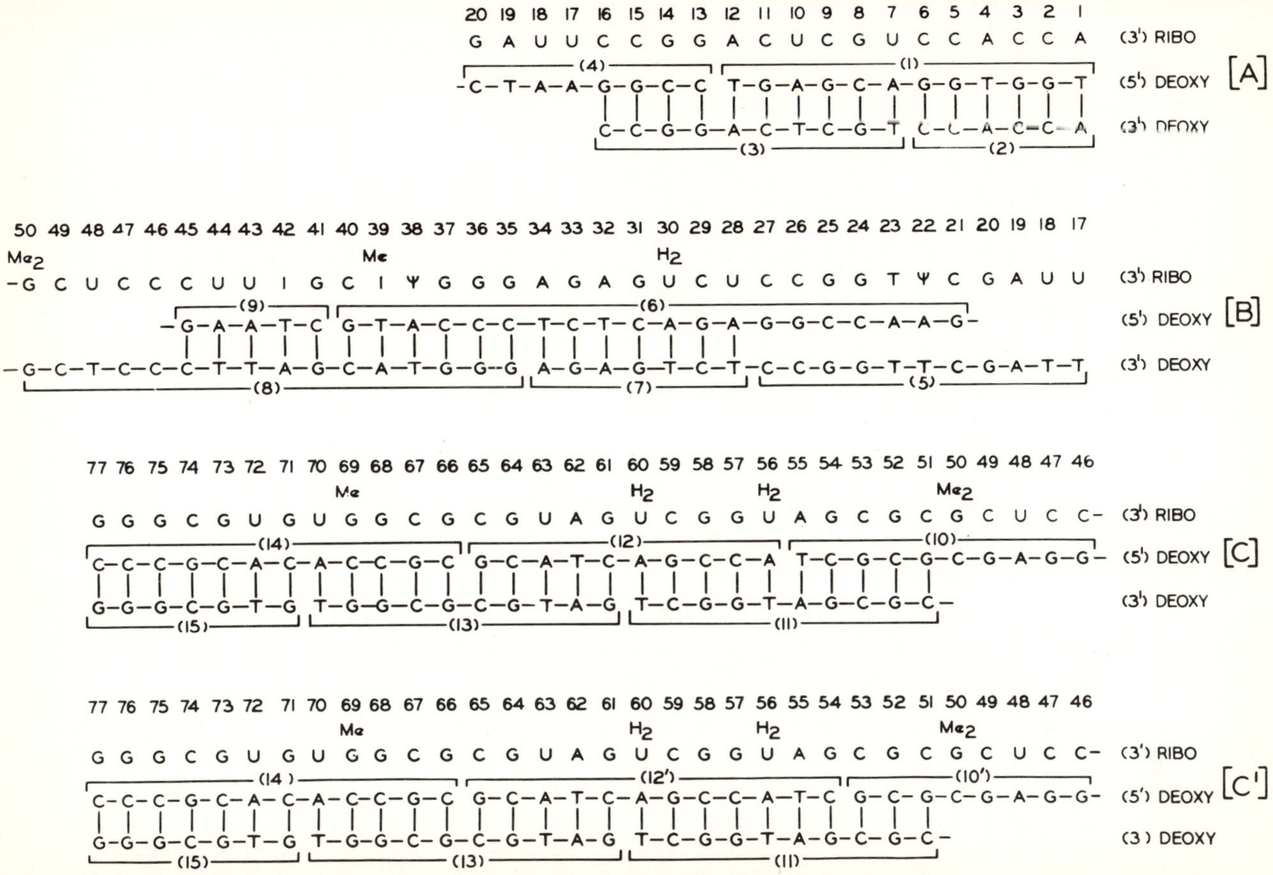

Fig. 3. Total plan for the synthesis of the yeast alanine tRNA gene. The chemically synthesized segments are in brackets, the serial number of the segment being shown within the brackets. A total of seventeen segments (including 10′ and 12′) varying in chain length from penta- to icosanucleotides were synthesized.

rather than the icosanucleotide shown in Fig. 3. The former was preferred because (1) the heptanucleotide (segment 7) could be prepared readily, and (2) the starting tetranucleotide block (CCGG) in the chemical synthesis of segment 5 could be used, in addition, for the synthesis of both segments 3 and 4 (part A of the gene). This left only the dodecanucleotide (segment 1) and the hexanucleotide, CCACCA (segment 2), to complete the A part of the gene.

Proceeding to the left of segment 6 (Fig. 3), it is seen that this region predominates in purine nucleotides which cause greater difficulty in chemical synthesis. It was therefore decided to use a pentanucleotide as segment 9 and then to go on to divide the remainder of the gene (part C) arbitrarily into decanucleotides (segments 10, 11, 12, and 13) and, finally, to provide the terminal sequences in the form of a dodecanucleotide (segment 14) and a heptanucleotide (segment 15). (The latter again predominates in the guanine nucleotide.) Because of the high G-C content in part C, problems of intramolecular secondary structure and possible self-aggregation were anticipated and, in particular, the self-complementarity of segment 10 was obvious. (Absorbance-temperature profiles described later, have borne this out, the T_m in 5 mM Mg^{2+} + 0.01 M Tris-HC1 buffer being 62°C.) To guard against the possibility of failure in joining of segment 10 to B or C or both simultaneously, a modified scheme (C′ of Fig. 3) for this part of the gene was also adopted. In this, the length of the segment 10′ was reduced to eight nucleotides while the adjoining segment 12′ was increased to a dodecanucleotide. No difficulty was in fact encountered in realizing synthesis according to plan C.

What I would like to do now is to first give a very brief oultine of the experimental methodology and then go on to describe the synthesis and characterization of the B part, then the A and the C parts, and then the joinings of these to form A-B and B-C and finally the total duplex.

Aspects of experimental methodology

The chemically synthesized segments were phosphorylated using the T4 polynucleotide kinase described previously. The reactions for enzymatic joining (on a 10–50 μl. scale) were set up in different conditions of Mg^{2+} concentration and temperature. The conditions used for preannealing the components varied, typical examples being given in the legends to figures. The kinetics of joining were followed by measuring resistance to bacterial alkaline phosphatase. The general conditions for assay on DEAE-paper strips have been described. The joined products were separated from the starting materials on Agarose or Sephadex columns and in all cases subjected to careful analysis, usually consisting of the following steps: (1) resistance to phosphatase; (2) degradation to 3′-nucleotides using the micrococcal and the spleen phosphodiesterase; and (3) hydrolysis to 5′-nucleotides using pancreatic deoxyribonuclease and venom phosphodiesterase. In a few cases, repair with DNA polymerase from

Escherichia coli was also used to characterize the product further. The mononucleotides were separated by solvent systems described previously and also by electrophoresis at pH 3.5.

Characterization of the joined products was aided by manipulation of the specific activity of γ-^{32}P-ATP used in phosphorylation reactions. Thus, radioactivity at the internal linkage points in A, B, and C was much weaker than that used at the termini which were to connect A to B and B to C. Further, in multiple joinings (for example, in C), stoichiometry of the joining of different segments was promoted by the concomitant use of ^{32}P and ^{33}P isotopes. Stoichiometric joining of A to B and B to C was similarly confirmed by using highly radioactive ^{32}P at the 5′-termini in B while the appropriate 5′-termini in A and C contained highly radioactive ^{33}P.

The synthesis of B

Of the three parts, the synthesis of this part was the most straightforward. Segments 5, 7, and 9 were phosphorylated at the 5′-termini using γ-^{32}P-ATP of the same specific activity, and the synthesis was carried out in two steps. Thus, at the start, the reaction mixture contained the four segments 6–9. After completion of this reaction, segment 5 was added and the incubation resumed as shown in Fig. 4. The joining of segment 5 to segment 7 was essentially complete. The resulting B was purified by passage through an agarose column and characterized.

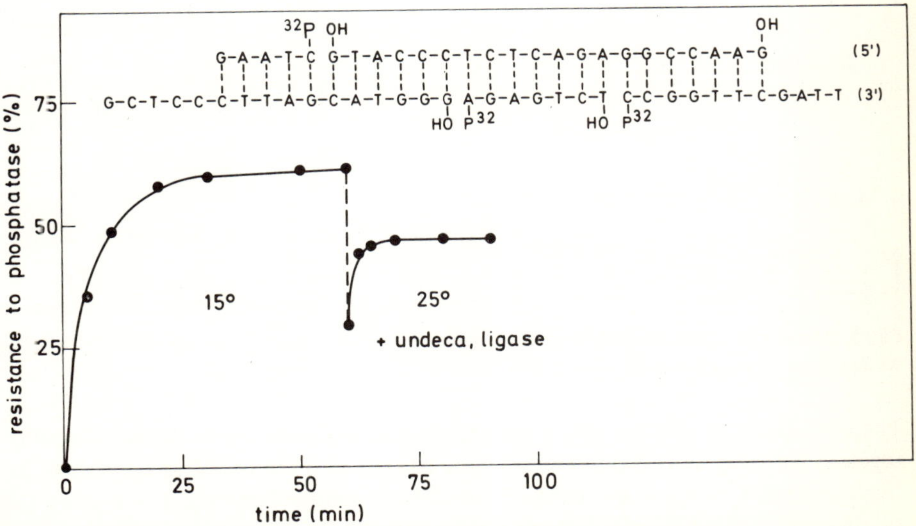

Fig. 4. Two-step synthesis of B. Segments 6–9 were joined first at 15 °C. The undecanucleotide (segment 5) was then joined at 25 °C. (For further details, see *Agarwal, K. L.* et al. [1], Fig. 5).

In various subsequent experiments on the joining of B to A and C, the two G units at the 5′-termini of B were phosphorylated with γ-^{32}P or ^{33}P-labeled ATP of much higher specific activity than that used at the internal positions.

The synthesis of A

In initial experiments, the synthesis of A was attempted by using 5′-^{32}P-labeled segments 2, 3, and 4 together with unphosphorylated segment 1 (top part of Fig. 5). While the joining of segment 2 to segment 3 went to completion, no joining of segment 4 occurred. Instead, the product formed was the dimer shown in the lower part of Fig. 5. This was clearly because of the self-complementary nature of the protruding CCGG single-stranded end in the duplex comprising segments 1, 2, and 3. To circumvent this, the synthesis of A was carried out in two steps. In step 1, segment 1 was joined to ^{32}P-labeled CCGGAATC (segment 4) in the presence of segment 3, the latter containing unphosphorylated 5′-end. This favored the kinetic rather than the thermodynamic control. The kinetics of the synthesis of the icosanucleotide (segments 1 + 4) are illustrated in Fig. 6. It is interesting that in the inital stages, a product was formed (dashed line in Fig. 6) which was identified as AMP-O-^{32}P-CCGGAATC. Ultimately, the formation of the required icosanucleotide was essentially complete as based on the amount of the segment 1. The icosanucleotide was isolated free from the unused segment 4 and the complementary segment 3 by passage through a column of Sephadex G-50 at room temperature with 0.05 M triethylammonium bicarbonate as the elutriator.

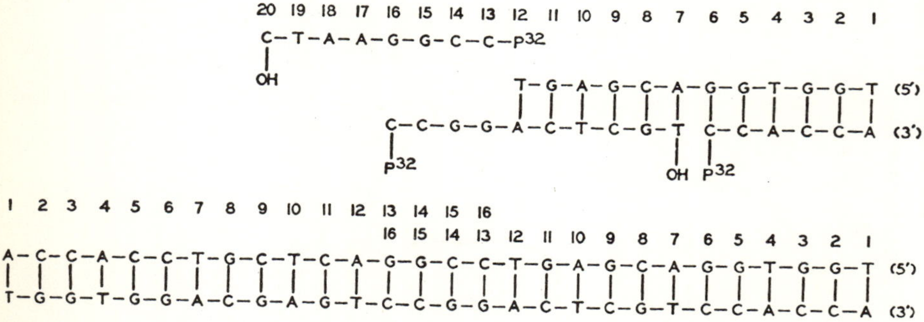

Fig. 5. Attempted synthesis of A. The four segments used are shown in the top part of the figure, while the product (a dimer derived from segments 1, 2, and 3) is shown below.

The synthesis of A was completed in a second step using the above icosanucleotide, together with segments 2 and 3 which were labeled with ^{32}P or ^{33}P at the 5′-end. In the "large scale" preparation of Fig. 7, the decanucleotide (segment 3) was phosphorylated with γ-^{33}P-ATP of very high specific activity. The same batch of ATP was also used for phosphorylation of the 5′-hydroxyls of segments 10 and 12. The preparative isolation of A was carried out as shown in Fig. 7.

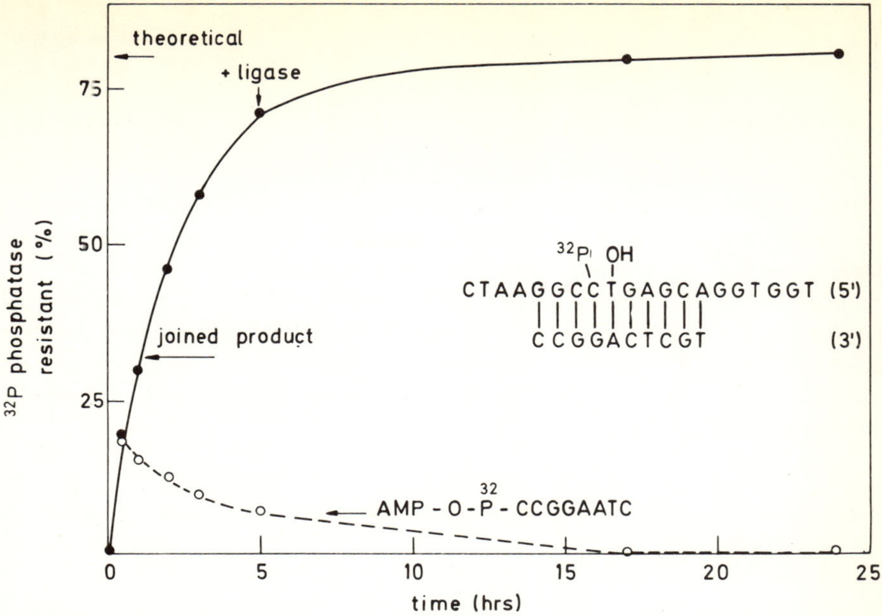

Fig. 6. Enzymatic joining of segment 1 to ^{32}P-labeled segment 4 in the presence of segment 3 (see also inset). (For details see [1].)

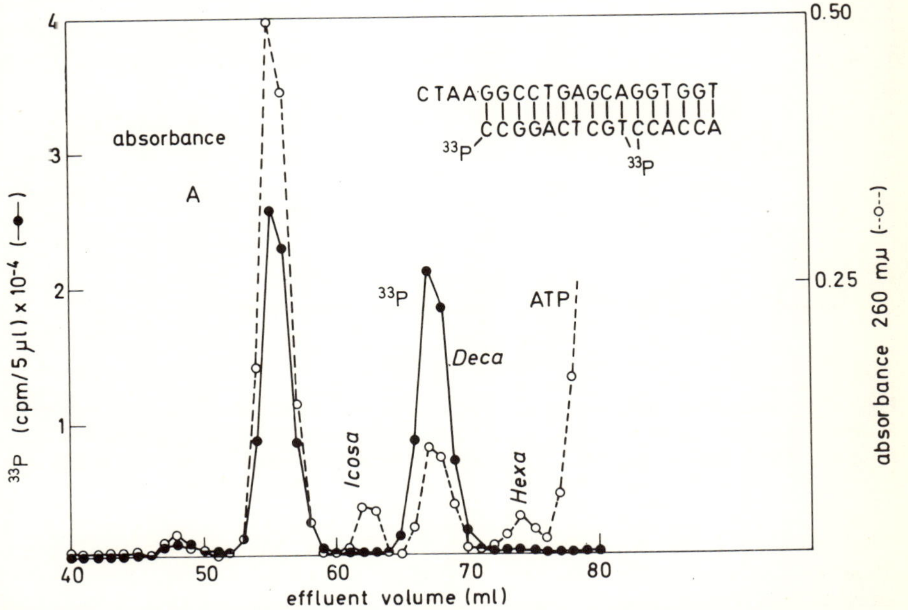

Fig. 7. The synthesis and purification of A. The labeling pattern of segments is shown in the inset. Bold ^{33}P refers to use of label of high specific activity relation to that in the hexanucleotide ^{33}P-C-C-A-C-C-A. (For further details see [1].)

The synthesis of C

The synthesis of this part presented problems of a different kind. Numerous attempts to perform joining reactions, using three component systems (for example, phosphorylated segment 11 + segments 12 and 13), in various combinations, either failed or showed moderate success. We ascribed this failure to the high G:C content of the components belonging to this part and believe that in all conditions previously tried, the formation of only a few G:C base-pairs stabilizes imperfectly matched duplexed structures so that it takes very prolonged periods (several weeks) for the components to align to form a perfect duplex. The procedure which was effective in the synthesis of C was as shown in Fig. 8.

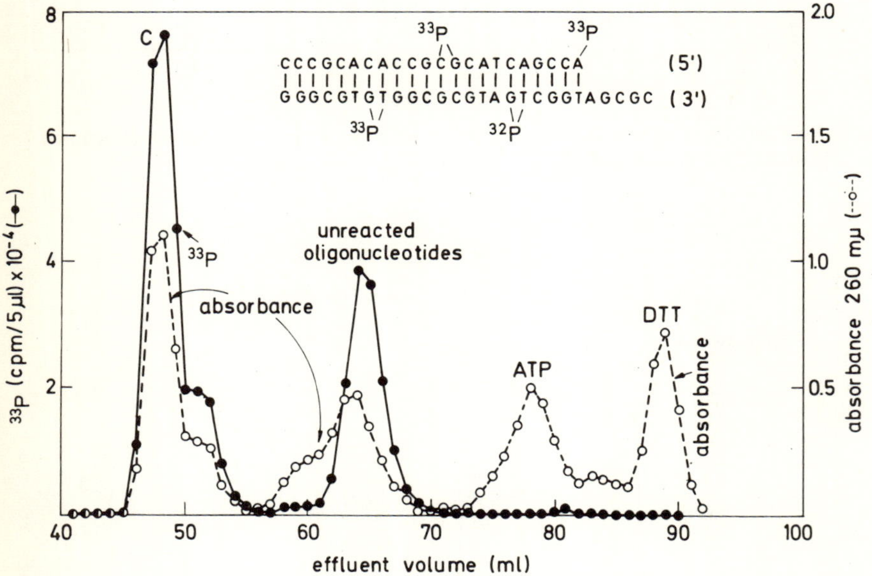

Fig. 8. The synthesis and purification of C. Four of the five components of the system contained 5′-labeled phosphate groups as shown in the inset, the ^{33}P group – in bold letter (segment 12) – being much more highly labeled than those used at the internal linkage points. (For details see [1].)

Thus, all the five components, four of which were suitably labeled (Fig. 8) were used all together. The yields in different experiments were 60–70 percent. The product was fully characterized. All of the data were as would be expected for C comprising the segments 11–15. It should be noted that segment 10 was deliberately excluded.

The joining of A to B and of B to C

The *joining of A to B* concerns the very weak tetranucleotide overlap comprising the nucleotide sequence 17–20. The joining reaction was therefore carried out at rather high Mg^{2+} concentration and low temperature (5°C). In the experiment shown in Fig. 9, A contained weak ^{32}P label, whereas the sample of B used was phosphorylated at its 5′-G ends using γ-^{33}P-ATP (the internal linkages in B contained insignificant ^{32}P label). The linkages which contain significant radioactive isotope are shown in the structure in the inset to Fig. 9. Also illustrated are the kinetics of the joining reaction as measured by the development of phosphatase-resistant ^{32}P and ^{33}P radioactivity. Both assays indicate that the joining occurs to the extent of about 60 percent. The product (A+B) was isolated as shown in Fig. 9. As expected, the first peak corresponding to the joined product contained both ^{32}P and ^{33}P labels in the expected ratio. Unreacted B containing only ^{33}P was eluted next, and unreacted A containing only ^{32}P was eluted last. Full characterization of A+B isolated from the first peak was carried out. All of the data are exactly was would be expected for the product.

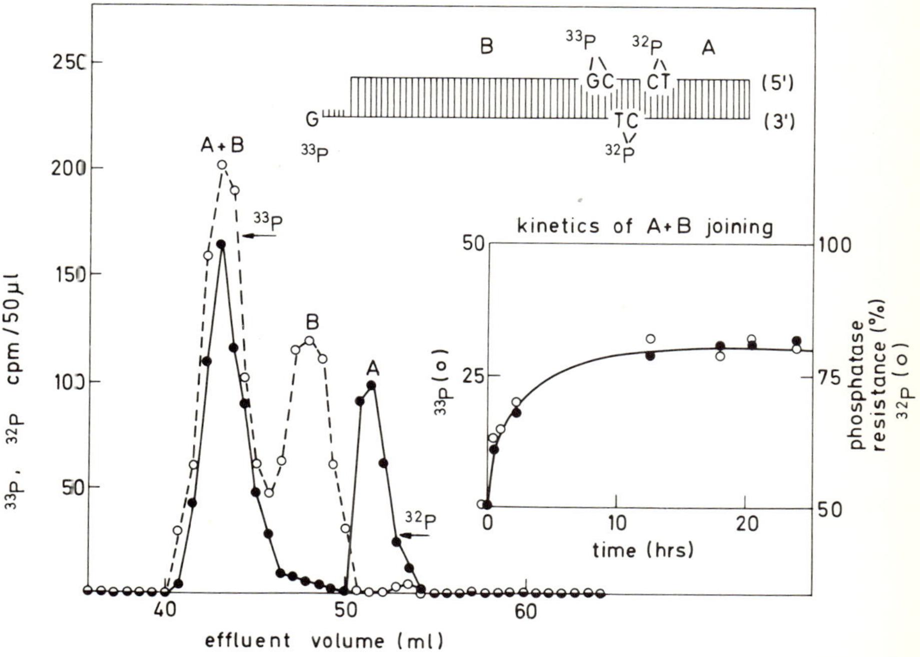

Fig. 9. The synthesis and purification of A + B. The points where significant radioactive label was present are shown in inset. The joining was performed at 5 °C using 18 mM Mg^{++}Cl. The kinetics of joining are shown. The separation was on an agarose 0.5 M column as in other experiments shown. (For further details see [1].)

Joining of B to C. The preparation of B used in this experiment carried very weak ^{32}P label in the internal linkages while the two 5′-G ends were labeled with high radioactive ^{32}P. The sample of C used contained very weak ^{32}P and ^{33}P labels in the internal linkages while the terminal A carried highly radioactive ^{33}P. Segment 10 also was labeled with the same highly radioactive ^{33}P. B and C were annealed separately, mixed, and segment 10 was added. Preincubation was now performed at 10 °C for 3 hours. The kinetics of joining are shown in the inset to Fig. 10. The reaction extent was 60–70 percent. The product (B+C) was isolated as shown in the same figure.

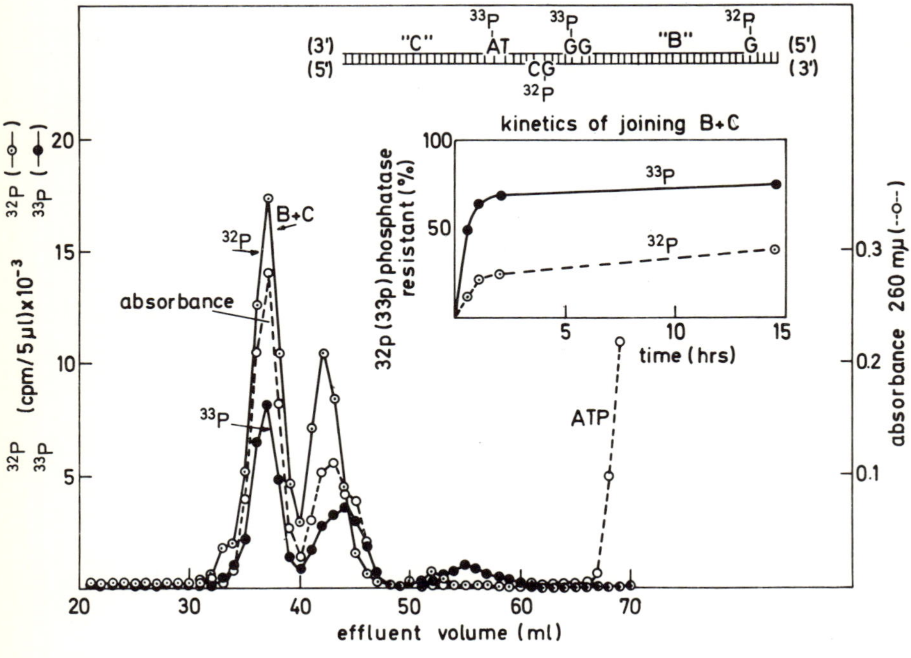

Fig. 10. The synthesis and purification of B + C.

The total synthesis of the 77 nucleotide long DNA

The joinings were performed in the two possible ways. In the first, preformed A+B was used in joining with C whereas in the second synthesis, preformed B+C was used in joining with A.

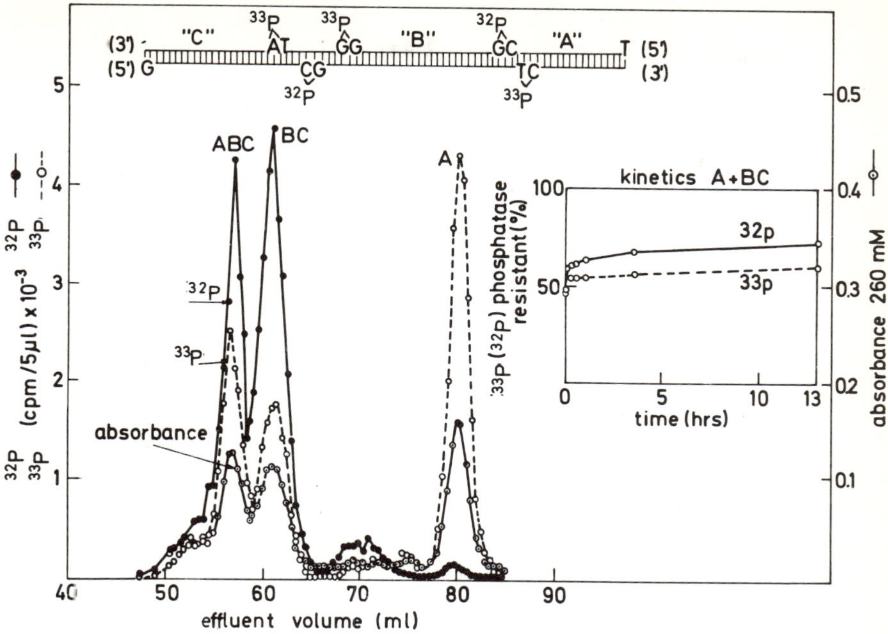

Fig. 11. The synthesis and purification of A + B + C by the joining of A to preformed B + C. The inset indicates that A contains a highly radioactive ^{33}P only on the 5′-nucleotide deoxycytidine. The joined product B + C contains a highly radioactive ^{32}P only on one of the 5′-nucleotides deoxyguanosine. Internally, as shown in the inset, B + C contains two highly radioactive ^{33}P labels and one ^{32}P. All other labeled internucleotide bonds are relatively less radioactive and have previously been characterized. The kinetics of joining are shown in the inset. Method of separation was as used in previous experiments.

Synthesis by joining preformed B+C to A. The preformed B+C was brought into reaction with A derived from the experiment of Fig. 7. The labeling pattern at the linkage points is shown in the abbreviated structure of the gene (inset to Fig. 11). The kinetics of the joining reaction is shown in the inset in the same figure. This reaction was carried out on a "large" scale such that ultraviolet absorbance of the different peaks could also be measured. The extent of the reaction (about 40 percent) was much lower than expected but all the properties (the ultraviolet absorbance to ^{32}P, ^{33}P ratios) of the first peak were as expected for the total gene.

Total synthesis by joining preformed A+B to C. A second synthesis of the gene is illustrated in Fig. 12. The preformed A+B (derived from experiment of Fig. 9) was brought into reaction with C (derived from experiment of Fig. 8) in the presence of the segment 10 containing weakly ^{32}P-labeled 5′-G end. The labeling pattern (inset to Fig. 12) shows the linkages where there was significant radioactivity, pronouncedly strong label being only at one site (the 5′-A unit in C). The kinetics of the joining are also shown in the inset in Fig. 12, the reaction having proceeded to about 65 percent. The profile obtained on separation of the products is shown in the figure. The first peak (excepting the inflexion at the right) corresponded to the total gene. Next came a small amount of unused A+B and finally the excess of C. Full characterization was performed and all the data were nicely consistent with theoretical expectations.

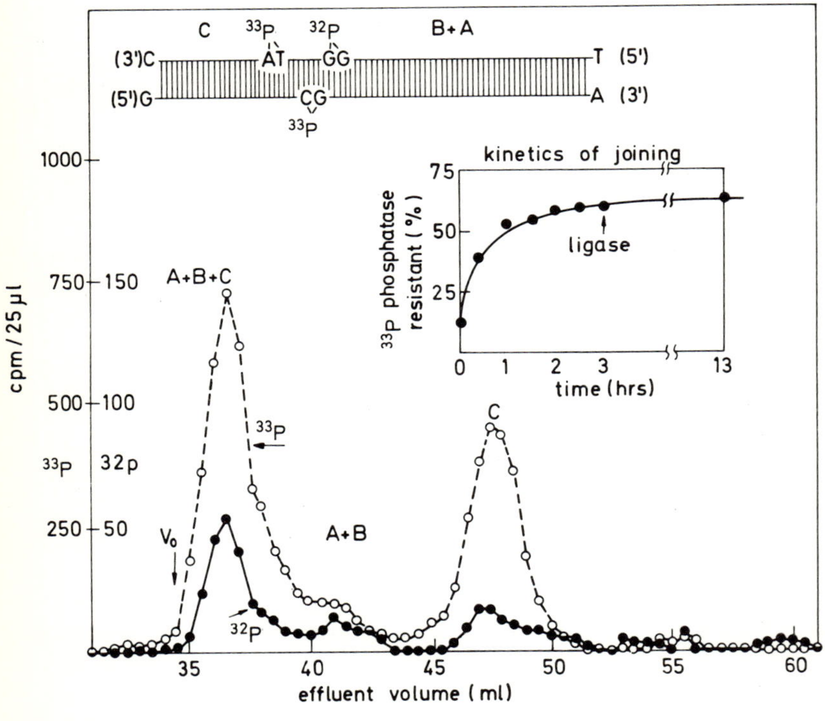

Fig. 12. The synthesis and purification of A + B + C by the joining of preformed A + B with C in the presence of segment 10.

Concluding remarks

The present successful synthesis of a seventy-seven nucleotide long DNA duplex shows that methodology is now available for the step-by-step synthesis of long bihelical DNA. The concepts and methods used are such that the synthesis can be controlled so as to be unambiguous. The first part consists of the chemical synthesis of oligonucleotide segments. Here the chain lengths involved are such that characterization can be unequivocal. The next phase involves the formation of duplexes from a few to several chemically synthesized segments. The joining steps in the synthesis of these duplexes can again be carefully monitored and the duplexes thoroughly characterized. The last step is the joining of the short duplexes to form longer duplexes. Here, again, in addition to the use of isotopic methods for characterization, the products can be readily separated and identified on the basis of size differences. Gel filtration methods proved to be very effective in all of the separation problems encountered in the present work and, hopefully, these can be extended for work with much longer duplexes.

Looking ahead, there are at least three directions that the present work can take. The first would be an attempt to include the transcription signals (initiation and termination) to the ends of the synthetic genes and in this way understand transcription in a precise way. The second direction would be the study of structure-function of tRNA genes by systematic alteration of the structural parts. Thirdly, an important direction lies in the addition of the synthetic genes to other genomes and introduction to living cells. One can see the chemical methods by which this is to be accomplished. One could in the near future, begin to undertake the synthesis of the genes for polypeptide hormones such as insulin, A.C.T.H. and these would be excellent candidates for insertion into the genomes of cells lacking these hormones.

References

[1] *Agarwal, K. L., Büchi, H., Caruthers, M. H., Gupta, N., Khorana, H. G., Kleppe, K., Kumar, A., Ohtsuka, E., RajBhandary, U. L., van de Sande, J. H., Sgaramella, V., Weber, H., and Yamada, T.* 1970. Nature **227**: 5253.

[2] *Holley, R. W., Apgar, J., Everett, G. A., Madison, J. T., Marquisee, M., Merrill, S. H., Penswick, J. R., and Zamir, A.* 1965. Science **147**: 1462.

Acknowledgements: This work has been supported by grants from the National Cancer Institute of the National Institutes of Health, U. S. Public Health Service (Grants No. 73675, CA05178), The National Science Foundation, Washington, D. C., (Grants No. 73078, GB–7484X), and The Life Insurance Medical Research Fund.

Discussion

Klingmüller: I recently learned that tRNA is made from longer precursors. If so, are the things you make really the genes for tRNA or must you add something? I should like you to comment on this a bit.

Khorana: The main point of my talk has been the development of methodology for the synthesis of bihelical DNA. In principle, we can make much longer chain lengths now. It is true that tRNA is made from a longer precursor, and we are actually in the process of synthesizing the DNA corresponding to that.

An Infectious and Replicating RNA of Low Molecular Weight: The Agent of the Exocortis Disease of Citrus

Heinz Ludwig Sänger

Arbeitsgruppe Pflanzenvirologie der Justus Liebig-Universität, 63 Gießen, West Germany

Summary: The transmissible agent of the exocortis disease of citrus (ExC) exhibits properties of an unusual free infectious RNA. It can be isolated from its herbaceous host *Gynura aurantiaca* DC by any method designed to extract and to concentrate undegraded host or viral RNA. The infectivity of ExC is found in nuclei and in association with the chromatin of the host cell. The agent of ExC is insensitive to different organic solvents, to proteolytic enzymes, and to DNase, but it is inactivated by RNase and by formaldehyde. In rate sedimentation experiments, it sedimented with about 6 to 8 s. Electrophoretic analysis in appropriate polyacrylamide gels revealed that infectivity of ExC is consistently associated with a monodisperse fraction of RNA with a molecular weight of ca. 50–60.000 daltons which would comprise molecules with a chain length of ca. 150–200 nucleotides.

Introduction

The exocortis disease of citrus (ExC) is an economically important disorder which is transmitted by budding and contaminated tools. In reactive citrus species, it causes scaling of bark, malformation of leaves, and dwarfing of whole trees, but in certain varieties no symptoms are ever expressed. Its incubation period can vary from several months to as much as 4–8 years. Transmission studies led one to assume that it was a virus disease [5]. The actual characterization of the proposed viral agent became possible when the host range of ExC could be extended to herbaceous host plants which produced typical symptoms a few weeks after inoculation [34]. Subsequent studies showed that no conventional virus is found in ExC infected plants and that the agent can only be extracted efficiently with phenol or alkaline-high salt solution. Its infectivity is susceptible to inactivation by RNase but not by DNase [28]. These properties together with its low sedimentation rate in sucrose gradient analysis and its behaviour in column chromatography suggested that it might be a conventionally infectious and double-stranded but free RNA molecule or possibly a RNA-DNA

Manuscript received: 2 March 1972

duplex [29]. The third possibility that the agent might be a single-stranded but unconventionally small infectious RNA was no reasonable alternative because the minimal molecular weight for infectious single-stranded RNAs is generally assumed to be in the range of ca. 1×10^6 daltons.

In this paper, I want to report the results of experiments which are capable of discriminating between these three alternatives. They substantiate that the infectivity of ExC is, in fact, associated with a fraction of small single-stranded RNA molecules of about twice the size of transfer RNA. Thus, next to the agent of the spindle tuber disease of potato [8, 30], the agent of ExC is the second member of a new and unusual class of minute infectious agents for which the name "viroids" has been proposed [8].

Material and methods

A Californian isolate of ExC was used which produces severe symptoms of epinasty and malformations of leaves in *Gynura aurantiaca* DC (Fig. 1) within 30–60 days after infection [34]. Inoculation was carried out by standardized needle puncture of the plant stem. ExC was propagated and tested in this host under greenhouse conditions at temperatures ranging between 20–25 °C. Infectivity is expressed in terms of "50 % infective dose" (ID_{50}), which is the dilution that would infect 50 % of the inoculated plants. It was calculated from three standard dilutions using five plants for each dilution [22]. Uninfected plants were always grown in parallel as a control.

Fig. 1. *Gynura aurantiaca* DC, the herbaceous host plant used for propagating and testing the infectious agent of the exocortis disease of citrus (ExC). Right: Symptoms of a systemic (generalized) ExC infection in the newly developed parts of the plant 6 weeks after inoculating the lower region of the stem (not visible) by needle puncture. Left: Healthy control plant.

Extraction of RNA

For routine work, the agent of ExC was extracted together with host nucleic acids from symptom-bearing Gynura plants in the presence of phenol, sodium dodecyl sulfate (SDS), and bentonite. Batches of 100 gr of leaf material were homogenized in the cold in a Sorvall Omnimixer in 400 ml of the BPSB extracting solution. It consisted of 200 ml TEMS buffer pH 7.5 (0.1 M Tris-HCl, 0.01 M EDTA, 0.01 M mercaptoethanol, 0.1 M NaCl), 200 ml TEMS-saturated phenol, 5 gr SDS, and 5 gr bentonite. The homogenate was clarified for 10 min at 10.000 x g, and the resulting supernatant was retreated with an equal volume of phenol and 10 mg bentonite per ml of the total volume. Upon clarification, the supernatant was retreated three more times by quick agitation with 5 mg, 2 mg, and 1 mg of bentonite, respectively. Nucleic acids were precipitated from the final clarified supernatant once with cetyltrimethylammoniumbromide [36] and then three more times in the cold with 2.3 volumes of 95 % ethanol containing 0.1 M sodium acetate. The final precipitate of RNA was dissolved in TEMS buffer and stored in portions at −30 °C until use. For comparison, RNA was also extracted from healthy Gynura plants grown under identical conditions.

Preparation of subcellular fractions

Nuclei were prepared according to the method of *Kuehl* [15] and chloroplasts according to the procedure of *Tewari* and *Wildman* [33]. The isolation of cytoplasmic ribosomes followed the method of *Clark* et al. [7], and mitochondria were isolated as described by *Wolstenholme* and *Gross* [35]. Chromatin was isolated from nuclei according to *Huang* and *Huang* [12] and directly from frozen tissue by the method of *Bekhor* et al. [4]. All procedures were followed as closely as possible. ExC was isolated from the subcellular fractions by BPSB extraction, concentrated 10 fold by ethanol precipitation and bioassayed as described.

Rate sedimentation analysis

Rate zonal centrifugation was accomplished in 5–20 % (w/v) linear sucrose gradients prepared with TEMS buffer in tubes of the SW 25.1 rotor at 4 °C. Fractionation was from the top by pumping 50 % glycerol into the bottom of the tube and monitoring the absorbancy at 254 nm with a Uvicord-UV-analyzer. The fifty fractions isolated from each gradient were appropriately diluted with TP-buffer and stored frozen until used for bioassay.

Electrophoresis

Molecular fractionation of nuclei acids was on polyacrylamidgels [16] in plexiglass tubes (10 cm long, 6 mm internal diameter) using gels with 5, 10, 15, or 20 % acrylamide and 2.5 % of bisacrylamide. The 2.5 % acrylamide gel contained 5 % bisacrylamide. The buffer had a pH of 7.6 and contained 36 mM Tris, 30 mM NaH_2PO_4, 1 mM EDTA, and 0.2 % SDS. About 50 μg of RNA in 25 μl electrophoresis buffer were applied to each gel, and electrophoresis was at room temperature at 50 V and 5 mA per gel. The gels were scanned at 260 nm in a Gilford Model 240 spectrophotometer equipped with a gel scanner and recorder. After fractionation into 2 mm slices, the corresponding slices of eight parallel gels from one run were pooled and kept frozen until use. For infectivity test, the slices were homogenized in pH 7.5 TP-buffer (0.05 M Tris, 0.05 M Na_2HPO_4), and all fractions were assayed at 10^{-1}, 10^{-2} and 10^{-3} dilution. Bromphenolblue and Pyronin G were used as marker dyes, the latter travelling with the buffer front in SDS-containing gels and buffer systems. Phenylalanin-specific transfer RNA (Boehringer RRAD) and a preparation of soluble RNAs from brewers' yeast (Boehringer RRAB) were used as marker RNAs. The size of ExC was estimated on the basis of the linear relationship between the log molecular weight and the relative electrophoretic mobility of marker RNAs [4, 16].

Results

Properties of ExC

Neither virus particles nor infectivity can be recovered from plants infected with ExC if conventional procedures for isolating plant viruses by differential centrifugation are used. However, highly infectious preparations are obtained by various methods and with additives used to extract undegraded cellular RNA (Fig. 2). It is interesting to note that a considerable amount of infectivity is sedimenting with the low speed pellet of pH 7.5 buffer homogenate from which it can be extracted with the aid of PSB. This would indicate a close association of ExC with cell debris and (or) subcellular components. The symptoms of ExC were only observed in those test plants inoculated with nucleic acid extracts derived from Gynura plants that had been previously infected with ExC and had developed characteristic symptoms of the disease. For routine isolation of ExC, the BPSB-procedure (Fig. 2, No. 7) as described in "Material and methods" was adopted.

For further characterization of ExC, infectious BPSB extracts were treated with various agents and bioassayed after the residual infectivity had been recovered and concentrated by ethanol precipitation. It was found that the agent of ExC present in such extracts is insensitive to DNase, to various proteolytic enzymes, to

	Extraction medium and additives	ID_{50} (– log 10)
1	0.1 M TEMS buffer pH 7.5	<1
2	+ 1 Vol chloroform + 1 Vol butanol	<1
3	+ 2.5 % Bentonite	2.80
4	+ 2.5 % Bentonite + 1 % SDS	3.60
5	+ 80 % Phenol	3.75
6	+ 80 % Phenol + 1 % SDS	3.80
7	+ 80 % Phenol + 1 % SDS + 4 x Bentonite treatment	4.20
8	0.3 M Na_2SO_3 pH 9.5 + 1 Vol chloroform + 1 Vol butanol	2.0
9	Low speed sediment of 1 extracted with 7	2.60
10	High speed sediment of 1 extracted with 7	2.0

Fig. 2. Infectivity in high speed supernatants of various extracts from infected Gynura plants

Treatment: Agent	Treatment: Concentration	ID_{50} (– log 10)
None	–	2.80
DNase	0.1 μg/ml	2.80
DNase	1.0 μg/ml	2.60
Trypsin	10.0 μg/ml	2.60
Chymotrypsin	10.0 μg/ml	2.60
Pronase	10.0 μg/ml	2.60
Urea	8.0 M	2.60
Guanidin-HCl	5.0 M	2.30
Dimethylsulfoxide	99.0 %	2.30
RNase	0.1 μg/ml	<1
RNase	1.0 μg/ml	<1
Formaldehyde	2.0 %	<1
Chloroform	50.0 %	2.40
Butanol	50.0 %	2.10
Ether	50.0 %	2.10

Fig. 3. Effect of different agents on the infectivity of ExC in purified extracts of total RNA from infected Gynura plants

dissociating agents like 8 M urea, 5 M guanidin-hydrochloride, and 99 % dimethylsulfoxide. ExC is also insensitive to treatment with organic solvents like chloroform, butanol, and ether. Infectivity of ExC is inactivated, however, by pancreatic RNase and by 2 % formaldehyde (Fig. 3). Taken together, all these results indicate that the agent of ExC is not a conventional virus but rather a free infectious RNA, as has been proposed previously [28, 29].

Association of ExC with nuclei and chromatin

A considerable part of the infectivity of ExC can be recovered from low speed pellets of TEMS buffer homogenates by BPSB extraction, although practically all infectivity is inactivated quite rapidly in the corresponding supernatant solution (Fig. 2, No. 9). This observation led to the investigation of the possible association of ExC with subcellular components of the host plant. Purified fractions of chloroplasts, ribosomes, and mitochondria prepared from infected plants were devoid of any infectivity as extractable with the BPSB procedure. However, infectivity of ExC could be recovered with BPSB in appreciable amounts from crude nuclear fraction as well as from nuclei that had been highly purified by a density gradient procedure (Fig. 4). The survival of infectivity during these procedures indicates that ExC must be located inside the nuclei, where it is well protected from enzymatic degradation. In fact, infectivity can be liberated by exposure of nuclei to 2 M NaCl, a procedure known to disrupt nuclei and to release nucleic acids from them.

These results suggested an association of ExC with the chromatin which is biochemically a complex of chromosomal DNA, nuclear RNA, histones, and proteins. When chromatin is isolated from purified nuclei or directly from plant tissue and subsequently treated with agents known to dissociate this complex, infectivity of ExC is released into the supernatant (Fig. 5). These results clearly show that the infectious RNA of ExC exists in a close association with the chromatin. This might explain the relative resistance of the agent in crude buffer homogenates and in low speed pellets. On the other hand, it might also have some bearing on its function and the possible mechanism of its replication.

Subcellular component	ID_{50} (−log 10)
Direct tissue extract	3.2
Chloroplasts	<1
Ribosomes	<1
Mitochondria	<1
Crude nuclear fraction	2.8
Highly purified nuclei	2.6
Chromatin isolated from nuclei	1.8
Chromatin extracted directly from infected tissue	2.1

Fig. 4. Infectivity of BPSB extracts from subcellular components isolated from infected Gynura plants

Treatment of chromatin	ID_{50} (−log 10)
None	<1
Freeze + thaw	1.2
Extraction with BPSB	2.8
Extraction with 1 % SDS	2.6
Extraction with 2 M NaCl	2.6
Extraction with 6 M Urea	2.6
Extraction with 2 M Guanidin-HCl	2.2

Fig. 5. Dissociation of ExC infectivity from isolated chromatin

Rate zonal centrifugation

Analysis of infectious BPSB extracts in 5–20 % sucrose gradients shows that their sedimentation profile is characteristic for preparations of total plant nucleic acids. The distribution of infectivity, however, is quite unusual since it bands in a region of RNA molecules with a sedimentation rate of 6–8 s (Fig. 6). The maximum of infectivity does not correlate with a distinct optical density peak, and there is no apparent difference in the absorbancy patterns of extracts from either healthy or infected plants. Under identical conditions the infectious, single-stranded RNA of tobacco mosaic virus (MW 2 x 10^6) sediments with about 30–32 s into the lower half of the gradient, travelling in the lower shoulder of the 28 s ribosomal RNA.

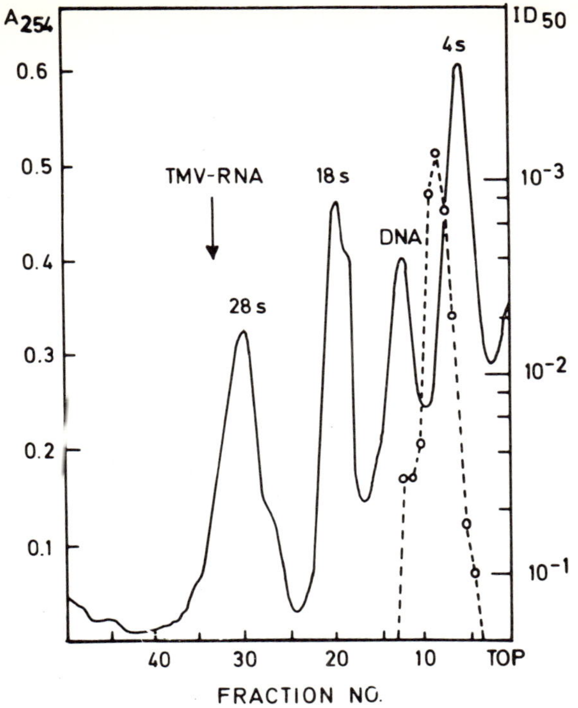

Fig. 6
UV-absorption profile (———) and infectivity distribution (○- - - -○) of a BPSB extract from ExC-infected Gynura analysed on a linear 5–20 % sucrose gradient after 18 hours at 24.000 rpm and 4 °C in a SW 25.1 rotor of the Spinco Model L centrifuge. The host cell nucleic acids are denoted by their sedimentation coefficients and the position of TMV-RNA under these conditions is shown by an arrow. A_{254} = Absorbancy at 254 nm; ID_{50} = Infectivity index given as "50 % infective dose."

Under the assumption that the agent of ExC might be a double-stranded RNA, several attempts were made to "melt" this proposed structure by a 10 min treatment at 100 °C and to keep the single strands segregated by rapid cooling. But there was neither an indication of any significant loss of infectivity nor of an alteration of its sedimentation behaviour. Its unusual properties led to analyse the actual dimensions of the agent by molecular fractionation in polyacrylamide gels.

Electrophoresis

Electrophoretic analysis of infectious BPSB extracts in polyacrylamide gels of different pore size revealed that the infectivity of ExC is banding in a region which is in between the location where the 4 s transfer RNA and the 7 s RNA are detectable by their optical density at 260 nm (Fig. 7). Under the usual conditions of fractionating conventional infectious viral RNAs with a molecular weight of 1–2 x 10^6 dalton gels with a polyacrylamide concentration of 2.1 to 2.5 % are used. It should be noted that under such conditions of electrophoresis, the infectious RNA of ExC might have migrated out of the gel. Accordingly, TMV-RNA barely enters the 2.5 %

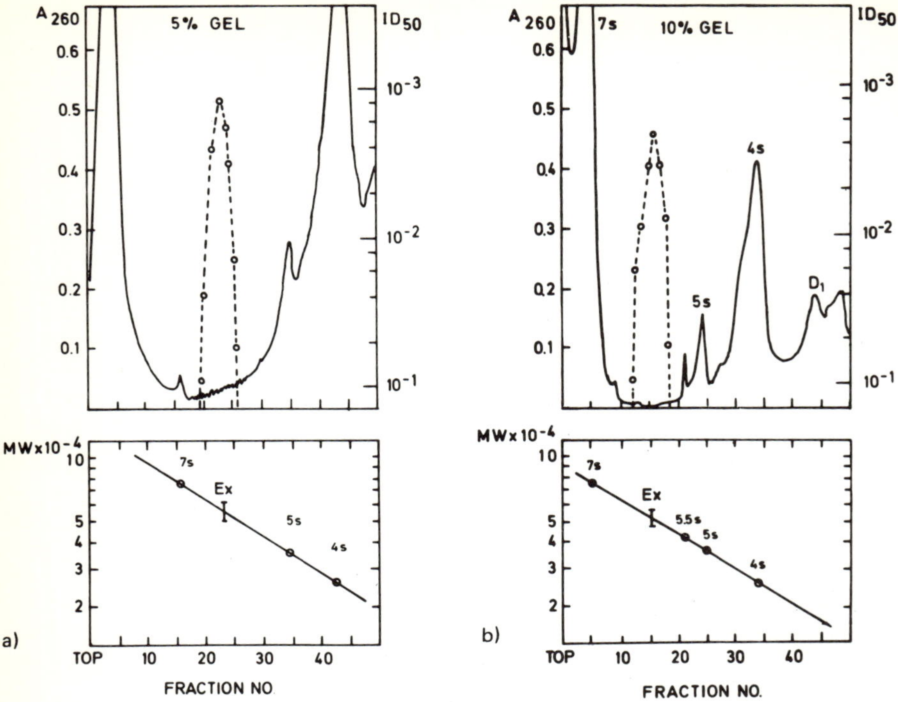

Fig. 7

gel under conditions where the infectivity of ExC is migrated into the lower half of this gel (not shown). In addition, gels with 5, 10, 15, and 20 % acrylamide and correspondingly smaller pores were used for analysis because they allow a better resolution of RNAs of low molecular weight. It was found (Fig. 7) that also under these conditions all infectivity of ExC entered the gels with a relative mobility in between that of t-RNA and 7 s RNA. As compared with these different marker RNAs of known size, the infectivity of ExC is migrating with a fraction of RNA molecules, the molecular weight of which is estimated to be in the range of 50.000 to 60.000 daltons.

Discussion

Based on the above results, I confirm earlier reports [28, 29] that the exocortis "virus" of citrus is a free infectious RNA with a low sedimentation coefficient. In addition to these results, my determinations of the actual molecular dimensions of

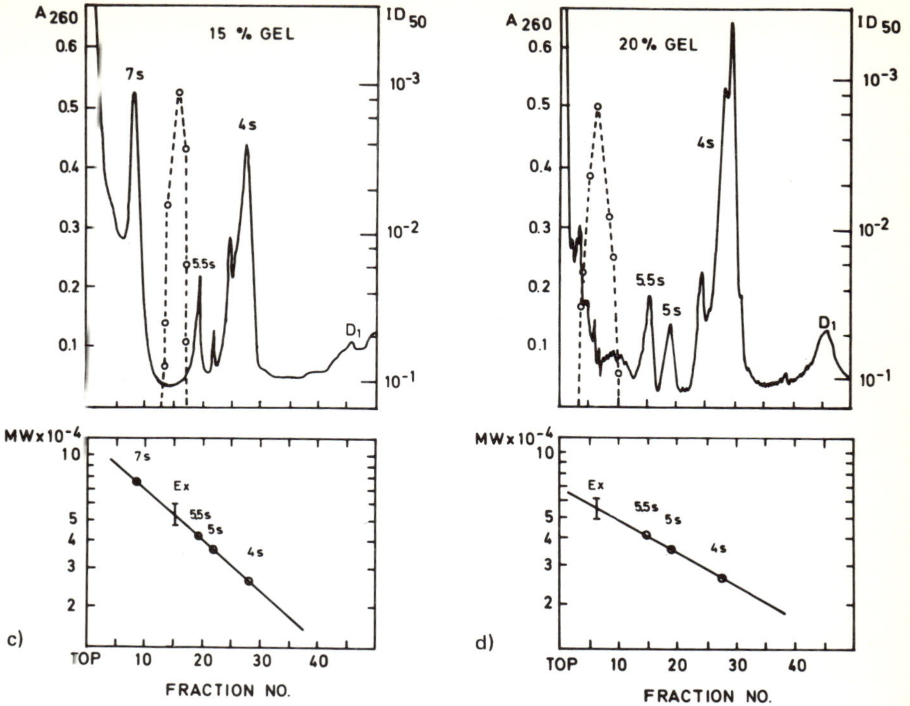

Fig. 7. Electrophoretic analysis of BPSB extracts from ExC-infected Gynura plants on 5, 10, 15, and 20 % polyacrylamide gels and relation between the relative mobility and the log molecular weight of RNAs. UV-absorption (———) and infectivity (o- - - -o) profiles. A_{260} = Absorbance at 260 nm; ID_{50} = Infectivity index given as "50 % infective dose"; D_1 = Bromphenolblue marker dye.

this RNA clearly show that this free infectious and replicating RNA has a molecular weight of 50.000 to 60.000 daltons which would indicate a chain length of about 150–200 nucleotides. Therefore, I conclude that ExC is the second member of the new class of minute infectious agents for which the term "viroids" has been proposed [8]. Viroids are small replicating free infectious RNA molecules which may cause pathological conditions in their host. The agent of the spindle tuber disease of potato (PSTV) was the first viroid discovered [8, 30].

Properties

If one compares the properties of the two viroids known at present, one finds that they have several biological properties in common. Both are characterized by long incubation periods, by a relatively high heat resistance, by their association with nuclei, and by their similar resistance to various agents in crude extracts as well as in a purified form. They differ, however, in their host range and in the type of symptoms they produce. Moreover, ExC is rather difficult to transmit mechanically by abrasion of leaves in the presence of sap. Elaborate transmission techniques like budding, grafting, or piercing the stem with razer slashes or needle punctures have to be used for efficient experimental transmission. In practice, the agent is spread in the course of the usual vegetative propagation of citrus by budding and grafting, and a low percentage of transmission can be obtained with contaminated pruning tools [1, 11]. PSTV, however, is transmitted easily even under natural conditions by simple contact of foliage or by machinery [17], and vertical transmission by seed has been reported for this type of disease [14]. It is interesting to note that both viroids have been discovered only because they induced symptoms of a viruslike disease in certain hosts of economical importance. But there are several host plants in which both agents are replicated efficiently without symptoms [19, 26]. In addition, I recently found an isolate of ExC from the Mediterranean area which produces severe symptoms in citrus and in petunia similar to the Californian isolate used in this study. In Gynura, however, this isolate is replicated without discernible symptoms [26] so that it would not have been detected in this test system.

There are also some interesting differences at the molecular level. As judged from the distribution of infectivity in polyacrylamide gels in one major band, the RNA of ExC seems to exist in a monodisperse state representing molecules with a chain length of about 150–200 nucleotides. The RNA of PSTV, on the other hand, seems to occur in various stages of aggregation which are obviously multiples of a smaller infectious monomer [8]. Some discrepancy still exists about the size of this monomer. In some experiments, its molecular weight appears to be 25.000 daltons which would give it the appearance of an infectious transfer RNA [30]. Other experiments indicate that its molecular weight is 50.000 [9], which is in the range of the size of the RNA of ExC as described here.

Possible mode of replication

One of the most interesting questions arising in this connection is how such small RNA molecules are replicated in their host cells. On the basis of their potential genetic information, at least five possible mechanisms of viroid replication are conceivable (Fig. 8) out of which three are rather unlikely to hold true. The infectivity dilution curve of ExC follows a theoretical single hit curve, and there is no indication whatsoever of a cooperative interaction in the process of infection [26]. This observation would exclude the possibility that the genome of ExC is polycistronic and that it exists in different segments which together could function cooperatively like a more or less conventional viral RNA. Therefore, the genome of

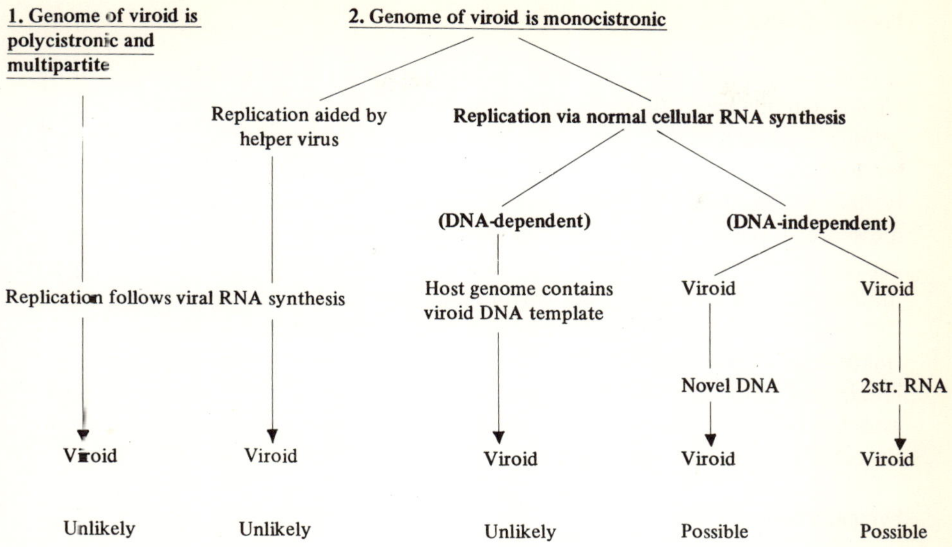

Fig. 8. Proposals for the possible mode of replication of viroid RNA

the viroid should be monocistronic. As a consequence, the limited genetic information encoded in this RNA can theoretically consist only of a cistron for a rather small protein, or it might even function only as recognition sites which alone might provide the basis for the replication of the molecule.

The single hit dependence of infection and the failure to isolate any other infectious agent from ExC infected tissue precludes the possibility of a helper virus being involved in the replication of the small infectious RNA. Thus, we are forced to assume that it is replicated by the RNA synthesizing machinery present in noninfected cells. Since it is rather unlikely that this new infectious RNA is transcribed from a part of the genome of the host which is normally suppressed, only two possible ways of hostmediated synthesis are left. The DNA-dependent mode of replication would require the existence in normal cells of a system which transcribes RNA into DNA. In fact, such a system has been claimed to exist in normal mammalian cells [20, 27]. The second alternative presumes that the replication of viroids is analogous to the replication of conventional RNA viruses. The existence of small double-stranded RNA species in apparently uninfected cells and tissues [18, 31], and the presence of a RNA-dependent RNA polymerase in healthy plant leaves [2] indicate that this mode of replication might also operate.

Possible Origin

Another interesting aspect of rather speculative nature is the possible origin of the viroids. On one hand, they might be considered to be either defective [23, 24], primitive, or regressed viruses which have found an alternative way of "surviving" without a protein coat and of being replicated in cells with only little genetic information introduced by themselves. Accordingly, they would have little or no messenger function and only exist in the replicating (vegetative) phase with the conventional virion phase lacking. In this sense, one is tempted to call them "naked mini-viruses". Alternatively, one might conjecture that viroids may represent the "missing link" between viruses and genes [32] or the postulated "oncogenes" of *Huebner* and *Todaro* [13], or they may even be carriers of extrachromosomal inheritance as has already been suggested by *Diener* [8]. Finally, because of their association with nuclei and in view of their size, viroids have some resemblance to the small RNA species recently found in association with nucleolar RNA [21].

Perspectives

The finding that two completely different plant diseases, regarded for a long time to be of viral origin, are actually caused by a replicating, chromatin-associated, low molecular weight RNA opens new perspectives for medical virology. Now the possibility has to be considered that certain infectious and (or) progressive viruslike diseases of man and animals might also be caused by viroids. It would be characteristic for such diseases that (1) all attempts to isolate a viral pathogen with conventional methods have failed so far; (2) the infectious agent is relatively insensitive to treatment with organic solvents, proteolytic enzymes, nucleases, UV-light and heat, especially if tested in crude tissue homogenates; (3) no specific antibodies are found in association with the disease in question because of the lack of virus-specific antigens, and (4) the expression of the disease is progressing rather slowly because viroids are primarily spread from cell to cell. It is conceivable that certain degenerative neurological diseases such as the scrapie disease of sheep, the Kuru disease and the Creutzfeld-Jacob disease of man [10], and even multiple sclerosis [3] might be caused by viroidlike agents.

In conclusion, the potential significance of the new class of small infectious RNA molecules discussed in this report is obvious for molecular biology and for genetics. They present a stimulating challenge for investigators of certain viruslike diseases of man and animals.

Acknowledgements: I thank Dr. *Weathers*, Riverside, for providing the Californian isolate of ExC, Dr. *T. O. Diener*, Beltsville, for helpful suggestions, and Miss *K. Ramm* and Mrs. *H. Muth* for capable technical assistance.

This work was supported by the Deutsche Forschungsgemeinschaft through a personal grant and through the Sonderforschungsbereich 47 – Virologie.

References

[1] *Allen, R. M.*, 1968. Survival time of exocortis virus of citrus on contaminated knife blades. Plant Disease Reporter **52**: 935–39.

[2] *Astier-Manifacier, S.*, and *Curnuet, P.* 1971. RNA-dependent RNA-polymerase in chinese cabbage. Biochim. Biophys. Acta **232**: 484–93.

[3] *Bauer, H. J.* 1970. Multiple Sklerose: Grundlagen und Hypothesen der modernen Ursachenforschung. Z. Neurol. **198**: 5–32.

[4] *Bekhor, I.*, *Kung, G. M.*, and *Bonner, J.* 1969. Sequence specific interaction of DNA and chromosomal protein. J. Mol. Biol. **39**: 351–64.

[5] *Benton, R. J.*, *Bowman, F. T.*, *Frazer, L.*, and *Kebby, R. G.* 1949. Stunting and scaly butt of citrus associated with Poncirus trifoliata rootstock. Agr. Gaz. N.S. Wales **60**: 521–26, 577–82, 641–45, 654.

[6] *Bishop, D. H. L.*, *Claybrook, I. R.*, and *Spiegelman, S.* 1067. Electrophoretic separation of viral nucleic acids on polyacrylamide gels. J. Mol. Biol. **26**: 373–87.

[7] *Clark, M. F.*, *Matthews, R. E. F.*, and *Ralph, R. K.* 1964. Ribosomes and polyribosomes in Brassica pekinensis. Biochim. Biophys. Acta **91**: 289–304.

[8] *Diener, T. O.* 1971. Potato spindle tuber "virus". IV. A replicating low molecular weight RNA. Virology **45**: 411–28.

[9] *Diener, T. O.*, and *Smith, D. R.* 1971. Potato spindle tuber viroid. VI. Monodisperse distribution after electrophoresis in 20 percent polyacrylamide gels. Virology **46**: 498–99.

[10] *Gajdusek, D. C.*, and *Gibbs, C. J.* 1971. Degenerative neurological diseases of viral etiology: Scrapie, Kuru and Creutzfeld-Jacob disease. *In* Proceedings of the Conference on Atypical Virus Infections, eds. *Christian, Phillips,* and *Williams,* pp. 167–77. The Arthritis Foundation Conference Series No. **15**, New York.

[11] *Garnsey, S. M.*, and *Jones, J. W.* 1067. Mechanical transmission of exocortis virus with contaminated budding tools. Plant Disease Reporter **51**: 410–13.

[12] *Huang, C. C.*, and *Huang, P. C.* 1969. Effect of protein-bound RNA associated with chick embryo chromatin on template specificity of the chromatin. J. Mol. Biol. **39**: 365–78.

[13] *Huebner, R. J.*, and *Todaro, G. J.* 1969. Oncogenes of RNA tumor viruses as determinants of cancer. Proc. Nat. Acad. Sci. U.S. **64**: 1087–94.

[14] *Hunter, D. D.*, *Darling, H. M.*, and *Beale, W. E.* 1969. Seed transmission of potato spindle tuber virus. Amer. Potato J. **46**: 247–50.

[15] *Kuehl, L.* 1964. Isolation of plant nuclei. Z. Naturforsch. B **19**: 525–32.

[16] *Loening, U. E.* 1967. The fractionation of high molecular-weight ribonucleic acid by polyacrylamide-gel electrophoresis. Biochem. J. **102**: 251–57.

[17] *Manzer, F. E.*, and *Merriam, D.* 1961. Field transmission of the potato spindle tuber virus and virus X by cultivating and hilling equipment. Amer. Potato J. **38**: 346–52.

[18] *Montagnier, L.* 1968. Présence d'un acide ribonucléique en double chaîne dans des cellules animales. Compt. Raid. Acad. Sci. Ser. D **267**: 1417–20.

[19] *O'Brien, M. J.*, and *Raymer, W. B.* 1964. Symptomless hosts of the potato spindle tuber virus. Phytopathology **54**: 1045–47.

[20] *Penner, P. E.*, *Cohen, L. H.*, and *Loeb, L. A.* 1971. RNA-dependent DNA polymerase in human lymphocytes during gene activation by phytohaemagglutinin. Nature, New Biology, London, **232**: 58–61.

[21] *Prestayko, A. W., Tonato, M.,* and *Busch, H.* 1970. Low molecular weight RNA associated with 28s nucleolar RNA. J. Mol. Biol. **47**: 505–15.

[22] *Reed, L. J.,* and *Muench, H.* 1938. A simple method of estimating fifty percent endpoints. Am. J. Hyg. **27**: 493–97.

[23] *Sänger, H. L.* 1968. Defective plant viruses. *In* Molecular genetics, eds. *H. G. Wittmann* and *H. Schuster,* pp. 300–36, 4. Wiss. Konf. Ges. Deutsch. Naturforsch. Ärzte, Berlin, 1967. Berlin: Springer-Verlag.

[24] *Sänger, H. L.* 1970. "Naked" plant viruses. Abstracts of the X. Int. Congr. for Microbiology Mexico 1970, p. 238.

[25] *Sänger, H. L.* in press. Association of the infectious agent of the exocortis disease of citrus with a low molecular weight RNA. Mol. Gen Genetics.

[26] *Sänger, H. L.* Unpublished work.

[27] *Scolnik, E. M., Aaronson, S. A., Todaro, G. J.,* and *Parks, W. P.* 1971. RNA dependent DNA polymerase activity in mammalian cells. Nature, London, **229**: 318–21.

[28] *Semancik, J. S.,* and *Weathers, L. G.* 1968. Exocortis virus of citrus: association of infectivity with nucleic acid preparations. Virology **36**: 326–28.

[29] *Semancik, J. S.,* and *Weathers, L. G.* 1970. Properties of the infectious forms of exocortis virus of citrus. Phytopathology **60**: 732–36.

[30] *Singh, R. P.,* and *Clark, M. C.* 1971. Infectious low molecular weight ribonucleic acid from tomato. Biochem. Biophys. Res. Comm. **44**: 1077–83.

[31] *Stern, R.,* and *Friedman, R. M.* 1970. Double-stranded RNA synthesized in animal cells in the presence of actinomycin D. Nature, London, **226**: 612–16.

[32] *Temin, H. M.* 1970. Malignant transformation of cells by viruses. Perspect. Biol. Med. **14**: 11–26.

[33] *Tewari, K. K.,* and *Wildman, S. G.* 1966. Chloroplast DNA from tobacco leaves. Science **153**: 1269–71.

[34] *Weathers, L. G.,* and *Greer, F. C.,* Jr. 1968. Additional herbaceous hosts of the exocortis virus of citrus. Phytopathology **58**: 1071.

[35] *Wolstenholme, D. R.,* and *Gross, N. J.* 1968. The form and size of mitochondrial DNA of the red bean, Phaseolus vulgaris. Proc. Nat. Acad. Sci. U.S. **61**: 245–52.

[36] *Ralph, R. K.,* and *Bellamy, A. R.* 1964. Isolation and purification of undegraded ribonucleic acid. Biochim. biophys. Acta (Amst.) **87**: 9–16.

Studies on Transcription and Ligation of RNA

Jonathan Leis, Robert Silber *, V. G. Malathi *, and Jerard Hurwitz **

* New York University School of Medicine, New York, N. Y., USA

** Department of Developmental Biology and Cancer. Albert Einstein College of Medicine, Bronx, New York, USA

The finding of an RNA-dependent DNA polymerase in RNA tumor viruses [5, 32] has prompted many to consider the flow of genetic information for viral RNA replication to be the reverse of the normal pathway. An important biological role for the viral DNA polymerase has been suggested by evidence indicating the existence of a DNA intermediate in viral replication. It has been proposed that this intermediate may explain the inhibition of virus production by actinomycin D and cytosine arabinoside [3, 10]. Further support for this concept was obtained by *Hanafusa* et al. [16] who described a variant of the Rous sarcoma virus (RSVα) which is deficient in RNA-dependent DNA polymerase activity and also is unable to infect or transform chick cells. These results do not distinguish between the direct role of the RNA-dependent DNA polymerase copying the entire viral RNA vs. a role in the transcribing of a selected portion of the viral genome which in turn activates the production of viral DNA by the host.

In the present communication, we report a study of the mechanism of action of the RNA-dependent DNA polymerase in an attempt to define the requirements of the enzyme. We also describe the properties of an enzyme obtained from T4-infected *Escherichia coli* capable of reacting with ends of RNA chains.

Jonathan Leis is a postdoctoral fellow of the Damon Runyon Foundation

Robert Silber is a recipient of a Guggenheim Fellowship.

V. G. Malathi is supported by USPHS Training Grant AM 5282

Manuscript received: 8 February 1972

A. Studies with viral RNA

The RNA-dependent DNA polymerase from AMV and Rauscher virions has been highly purified and is primed by RNA as well as DNA. The investigation of several characteristics of the viral DNA polymerase reaction conducted in our laboratory can be summarized as follows. [17, 23].

Priming of polymerases with DNA

The presence of DNA-dependent DNA polymerase in RNA tumor viruses was first demonstrated by *Spiegelman* et al. [30]. These workers showed that a variety of exogeneously supplied DNA preparations supported deoxynucleotide incorporation; they also noted a marked preference for double-stranded DNA and suggested that this feature distinguished the viral polymerase from previously described DNA-dependent DNA polymerases.

We have reinvestigated this problem and have found, as they reported, that the polymerases exhibit a marked preference for native DNA. It was shown that this preference is due to the requirement for a 3′-OH containing primer strand hydrogen bonded to an overlapping template strand. This characteristic is true for all DNA polymerase activities so far studied. The only exception to this general requirement in primed deoxynucleotide incorporation is the terminal deoxynucleotidyl transfer enzyme studied by *Bollum* and coworkers [20].

Alterations of templates and their influence on priming efficiency

The structural modification required to activate native DNA as template was studied. Treatment of native thymus DNA with exonuclease III or with relatively large amounts of pancreatic DNase (Fig. 1, line 5) activated it as a template. The following treatments were without effect: incubation with Neurospora nuclease or micrococcal nuclease, heat denaturation, and introduction of single-strand breaks with pancreatic DNase.

The low levels of DNase employed (Fig. 1, line 4) optimally activate calf thymus DNA as a template for DNA polymerase I [2]. The single-strand breaks introduced by this procedure do not activate DNA as a primer for the Rauscher or avian polymerases (not shown). Since DNA polymerase I utilizes these single-strand breaks by "nick translocation" due to the presence of the 5′ → 3′ exonuclease [22], these observations suggest that the Rauscher and avian polymerases lack such an exonuclease activity. This interpretation was substantiated by the following experiments. DNA containing 5′^{32}P-labeled ends was not attacked by Rauscher polymerase preparations. In addition, limit addition products formed with DNA polymerase I

Treatment of Native DNA	Incorporation of ^{3}H-dCMP pmoles/30 min
1. None	0.4
2. Sonication	0.2
3. Neurospora nuclease	<0.1
4. Pancreatic DNase (0.18 mμg)	0.3
5. Pancreatic DNase (0.18 μg)	11.4
6. Micrococcal nuclease (0.015–1.5 μg)	0.4–0.1
7. Exonuclease III	6.9
8. None – heated at 100 °C for 3 min	<0.1
9. As in 5 – followed by heating at 100 °C for 3 min	1.1

Fig. 1. Template requirements of Rauscher DNA polymerase

The different DNA treatments were as follows:

A. Sonication – Calf thymus DNA (0.97 μmole/ml) in 0.05 M NaCl, 0.001 M Tris-HCl, pH 8.0, 0.005 M EDTA was sonicated for 45 seconds at 20 Kc in the Branson sonifier.

B. Neurospora nuclease – reaction mixtures (0.25 ml) containing 10 μmoles of Tris-HCl, pH 7.5, 1 μmole of $MgCl_2$, 448 mμmoles of calf thymus DNA, and 0.4 unit of Neurospora nuclease were incubated for 60 min. Reaction mixtures were heated at 65 °C for 10 min to inactive the nuclease.

C. Pancreatic DNase – Reaction mixtures were as described in Fig. 3.

D. Micrococcal nuclease – Reaction mixture (0.25 ml) contained 15 μmoles of Tris-HCl, pH 8.6, 224 mμmoles of calf thymus DNA, 0.25 μmole of $CaCl_2$, and the indicated amount of micrococcal nuclease. After 30 min at 37 °C, reaction mixtures were treated with 5 μmoles of EDTA, 5 μmoles of 2-mercaptoethanol and heated at 65 °C for 15 min to inactivate the endonuclease.

E. Exonuclease III – Reaction mixtures (0.25 ml) containing 15 μmoles of Tris-HCl, pH 8.0, 1 μmoles of $MgCl_2$, 5 μmoles of dithiothreital, 448 mμmoles of calf thymus DNA, and 12 units of exonuclease III were incubated for 30 min at 37 °C and then heated at 65 °C for 10 min.

containing ^{3}H-dTMP at the 3′-OH ends were not attacked. These results indicate that the Rauscher polymerase is devoid of both the 5′→3′ and 3′→5′ exonuclease activities toward native DNA. These experiments were not carried out with the avian polymerase preparation.

Nature of the primer and product

Exonuclease III treatment or extensive nuclease action activates native DNA templates. These observations suggest that the polymerases catalyze incorporation of deoxynucleotides at the 3′-OH end of one strand and utilize the single-stranded

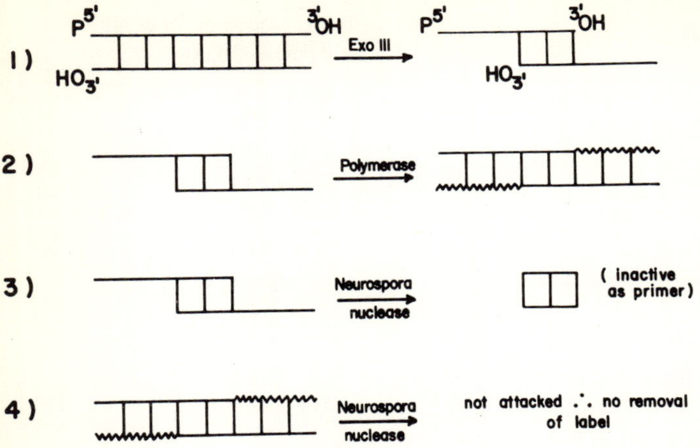

Fig. 2. Proposed action of polymerase and neurospora nuclease on DNA primers and products. The structures proposed above are those expected by the action and specificity of the enzymes indicated.

regions of the complementary strand as template. If this model is correct, structures resembling those produced by the action of exonuclease III (Fig. 2) would be active templates. Also, since Neurospora nuclease specifically attacks single-stranded regions in DNA as well as RNA, such treatment should destroy priming efficiency of any active DNA (or RNA). The duplex structure generated by the "repair" action of the polymerase should also be resistant to the digestion by the Neurospora nuclease.

As shown in Fig. 3, the addition of the Neurospora nuclease to reaction mixtures containing DNA activated by the action of Rauscher endonuclease, pancreatic DNase, or exonuclease III significantly reduced deoxynucleotide incorporation. Polymers such as poly (dI)·poly (dC) which specifically lead to dGMP incorporation were also inactivated by Neurospora nuclease. Thus, all active substrates possess a primer template structure, and the template must contain single-strand regions which are susceptible to Neurospora nuclease.

The expected product of repair synthesis is a duplex structure resistant to Neurospora nuclease. Results in accord with this mechanism are presented in Fig. 4. As shown, the products formed in the polymerase reaction are resistant to the action of Neurospora nuclease but are completely susceptible after heat denaturation. When products formed in reactions primed with poly (dI)·poly (dC) were exposed to the same treatment, denaturation hardly increased the susceptibility of the product to Neurospora nuclease action. This result would be expected for a product which can undergo reversible denaturation-renaturation.

Experiment	Additions	dTMP Incorporation pmoles/30 min
1	A. Complete System (DNase I treated)	44.4
	B. as in A + Neurospora nuclease	6.2
	C. Complete System (DNA Exo III treated)	29.4
	D. as in C + Neurospora nuclease	1.9
		dGMP Incorporation pmoles/30 min
2	A. poly (dI) · poly (dC)	180
	B. as in A + Neurospora nuclease	10

Fig. 3. Influence of neurospora nuclease on priming activity of DNA templates

Reaction mixtures (0.05 ml) were, as described previously [17] with DNA. In experiment 1, 10 mμmoles of DNA were present while 0.04 unit of Neurospora nuclease was added where indicated. In experiment 2, 2 mμmoles of poly (dI) · poly (dC) were added with α-^{32}P-dGTP (1.5 mμmoles).

DNA Primer Added	Heat Treatment	Acid-insoluble DNA pmoles
E. coli treated with Exo III	55 °C or 100 °C	35
"	55 °C + Nuclease	28
"	100 °C + Nuclease	0.3
poly (dI) · poly (dC)	55 °C or 100 °C	301
	55 °C + Nuclease	286
	100 °C + Nuclease	196

Fig. 4. Susceptibility of product to Neurospora nuclease

Products were formed with Rauscher DNA polymerase as previously described [17]. After 30 min incubation, the products were either heated at 55 °C for 10 min to inactivate the polymerase or heated at 100 °C for 5 min to denature the DNA product as well. Aliquots were added to reaction mixtures (0.1 ml), as described previously [17] for Neurospora nuclease action, and treated with 0.08 unit of Neurospora nuclease for 30 min at 37 °C. The amount of acid-insoluble radioactivity was measured as previously described.

Direct evidence for the incorporation of nucleotides at the 3′-OH end of exonuclease III treated DNA has been obtained. Deoxynucleotides (37 pmoles of ^{32}P-dGMP) incorporated with ^{3}H-λ DNA were quantitatively removed by the action of exonuclease III while less than 4 pmoles of ^{3}H-λ DNA was rendered acid-soluble during the reaction.

Covalent linkage of newly synthesized DNA to primer DNA

The data presented above suggests that there is incorporation of deoxynucleotides at the 3′-OH ends of duplex structures rather than the initiation of new DNA chains. Proof of this mechanism was obtained by demonstrating covalent attachment of the newly synthesized product to the primer DNA. In these experiments, ^{3}H-d(AT) copolymer was used as a primer and synthesis was carried out with ^{32}P-dATP and unlabeled TTP. After 15 % synthesis of ^{32}P-d(AT) copolymer, the sedimentation profile of the ^{3}H-d(AT) copolymer primer and the ^{32}P-d(AT) copolymer product was analyzed in alkaline sucrose gradients (Fig. 5 A). Both the ^{32}P and ^{3}H-labeled polymers cosedimented almost identically; the peak of ^{32}P-d(AT) product, however, sedimented slightly slower than the ^{3}H-d(AT) copolymer primer. This difference can be explained by preferential utilization of smaller species of ^{3}H-d(AT) copolymer as primer.

Further evidence supporting covalent linkage between the primer and product was obtained using α-^{32}P-dATP and dBrUTP and ^{3}H-d(AT) copolymer as primer and analyzing the products of the reaction by banding in alkaline CsCl solution (Fig. 5B). If new chains are initiated during DNA synthesis, ^{32}P-labeled product would be found at the density of heavy d(ABu) copolymer. If, however, dBuMP and ^{32}P-dAMP were covalently attached to the ^{3}H-d(AT) copolymer, the banding density would depend upon the contribution of primer material to the product formed. As shown, under conditions of only 3–5 % synthesis of ^{32}P-product (relative to input primer), both the ^{32}P product and the ^{3}H-primer band at a light density. There is no indication of any completely dense material (expected near the bottom of the gradient).

Extent of repair synthesis in the polymerase system

The mechanism of synthesis outlined above suggests that the tumor virus polymerases catalyze only repair synthesis. We, therefore, studied the extent of single-stranded regions of exonuclease III treated DNA which the viral polymerase could repair. For this purpose, ^{3}H-λ DNA was sonicated to a size corresponding to a molecular weight of 10^{6}, then treated with exonuclease III, and the extent of nucleotides removed was measured. Preparations in which up to 50 nucleotides had been removed were quantitatively repaired (Fig. 6). At the time indicated (30 minutes) more enzyme or

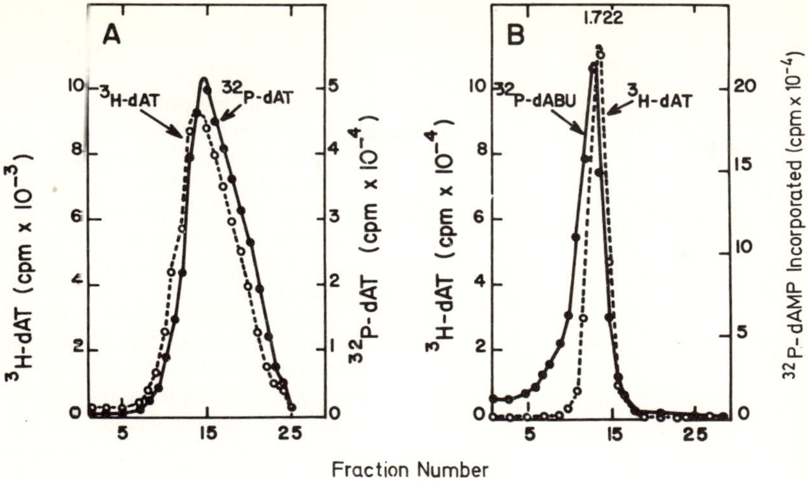

Fig. 5A. Sedimentation profile of ^{3}H-d(AT) copolymer primer and ^{32}P-d(AT) product in alkaline sucrose.

Reaction mixtures (0.05 ml) were as previously described [17] with 2.0 mμ moles of ^{3}H-d(AT) copolymer (4.2 x 10^7 cpm/μ mole), 2 mμ moles of dTTP, 1 mμ mole of α^{32}P-dATP (2000 cpm/pmole), and 0.54 unit of Rauscher polymerase. After 30 minutes at 38°, approximately 15 % synthesis of ^{32}P-d(AT) copolymer was detected. Reaction mixtures were treated with 0 02 ml 1 M EDTA, 0.13 ml 2 M NaOH; the entire reaction mixture was layered on to a 5–20 % sucrose gradient containing 0.3 M NaOH, 0.7 M NaCl, and 0.005 M EDTA and centrifuged for 120 minutes at 50.000 rpm in the SW50 rotor at 4°. Fractions were collected and the acid-insoluble ^{3}H- and ^{32}P-measured. The recovery of input ^{3}H and ^{32}P was greater than 90 % in both cases. The sedimentation profile of ^{3}H-d(AT) copolymer presented above was identical to the sedimentation profile observed with copolymer unexposed to Rauscher polymerase preparations in keeping with the absence of endonuclease activity in these fractions.

Fig. 5B. Equilibrium sedimentation in CsCl solution of ^{3}H d(AT) primer and ^{32}P-d(ABu) containing product.

Reaction mixtures (0.1 ml) contained 21 mμ moles of ^{3}H-d(AT), 3 mμ moles of dBrUTP, 2.5 mμ moles of α^{32}P-dATP (1850 cpm/pmole), and 0.40 unit of Rauscher polymerase (phosphocellulose eluate). All other conditions were the same as described before. Incubation was for 30 minutes at which time 650 pmoles of ^{32}P had been incorporated into an acid-insoluble form. Reaction mixtures lacking dBrUTP incorporated less than 15 pmoles of ^{32}P into an acid-insoluble form. Reaction mixtures were treated with 0.02 ml of 1 M EDTA, 0.2 ml 2 M NaOH, 1.88 ml H_2O, and 2.73 gm of CsCl and then centrifuged at 40.000 rpm at 25° in polyallomer tubes for 44 hours. The density was measured on samples collected after piercing the bottom of the tubes, and the amount of acid-insoluble radioactivity determined. The banding density of the ^{3}H-d(AT) corresponds to 1.722, and the recovery of ^{32}P in acid-insoluble material was 91 %.

The experiments described in Figs. 5A and 5B have been performed with the purified avian polymerase fraction with identical results.

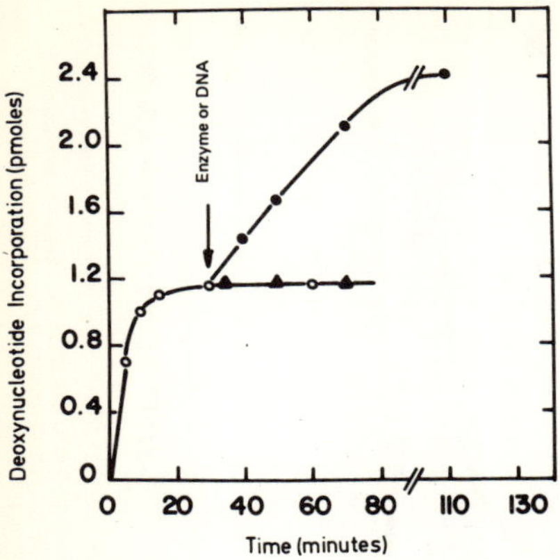

Fig. 6

Extent of deoxynucleotide incorporation with Exonuclease III treated DNA.

Reaction mixture (0.05 ml) contained 5 mμ moles each of dATP, dGTP, and dTTP; 1 mμ mole of α^{32}P-dCTP (2100 cpm/pmole); 5 μg of albumin; 2.5 μmoles of Tris · HC1, pH 8.0; 0.5 μmole of $MgCl_2$; 0.25 μmole of dithiothreitol; 0.25 μmole KCl; 216 pmoles of sonicated λ DNA treated with Exonuclease III which had removed 2 % of the total nucleotides, and 0.39 unit of Rauscher DNA polymerase (phosphocellulose eluate). Reactions were incubated for the indicated length of time; after 30 minutes additional enzyme of DNA (216 pmoles) was added. The incorporation refers only to labeled nucleotide and should be multiplied by 4 to calculate total nucleotide incorporation. (– ○ –) deoxynucleotide incorporation; (– ● –) deoxynucleotide incorporation after addition of DNA; (— ▲ —) after addition of enzyme.

DNA was added and, as shown, resumption of deoxynucleotide incorporation recurred only after DNA addition. Thus, this DNA preparation contained a limited ability to accept nucleotides; once these nucleotides were added, the DNA was no longer active as primer.

The ability of Rauscher polymerase to fill larger regions of sonicated DNA treated with exonuclease III is shown in Fig. 7. In the presence of Mg^{++} as the only divalent cation, limiting amounts of substrate and large excess of polymerase, the extent of repair exactly matched the amount of nucleotides removed by exonuclease III up to about 100 nucleotides. When 250 nucleotides had been removed, the repair was only 50 % of that expected, and with 750 nucleotides removed, the activity was greatly reduced. As noted by *Scolnick* et al. [28], the activity of a template such as poly (A) · poly (U) is greatly increased in the presence of Mn^{++}. A similar observation was noted in the repair of DNA with extensive single-stranded stretches.

The above information has defined the type of priming and product formed with DNA. The following section deals specifically with the priming of reactions with RNA.

Specificity of DNA polymerase for RNA templates

A repair-like synthesis has been found using RNA primers in the reaction. A number of RNA preparations supported deoxynucleotide synthesis with purified polymerase preparations in agreement with the results reported by *Duesberg* et al. [8]. Viral

DNA rendered acid-soluble by Exonuclease III (%)	Yield of Nucleotide Incorporation (%)
1. 0.21	0.20
2. 0.86	0.88
3. 2 04	2.02
4. 4.0	4.2
5. 10	5.3 10.3 (+ Mn^{++})
6. 30	2.3 21.3 (+ Mn^{++})

Fig. 7. Repair of sonicated λ DNA treated with exonuclease III

Reaction mixtures were as described in Fig. 3 with α^{32}P-dATP (660 cpm/pmole). The amount of DNA added was different in each case, e.g., (1) 3.82 mμmoles; (2) 0.47 mμmole; (3) 0.226 mμmole; (4) 0.08 mμmole; (5) 0.1 mμmole; (6) 0.181 mμmole. In reaction mixtures designated + Mn^{++}, 0.05 μmole of $MnCl_2$ replaced $MgCl_2$. Incubation in all cases was for 60 min at 37 °C. In experiments with extensively digested DNA (Nos. 5 and 6), the addition of more enzyme (0.39 unit) or extended incubation periods (up to 180 min) did not change the extent of nucleotide incorporation.

RNA isolated from Rous sarcoma and Rauscher leukemia viruses stimulated dTMP incorporation to about the same extent as AMV RNA.

Several RNA preparations from sources other than RNA tumor viruses were tested. Of these, f2, MS2, Qβ RNAs and bulk *E. coli* tRNA preparations were active templates for the synthesis of DNA in the presence of Mn^{++} and Mg^{++}, while a purified preparation of *E. coli* $tRNA_f^{Met}$ was not. The efficiency of priming with these RNA preparations, however, was markedly lower than that observed with tumor viral RNAs.

The native conformation of the tumor viral RNA preparations is disrupted by heating at 80° in low ionic strength buffers [4, 7, 12]. This treatment results in an irreversible change in the sedimentation coefficient of the RNAs from 60S to 30S and, as reported by *Duesberg* et al. [8], a marked decrease in template activity. RSV and AMV RNAs were tested after heat treatment and found to be active. Ten to 30 % of the template activity was retained when assayed in the presence of Mg^{++}, while 50 to 75 % of the template activity was observed when assayed in the presence of Mn^{++} and Mg^{++}.

Extent of DNA synthesis

The extent of DNA synthesis with AMV RNA as primer was investigated (Fig. 8). As shown, the yield of deoxynucleotide incorporation was equivalent to 3 % of the

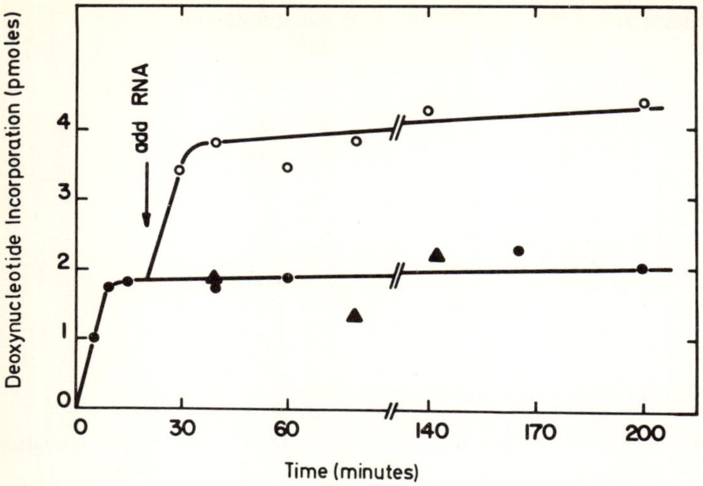

Fig. 8. Kinetics of deoxynucleotide incorporation with AMV RNA and DNA polymerase. Assay conditions were as described in "Methods" [22] with ^{3}H-dCTP (4460 cpm/pmole), 67 pmoles of AMV RNA, and 0.375 unit of purified polymerase II added. After 20 minutes of incubation, an additional 67 pmoles of AMV RNA or 0.15 unit polymerase pII was added. (— ● —), deoxynucleotide incorporation; (— ○ —), deoxynucleotide incorporation after addition of RNA at 20 minutes; (— ▲ —), deoxynucleotide incorporation after addition of enzyme at 20 minutes.

added RNA primer. The observed plateau was not due to inactivation of enzyme activity during the incubation since the addition of more enzyme at 20 min did not further increase deoxynucleotide incorporation. The addition of more RNA, on the other hand, resulted in an immediate resumption of deoxynucleotide incorporation which again reached the plateau corresponding to 3 % of the added RNA. These results show that only a limited amount of deoxynucleotide can be incorporated with a given amount of RNA. In other experiments (not shown) conducted under conditions described in Fig. 8, the addition of RNase (2 μg) at 20 min had no effect on the yield of DNA synthesis.

Susceptibility of Rauscher RNA to nuclease and its effect on template activity

The RNA from RNA tumor viruses sediments in neutral sucrose gradients with a sedimentation coefficient of 60 to 70S [9, 26, 27]. Denaturation by treatment with dimethyl sulfoxide, urea or heat [4, 7, 12] reduces the sedimentation coefficient to 20–30 S and smaller materials, suggesting that the native RNA is composed of segments hydrogen-bonded together. Earlier support for such a structure has been

RNA	treatment	acid-insoluble RNA %
^{3}H-RLV RNA	RNase (4 μg)	<5
^{3}H-RLV RNA	*Neurospora* nuclease (0.02 unit)	56
^{3}H-RLV RNA	*Neurospora* nuclease (0.04 unit)	43
^{3}H-RLV RNA	Mung bean nuclease I (0.2 unit)	51
^{3}H-RLV RNA	Pancreatic DNase (10 μg)	93
^{3}H-RLV RNA	RNA, heat at 100 °C for 3 min; then add *Neurospora* nuclease (0.02 unit)	13
^{3}H-poly (U)	RNase (4 μg)	<2
^{3}H-poly (U)	*Neurospora* nuclease (0.04 unit)	<2
^{3}H-poly (U)	Mung bean nuclease I (0.2 unit)	<2
^{3}H-poly (U)	Poly (A) + RNase (4 μg)	7
^{3}H-poly (U)	Poly (A) + *Neurospora* nuclease (0.04 unit)	>98
^{3}H-poly (U)	Poly (A) + Mung bean nuclease I (0.2 unit)	>98

Fig. 9. Susceptibility of Rauscher RNA and poly (U) to enzymatic digestion

Tritium labeled RLV RNA (80 pmoles, 2340 cpm) was incubated in reaction mixtures (0.1 ml) containing 5 μmoles Tris · HCl, pH 7.5, 0.5 μmole $MgCl_2$, and nucleases as indicated. ^{3}H-poly (U) (7 mμmoles, 62,000 cpm) and poly (A) (10 mμmoles) were added where indicated, and reaction conditions were the same as those described with RLV RNA. In reaction mixtures treated with mung bean nuclease I, the polynucleotides as indicated were added to solutions containing $MgCl_2$. After mixing, potassium acetate buffer pH 5.1 (2 μmoles) was added and then enzyme. All other conditions were as described above.

presented by *Bader* and *Steck* [4] who observed that the 60S to 70S Rauscher RNA remains intact after action of endogenous viral RNase; upon denaturation, small RNA species (< 20–30 S) were detected.

We investigated the structure of the Rauscher RNA by studying its susceptibility to attack by several nucleases. Tritium labeled RLV RNA was rendered acid-soluble after incubation with pancreatic RNase (Fig. 9). This result suggests that the secondary structure of the RNA does not involve long stretches of complementary sequences such as those found with the duplex reo virus RNA. In addition, complex structures such as poly (A) · poly (U) are also susceptible to pancreatic RNase attack under the conditions employed (Fig. 9).

The Neurospora and mung bean I nucleases [19, 25] specifically degrade single-stranded but not duplex structures of DNA. These enzymes also attack single-stranded regions of RNA in RNA duplexes, or single-stranded regions of RNA and DNA in RNA · DNA hybrids. As shown in Fig. 9, ^{3}H-poly (U) is completely susceptible to the action of Neurospora and mung bean nucleases; after annealing to poly (A), ^{3}H-poly (U) was completely resistant to the nucleases. In other experiments, RNA · DNA hybrids were formed with fd DNA and RNA polymerase and isolated by gradient density banding in Cs_2SO_4. The RNA in such structures is resistant to Neurospora nuclease; heat denaturation of the hybrids renders the RNA susceptible to this nuclease preparation.

When ^{3}H-RLV RNA was treated with either mung bean or Neurospora nucleases, approximately 50 % of the RNA remains acid-insoluble. These results suggest that there is considerable internal structure of the RNA, in accord with the previously reported observations (cited above). Upon heat denaturation, 90 % of the RNA was susceptible to Neurospora nuclease action, suggesting a marked disruption of the internal structure of the RNA. The interpretation of the latter finding is complicated by the presence of nicks in the ^{3}H-RLV RNA. Approximately 50–60 % of the ^{3}H-RLV RNA in our preparations sedimented slower than 20 S after heat denaturation. Nicks present in the RNA would yield structures incapable of reversible denaturation if, for example, hairpin structures existed in intact RNA. A diagrammatic representation would be as follows (Fig. 10).

The single-stranded regions of the AMV RNA were shown to be essential for template activity by use of the Neurospora nuclease. The nuclease was preincubated with AMV RNA for 10 min at 38° to degrade single-stranded regions. After heating at 65° for 4 min to inactivate the nuclease activity, deoxynucleoside triphosphates and polymerase were added and the rate of DNA synthesis measured. AMV RNA which

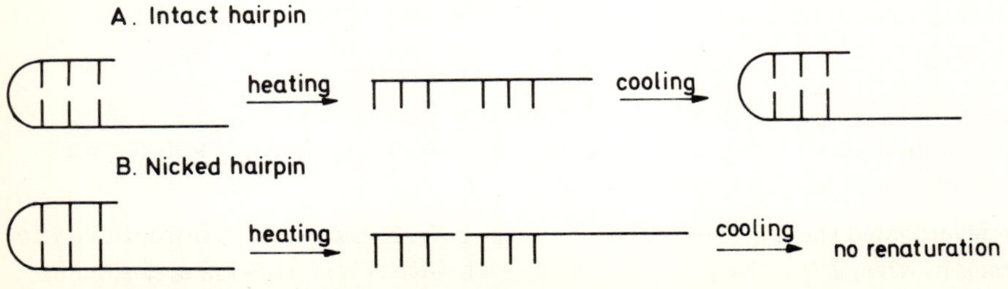

Fig. 10. The region designated with a bar represents complementary sequences.

was preincubated without the nuclease retained 70 % of its activity compared to nonpreincubated controls. On the other hand, RNA preincubated with the nuclease did not prime DNA synthesis. Similar structural requirements for single-stranded regions have been demonstrated for DNA primed DNA synthesis as described above.

Characteristics of DNA polymerase product

The nature of the product of the AMV DNA polymerase and AMV RNA was examined by treatment with various chemical and enzymatic reagents. As reported by others, the labeled product possessed characteristics expected of DNA. It was resistant to digestion by RNase and quantitatively attacked by DNase. After treatment with alkali, the labeled product banded at a density of 1.45 in Cs_2SO_4 solution characteristic of DNA. The alkali treated product had a sedimentation coefficient of 6 to 7 S measured in alkaline sucrose gradients using single-stranded ^{14}C labeled fd DNA as a reference marker. This represents a molecular weight of approximately 1.7–2.6 x 10^5. The size of the DNA was the same whether it was synthesized in the presence of Mg^{++} and Mn^{++} or Mg^{++} alone.

Product is an RNA · DNA hybrid

Duesberg et al. [8] and *Faras* et al [14] have reported that newly synthesized DNA formed with purified polymerase and RSV RNA cosediments with 60 to 70 S viral RNA in neutral sucrose gradients. We have repeated this experiment with ^3H-RLV or AMV RNA preparations with similar results. A small portion of the ^{32}P-labeled DNA was found at the top of the gradient presumably due to the presence of RNase in the purified AMV polymerase preparation used. These results suggest that the first polymerase product is an RNA · DNA hybrid. This has been shown to be the case by the following observations: (1) the majority of the ^3H-dTMP-labeled product was found to be resistant (55–80 %) to digestion by Neurospora or mung bean I nucleases. After the product was treated with alkali to digest RNA, between 5–10 % of the DNA was resistant to Neurospora or mung bean I nucleases; (2) the dTMP-labeled product banded at a density of 1.835 in CsCl solution which is characteristic of RNA, while less than 2 % banded as free DNA. These experiments indicate that the product, under the conditions employed, is mostly, if not exclusively, an RNA · DNA hybrid.

Covalent attachment of the RNA to DNA

As described above, Rauscher and AMV DNA polymerases are incapable of initiating synthesis of new chains with poly d(AT) copolymer as template. Deoxynucleotide incorporation occurred at the 3′-OH end of a primer strand and the newly synthesized product was covalently attached to the poly d(AT) copolymer primer. We therefore investigated the nature of the linkage between the AMV RNA and the DNA in the hybrid product. If RNA is covalently attached to the DNA, then the two components should be inseparable by density gradient centrifugation after denaturation in the presence of formaldehyde. A ^{3}H-dTMP-labeled product was prepared and divided into two equal portions. One half was heated at 85° for 10 min in 0.25 M formaldehyde while the other sample was unheated. Both products were then analyzed by Cs_2SO_4 solution equilibrium density gradient centrifugation with ^{3}H-poly (U) and ^{14}C-T7 DNA as density markers. The untreated product banded as a broad peak at densities intermediate to that of free DNA and RNA (Fig. 11 A). This heterogeneity would be expected for RNA · DNA hybrids possessing varying amounts of the two components. The heat denatured product in formaldehyde also banded at a hybrid density (Fig. 11B); no detectable product banded at the density of free DNA indicating that, under

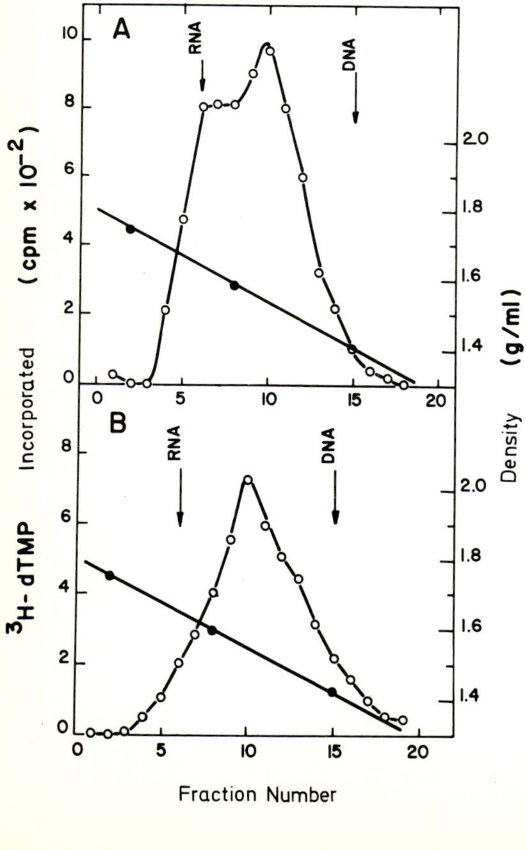

Fig. 11

Equilibrium Cs_2SO_4 solution gradient centrifugation of DNA polymerase product.

(A) A polymerase product (2.1 x 10^4 cpm) was prepared with AMV RNA (206 pmoles) and 0.2 unit purified polymerase dialyzed against SSC and extracted with phenol. An aliquot was sedimented to equilibrium in Cs_2SO_4 solution for 44 hours at 123,000 x g as described in "Methods." The recovery of ^{32}P was 71 %.

(B) An aliquot of the polymerase product was heated in 0.25 M formaldehyde at 85° for 10 minutes and sedimented to equilibrium in formaldehyde Cs_2SO_4 solution as above. The recovery of ^{32}P was 66 %. (– ○ –), ^{32}P-dTMP labeled product; (– ● –), density.

the conditions employed, the entire DNA product was covalently linked to RNA. The shift in density observed in formaldehyde heated preparations is most likely due to the single-strand breaks present in the 60–70 S AMV RNA primer which results in products containing a higher proportion of DNA.

The covalent linkage of the DNA product to the RNA is not limited to the viral RNA template. Similar observations were made with the synthetic polynucleotides poly (A) · poly (U), as well as with phage RNA f2. ^{32}P-dTMP-labeled polymerase products were prepared with ^{3}H-f2 RNA in the presence of Mn^{++}. One half of the product was heated at 65° for 15 min in 3 % formaldehyde and both products sedimented through a neutral-SDS sucrose gradient. Formaldehyde was included in the gradient containing the heated product. The ^{32}P-labeled product cosedimented with the ^{3}H-labeled f2 RNA before and after the formaldehyde treatment (Fig. 12)

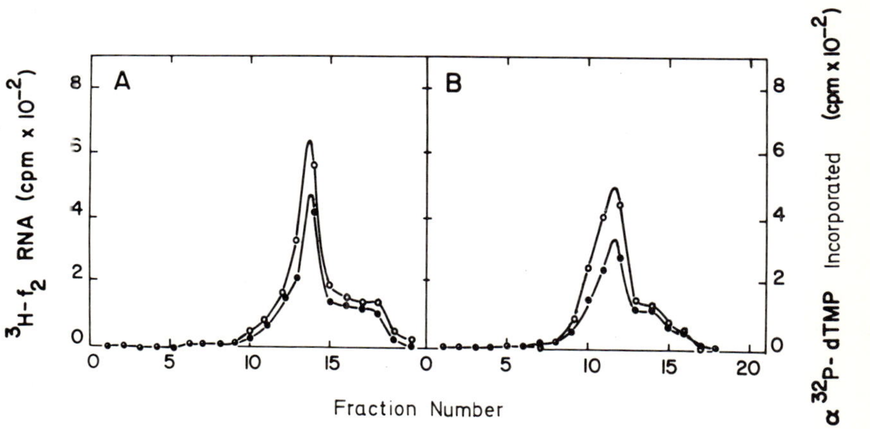

Fig. 12. Gradient centrifugation of DNA polymerase product formed with f2 RNA as primer.

(A) A DNA polymerase product was prepared in 0.1 ml with 76 mμmoles of ^{3}H labeled f2 RNA (338 cpm/mμmole), α^{32}P-dTTP (5000 cpm/pmole), 0.5 mM $MnCl_2$, and 0.28 unit purified polymerase as described in "Methods." [22] After 15 minutes incubation at 38°, the reaction mixture was brought to a final concentration of 0.1 M EDTA and 0.4 % SDS in a volume of 0.25 ml, layered onto a 5 ml 5 to 20 % linear sucrose gradient in 0.2 % SDS, 20 mM sodium phosphate, pH 7, 1 mM EDTA, and 0.1 M NaCl, and centrifuged at 180.000 x g for 80 minutes at 20° in a Spinco SW65 rotor. Thirty drop fractions were collected by gravity after piercing the bottom of the centrifuge tube, precipitated with 5 % TCA, collected on glassfiber filters, and counted in a scintillation counter. Untreated tritium labeled f2 RNA (30S) sedimented in a separate gradient at the same rate as the treated f2 RNA.

(B) A DNA polymerase product prepared as above was heated to 65° for 15 minutes in 3 % formaldehyde after addition of EDTA and SDS. The product was analyzed as above except that the sucrose gradient contained 3 % formaldehyde and was centrifuged at 300.000 x g for 240 minutes. The formaldehyde treated f2 RNA sedimented with a sedimentation coefficient of 13S. (—○—), ^{3}H-f2 RNA; (—●—), ^{32}P-dTMP labeled product.

The ratio of ^{32}P to ^{3}H for each experiment is summarized in the inserts in the above figures.

indicating that the DNA was covalently linked to the f2 RNA. Similar experiments using ^{3}H-RLV RNA as template were performed. There was general correspondence between the sedimentation profile of ^{32}P-DNA and ^{3}H-RNA. A sizeable amount of the RNA, however, did not cosediment with the ^{32}P DNA after formaldehyde denaturation most likely due to the presence of nicks in the 60–70 S RNA. In the absence of formaldehyde treatment, ^{3}H- and ^{32}P-labeled materials cosedimented as previously discussed.

The synthetic homopolymers, poly (A) · poly (U) in the presence of Mn^{++} specifically support the synthesis of poly (dT) as reported by *Scolnick* et al. [27]. The product poly (dT) formed in such reaction mixtures banded in Cs_2SO_4 solution as a sharp peak at the density of 1.52; ^{3}H-poly (U) and ^{3}H-poly (dT), added to separate gradients as standard markers, banded at a density of 1.66 and 1.43, respectively. After heat denaturation in formaldehyde, the ^{32}P product still banded at 1.520 indicating covalent attachment of the poly (dT) to the RNA primer. The covalent attachment of ^{32}P-poly (dT) to poly (U) was further demonstrated by subjecting ^{32}P-poly (dT) products to alkaline hydrolysis. If ^{32}P-dTMP is covalently attached to poly (U), alkaline hydrolysis should result in the transfer of ^{32}P to UMP residues yielding ^{32}P-labeled 2′ (3′) UMP. A poly (dT) product was prepared with several enzymes: DNA polymerase I from *E. coli*, AMV and RLV DNA polymerases. After alkaline hydrolysis, carrier 2′ (3′) mononucleotides were added and separated by high voltage electrophoresis. The area of the electropherograms corresponding to 2′ (3′) UMP contained 86, 90, and 92 % respectively of ^{32}P label in the mononucleotide regions (Fig. 13). Further substantiation that the ^{32}P product present in these

	DNA polymerase system used		
2′ (3′) mononucleotide	DNA polymerase I	AMV	Rauscher
	% distribution of ^{32}P		
Up	86	90	92
Gp	<2	<2	<3
Ap	12	8	4
Cp	<2	<2	<2

Fig. 13. Distribution of ^{32}P in mononucleotides after alkaline hydrolysis

A polymerase product was prepared in reactions primed with poly (A) · poly (U) as described in "Methods" [22] labeled with 5′ α^{32}P-dTTP. The product was hydrolyzed with alkali and the mononucleotides separated by high voltage electrophoresis in 0.05 M sodium citrate, pH 3.5. The amount of ^{32}P counted corresponded to 4400 cpm (DNA polymerase I), 450 cpm (AMV polymerase), and 1120 cpm (Rauscher polymerase). Control experiments were carried out with ^{32}P-labeled poly (dT) synthesized in reaction mixtures primed with poly (A) · poly (dT). The product (1.09 x 10^5 cpm) carried through the above procedure did not lead to the isolation of detectable labeled nucleotides (<80 cpm).

regions contained 2′ (3′) phosphomonoester structure was obtained; all of the ^{32}P was converted to a Norit-nonadsorbable form with alkaline phosphatase. In contrast, 5′ nucleotidease (crude snake venom) did not effect the Norit-adsorbable property of the ^{32}P product. The radioactivity detected in the Ap region was not further characterized and may represent aberrant TMP incorporation onto the poly (A) chain.

In the presence of four deoxynucleoside triphosphates and Mg^{++}, the purified viral enzymes catalyze repair-like reactions on RNA, DNA or RNA · DNA hybrids. Polymerase activities primed by these different templates were always found in the same fractions during enzyme purification and were inactivated by heat at the same rate suggesting that one enzyme is capable of using these different templates for DNA synthesis. A similar conclusion has been reached by *Duesberg* et al. [8]. The incorporation of deoxynucleotides occurred at approximately the same rate with an equal amount of either DNA or RNA as template with the AMV polymerase.

The mechanism of deoxynucleotide incorporation with RNA primers appear to be same as that found with DNA primers as well as with synthetic RNA · DNA hybrids. Evidence has been presented that deoxynucleotide incorporation occurs at the 3′-OH end of primer strands attached to template strands. These structures must possess single-stranded regions to direct synthesis represented as follows (Fig. 14).

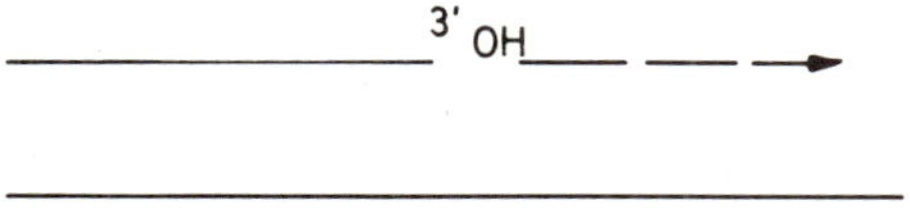

Fig. 14
The arrow denotes direction of deoxynucleotide incorporation.

Thus when single-stranded regions of AMV RNA were degraded with Neurospora nuclease, this RNA no longer supported deoxynucleotide incorporation. As expected, the DNA synthesized with AMV RNA was largely resistant to the action of Neurospora and mung bean nucleases indicating its duplex structure. In reactions primed with poly (A) · poly (U) (in the presence of Mn^{++}), the rate of dTMP incorporation was markedly increased by the use of poly (U) previously nicked with micrococcal nuclease followed by alkaline phosphatase treatment. Decreasing the poly (U) average chain length from 200–300 nucleotides to 50, increased the rate of dTMP incorporation 10–20-fold. Furthermore, following alkaline hydrolysis, experiments presented in Fig. 13 indicate that the 3′-OH end of the poly (U) chain is the site of dTMP incorporation.

The DNA formed in the presence of AMV RNA is an RNA · DNA hybrid whose polynucleotide content is mostly RNA. It bands in CsCl and Cs_2SO_4 solution at densities expected of products rich in RNA, and no labeled product was detected banding at a density of free DNA.

As observed in reactions primed with DNA polymers, the polymerase does not initiate synthesis of new chains *de novo* with RNA primers, but incorporates deoxynucleotides onto the 3′-OH ends of the primer strands. Thus, the newly synthesized DNA is covalently linked to the RNA primer in a hybrid structure. This attachment was demonstrated by the bonding of AMV RNA-polymerase products in Cs_2SO_4 solution in the hybrid density regions after treatment with formaldehyde. Similarly, ^{32}P dTMP-labeled products cosedimented with ^{3}H-f2 RNA primer-template in neutral sucrose SDS gradients both before and after treatment with formaldehyde.

Only about 3 % of the AMV RNA added as primer is transcribed into DNA. The size of synthesized DNA was determined to be 6–7 S by sedimentation in alkaline sucrose gradients, equivalent to a molecular weight of approximately $1.7–2.6 \times 10^5$. DNA isolated after incubation of partially permeable virus with labeled deoxynucleoside triphosphates was found to have similar sedimentation coefficients [11]. At present, it is not known whether the DNA synthesized in our system represents a heterogeneous DNA population complementary to the entire RNA genome as reported by *Duesberg* et al. [11] for virion generated DNA.

Since viral RNA primers support a limited extent of DNA synthesis, it appears that the DNA polymerase is unable to denature the secondary structure of RNA. This restricted activity is similar to that found with the T4 induced DNA polymerase and DNA polymerase II of *E. coli.* With these two enzymes, specific proteins have been shown to play an important role in repair reactions. The T4 gene 32 protein specifically stimulates the T4 induced DNA polymerase [1], and a protein recently isolated from *E. coli* specifically stimulates DNA polymerase II in repair of DNA [15]. Similar proteins may be present either in virions or in infected cells and could facilitate viral polymerase transcription of RNA.

We have recently prepared crude extracts of the AMV with ammonium sulfate in which 90–95 % of the RNase activity towards poly (U) has been removed. These extracts contain DNA polymerase and can be primed with exogeneously added AMV RNA. The extent of deoxynucleotide incorporation under these conditions is 4 to 6 times greater than with a comparable amount of purified AMV DNA polymerase alone. All of the deoxynucleotide incorporation is sensitive to the addition of pancreatic RNase. The stimulatory activity present in these extracts is inactivated by heat treatment and is nondialyzable. The DNA products formed under these conditions sediment at 6–7 S in alkaline sucrose gradients.

Manly et al. [24] and *Fanshier* et al. [13] have reported that both single- and double-stranded DNA, free of RNA, are produced by detergent treated virions. As of yet, we have not detected free DNA products in our purified system in vitro but only RNA · DNA hybrids. Thus, we believe that factors present in the virion participate in the formation of free DNA. Experiments with the crude ammonium sulfate extracts described above support this concept. The DNA products formed with exogenously added AMV RNA to these crude extracts are increasingly sensitive to digestion with mung bean nuclease. As a function of the incubation period, e.g., 4 %, 17 %, 28 %, and 41 % of the DNA produced is hydrolyzed by mung bean nuclease after 10, 30, 60, and 120 min of synthesis, respectively. In addition, 30–40 % of the DNA product at 60 min bands at a density of 1.72 in CsCl, characteristic of free DNA. Possible mechanisms yielding free and double-stranded DNA from RNA·DNA hybrids might include the following: (1) the covalent linkage between DNA and RNA could be hydrolyzed by an endonuclease in the virion. This cleavage would be followed by denaturation of the DNA chain from the RNA by a concerted process involving transcription and peeling as postulated by *Manly* et al. [24].

Alternatively, direct denaturation may be facilitated by replication of double-stranded DNA. In either case, both models require localized denaturation which could be catalyzed by a T4-gene 32-like protein. In one case, single-stranded DNA is generated while in the second case, double-stranded DNA is generated. It is equally possible that both mechanisms occur concomitantly. (2) Inherent in the model of the primer-template role that viral RNA plays is the importance of hairpin structures. If the polymerase can only elongate chains in repair-like reactions, it would be unable to transcribe the regions of the RNA involved in the hairpin structure. This loss of genetic information can be obviated if covalently linked DNA can form hairpins intermolecularly. Evidence for such hairpin structures has been obtained with AMV DNA prepared by alkali digestion of the RNA · DNA hybrid formed with the DNA polymerase and AMV RNA. The alkali treated DNA supports deoxynucleotide incorporation with the purified AMV DNA polymerase and purified *E. coli* DNA polymerases I and II. The mechanism of action of these enzymes requires that the DNA template must possess a primer-template structure. The newly synthesized DNA prepared with *E. coli* DNA pol I, under conditions leading to 25 % synthesis (compared to DNA added), was found to be completely resistant to digestion by mung bean nuclease, thus indicating its duplex structure. After heating at 100° for 3 min. 86 % of the newly synthesized DNA was still resistant to mung bean nuclease action, indicative of reversible denaturation. These results suggest that the AMV DNA can form intramolecular hairpin structures. Alternatively, in vivo initiation may involve other mechanisms. It is possible that small 3′-OH terminated polynucleotides can serve as nucleation points for DNA synthesis, or additional factors may operate to facilitate initiation of DNA on viral RNA without the direct participation of a primer strand. To date, however, no known DNA polymerase can initiate new chains in DNA directed reactions. If viral DNA is an intermediate in replication, it is necessary to transcribe this DNA into RNA progeny. Whether this transcription involves a host cell or viral RNA polymerase is unknown. At present, we, as well as others, have not detected ribonucleotide incorporation with virions or crude extracts under a wide variety of conditions. Our understanding of the RNA oncogenic viruses would be facilitated by the discovery of the mechanism of RNA replication.

B. Reaction of RNA terminus

It is evident that information about the nature of the breaks in RNA and the possible repair of such breaks is of importance in understanding the behavior of RNA tumor viruses. During the course of studies concerned with DNA polynucleotide ligase of T4-infected *E. coli,* it was noticed that relatively crude enzyme fractions, in addition to sealing single-strand breaks of DNA, reacted with a variety of 5′-^{32}P-labeled RNAs to render the 5′-phosphate terminus resistant to bacterial alkaline phosphatase (BALP) [6]. The RNA-ligase type activity has now been purified approximately 300-fold from T4-infected *E. coli* [29]. It is free of detectable DNA ligase, poly (A) polymerase and completely free of polynucleotide phosphorylase.

A variety of RNA preparations labeled at the 5′-phosphate end serve as substrate in this reaction. As shown in Fig. 15, tRNA and poly (A) are both active and in each

Exp.	Additions	BALP Resistant Activity pmoles/30 min
A	1. tRNA + ATP + Mg^{++}	5.91
	2. omit ATP or Mg^{++}	<0.1
	3. (1) + RNase (5 μg)	<0.1
	4. (1) + DNase (5 μg)	5.70
B	1. Poly (A) + ATP + Mg^{++}	47.2
	2. Omit ATP or Mg^{++}	<0.1
	3. (1) + RNase (5 μg)	39.4
	4. (1) + DNase (5 μg)	46.4
	5. (1) + Micrococcal nuclease (1 μg) + Ca^{++}	2.1
	6. (1) + pCMB (10^{-4} M)	<0.1
	7. (1) + 10^{-3} M P_i	50.0
	8. (1) + 10^{-4} M PP_i	<0.1

Fig. 15. Requirements of system

All reaction mixtures (0.1 ml) contained the following: Tris-HCl, pH 7.5, 5 μmoles; $MgCl_2$, 0.5 μmole; DTT, 5 nmoles; ATP, 5 nmoles; bovine serum albumin, 3 μg. The substrate in Experiment A was 5′^{32}P-tRNA (*E. coli,* 18 mμmoles, containing 0.38 mμmole of 5′-^{32}P termini specific activity 425 cpm/pmole) and in Experiment B poly (A) 14 mμmoles (containing 0.44 mμmole of ^{32}P, specific activity 280 cpm/pmole). In Experiments B, where P_i and PP_i were used, the reaction was terminated by the addition of trichloroacetic acid and the precipitate washed with 1 % TCA. The precipitate was neutralized with dilute NH_4OH, adjusted to pH 8.0 with Tris buffer and incubated with 1 unit of BALP for 35 min at 80 °C. One unit of enzyme is defined as the activity which renders 1 nmole of substrate resistant to BALP over a 30 min incubation.

case the requirements for the reaction are the same i. e., ATP and Mg^{++} are essential. The reaction with tRNA is sensitive to RNase but insensitive to DNase while the activity with poly (A) is resistant to both of these nucleases but sensitive to micrococal nuclease. The presence of pHMB or PP_i inhibited the reaction with either substrate while P_i had no effect. When ^{3}H-ATP (1100 cpm per pmole) was used in the reaction, the extent of incorporation was less than 10 % of the termini rendered alkaline phosphatase resistant. This finding indicates that the addition of ATP to the 5′-OH terminus cannot explain the observed resistance to BALP. The following had no effect on the ligase reaction with tRNA or poly (A) as substrate: all 4 ribonucleoside triphosphates (10^{-3}M of each), AMP (10^{-3} M), ADP (10^{-3} M), and RNA polymerase (10 units).

Stoichiometry of the reaction

The products formed from ATP in reaction mixtures with poly (A) appear to be AMP and PP_i in amounts equivalent to the ends rendered resistant to alkaline phosphatase (Fig. 16). A small amount of ADP was also found (not shown). The ADP formed corresponded to the amount of P_i detected. This activity, as well as that leading to PP_i formation in the absence of poly (A), reflects the presence of an ATPase present as a contaminant in the enzyme preparation. No further characterization of this activity has been undertaken.

The RNA ligase preparation catalyzed a $^{32}PP_i$ exchange reaction into ATP which was independent of poly (A) addition. When ATP was replaced by dATP in the routine measurement of RNA-ligase activity, the reaction proceeded at only 10 % of the rate observed with ATP.

Additions	P_i	PP_i	AMP pmoles	BALP Resistant ^{32}P
1. Poly (A) + enzyme + ATP	68	244	259	197
2. (1) – omit Poly (A)	64	71	63	–
	Δ =	173	196	197

Fig. 16. Stoichiometry of reaction

Three reaction mixtures were prepared as routinely used except for the following differences; one contained 5′-^{32}P-poly (A) (420 cpm/pmole) and cold ATP; the second contained ^{3}H-ATP (2100 cpm/pmole) and 5′-^{32}P labeled poly (A); while the third contained γ-^{32}P-ATP (422 cpm/pmole) and cold poly (A). All contained 0.37 unit of enzyme and 2.1 mμmoles of ATP and were incubated for 60 minutes at 38 °C. The amount of BALP resistant material was measured when possible; the amount of $^{32}PP_i$ and $^{32}P_i$ was measured by electrophoretic separation on paper (3500 volts for 1.5 hous) of the acid-soluble, norit-nonadsorbable material. The amount of AMP produced was measured in reaction mixtures containing ^{3}H-ATP after chromatography of the entire reaction mixture on Dowex-1-Cl.

Activity with other polynucleotides

A large variety of 5′-^{32}P-terminated polyribonucleotides are active in the above system (Fig. 17), if the synthetic polynucleotides, the polypurines (I and A), are most effective while the polypyrimidines are only slightly active.

Additions	5′-^{32}P-terminus added pmoles	BALP Resistant pmoles/30 min
tRNA (34.7)	167	3.9
Poly (A) (1.4)	50	7.8
Poly (I) (21.5)	185	2.1
Poly (C) (30)	270	0.1
Poly (U) (23)	185	0.07

Fig. 17. Comparison of various 5′-^{32}P-polyribonucleotides

Reaction mixtures (0.1 ml) contained 5 μmoles of Tris buffer pH 7.5, 0.5 μmole of Mg^{++}, 5 mμmoles of ATP, 5 mμmoles of dithiothreitol, 0.6 μg of protein (Sephadex fraction) and polynucleotides as indicated. The numbers in parentheses indicate the nmoles of nucleotides added. Incubation was for 30 min at 38 °C.

Nature of product: chemical characterization

The products formed from 5′-^{32}P terminated RNAs have been investigated. All alkaline phosphatase resistant products formed are rendered acid-soluble when hydrolyzed with alkali. After alkaline hydrolysis, all the ^{32}P is norit-adsorbable and sensitive to BALP action. These observations have been carried out with tRNA, rRNA, and all synthetic polynucleotides used. As shown in Fig. 18, the BALP resistant products formed with poly (A) or poly (I) lead to 2′(3′) AMP or IMP after alkaline hydrolysis. Although ATP is essential in the RNA ligase reaction, it is evident that AMP has not been added at the 5′-phosphate ends of poly (I) since the expected products would then be A_p^*. These results indicate that 5′-^{32}P termini are converted into internal phosphodiester bonds during the course of the reaction.

Physical characterization

The reaction proceeds best with poly (A), and for this reason this material has been used to measure the fate of polymers during the ligase action. No evidence was

Material Hydrolyzed	% Product Recovered as			
	pAp	Ap	pIp	Ip
1. 5′-^{32}P-poly (A) (1.54 x 10^4 cpm)	>95	<5	<5	<5
2. Poly (A); BALP Resistant (8,3 x 10^3 cpm)	4	93	<2	<2
3. 5′-^{32}P-Poly (I) (1.12 x 10^4 cpm)	<5	<5	>95	<5
4. Poly (I); BALP Resistant (5.1 x 10^3 cpm)	<3	<3	3	92

Fig. 18. Characterization of poly (A) and poly (I) products

Reaction products were prepared as described in Fig. 15. The substrate poly (A) or poly (I) and their respective ligase products were hydrolyzed in 0.5 M KOH at 38 °C for 18 hours. The alkali was carefully neutralized with Dowex 50 H^+, and aliquots were spotted on Whatman 3 MM paper and subjected to electrophoresis for 2.5 hours at 5000 volts. Markers corresponding to the indicated nucleotides were included. The paper was cut into 1 cm strips and counted in toluene-POPOP in a Packard scintillation counter.

obtained for a marked increase in the size of the poly (A) products. Certain properties of the product are in keeping with the formation of circular structures during the reaction.

Products centrifuged in HCHO-sodium dodecyl sulfate gradients (Fig. 19) are almost identical in sedimentation profiles compared to the starting poly (A). Fractions were pooled and analyzed for BALP resistance. It was found that while fractions 4–9 from Fig. 19 contained less than 2 % of the total radioactivity, all of the ^{32}P was resistant to the action of BALP. Pooled aliquots corresponding to fractions 15–19 contained 40 % of the ^{32}P and were 76 % resistant to BALP. Fractions corresponding to 25–33 contained 7 % of the ^{32}P and were 30 % resistant to BALP. Thus the poly (A) in the lower part of the gradient is more resistant to BALP than the poly (A) in the upper region of the gradient. In order to evaluate the selective utilization of longer poly (A) chains as ligase substrates, the experiment was repeated using poly (A) isolated from regions corresponding to fractions number 24–29. These fractions were dialyzed and subjected to the action of RNA ligase. Approximately 40 % of the termini added were rendered BALP resistant during the incubation. The profile of the product and starting material in HCHO-sucrose gradient is summarized in Fig. 20. No increase in the size of the product was noted over the initial substrate, but again a marked skewing to the lower region was detected for the BALP resistant material. These observations might be expected if the product of the reaction is circular in nature. If the RNA product formed is a circular material, there should be no 3′-OH or 5′-phosphate ends available for exonuclease attack. Experiments with venom diesterase, polynucleotide poshorylase, and RNase II indicate that the RNA ligase product is resistant to the action of these enzymes in contrast to the initial substrate.

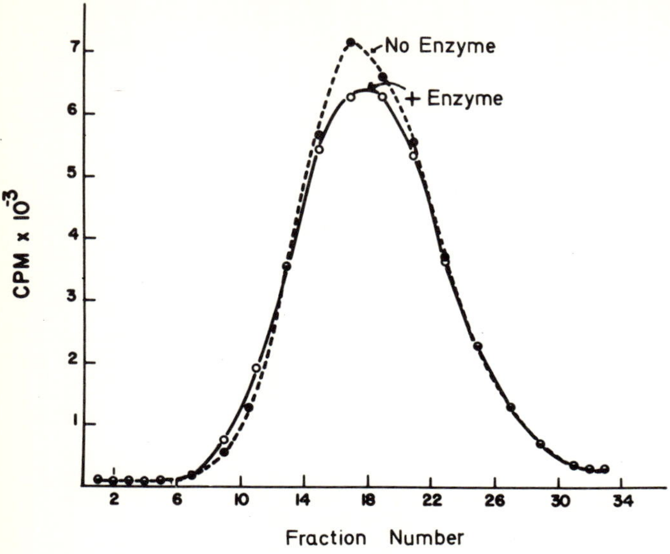

Fig. 19. Sedimentation properties of poly (A) and ligase product.
Reaction mixture (0.1 ml) contained 14 mμmoles of ^{32}P-Poly (A), 0.5 μmoles of $MgCl_2$, 10 mμmoles of ATP, 5 μmoles of Tris buffer, pH 7.5, and 3.2 units of G-200 Sephadex fraction III. After 90 minutes at 30°, approximately 70 % of the input ^{32}P were resistant to BALP. Reactions were halted by the addition of 0.04 ml of a solution containing 5.5 M HCHO, 1 M sodium phosphate buffer, pH 7.0, and 10 % sodium dodecyl sulfate, and the mixture was heated at 65° for 10 minutes. The entire solution was added to 5.2 ml of 5–25 % SDS-sodium phosphate-HCHO sucrose gradient which was centrifuged at 50.000 rpm for 36 hours at 20° in an SW65 rotor. Fractions were collected and aliquots counted in Bray's scintillation fluid.

Nature of the secondary structure required

The influence of various polymers on the reaction has been studied. The results presented in Fig. 21 suggest that some polynucleotides containing a duplex secondary structure are inhibitory. This is evident in reaction mixtures in which 5′-^{32}P-poly (U) and poly (A) or 5′-^{32}P poly (1) and poly (C) or 5′^{32}P poly (C) and poly (1) are combined. Surprisingly, the mixture 5′-^{32}P poly (A) · poly (U) possesses the same activity as 5′-^{32}P poly (A) alone.

The system described above is capable of utilizing a variety of RNA substrates and converting ends into internal phosphodiester bonds. Recent studies [21, 33] have indicated that T4-DNA ligase is also capable of utilizing DNA-RNA hybrid substrates. The data reported here indicates that the observed ligase action with poly (R) is not due to DNA ligase. The evidence supporting this conclusion can be summarized as

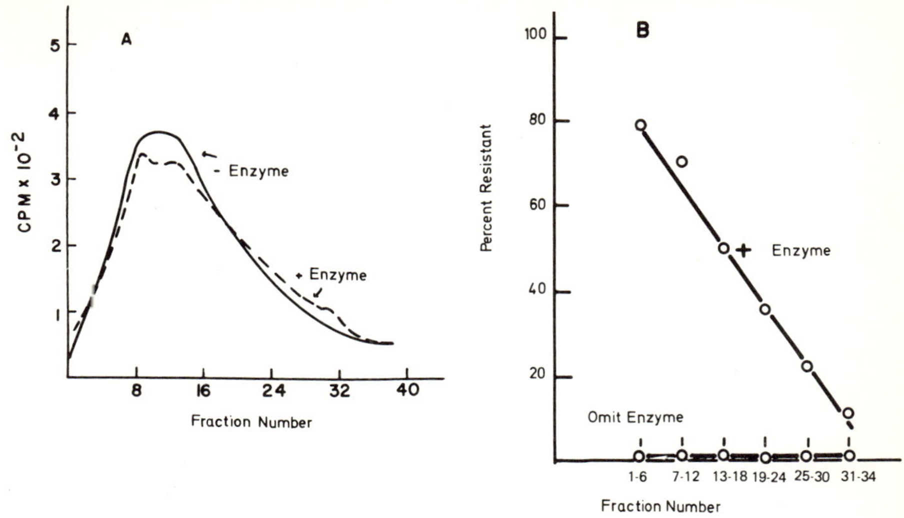

Fig. 20 A und B. Sedimentation properties (A) and distribution of alkaline phosphatase resistant material (B).

Poly (A) obtained from gradients as described in Fig. 19 (fractions 24–29) were dialyzed against 0.05 M NaCl for 24 hours to remove HCHO-sucrose and SDS and then against 0.02 M NaCl for an additional 24 hours. Reaction with RNA ligase was carried out as described in Fig. 19 with 1.13×10^5 cpm of poly (A). After 60 minutes at 30°, approximately 40 % of the ^{32}P was resistant to alkaline phosphatase. The material was treated as described in Fig. 19 and centrifuged for 63 hours. Aliquots were counted in Bray's scintillator fluid. Fractions corresponding to 1–6, 7–12, etc., as indicated were pooled and dialyzed for 48 hours against 0.02 M NaCl. These fractions were then assayed for the % of total ^{32}P-resistant to alkaline-phosphatase. Curves labelled "no enzyme" indicate that RNA ligase was omitted during the above reaction.

$5'$-^{32}P-labeled polymers	Unlabeled Polymer Added				
	NONE	poly (I)	poly (C)	poly (A)	poly (U)
	pmoles/30 min				
$5'$-^{32}P-poly (A)	115	84	106	–	102
$5'$-^{32}P-poly (U)	8.2	7.2	8.6	2.0	
$5'$-^{32}P-poly (I)	14	–	1.0	4.3	11
$5'$-^{32}P-poly (C)	7.7	2.4	–	5.6	8.1

Fig. 21. Influence of combination of polymers

Reaction mixtures were as described in Fig. 15 with the following specific addition: $5'$-^{32}P-poly (A), 14 mμmoles (0.52 mμmole of terminus); $5'$-^{32}P-poly(I), 56 mμmoles; $5'$-^{32}P-poly (C), 52 mμmoles. The unlabelled polymers were commercial polymers added in equal amounts to the $5'$-^{32}P labelled polymers. Mixtures were incubated in 0.02 M Tris buffer, pH 7.5 at 4 °C for 48 hours; they were then used as standard ligase mixtures.

follows: (1) to date only T even phages induce the RNA-ligase activity while T1, T3, and T7 are capable of inducing DNA ligase in *E. coli* [18]. (2) Highly purified DNA phage ligase from *E. coli* or T4 does not substitute for the RNA ligase with poly (A) or tRNA. (3) The purification of T4 RNA ligase removes all detectable DNA-ligase activity. More definitive experiments such as the use of T4 mutants which do not induce DNA-ligase remain to be done.

At present, the secondary structure required for the RNA-ligase reaction is unknown. In contrast to the action of DNA-ligase, duplex structures are not required. It is possible that the intramolecular stacking of purine polymers can explain how ends of polymers can be brought near one another. It is also possible that an enzyme capable of repairing RNA does not require a rigid duplex structure as expected of a DNA repair system. More work is essential to answer this question. Studies are presently underway to repair nicked tRNAs.

Acknowledgements: This work was supprted by grants from the National Cancer Institute (NIH-NCI 71-2251) and United States Public Health Service Grant (CA06657).

References

[1] *Alberts, B.,* and *Frey, L.* 1970. Nature **227**: 1313.

[2] *Aposhian, H. V.,* and *Kornberg, A.* 1962. J. Biol. Chem. **237**: 253.

[3] *Bader, J. P.* 1965. Virology **26**: 253.

[4] *Bader, J. P.,* and *Steck, T. L.* 1969. J. Virology **4**: 454.

[5] *Baltimore, D.* 1970. Nature **226**: 1209.

[6] *Becker, A.* 1966. Fed. Proc. **25**: 276.

[7] *Duesberg, P. H.* 1968. Proc. Natl. Acad. Sci. **60**: 1511.

[8] *Duesberg, P., Canaani, E.,* and *Helm, K. U. D.* 1971. Proc. Natl. Acad. Sci., U.S. **68**: 747.

[9] *Duesberg, P. H.,* and *Robinson, W. S.* 1966. Proc. Natl. Acad. Sci., U.S. **55**: 219.

[10] *Duesberg, P. H.,* and *Vogt, P. K.* 1969. Proc. Natl. Acad. Sci., U.S. **64**: 939.

[11] *Duesberg, P. H., Vogt, P. K.,* and *Canaani, E.* 1971. Structure and Replication of Avian Tumor Virus. Ed. *L. G. Silvestri,* pp. 154–66. 2nd Lepetit Colloquia on the "Biology of Oncogenic Viruses", Amsterdam-London: No. Holland Publishing Co.

[12] *Erikson, R. L.* 1969. Virology **37**: 124.

[13] *Fanshier, L., Garapin, A. C., McDonnell, J., Faras, A., Levinson, W.,* and *Bishop, J. M.* 1971. J. Virol. **7**: 77.

[14] *Faras, A., Fanshier, L., Garapin, A., Levinson, W.,* and *Bishop, J. M.* 1971. J. Virol. **7**: 539.

[15] *Gefter, M.,* and *Kornberg, T.* 1971. Fed. Proc. **30**: 1110.

[16] *Hanafusa, H., Hanafusa, T.,* and *Miyanoto, T.* 1971. p. 170. 2nd Lepetit Colloquia on the "Biology of Oncogenic Viruses", Amsterdam-London: No. Holland Publishing Co.

[17] *Hurwitz, J.*, and *Leis, J.* 1972. J. Virol. **9**: 116.

[18] *Hurwitz, J., Lyn, G., Gefter, M.*, and *Becker, A.* 1967. Proc. Natl. Acad. Sci., U.S. **58**: 1996.

[19] *Johnson, P. H.*, and *Laskowski, M.* Sr. 1970. J. Biol. Chem. **245**: 891.

[20] *Kato, K., Goncalves, J. M., Houts, G. E.*, and *Bollum, F. J.* 1967. J. Biol. Chem. **242**: 2780.

[21] *Kleppe, K., van de Sande, J. H.*, and *Khorana, H. G.* 1970. Proc. Natl. Acad. Sci., U.S. **67**: 68

[22] *Kornberg, A.* 1969. Science **163**: 1410.

[23] *Leis, J.*, and *Hurwitz, J.* 1972. J. Virol. **9**: 130.

[24] *Manly, K. F., Smoller, D. F., Bromfield, E.*, and *Baldimore, D.* 1971. J. Virol. **7**: 106.

[25] *Rabin, E. Z., Mustard M.*, and *Frazer, M. J.* 1968. Can. J. Biochem. **46**: 1285.

[26] *Robinson, W. S.*, and *Baluda, M. A.* 1965. Proc. Natl. Acad. Sci., U.S. **54**: 1686.

[27] *Robinson, W. S., Pitkamen, A.*, and *Rubin, H.* 1965. Proc. Natl. Acad. Sci., U.S. **54**: 137.

[28] *Scolnick, E., Raub, E., Aaronson, S. A.*, and *Todaro, G. J.* 1970. Proc. Natl. Acad. Sci., U.S. **67**: 1789.

[29] *Silber, R., Malathi, V. G.*, and *Hurwitz, J.* Fed. Proc. 1972. **31**: 471

[30] *Spiegelman, S., Burny, A., Das, M. R., Keydar, J., Schom, J., Travnicek, M.*, and *Watson, K.* 1970. Nature **227**: 1029.

[31] *Temin, H. K.* 1964. Virology **23**: 486.

[32] *Temin, H.*, and *Mizutani, S.* 1970. Nature **226**: 1211.

[33] *Weiss, B., Jacqemin-Sablon, A., Live, T. R., Fareed, G. C.*, and *Richardson, C. C.* 1968. J. Biol. Chem. **243**: 4543.

Discussion

Bauer: Did I understand you correctly that the existence of a single-stranded DNA product of the reverse transcriptase results from the liberation of this DNA by a new enzyme which acts on the hybrid?

Hurwitz: The easiest way of envisaging the generation of single-stranded DNA is to assume that there is something which displaces the DNA from the path of the enzyme motion.

Bauer: I would like to mention an alternative possibility. We have been asking whether single-stranded DNA might be generated by digestion of the RNA from the RNA-DNA hybrid. *Hausen* and *Stein* first described such an enzyme two years ago, and in collaboration with *Hausen's* group in Tübingen we have been examining avian myeloblastosis virus (AMV) for such enzyme acitvity. We have found that purified reverse transcriptase preparations do contain an activity which selectively digests the RNA in an RNA-DNA hybrid. We, therefore, assume that this is the way in which the single-stranded DNA develops. Would you agree with our conclusion?

Hurwitz: I am quite familiar with your work. Dr. *Bolognesi* was kind enough to send us a preprint of your paper. While I am just guessing, I would answer, "no, I don't believe your conclusion is correct." For one thing, the enzyme preparations that we have are completely free of ribonuclease activity, as far as we can tell. Secondly, the fact that the enzyme is present in virions does not convince me that it plays any role with regard to the replication of RNA or generation of DNA. Since a large number of enzymes are detected in the virions, it is pretty hard to tell right now what is for real and what is not. One must remember that the virions are quite open during the period of their generation, during the time when they are budding off from cells. One thing which I noticed in your preprint was that there appeared to be no clear-cut quantitative relationship between the RNA attacked by ribonuclease-H activity and the deoxynucleotide incorporation data.

Bauer: That is right, and we don't understand it yet either. I am, however, surprised that you don't find this enzyme activity, because we can't separate it from the reverse transcirptase complex.

Hurwitz: Perhaps we inactivate it during purification of the polymerase. That is a possibility.[1)]

Duesberg: If your model is correct, the RNA hairpin of the AMV RNA would not be transcribed into DNA. Do you feel that the base sequence of the hairpin is genetically not essential for the viral RNA, and that it is added to the RNA after it is transcribed from a viral DNA or RNA plate by an enzyme?

Hurwitz: The question is whether the hairpin requirement in my model excludes the possible transcription of some of the genetic information stored in the RNA. That is certainly something which one cannot explain by a model such as this. On the other hand, there is the potential of something like a "T_4" phage protein, which would be capable of causing denaturation of a structure. For example once DNA synthesis starts, it may be immediately melted. If there is redundancy in that region, remelting would cause the DNA pieces to hybridize to the RNA and result in the uncovering of this region. I should add that one other major problem which remains to be elucidated is the nature of the multiple pieces of RNA which appear to constitute the viral 70 s RNA genome. Since this complex may be composed of 4–5 30S pieces of RNA, hydrogen bonded together, the question arises whether or not these 30S RNA molecules are the same or different.

---

1) Information received by the author at proof stage:

"We have recently confirmed Dr. Bauer's results. Our original experiments utilized ^{3}H-poly(U)-poly d(A); this hybrid structure is not attacked in contrast to ^{3}H-poly(A)-poly d(T) which is attacked by the nuclease activity which copurifies with the polymerase."

RNA Tumor Virus Replication: Facts and Fancy

Peter H. Duesberg

Department of Molecular Biology and Virus Laboratory, University of California, Berkeley, California 94720, USA

Summary: Indirect evidence suggests that the replication of RNA tumor viruses differs from that of cytocidal RNA viruses in two distinctive properties: a requirement for early DNA synthesis and one for permanent DNA dependent RNA synthesis. These requirements led to the provirus hypothesis of *Temin* which states that there is a DNA intermediate in RNA tumor virus replication. The provirus hypothesis has been strengthened by the discovery of virus-associated RNA dependent DNA polymerase. Yet, neither provirus hypothesis nor the virus-associated RNA dependent DNA polymerase explain the peculiar requirement of RNA tumor virus replication for cellular DNA synthesis.

Numerous recent observations suggest that some RNA tumor virus genes, in some cases even all viral genes, preexist in normal cells. If these genes exist in phenotypically normal cells as unexpressed DNA, it is conceivable that "cellular" DNA synthesis is required in RNA tumor virus replication to derepress these genes. In this case the genetic role of the virus or of the viral RNA in RNA tumor virus infection becomes a philosophical question.

Introduction

In early 1970 it was still possible to consider RNA tumor viruses as just another class of RNA containing viruses distinguished only by their capacity to cause tumors [3]. It was claimed [3] that RNA tumor viruses replicate by the orthodox model of cytocidal RNA viruses [16, 44]. This rather simple model says that in the cell the RNA of a given virus is converted to a double-stranded RNA or replicative form (RF) by a virus-specific replicase. Subsequently, the replicase uses the RF as a template to make more viral RNA. The model has two important consequences: Almost all cytocidal RNA viruses including tobacco mosaic virus, Qβ-phage, and polio can do without DNA synthesis and without DNA dependent RNA synthesis. That is, RNA of a typical cytocidal RNA virus is a self-replicating mRNA using the cell more or less as a convenient "in vitro" system. Evidence available today, however, makes it unlikely that RNA tumor viruses replicate according to the orthodox model.

Manuscript received: 17 January 1972

Requirements for RNA tumor virus replication

Indirect evidence suggests that the mechanism of RNA tumor virus replication differs from that of cytocidal RNA viruses in two or three distinct properties: There is a permanent requirement of DNA dependent RNA synthesis, and an early requirement of DNA synthesis. This early requirement for DNA synthesis appears to be dual; including, possibly requirements for both virus-specific and cellular DNA synthesis. I will now discuss some evidence for these peculiar requirements which have kept RNA tumor virology – until not very long ago – the work of a heretical and suspected minority.

1. Evidence for a requirement of DNA dependent RNA synthesis: In 1963 and later, it was found by a number of RNA tumor virologists that addition of actinomycin D to infected cells inhibits tumor virus production and RNA synthesis at any time during the growth cycle [4, 15, 22, 28, 51]. By contrast, actinomycin D which specifically binds to DNA thereby inhibiting DNA dependent RNA synthesis has no effect on the replication of most cytocidal RNA viruses, whose replication involves RF. Moreover, extensive search for a RF in tumor virus infected cells gave negative results [40]. Positive evidence for a RF of an RNA tumor virus was reported in one case [3] but was not confirmed in other cases.

2. Evidence for a requirement of DNA synthesis: One possible interpretation of the actinomycin D sensitivity of RNA tumor virus replication and of the failure to find RF was that DNA instead of RNA was involved in replication. To see if any DNA is synthesized as a result of infection, the effects of various inhibitors of DNA synthesis on tumor virus replication were studied. It was found by *Temin* [54], *Vigier* and *Goldé* [62], and *Bader* [5, 6] that inhibitors of DNA synthesis including BUDR, FUDR, cytosine arabinoside, thymidine, and others prevent avian tumor virus replication but only during the first 8–12 hours p. i. Later virus replication becomes refractory to inhibitors of DNA synthesis.

The provirus hypothesis

The evidence for requirements of early DNA synthesis and permanent DNA dependent RNA synthesis in tumor virus replication led *Temin* in 1964 to postulate a DNA intermediate in RNA tumor virus replication, which he called the "provirus" [52, 53]. The provirus was postulated to arise by reverse transcription from viral RNA, to serve as a template for synthesis of viral RNA, and to be integrated in the cellular chromosome [52].

The provirus hypothesis – radical as it was – was nevertheless a love at first sight for many, though not all, scientists studying RNA tumor virus replication. In particular, the implication of the provirus hypothesis that viral RNA could be integrated in the celluclar chromosome after transcription to DNA was well received for the following reason: RNA tumor viruses, unlike cytocidal RNA viruses, do not kill the host cell, and RNA tumor virus infected or transformed cells growing in culture almost never revert to the normal state. Therefore, the number of RNA genomes and viral RFs in the transformed cell must either be very large and self-replicating or the viral genome must be integrated with the chromosome of the host cell to ensure that the viral genome is not segregated out during mitosis. Since RFs could not be found in virus producing cells and virus-transformed cells, the search for "integrated" viral DNA à la *Temin's* provirus hypothesis became a favorite experiment in the field since 1964.

Search for virus-specific DNA

To examine the possibility that a virus-specific "DNA provirus" was present in infected cells, *Temin* [52] first employed molecular hybridization of cellular DNA with viral RNA. Unfortunately, the difference in annealing efficiencies between viral RNA and DNA from infected or uninfected cells was not more than a few counts per minute, hardly enough to support the provirus hypothesis. Since then, many tumor virologists have carried out similar hybridizations with either viral RNA or DNA, made by the viral reverse transcriptase, with similar experimental results but alternating interpretations. Some interpreted their results to indicate that DNA complementary to viral RNA is indeed only present in infected cells [8, 13, 14, 52]. Others, by now a majority, concluded that some DNA complementary to viral RNA is detectable not only in infected but also in uninfected cells [31, 34, 47, 60, 65, 66]. Yet none of these experiments provide convincing proof for the provirus hypothesis particularly because of the low annealing efficiencies obtained with both infected and uninfected cells. These poor efficiencies may be explained in a positive way, that is, as evidence for the ubiquity of RNA tumor virus genes in all cells (see below). Alternatively, it could be an artefact due to contamination [46] or association [29] of viral RNA with cellular RNA. Finally, it may be due to limited nucleic acid sequences common to virus and cell (common initiation sequences for protein synthesis for example) rather than being evidence for an "integrated" complete genome of the virus.

Perhaps the most compelling, although indirect, experiment to prove that genetically specific viral DNA may be required for virus replication is the finding by *Peter Vogt* and myself that preinfection of chicken cells with a given avian tumor virus does not eliminate the requirement of early DNA synthesis for other avian tumor viruses superinfecting the same cell [23]. This finding is compatible with the idea that at least some of the early DNA is genetically specific for each virus strain.

Many other experiments have been done to trace the evasive provirus. For example, RNA tumor virus replication was shown to become light sensitive if early replication occurs in the presence of BUDR, a light-sensitive analog of the DNA precursor thymidine, implying that a virus-specific, BUDR-labeled DNA was made [10, 17]. Yet the possibility that cellular DNA was BUDR labeled and inactivated by light was not convincingly ruled out. Further BUDR was shown to be mutagenic for RNA tumor viruses [9], implying again a DNA intermediate in virus replication. But again it was not excluded that these mutants have arisen spontaneously [43a, 62a] or that they were due to other effects of BUDR [38, 42].

Despite these numerous and in part ingenious experiments demonstrating the essential role of DNA synthesis in RNA tumor virus replication, the "desired" DNA has never been isolated from an infected cell. Thus from an experimentalist's point of view, the story of the provirus DNA is still reminiscent of the story of the *Emperor's new clothes.* It follows that "integration" of the provirus DNA has yet to be demonstrated.

Is RNA dependent DNA polymerase involved in tumor virus Replication?

Fascinated by the challenge to solve RNA tumor virus replication but frustrated by the results of their in vivo experiments, tumor virologists were waiting for the coming of the "golden age." It appeared to arrive about 18 months ago when an RNA dependent DNA polymerase was discovered in avian and murine RNA tumor viruses by *Baltimore* and *Temin* [11, 57]. Within a few weeks after its discovery, this enzyme became the darling of numerous molecular biologists for both scientific and financial reasons. The enzyme could transcribe viral RNA readily to, presumably, the DNA [21, 31, 32, 59] that had been postulated and implicated as an intermediate in virus replication. Moreover, it came as a surprise that this enzyme, which most everybody anticipated in the cell [5], was associated with mature virus particles. The heresy of the *Temin* model, the reverse transcription from viral RNA to DNA could now be tested at least by in vitro experiments. On this occasion even the ultimate authority of molecular biology, *Francis Crick,* thought it was necessary to reinterpret the central dogma [18]. RNA dependent DNA synthesis, now called "Teminism" was officially legalized.

However, the excitement about this enzyme derives primarily from the assumption that it actually transcribes in vivo viral RNA to DNA as predicted by the provirus hypothesis. Such DNA would then function as a template for the replication of viral RNA. In addition, it is assumed that this DNA would be integrated into the cellular chromosome to cause stable transformation of the cell [52].

After more than a year of a passionate romance with the polymerase by both veteran and proselyte tumor virologists, the "love-is-blind" phase now appears to be over, and we can ask and answer some academic questions about this virus-associated DNA polymerase. The key question is whether the enzyme is the viral replicase or is at least involved in virus replication. Several properties of the enzyme have become known which favor its putative role of a virus-specific transcriptase involved in replication. These include: (1) the ubiquity of the enzyme in all known infectious forms of RNA tumor viruses [30], (2) a tenfold preference of this enzyme among natural RNA templates for 60–70 S tumor virus RNA [26, 27], (3) the ability of the unpurified enzyme to transcribe most or all viral RNA to DNA [21, 31, 32, 59]. However, the DNA produced under conditions where almost all viral RNA is transcribed is 5–8 S or about 100–200,000 daltons [21, 32], only a small fraction of the largest viral RNA subunit whose molecular weight is 3–4 $\times 10^6$ [20]. Furthermore, the 5–8 S DNAs are not a proportionate representation of the sequences present in viral RNA. A majority of the small DNAs are transcribed from an approximately 25 % fraction of the total viral RNA [21, 31, 59, 60].

Quantitative comparisons performed in my laboratory using several different RNA species as templates indicate that the relatively high template activity of 60–70 S tumor virus RNA is due to its aggregate structure [27]. The 60–70 S tumor virus RNA probably consists of several single-stranded major and minor RNA subunits held together by limited base pairing [20]. After heat dissociation of 60–70 S tumor virus RNA into subunits, the high template activity for enzyme is lost [26, 27]. It seemed possible then that the tumor virus enzyme is in vitro analogous to the Kornberg DNA polymerase requiring one strand of RNA as template and another strand, partially base paired to the template, as primer. An obligatory primer requirement for tumor virus DNA polymerase has been independently demonstrated in the laboratories of *Baltimore* [12, 61] and *Hurwitz* [37, 39]; they have further shown that RNA or DNA primers are covalently linked by the enzyme to the DNA product [12, 37, 39, 61]. We have also obtained evidence that some of the DNA produced by the enzyme in presence of 60–70 S Rous sarcoma virus (RSV) RNA as template is covalently linked to RNA. About 20 % of the DNA produced during a 15 minute polymerization reaction in the presence of 60–70 S [^{32} P] RSV RNA as template had a slightly higher sedimentation coefficient and a higher buoyant density in 1.1 M formaldehyde than KOH-treated reaction product indicating that some of this DNA was covalently linked to some RNA primer. The small size of the RNA linked to the DNA product suggested that the RNA primers of 60–70 S RSV RNA were considerably smaller than the largest RNA subunits [17a]. Some DNA did not appear to be linked to RNA or was linked to very small primers [17a].

If a primer requirement is obligatory for the enzyme to work, then it is not an independently initiating nucleic acid polymerase of the Qβ-type [49]. Such a primer-requiring enzyme would not fill the bill of the sole catalyst for self-replication of viral RNA via DNA. Additional factors to initiate DNA synthesis and to transcribe the primer sequences of the viral RNA would be required. The relative template specificity of RSV DNA polymerase among RNAs for 60–70 S tumor virus RNA [27] appears then to be due to the super-structure of the viral RNA which supplies both primers and templates simultaneously for the enzyme [17a, 26, 27]. This relative template specificity of RSV DNA polymerase is only compatible with, but not proof for the view that the enzyme is involved in virus replication. There are other properties of the enzyme which suggest that it may not be specific for virus replication. For example, very different kinds of nucleic acids including RNAs and DNAs, and synthetic nucleic acids can serve as templates for virus-associated DNA polymerase [26].

Role of cellular DNA

There is good evidence to suggest that cellular DNA synthesis is also necessary to initiate RNA tumor virus replication. This evidence is almost prehistoric compared to the recent explosive evolution of the field. Back in 1959, *Rubin* and *Temin* [48] found that the ability of cells to initiate RSV replication is several orders of magnitude more sensitive to X- or UV-irradiation than the ability of the same cell to replicate a typical cytocidal virus (NDV).

Although much harder to interpret than the inhibitor data, these early observations by *Rubin* and *Temin* suggested that the unirradiated integrity of cellular chromosomes or DNA was necessary for virus replication. It was also shown by *Rubin* and *Temin* that the capacity of the cell to divide was as sensitive to radiation as its ability to initiate RSV replication. It was the correlation between this distinctive cellular x-ray sensitivity of the mitosis and ability to propogate RSV which led *Rubin* and *Temin* to commit together the "original sin" against the orthodox model of RNA virus replication. They suggested that the "genome of the Rous sarcoma virus must be integrated with that of the cell before virus production can begin" [48].

The *Rubin* and *Temin* proposal was obviously fashioned after the λ-model, and that may be the reason why it was so readily accepted in 1959. Its heretical implications regarding the central dogma seem to have been ignored by the readers of *Virology* in 1959, probably because the central dogma had only been enunciated in 1958 [18] and was not yet canonical knowledge at the time. Neither was it clear in 1959 whether RSV was a RNA or a DNA virus.

Others have since confirmed the idea that cellular DNA synthesis [6, 7, 45, 55] and possibly mitosis [7, 35, 55] are necessary. It is conceivable that cellular DNA synthesis may play an indirect role in tumor virus replication, that is to say it may be necessary to derepress a cellular function instrumental in virus replication. Alternatively cellular DNA synthesis may be necessary to *"integrate"* incoming viral genes or to *"excise"* preexisting viral genetic information. These fancy possibilities presumably raise more questions than they answer.

Nevertheless the following data are compatible with the idea that the genetic material of certain types of RNA tumor viruses preexists in all animal cells: (1) There is evidence, already discussed above, for homology between viral and cellular nucleic acids. Although inconclusive, this evidence implies that some nucleic acid sequences may be common to virus and cell. (2) Antigens have been found reproducibly in some "virus-free" chicken cells which are indistinguishable from the group-specific antigen of RNA tumor viruses [19], the major structural protein of RNA tumor viruses [20]. (3) *Altaner* and *Temin* [2] have discovered that certain avian RNA tumor viruses (B77) after passage in mammalian cells could pick up new genetically stable properties. It was suggested that this phenomenon was due to recombination of the virus with viral genetic information preexisting in the host cell rather than due to selection of a variant. A similar phenomenon was described for murine leukemia virus [1]. (4) Certain "virus-free" cells can help defective Rous sarcoma virus to replicate and to produce complete virus suggesting that these cells contain RNA tumor virus genes. Such cells have been called "helper-cells" by *Weiss* [63]. (5) Finally, and perhaps most conclusively, it has been shown this year by several groups, in particular by *R. Weiss* [64], *W. Rowe* [42], *Klement* [38] and collaborators, that probably all "virus-free" animal cells in culture can be induced by certain manipulations to release nontransforming RNA tumor viruses. The induction of these viruses has since become a new bandwagon in RNA tumor virology although induction of leukemia viruses in animals has been shown by *Lieberman* and *Kaplan* [41] many years ago. The most efficient inducer in culture of nontransforming tumor viruses turned out to be BUDR and IUDR, but others such as radiation or mitomycin C were also effective.

Clearly these results suggest that RNA tumor viruses in a broad sense are "integrated" in the cell. Whether this is integration in a molecular sense, that is, viral DNA integrated in cellular DNA, or whether it is a DNA plasmid or a latent virus replicating by a variation of the orthodox model cannot be answered at present.

The differential capacity of cells to express some or all genes of endogenous RNA tumor viruses, or to complement an incoming defective virus may be compatible with our recent observation that the RNA of RNA tumor viruses exists in subunits or subgenomic fragments. In particular, the capacity of sarcoma viruses to transform seems to be correlated with a RNA subunit absent from nontransforming viruses [24, 43a].

The possibility that some or all genetic information of RNA tumor viruses preexists in normal cells raises important philosphical questions about the role of the virus and its RNA as an infectious agent. Does the classical definition of an infectious agent apply to RNA tumor viruses? What is left for the exogenous virus to do, if most or all viral genes preexist endogenously in the cell – in other words, if the cell is already infected? As an answer to these questions, I would like to briefly describe two hypotheses which provide a philosphy of the problems of RNA tumor virus replication and integration.

Oncogene and protovirus hypotheses

The oncogene hypothesis by *Huebner* and *Todaro* [36] postulates that some, perhaps all, viral genes ("virogenes and oncogenes") are present in animal cells and are transmitted from animal to progeny and from cell to cell in a latent form. Carcinogens, irradiation, ageing, and excitement all favor the partial or complete activation of these genes. This theory is based mainly on "seroepidemiological" [36] evidence which leaves many questions unanswered. What is the role of the virus or the viral nucleic acid if radiation, ageing, and excitement are enough to "switch on" the oncogene? Although direct and reproducible evidence for infectious tumor virus RNA is not available, there is the following indirect evidence that viral RNA carries genetic information. Successful radioinactivation [20, 48] or mutagenization [43, 58] of RNA tumor virus suggests that the viral nucleic acid plays a genetic role and is not just an "inducer." The appearance of the group-specific antigen of avian tumor viruses in mammalian cells after transformation with avian tumor viruses [50] further indicates that viral RNA may be the messenger of the viral group-specific antigen. Also, correlation of the ability of avian tumor viruses to transform with the presence of a distinct viral RNA subunit (termed *a* subunit [24]) suggests a relation between RNA structure and biological function of the virus [24].

The oncogene hypothesis might explain, however, why mitosis or an event leading to mitosis may be necessary for virus replication. It may be required to derepress switched-off genetic information preexisting in the cell. This may be analogous to the processes which are thought to lead to differentiation of somatic cells [33].

As a variant of the oncogene hypothesis, *Temin* has recently proposed the "protovirus hypothesis" [56]. The protovirus hypothesis postulates that the germ line contains directly the potential for somatic *de novo* evolution of the information for cancer. But this is not the normal role of this potential. Normally the "protoviruses" as part of the germ line are thought to evolve in somatic cells to become specific genomes of differentiated cells and thus account for stable differentiation, creation of antibody diversity, and stable memory. According to *Temin,* the genetic change which might occur during differentiation or transformation of a cell happens at the RNA level. The protovirus is first transcribed to RNA. It may then be altered and transcribed back to DNA involving reverse transcriptase. Further, it is proposed that exchange occurs between cells at the RNA level [56].

An abnormal, that is a transforming, protovirus gene could then lead to the evolution of an RNA tumor virus. A suitable "oncogene" and an "envelope gene" may join in the host cell and the two together could form a virus, leave the cell, and enter new cells from outside. It is a serious shortcoming of the protovirus hypothesis, however that it provides also a potential mechanism for inheritance of acquired properties.

Without a corollary mechanism to protect the germ line of an organism against the acquisition of new or altered genes, the protovirus hypothesis implicates a potentially dangerous extension of the theories of *Lamarck* and *Lysenko*.

Acknowledgement: I would like to thank Drs. *L. Lewandowski, M. Lai, E. Canaani,* and *K. v. d. Helm* for their reviews of the manuscript and Dr. *P. H. Hofschneider* for suggesting a major fraction of the title.

Supported by U.S. Public Health Service Research Grant CA 11426 from the National Cancer Institute and by Contract No. 71-2173 within the Special Virus Cancer Program of the National Cancer Institute.

References

[1] *Aaronson, S. A.* 1971. Common genetic alterations of RNA tumor viruses grown in human cells. Nature **230:** 445–47.

[2] *Altaner, C., and Temin, H. M.* 1970. Carcinogenesis by RNA sarcoma viruses XII. A quantitative study of infection of rat cells in vitro by avian sarcoma viruses. Virology **40:** 118–34.

[3] *Anonymous* 1970. Chasing a minus strand. Editorial, Nature **225:** 690.

[4] *Bader, J. P.* 1964. The role of deoxyribonucleic acid in the synthesis of Rous sarcoma virus. Virology **22:** 462–68.

[5] *Bader, J. P.* 1966. Metabolic requirements for infection by Rous sarcoma virus I. The transient requirement for DNA synthesis. Virology **29:** 444–51.

[6] *Bader, J. P.* 1966. Metabolic requirements for infection by Rous sarcoma virus II. The participation of cellular DNA. Virology **29:** 452–61.

[7] *Bader, J. P.* 1967. Metabolic requirements in Rous sarcoma virus replication. *In:* The molecular biology of viruses, (eds.) *Colter* and *Paranchych*, pp. 697–708. New York: Academic Press.

[8] *Bader, J. P.* 1967. Metabolic requirements for infection by Rous sarcoma virus. *In:* Subviral carcinogenesis: Monograph of First International Symposium on Tumor Viruses, pp. 144–55. Japan: Nagoya.

[9] *Bader, J. P., and Brown, N. R.* 1971. Induction of mutations in an RNA tumor virus by an analogue of a DNA precursor. Nature New Biology **234:** 11–12.

[10] *Balduzzi, P., and Morgan, H. R.* 1970. Mechanism of oncogenic transformation by Rous sarcoma virus I. Intracellular inactivation of cell-transforming ability of Rous sarcoma by 5-bromodeoxyuridine and light. J. Virol. **5:** 470–77.

[11] *Baltimore, D.* 1970. Viral RNA-dependent DNA polymerase. Nature **226:** 1209–111.

[12] *Baltimore, D., and Smoler, D.* 1971. Primer requirement and template specificity of the DNA polymerase of RNA tumor viruses. Proc. Nat. Acad. Sci. **68:** 1507–111.

[13] *Baluda, M. A., and Markham, P.* 1971. Nucleotide composition of RNA hybridized to homologous DNA from cells transformed by avian tumor viruses. Nature New Biology **231:** 90–91.

[14] *Baluda, M. A., and Nayak, D. P.* 1970. DNA complementary to viral RNA in leukemic cells induced by avian myeloblastosis virus. Proc. Nat. Acad. Sci. **66:** 329–36.

[15] *Bather, R.* 1963. Influence of 5-fluorodeoxyuridine and actinomycin D on the reproduction of Rous sarcoma virus in vitro. Proc. Am. Ass. Cancer Res. **4:** 4.

[16] *Blair, C. D., and Duesberg, P. H.* 1970. Myxovirus ribonucleic acids. Ann. Rev. Microbiol. **24:** 539–74.

[17] *Boettiger, D., and Temin, H.* 1970. Light inactivation of focus formation by chick embryo fibroblasts infected with avian sarcoma virus in the presence of 5-bromodeoxyuridine. Nature **228:** 622–24.

[17a] *Canaani E., and Duesberg, P. H.*, (1972). Role of subunits of 60–70 S avian tumor virus RNA in its template activity for the viral DNA polymerase. J. Virol., July issue.

[18] *Crick, F.* 1970. Central dogma of molecular biology. Nature **227:** 561–63.

[19] *Dougherty, R. M., Di Stefano, H. S., and Roth, R. K.* 1967. Virus particles and viral antigens in chicken tissues free of infectious avian leukosis virus. Proc. Nat. Acad. Sci. U.S. **58:** 808–17.

[20] *Duesberg, P. H.* 1970. On the structure of RNA tumor viruses. Current Top. Microbiol. and Immunol. **51:** 79–104.

[21] *Duesberg, P. H., and Canaani, E.* 1970. Complementarity between Rous sarcoma virus (RSV) RNA and the in vitro-synthesized DNA of the virus-associated DNA polymerase. Virology **42:** 783–88.

[22] *Duesberg, P. H., and Robinson, W. S.* 1967. Inhibition of mouse leukemia virus (MLV) replication by actinomycin D. Virology **31:** 742–46.

[23] *Duesberg, P. H., and Vogt, P. K.* 1969. On the role of DNA synthesis in avian tumor virus infection. Proc. Nat. Acad. Sci. **64:** 939–46.

[24] *Duesberg, P. H., and Vogt, P. K.* 1970. Differences between the ribonucleic acids of transforming and nontransforming avian tumor viruses. Proc. Nat. Acad. Sci. U.S. **67:** 1673–80.

[25] *Duesberg, P. H., Vogt. P. K., and Martin, G. S.* 1971. The 60–70 S RNA of avian sarcoma and leukosis viruses: Distribution of class *a* and *b* subunits. *In:* Comp. leukemia research Bibl. haemat., ed. R. M. Dutcher. Basel; Karger: in press.

[26] *Duesberg, P. H., von der Helm, K., and Canaani, E.* 1971. Comparative properties of RNA and DNA templates for the DNA polymerase of Rous sarcoma virus. Proc. Nat. Acad. Sci. **68:** 2505–9.

[27] *Duesberg, P. H., von der Helm, K., and Canaani, E.* 1971. Properties of a soluble DNA polymerase isolated from Rous sarcoma virus. Proc. Nat. Acad. Sci. **68:** 747–51.

[28] *Eidenoff, M. L., Bates, B., Perez, A., and De la Sierra, A.* 1963. Effects of inhibitors of nucleic acid synthesis on the cell culture system; chick embryo cell-Rous sarcoma virus (RSV). Proc. Am. Assoc. Cancer Res. **4:** 18.

[29] *Erikson, E., and Erikson, R. L.* 1971. Association of 4 S ribonucleic acid with oncornavirus ribonucleic acids. J. Virol. **8:** 254–56.

[30] *Gallo, R. C.* 1971. Reverse transcriptase, the DNA polymerase of oncogenic RNA viruses. Nature **234:** 194–98.

[31] *Gelb, L. D., Aaronson, S. A., and Martin, M. A.* 1971. Heterogeneity of murine leukemia virus in vitro DNA; detection of viral DNA in mammalian cells. Science **172:** 1353–55.

[32] *Green, M.* 1971. Transcription of viral genes in cells transformed by DNA and RNA viruses. *In:* Third Miami PCRI – Winter Symposium, eds. D. W. Ribbons, J. F. Woessner, and J. Schultz, pp. 380–418. Amsterdam: North-Holland Publishing Co.

[33] *Gurdon, J. B., and Woodland, H. R.* 1970. On the long-term control of nuclear activity during cell differentiation. *In:* Current topics in developmental biology. **5:** 39–70. New York: Acedemic Press.

[34] *Harel, L., Harel, J., Lacour, F., and Huppert, J.* 1966. Homologie entre génome du virus de la myéloblastose aviaire (AMV) et génome cellulaire. C. R. Acad. Sc. Paris **263:** 616–19.

[35] *Hobom-Schnegg, B., Robinson, H. L., and Robinson, W. S.* 1970. Replication of Rous sarcoma virus in synchronized cells. J. Gen. Virol. **7:** 85–93.

[36] *Huebner, R. J., and Todaro, G. J.* 1969. Oncogenes of RNA tumor viruses as determinants of cancer. Proc. Nat. Acad. Sci. **64:** 1087–94.

[37] *Hurwitz, J., and Leis, J. P.* 1972. Studies on the RNA dependent DNA polymerase activity of RNA tumor viruses, I. Directing influence of DNA in the reaction. J. Virol. **9,** 116–129.

[38] *Klement, V., Nicolson, M. O., and Huebner, R. J.* 1971. Rescue of the genome of focus forming virus from rat non-productive lines by 5-bromodeoxyuridine. Nature New Bioloy **234:** 12–14.

[39] *Leis, J. P., and Hurwitz, J.* 1972. Studies on the RNA-dependent DNA polymerase activity of RNA tumor viruses. II. Directing influence of RNA in the reaction, J. Virol. **9,** 130–142.

[40] *Levinson, W., Bishop, M. J., Quintrell, N., Jackson, J., and Fanshier, L.* 1970. Synthesis of RNA in normal and Rous sarcoma virus-infected cells: Effect of bromodeoxyuridine. Virology **42:** 221–24.

[41] *Lieberman, M., and Kaplan, H. S.* 1954. Leukemogenic activity of filtrates from radiation-induced lymphoid tumors of mice. Science **130:** 387–88.

[42] *Lowy, D. R., Rowe, W. P., Teich, N., and Hartley, J. W.* 1971. Murine leukemia virus: High-frequency activation in vitro by 5-fluorodeoxyuridine and 5-bromodeoxyuridine. Science **174:** 155–56.

[43a] *Martin, G. S., and Duesberg, P. H.* 1972. The a-subunit in the RNA of transforming avian tumor viruses. I. Occurrence in different virus strains. II. Spontaneous loss resulting in nontransforming variants. Virology **47,** 494–497.

[43] *Martin, G. S.* 1970. Rous sarcoma virus: a function required for the maintenance of the transformed state. Nature **227:** 1021–23.

[44] *Montagnier, L.* 1968. The replication of viral RNA. In: The Molecular Biology of Viruses, Symp. Soc. Gen. Microbiol. **18:** 125–48.

[45] *Nakata, Y., and Bader, J. P.* 1968. Studies on the fixation and development of cellular transformation by Rous sarcoma virus. Virology **36:** 401–10.

[46] *Obara, T., Bolognesi, D. P., and Bauer, H.* 1971. Ribosomal RNA in avian leukosis virus particles. Int. J. Cancer **7:** 535–46.

[47] *Rosenthal, P. N., Robinson, H. L., Robinson, W. S., Hanafusa, T., and Hanafusa, H.* 1971. DNA in uninfected and virus-infected cells complementary to avian tumor virus RNA. Proc. Nat. Acad. Sci. **68:** 2336–40.

[48] *Rubin, H., and Temin, H. K.* 1959. A radiological study of cell-virus interaction in the Rous sarcoma. Virology **7:** 75–91.

[49] *Stavis, R., and August, T.* 1971. The biochemistry of RNA bacteriophage replication. Ann. Rev. Biochem. **39:** 527–60.

[50] *Svoboda, I., and Hlozanek, I.* 1970. Role of cell association in virus infection and virus rescue. Advances in Cancer Research **13:** 217–69.

[51] *Temin, H. K.* 1963. The effects of actinomycin D on growth of Rous sarcoma virus in vitro. Virology **20:** 577–82.

[52] *Temin, H. K.* 1964. Homology between RNA from Rous sarcoma virus and DNA from Rous sarcoma virus-infected cells. Proc. Nat. Acad. Sci. **52:** 323–29.

[53] *Temin, H. K.* 1964. Nature of the provirus of Rous sarcoma. Natl. Cancer Inst. Monograph **17:** 557–70.

[54] *Temin, H. K.* 1964. The participation of DNA in Rous sarcoma virus production. Virology **23:** 486–94.

[55] *Temin, H. K.* 1967. Studies on carcino-genesis by avian sarcoma viruses. V. Requirement for new DNA synthesis and for cell division. J. Cell. Physiol. **69:** 53–69.

[56] *Temin, H. K.* 1971. The protovirus hypothesis: Speculations on the significance of RNA-directed DNA synthesis for normal development and for carcino-genesis. J. Nat. Canc. Inst. **46:** 3–7.

[57] *Temin, H. K., and Mizutani, S.* 1970. RNA-dependent DNA polymerase in virions of Rous sarcoma virus. Nature **226:** 1211–13.

[58] *Toyoshima, K., and Vogt, P. K.* 1969. Temperature sensitive mutants of avian sarcoma virus. Virology **39:** 930–31.

[59] *Varmus, H. E., Levinson, W. E., and Bishop, M. I.* 1971. Extent of transcription by the RNA-dependent DNA polymerase of Rous sarcoma virus (RSV). Nature New Biology **233:** 19–21.

[60] *Varmus, H. E., Weiss, R. A., Friis, R., Levinson, W., and Bishop, M. J.* 1972. Detection of avian tumor virus-specific nucleotide sequences in avian cell DNAs. Proc. Nat. Acad. Sci. U.S. **69,** 20–24.

[61] *Verma, I. M., Meuth, N. L., Bromfeld, D., Manly, K. F., and Baltimore, D.* 1971. Covalently linked RNA-DNA molecule as initial product of RNA tumor virus DNA polymerase. Nature New Biology **233:** 131–34.

[62a] *Vogt, P. K.* 1971. Segregation of nontransforming viruses from cloned sarcoma viruses. Virology **46:** 939–46.

[62] *Vigier, P., and Goldé, A.* 1964. Effects of actinomycin D and mitomycin C on the development of Rous sarcoma virus. Virology **23:** 511–19.

[63] *Weiss, R. A.*, 1972. Helper viruses and helper cells. *In:* RNA viruses and host genome in oncogenesis, eds. P. Emmelot, and P. Bentvelzen. Amsterdam: North-Holland Publishing Co. pp. 117–135.

[64] *Weiss, R. A., Friis, R. R., Katz, E., and Vogt, P. K.* 1971. Induction of avian tumor viruses in normal cells by physical and chemical carcinogens. Virology **46:** 920–38.

[65] *Wollmann, R. L., and Kirsten, W. H.* 1968. Cellular origin of mouse leukemia viral ribonucleic acid. J. Virol. **2:** 1241–48.

[66] *Yoshikawa-Fukada, M., and Ebert, I. D.* 1971. Further characterization of the cellular DNA hybridizing with the RNA of Rous sarcoma virus. Proc. Nat. Acad. Sci. **68:** 743–46.

Discussion

Sinsheimer: Is there a critical distinction between the oncogene hypothesis and the protovirus hypothesis?

Duesberg: Not really. Only that *Temin* has postulated that the protovirus is, in fact, cellular DNA and that it is not directly the information for the viral RNA. That is, it does not directly encode the information for the viral RNA but can evolve to become viral RNA. He postulates that some polynucleotide chain of the cellular DNA which is capable of causing transformation can transcribe this RNA. This RNA then becomes modified and arranged in the cell. It may somehow meet a gene which provides an envelope. The two then get together and form a tumor virus. *Temin* does not say that the protovirus is directly the information of the virus, but rather a genetic potential whose main role is creating antibody diversity, differentiation, stable memory, and so on.

Hill: I would like to stress that we have rescued a Sarcoma virus with DNA treatment of permissive cells.

Duesberg: You have infectious DNA?

Hill: Yes.

Duesberg: Did you get the same virus back which originally transformed the cell? Can you exclude that it is induction of a virus preexisting in recipient cells?

Hill: Yes, the virus which is rescued is antigenically identical to the original transforming virus.

The Presence of Infectious DNA in Rous Sarcoma Virus Transformed Cells

Miroslav Hill and Jana Hillova

Groupe de Recherche No 8 du C.N.R.S., Institut Gustave-Roussy, (94) Villejuif, France

Summary: The DNA of XC cells transformed by PR-RSV gives rise to a sarcoma virus when introduced into the chick embryo cells. Since the rescued virus resembles the virus which originally transformed the XC cells, the DNA involved in the rescue appears to contain a copy or copies of the RSV genome which are expressed in the permissive cells.

Chick embryo cells are capable of taking up a considerable amount of foreign DNA which has been dissolved in culture medium. It was demonstrated recently [3] that a portion of this DNA amounting to about 0.2 percent of the host cell genome is integrated as single-stranded pieces into the newly replicated DNA strands. Another portion of foreign DNA (representing about 2 percent of the host cell genome) persists in the host cell without being bound to the cellular genome [4]. These results permit us to speculate that the genetic information transferred in such a way into the host cell might lead to a stable change in the phenotype.

To test the validity of this speculation, the information for Rous sarcoma virus (RSV) was chosen as a marker, and assayed for DNA mediated transfer from non-permissive rat Rous sarcoma cells to permissive chick embryo cells. This choice short-circuited the current uncertainty concerning the presence of copy(ies) of RSV genome in the DNA of Rous sarcoma cells [11]. Preliminary results of this work were reported elsewhere [2].

The DNA used to treat chick embryo cells was extracted from an established line of rat XC cells (kindly supplied by Dr. *J. K. Youn)* which were transformed about ten years ago by the Prague strain of RSV (PR-RSV) [9]. These cells do not produce

Manuscript received: 30 December 1971

viral particles [10, 12] but yield infectious virus upon fusion with normal chick embryo fibroblasts. The DNA was extracted essentially by the Marmur's procedure [7] with pronase digestion (500 μg/ml; 4 hours at 37 °C) following a treatment with RNase. The DNA fibres after the isopropanol precipitation step were sterilized in 75 percent alcohol overnight, dissolved in sterile 0.15 M NaCl – 0.015 M Na citrate, and kept at 4 °C before use. The analytical sedimentation profiles showed heterogeneity in the molecular weight of the DNA. The average molecular weight calculated according to *Studier* [8] was 8.9×10^6 daltons, but about 25 percent of the molecules were larger having 35×10^6 to 44×10^6 daltons. DNA was also extracted in a similar manner from thymocytes of healthy Wistar rats. The average molecular weight of this DNA was 17.5×10^6 daltons; about 25 percent of the molecules had 21.1×10^6 to 45×10^6 daltons.

Chick embryo cells were obtained from 11-day-old embryos from a controlled lymphomatosis-free flock of brown Leghorns (J. Carr strain), and cultivated in Falcon plastic flasks in a modified Eagle's medium [6] supplemented with 10 percent tryptose phosphate broth (DIFCO) and 5 percent preheated (56 °C, 30 minutes) calf serum. At various time intervals the cell monolayers were trypsinized, and the cells were equally distributed into two flasks as indicated in Figs. 1 and 2. One of these flasks was used for further passage of cells while the other was examined each day for the presence of foci of transformed cells and discarded after 3–4 days.

About 12–24 hours after seeding, the cell monolayer was twice washed with 0.14 M NaCl – 0.05 M Tris-HCl, pH 7.4, and treated first with DEAE-dextran (100 μg/ml; 15 minutes at 37 °C) [5] and then with DNA (17 μg/ml; 15 minutes at 37 °C) in the same buffer (lines 45 and 46). Some cell lines (lines 28A and 29A) were treated with DNA preincubated with DNase (Worthington, RNase free; 10 μg DNase/ml, 1 hour at 37 °C) and others (lines 26 and 27, and sublines 28B and 29B) with DNA preincubated alone.

Figures 1 and 2 show that repeated treatments of cells with XC cell DNA resulted on each occasion in the appearance of foci of transformed cells (Fig. 3) whereas no cell transformation occurred in cell lines treated with DNase digested DNA. This transformation was also not observed when the DNA from thymocytes of healthy Wistar rats was applied instead of the DNA from rat XC cells (Fig. 2).

Cell monolayers with transformed cells were passed two or three times to obtain completely transformed cell cultures. These cells were examined by electron microscopy and found to contain viral particles characteristic for avian tumor viruses (Fig. 4) which budded from the cell surface (Fig. 5). Culture medium was filtrated through millipore HA filter (pore diameter 0.45 μ) and subsequently assayed for the presence of infectious virus. It was found in such a way that each transformed culture produced viral particles giving rise to microsarcomas when injected on

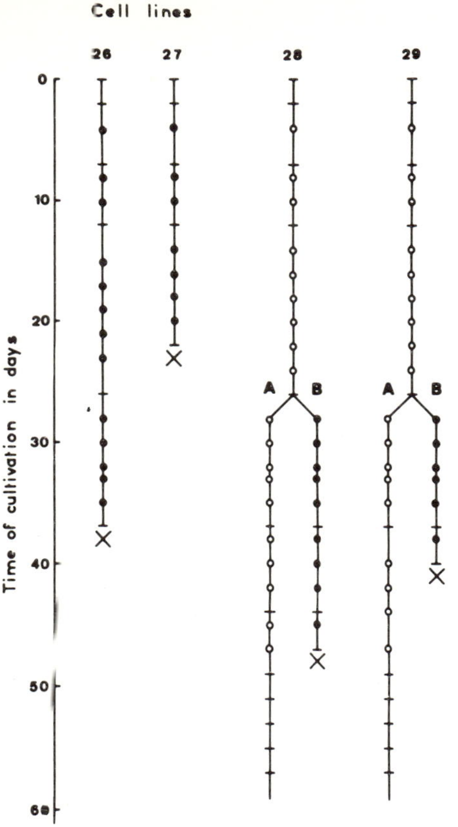

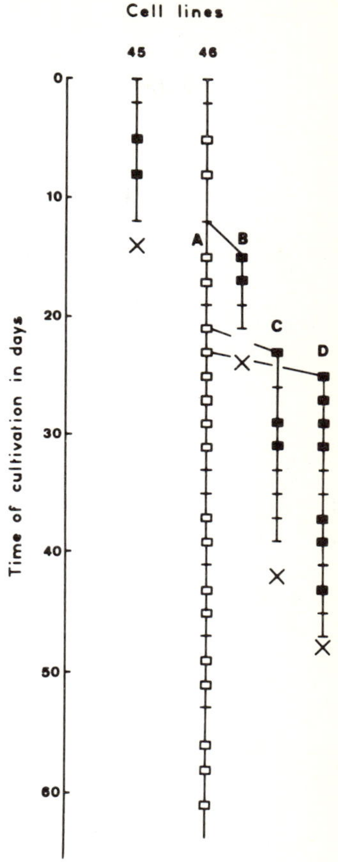

Fig. 1

Schematic representation of virus rescue in chick embryo cells treated with XC cell DNA. An appropriate amount of cells obtained from one chick embryo was put into each of four flasks (lines 26–29) at the day 0 and then serially propagated in vitro. Explanation of symbols: (–) passage of cells without subsequent DNA treatment; (●) passage followed by a DNA treatment of cells 12–24 hours later (DNA was preincubated alone for 1 hour at 37 °C); (○) passage followed by a treatment of cells with the DNase digested DNA 12–24 hours later; (X) appearance of foci of transformed cells.

Fig. 2

Schematic representation of virus rescue after several treatments of chick fibroblasts with rat XC cell DNA (■), and the failure to rescue the virus if rat thymus DNA (□) was pplied. Other conditions as in Fig. 1.

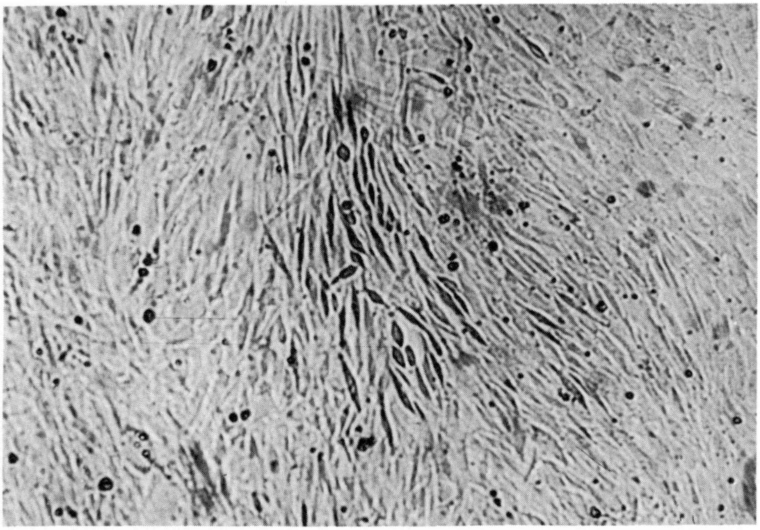

Fig. 3. Typical appearance of one of the first foci of transformed cells in a chick cell monolayer previously treated with XC cell DNA. Photographed in the living state with bright field optics. × 430.

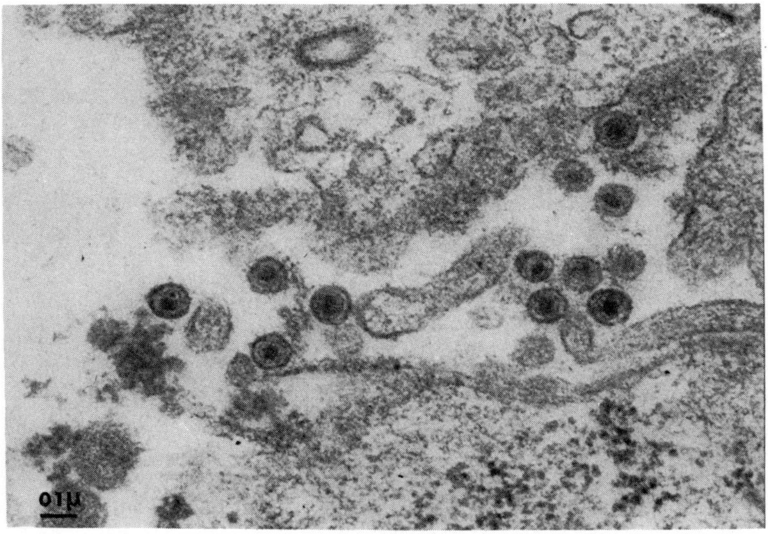

Fig. 4. Free virus particles rescued by the treatment of chick embryo fibroblasts with XC cell DNA.

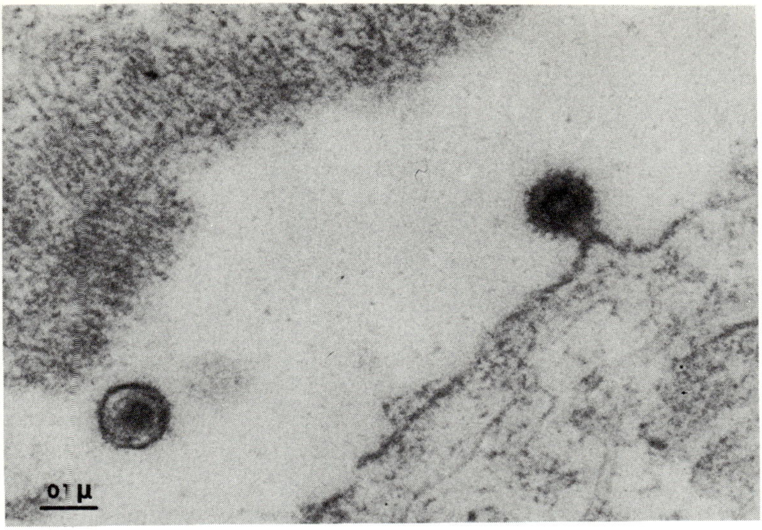

Fig. 5. Virus particle budding at the cell surface in the same culture as in Fig. 4.

chorio-allantoic membrane, and to foci of transformed chick embryo cells when plated under agar medium. Completely transformed cultures contained about 10^4 focus-forming units per ml of culture medium.

The morphology of the foci and of the transformed cells themselves were the same for all the viruses rescued by XC cell DNA treatment, and resembled those characteristic for PR-RSV.

Further experiments were performed to compare the DNA rescued virus with that which was rescued by fusion of XC cells with chick fibroblasts. Both viruses were incubated with antisubgroup A, B, and D sera for 1 hour at 37 °C and titrated by the focus formation method. As can be seen from Fig. 6, the same patterns of neutralization were obtained. Both viruses were very slightly neutralized with anti-A and anti-B sera and a little more with a mixture of anti-A and anti-B sera. This observation could be explained either by phenotypic mixing of envelope antigens as was already found by *Duff* and *Vogt* [1] in a crude stock of PR-RSV, or by genetic recombination taking place in the course of virus rescue. By aspirating a focus formed under agar, cloning of virus was possible. A virus was considered cloned after three isolations under agar medium. These cloning experiments have supported the second possibility because the cloned viruses so far tested show the same neutralization patterns as the uncloned ones.

Antiserum specificity*)	Fraction of virus survivors**)			
	28B	Clone 11 of 28B	Clone 15 of 28B	1R
Anti-A	0.49	0.65	0.43	0.46
Anti-B	0.43	0.60	0.31	0.52
Anti-A + Anti-B	0.32			0.26
Anti-D	0.94	0.92	0.88	1.06
Control***)	1.00	1.00	1.00	1.00

Fig. 6. Effect of neutralizing antisera on viruses rescued either by DNA treatment or cell fusion

---

*) The antisera listed were prepared in chickens against RAV-1 (subgroup A), RAV-2 (subgroup B), and RAV-50 (subgroup D), and kindly supplied by Dr. *Ph. Vigier.* A 1 : 10 dilution of serum was incubated with culture medium containing the virus for 1 hour at 37 °C, and the surviving virus was titrated by the focus formation method. Fraction of survivors was determined in four plates by counting foci under agar medium.

**) Virus 28B was rescued in chick fibroblasts treated with XC cell DNA (Fig. 1). The clones of this virus were isolated by aspiration of a single focus through the undisturbed agar overlay. The cloning procedure was repeated three times. Virus 1R was rescued by fusion of XC cells with chick fibroblasts.

***) Calf serum heated at 56 °C for 30 minutes prior to use.

Addendum: Soon after the return from this conference, experiments were completed with a virus rescued in chick fibroblasts treated with the DNA from nonpermissive hamster cells already transformed in Dr. *Vigier's* laboratory with a thermosensitive mutant of Schmidt-Ruppin RSV, subgroup D. The rescued virus gave rise to foci characteristic for Schmidt-Ruppin RSV which were quite different from those formed by PR-RSV. Furthermore, this virus is thermosensitive and by neutralization studies belongs to the subgroup D as is the case of the original virus. These results furnish evidence that three genetic markers were transferred by the infectious DNA: (1) morphology of foci, (2) thermosensitivity, and (3) envelope antigenicity.

Viruses rescued by XC cell DNA treatment or by fusion of XC cells with chick fibroblasts were both neutralized (surviving fraction 0.02) by anti-PR-RSV serum kindly supplied to us in January 1972 by Dr. *P. K. Vogt.*

Acknowledgements: The authors wish to thank Miss *D. Widmann* for electron microscopic examinations, and Dr. *Ph. Vigier* for furnishing us the neutralizing antisera and the clone of hamster cells transformed by a thermosensitive mutant of RSV.

References

[1] *Duff, R. G., and Vogt, P. K.* 1969. Virology **39:** 18.

[2] *Hill, M., and Hillova, J.* 1971. Comptes Rendus Acad. Sci. Ser. D **272:** 3094.

[3] *Hill, M., and Hillova, J.* 1971. Nature New Biol. **231:** 261.

[4] *Hill, M., and Hillova, J.* 1971. *In:* Informative molecules in biological systems, ed. *L. G. H. Ledoux,* p. 113. Amsterdam: North-Holland Publishing Co.

[5] *Howard, B. V., Estes, M. K., and Pagano, J. S.* 1971. Biochim. Biophys. Acta **228:** 105.

[6] *Macpherson, I., and Stoker, M.* 1962. Virology **16:** 147.

[7] *Marmur, J.* 1961. J. Mol. Biol. **3:** 208.

[8] *Studier, F. W.* 1965. J. Mol. Biol. **11:** 373.

[9] *Svoboda, J.* 1964. Nat. Cancer Inst. Monogr. **17:** 277.

[10] *Svoboda, J., Chýle, P., Šimkovič, D., and Hilgert, I.* 1963. Folia Biol. Prague **9:** 77.

[11] *Temin, H. M.* 1964. Nat. Cancer Inst. Monogr. **17:** 557.

[12] *Vigier, Ph., and Svoboda, J.* 1965. Comptes Rendus Acad. Sci. Ser. D **261:** 4278.

Discussion

Glick: Would you care to expand on the source of your DNA preparations?

Hill: The DNA came from XC cells transformed 10 years ago by the Prague strain of RSV. In recent experiments, the source of the Rous DNA were cells transformed with Schmidt-Ruppin strain of RSV, and in this case the resulting virus produced foci whose morphology was characteristic for Schmidt-Ruppin RSV.

Glick: Could your results be consistent with the human "oncogene" hypothesis? What I am asking is whether or not any DNA could induce the expression of virus universally present in the cells in a latent form? Have you ever used any other source for DNA?

Hill: Yes, many times, and each time the results with DNA from normal rat cells were negative.

Axelrod: What were the size characteristics of the DNA used in these experiments?

Hill: The average molecular weight was about 10 million. The molecular weight of 25 % of molecules was between 35 and 44 million daltons.

Axelrod: Was this DNA treated with RNase as well, and were the samples treated with protease prior to the nuclease digestions?

Hill: During the extraction, the DNA was treated first with RNase and then with pronase. The DNase treatment of the purified DNA eliminated its transforming activity.

Friis: Dr. *Hill,* I did not really understand whether the envelope properties of the induced virus were actually identical with virus that can be isolated from XC cells, for example, with Sendai virus.

Hill: We have tested the rescued virus with neutralizing antisera against subgroup A, B, and D. The reaction patterns were the same as those of the original virus rescued by fusion of XC cells and chick fibroblasts. Thus, the envelope properties of both viruses seem to be identical.

Expression and Integration of Viral DNA in Animal Cells

Bernhard Hirt and Chantal Widmer-Favre
Swiss Institute for Experimental Cancer Research, Lausanne, Schwitzerland

Summary: Evidence for physical integration of SV 40 DNA in transformed 3T3 cells is summarized.

Attempts were made to introduce the genes for thymidine kinase and inosinic pyrophosphorylase into deficient animal cells using polyoma pseudovirions. The results were negative.

Metabolic cooperation (intercellular exchange of nucleotides) could be observed for BHK cells, but was absent in L cells and in mouse mast-cells.

Transformation by DNA tumor viruses

DNA tumor viruses can transform cells. The transformed cells divide and the viral DNA replicates with them, but no virus growth is observed [1]. The transformed cells so far could not be „cured", i. e., at least part of the viral genome remains stably associated with the cell.

The different known classes of RNA tumor viruses are similar in structure. They all contain a 70S RNA. Among the different DNA tumor viruses, however, there is little similarity in structure: polyoma and SV40 contain circular DNA with a molecular weight of 3 million daltons. The DNA of adeno viruses is linear with a molecular weight of 20 million. Some herpes viruses seem also to be carcinogenic. They contain linear DNA molecules with a molecular weight of 100 million.

Cells transformed by DNA tumor viruses contain virus specific DNA and RNA and a virus specific nuclear antigen, the T-antigen. These characteristics are genetically stable. The manifestation of cell transformation seems to be similar for the different DNA viruses despite the variety of size and structure of the viral genomes. The question now is: How is the viral genome linked to the genome of the cell?

Manuscript received: 9 February 1972

For one type of transformed cells, SV-3T3, it is known that the viral DNA is covalently linked to the DNA of the cell. SV-3T3 are mouse cells transformed by the monkey virus SV40. Mouse cells are not permissive for SV40, i.e., they support neither viral growth nor replication of viral DNA. Therefore, virus cannot be induced to grow in SV-3T3 cells. Virus can, however, be rescued by cocultivation or fusion of SV-3T3 with permissive monkey cells. Because of the lack of permissiveness of 3T3 cells for SV40, it is very hard to consider lysogenic bacteria as a model system for SV-3T3 cells.

The reports from two groups have shown that viral DNA is covalently linked to the cellular DNA in SV-3T3 cells, and we shall summarize them here:

Sambrook et al. [7] made radioactive SV40 specific RNA in vitro using *E.-coli* polymerase. This RNA hybridizes to the viral DNA in transformed cells. By sedimentation through an alkaline sucrose gradient, the cellular DNA can be fractionated according to size. Sambrook and co-workers could show that the viral RNA hybridizes exclusively to DNA of molecular weight greater than 100 million. From this they concluded that the viral DNA is covalently linked to large cellular DNA. So far nothing is known about the mechanism or the site of attachment of the viral DNA.

Wall and *Darnell* [10] have studied the virus specific messenger RNA in SV-3T3 cells. They labeled the cells with radioactive uridine, extracted the RNA, and hybridized it to SV40 DNA. They chose hybridization conditions mild enough to leave the RNA intact, and they did not treat the hybrids with RNAase. The RNA which hybridized to SV40 DNA was recovered and hybridized again to SV40 DNA and to 3T3 DNA with the result that it hybridized to both. The conclusion is that one molecule of RNA contains both SV40 message and 3T3 message. It would originate therefore from a cellular DNA molecule containing covalently linked SV40 DNA.

There is one problem however: even in lytically infected cells, the viral messenger is found in pieces which are larger than a viral genome [9]. It would be a good control to do the *Wall* and *Darnell* experiment with this large messenger from lytic infection, which is supposed to be copied from viral DNA exclusively.

Alternatively, the data of *Wall* and *Darnell* could involve the joining of different messengers by an RNA ligase (*Hurwitz* [2]).

We can conclude that for the specific case of SV40 transformed 3T3 cells, the virus DNA is covalently linked to the celluar DNA. No data exist on all the other cells which are transformed by DNA viruses. Very little is known, therefore, about the physical integration of viral DNA in transformed cells. But we know, and this is important, that the viral genome is present in the transformed cells, that it cannot be removed, and that it is inherited by the daughter cells.

Transduction experiments with polyoma pseudovirions

During our work with polyoma virus, we found some virus preparations with a very high content of pseudovirions [4]. These are polyoma capsids containing mouse DNA instead of the viral genome. We tried to use these pseudovirions to transfer a genetic marker from one host cell to another cell deficient in this marker. Thymidine kinase (TK) and inosinic pyrophosphorylase (IPP) were used as markers. Cells deficient in TK cannot incorporate thymidine, and IPP deficient cells do not take up hypoxanthine. The deficient cells do not grow in a medium containing amethopterin, hypoxanthine, and thymidine, while wild-type cells grow normally.

We infected deficient cells with large amounts of pseudovirions harvested from primary mouse kidney cells. Two methods were used to look for a gene transfer:
1) autoradiography after labelling with tritiated thymidine or hypoxanthine;
2) colony formation in medium restrictive for deficient cells.

All the results were negative [11].

It is clear that the probability of having a specific cellular gene contained in a pseudovirion is very small. The experiment was performed anyway because it was hoped that, since the viral coat increases the infectivity of viral DNA approximately 1000-fold, this might also be true for the expression of a nonviral gene. We estimate that we could introduce one TK or IPP gene into approximately one cell out of 100. We observed no gene expression in the 10^7 cells scanned. Such a negative result could have been obtained for a number of reasons.

Metabolic cooperation

Although gene transfer with pseudovirions did not work, we obtained interesting results from the control experiments.

We mixed IPP deficient BHK cells (hamster) with wild-type cells and labeled them with tritiated hypoxanthine. The result is illustrated by Fig. 1: not only was the wild-type cell labeled, but all the cells which are in contact with it as well. This phenomenon has been described by *Subak-Sharpe* et al. [8] and was called metabolic cooperation. It is probably due to transfer of nucleotides [6]. Nucleotides normally cannot pass the cell membranes.

When the same experiment was done with L cells (mouse), no metabolic cooperation was observed (Fig. 2) [12]. Metabolic cooperation was also absent in TK deficient L cells, in a line of TK deficient BHK cells, and a line of mast-cells from mouse origin lacking IPP. The absence of metabolic cooperation in spite of close contact between the cells has not been observed before.

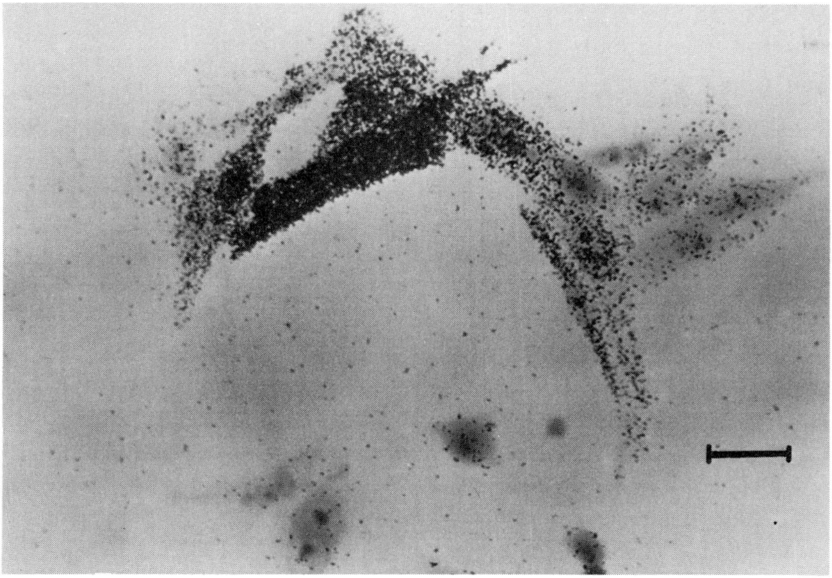

Fig. 1. Metabolic cooperation between BHK-IPP deficient cells and BHK cells. The deficient cells were mixed with BHK cells in a ratio 1 : 100. Cells were labeled with tritiated hypoxanthine for 12 hours. The autoradiography shows that deficient cells in contact with BHK cells incorporate radioactive precursor. Bar : 20 microns.

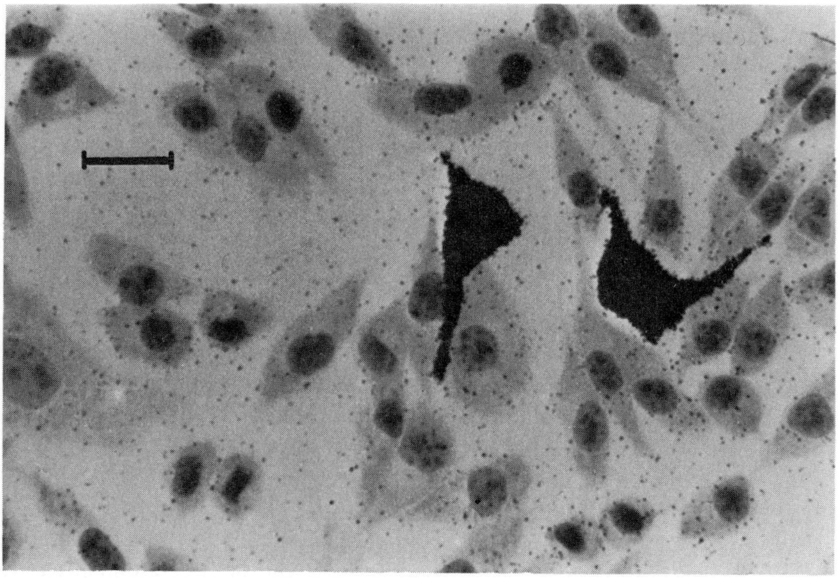

Fig. 2. Absence of metabolic cooperation between L cells deficient in IPP and normal L cells. Compare to Fig. 1. Bar : 20 microns.

Expression of viral genes in animal cells

Since we were not able to change the genome of a deficient cell with pseudovirions, we tried a different approach: Herpes viruses contain the gene for thymidine-kinase. Purified herpes DNA was sheared and put on TK deficient cells. We estimated that such cells would take up approximately 1000 TK genes. The expression of the gene could be followed by autoradiography after labeling with tritiated thymidine. In the best case, less than one cell out of a thousand was labeled. If the cells are infected with complete virus, most of them incorporate thymidine. The low frequency of gene expression is interesting in the context of the expression of bacterial genes in animal cells [3].

So far we have not been able to isolate a cell colony which would contain the viral TK gene. These results are preliminary and further work is needed.

We would like to mention here that cells transformed with U.V.-inactivated herpes contain the viral gene [5].

Acknowledgement: Supported by the "Fonds national Suisse de la Recherche scientifique."

References

[1] *Green, M.* 1970. Oncogenic viruses 765. Ann. Rev. Biochem. **39**: 701.

[2] *Hurwitz, J.* This volume, p. 117 ff.

[3] *Merril, C.R., Geier, M.R.,* and *Petricciani, J.C.* 1971. Bacterial virus gene expression in human cells. Nature **233**: 398-400.

[4] *Michel, M.R., Hirt, B.,* and *Weil, R.* 1967. Mouse cellular DNA enclosed in polyoma viral capsids (pseudovirions). Proc. Natl. Acad. Sci., U.S. **58**: 1381-88.

[5] *Munyon, W., Kraiselburd, E., Davis, D.,* and *Mann, J.* 1971. Transfer of thymidine kinase to thymidine kinaseless L cells by infection with ultraviolet-irradiated herpes simplex virus. J. Virology 7: 813-20.

[6] *Pitts, J. D.* 1971. Molecular exchange and growth control in tissue culture. *In* Growth control in cell cultures, eds. Wolstenhome and Knight, Ciba found. Symp., pp.89–98. Edinburgh and London: Churchill and Livingstone.

[7] *Sambrook, J., Westphal, H., Srinivasan, P.R.,* and *Dulbecco, R.* 1968. The integrated state of viral DNA in SV40 transformed cells. Proc. Natl. Acad. Sci., U.S. **60**: 1288-95.

[8] *Subak-Sharpe, H., Bürk, R.R.,* and *Pitts, J.D.* 1969. Metabolic cooperation between biochemically marked mammalian cells in tissue culture. J. Cell. Sci. **4**: 353-67.

[9] *Tonegawa, S., Walter, G., Bernardini, A.,* and *Dulbecco, R.* 1970. Transcription of the SV40 genome in transformed cells and during lytic infection. *In* Transcription of genetic material. Cold Spring Harbor Symp. on Quant. Biology Vol. **35**: 823-31.

[10] *Wall, R.*, and *Darnell, J.E.* 1971. Presence of cell and virus specific sequences in the same molecules of nuclear RNA from virus transformed cells. Nature N.B. **232**: 73-76.

[11] *Widmer-Favre, C.* 1971. Essais de transduction par les pseudovirions de polyome. Ph. D. thesis, University of Lausanne, Switzerland.

[12] *Widmer-Favre, C.* Absence of metabolic cooperation between certain types of mammalian cells in culture. J. Cell Sci. In press.

Discussion

Axelrod: Haven't *Martin, Gelb,* and *Cohn* suggested that normal green monkey cells contain as much as a half-genome equivalent of SV 40 DNA?

Hirt: Yes, by hybridization.

Axelrod: Would not that make it very difficult to interpret the data you have on covalent linkage? Would not that suggest that it might easily be explained by the oncogene hypothesis?

Hirt: No, this half-genome is in untransformed cells. And *Sambrook* was smart enough to use the proper control. He did not use untransformed cells as a control, but polyoma transformed cells.

Axelrod: But I was wondering about the SV 40 system. Would not that make it difficult to interpret the data you suggested on the size of the messenger being covalently linked.

Hirt: No, there again you have an effect that is definitely above background, and a proper control done with uninfected, untransformed cells.

Hilz: You mentioned the hybridization experiments of *Darnell.* How much could sequences of poly-A or other homologous sequences interfere with the specificity of such a binding? I think poly-A sequences could introduce an unspecificity.

Hirt: I think that is a good objection, but the proper control there would be to take the big RNA from lytically infected cells, and that was not done. Sofar, probably because of the difficulties of the analyses, both the studies of *Sambrook* et al. and *Wall* and *Darnell* were not repeated by other laboratories.

Doerfler: I should like to remind you, since I am the only representative at this meeting of the groups working on adenoviruses, that in hamster cells abortively infected with adenovirus type 12, we have shown that viral genomes are covalently linked to cellular DNA. However, covalent linkage between viral and host cell DNA has not yet been shown in hamster cells transformed by adenovirus type 12.

Hirt: May I correct you? You said "some of the viral genomes"; I would say "pieces" of viral genomes.

Doerfler: Yes, fragments. But still, viral DNA.

Hirt: Yes, I agree with that, and I think any DNA you throw on the cells will be partially integrated, but that is a guess. There are unpublished data from Weil's Laboratorie indicating that in lytic infections with polysoma virus, some of the DNA is integrated, but still we don't know if that has anything to do with the lytic infection.

Doerfler: I was talking only about cells abortively infected with adenovirus type 12.

Bodmer: Do the 3T3 cells show metabolic cooperation with respect to the TK minus marker?

Hirt: Unpublished work of Pitts indicates that they do. We were using BHK and L cells only.

Bodmer: So you were not using the 3T3 cells for your pseudovirion experiment.

Hirt: No, we did not use 3T3 because they are permissive, and our pseudovirion preparation contained infectious virus.

Yarmolinsky: Little pieces of DNA can get integrated also in lambda. Jacob and Wollmann isolated a petite mutant of lambda which has a small insertion, and quite recently a similar mutant was isolated in a very different way (about 6000 miles away and 16 years later) and shown to include a segment of DNA perhaps derived from another phage. The DNAs of these two mutants were hybridized with each other and shown to contain the same little alien piece. So we have analogies, even in "lambdology", to such peculiar events as the uptake of pieces of viral genomes, here by another virus.

Siniscalco: *Mary Weiss* [PNAS, 1970, **66**, 79–86] has published a series of papers where she shows that a somatic cell hybrid between a human SV 40-transformed line and the 3T3 mouse line deficient in TK had lost the transformed phenotype but expressed the T-antigen coded by the viral genome. Later on the T-antigen was also lost in most of the clones, and this loss was correlated with the progressive loss of the human chromosomes which is characteristic for this type of interspecific hybrid. Dr. *Weiss* thinks that her results not only support the hypothesis that the viral genome is associated with single chromosomes of the transformed cells, but also that the site of "integration" is multiple. I would like to know how far this kind of evidence is acceptable to specialists in viral oncology.

Hirt: This is completely acceptable. Dr. *Weiss* used human cells transformed with SV 40. As I mentioned before, SV-3T3 are the only cells where we know something about physical integration of the viral DNA, and even there we know nothing about the site of integration.

Duesberg: Isn't that in contradiction with what you mentioned yourself and with *Martin*'s [*Gelb, Kohne,* and *Martin,* 1971, J. Mol Biol. **57**, 129–45] experiment, that, at most, half an SV 40 genome is present in a transformed cell? If they fractionate chromosomes and find it in every chromosome, somebody must be wrong.

Hirt: The work of Gelb et al. was done on SV 40 transformed mouse and hamster cells. In four lines, they found one genome equilvalent of viral DNA, in one line, however, they found three genomes. It would, therefore, not be surprising if a transformed human line had more than one genome.

Duesberg: And I have one more sceptical question, you might say. Why is it that you always find integrated DNA which is easily detected? If DNA of tumor virus transformed rat cells is put on a chick cell, the only marker you seem to find expressed is the tumor virus genome. Yet rat DNA has a molecular, weight of 10^{12}, and viral DNA is at best 10^{7}. So it's only one 10^{-5} of the DNA which is integrated. Why is no other marker of the rat detectable in the chicken cell? Moreover, if you do these thymidine kinase transformation experiments, it always seems to work with an enzyme that is nonessential. The cell can live without that enzyme. How about a more critical marker?

Hirt: The advantage of the marker is simply that one can easily follow it, and that there exists a selective medium for it. There are not so many of these markers. Inosinic pyrophosphorylase is essential for humans. Lack of IPP causes an illness called Lesch-Nyhan syndrome. Children with this syndrome show psychomotor retardation and selfmutilatory behaviour.

Duesberg: But these cell hybrids exist. Can one use such hybrids? That is, more essential markers could theoretically be integrated, but they never are, or at least I have never heard of anthing like that.

Hirt: One can work on hybrid cells, it's true, but they are not stable. They tend to loose chromosome and markers. Dr. *Klein* will tell you about this.

Sinsheimer: Things are confused enough, but isn't there one other point that can add to the uncertainty, and that is that your standard here is always the DNA you extract from virions; and it is difficult to exclude the possibility that the virion DNA includes, even in closed rings, some host DNA.

Hirt: The viral DNA molecules could contain some host DNA, and this would explain the background in hybridization of untransformed cells. One thing which comes out of *Wall's* and *Darnell's* data is that the viral messenger would be homologous to highly repetitious cellular DNA and not to unique sequences. From hybridization data, one can conclude that there is no highly repetitious cellular DNA integrated into viral DNA.

Sinsheimer: There have been some recent experiments in *Vinograd's* laboratory using hetero-duplex hybridization showing that particles, particularly when grown under high multiplicity conditions, have regions of DNA in closed circles which are not homologous to those in other closed circles.

Interaction of Pseudovirions with Infectious Virions of Papova Viruses in Susceptible Host Cells

David Axelrod and David Trilling

Division of Laboratories and Research, New York State Department of Health, Albany, New York, USA

Summary: Examination of the population of virions of a productive infection with Simian virus 40 (SV40) was undertaken and methods were developed for the quantitation of the various kinds of DNA encapsidated. Since the potential for generalized transduction by SV40 pseudovirions had previously been reported by us, studies were undertaken to demonstrate phenotypic alterations produced by gene transfer from one host cell to another host cell of a different animal species. Polyoma virus was grown on SV40-transformed mouse cells, and pseudovirions obtained thereby were used to infect primary monkey tissues at high multiplicities. Immunofluorescent staining, carried out to identify the expression of the transfer of the SV40-T-gene, failed to establish the presence of transduced cells with the SV40-T-gene.

Previous experiments have demonstrated the appearance of three distinct populations of particles, following infection of susceptible cells with Simian virus 40 (SV40). The populations consist of mature plaque-forming particles containing circular DNA molecules, a defective particle with a somewhat smaller DNA [27], and a third particle, the pseudovirus, containing random linear host DNA fragments in a SV40 capsid [9,23]. Methods were devised for the ready quantitation of these three species of DNA in a population of SV40.

Although a number of workers have now demonstrated the ability of several of the papova viruses to encapsidate host DNA [9, 15, 17, 26] and to introduce the DNA into the nucleus of an infected cell [19, 7], no evidence has been presented for the expression or replication of this genetic material. The very elegant experiments recently reported by Merril et al. provide the first evidence for the transduction of human cells, albeit by a bacteriophage [14]. Our experiments were designed to provide a technique for the quantitation of the components of a population of papova particles and to demonstrate the transduction of functional genes from one host cell to another by pseudovirions. Several methods for the detection of functional genes were investigated. The method to be described depends on the immunofluorescent demonstration of the transfer of SV40-associated tumor antigens by polyoma pseudovirions from mouse cells to African green monkey cells.

Manuscript received: 11 December 1971

Materials and methods

Characterization and quantitation of viral DNA

African green monkey kidney (AGMK) cells were infected at a multiplicity of 100 with a small plaque SV40, plaque-purified in Vero cells, a continuous AGMK cell line. Cells were harvested at maximum cytopathic effect, disrupted by three cycles of freezing and thawing, and then treated with deionized water overnight at 5 °C [8] Virus was purified from the supernatants by zone sedimentation onto a CsCl cushion (1.40 g/cm^3) followed by three cycles of equilibrium density centrifugation at 35,000 rpm for 20 hours at 5 °C in 4.0 ml CsCl with a mean density of 1.33 g/cm^3 in the Spinco SW50.1 rotor. The two discrete lightscattering bands were collected separately and dialyzed against phosphate buffered saline, pH 7.6, and treated with DNase and RNase [23].

To label mature and defective virus particles isotopically [^{3}H] thymidine ([^{3}H] Tdr) was added 24 hours after infection of confluent stationary cells. To label pseudovirus, [^{3}H] Tdr was added to nonconfluent primary AGMK cells; the cells were allowed to reach confluence, and the labeling medium was removed. Monolayers were washed with medium containing 5×10^{-6} M nonlabeled Tdr and incubated with the Tdr-containing medium for an additional 20 hours before cells were infected. The medium was again changed 24 hours after infection.

The total DNA encapsidated by the virions was extracted by lysing the virus in 1 % sodium dodecyl sulfate (SDS) at 50 °C for 30 minutes, cooling to 25 °C, and extracting with phenol. The residual phenol was removed with ether, and the DNAs were passed through a G-100 Sephadex column equilibrated with 1.5 mM NaCl–0.15 mM sodium citrate (0.1 × SSC). For native DNA, an A_{260} of 0.20 was assumed to be equal to 10 μg/ml; for denatured DNA, an A_{260} of 0.24 to be equal to 10 μg/ml.

To prepare supercoiled, double-stranded circular DNA (DNA-I) of the mature virus, labeled with [^{14}C] thymidine, Vero cells were infected at a multiplicity of 0.5–1.0 in the presence of [^{14}C] thymidine [11], the progeny virus was lysed in 1 % SDS at 50 °C for 30 minutes, and the double-stranded, closed, circular DNA molecules were separated from linear and nicked molecules by equilibrium density centrifugation in CsCl containing ethidium bromide [20]. At pH 12.5 in a sucrose density gradient, the SV40 [^{14}C] DNA-I cosedimented with polyoma virus DNA-I, which has a 53S value under these conditions [25]. Mature, unlabeled viral DNA-I and defective, unlabeled viral DNA-I were extracted and purified from upper and lower band virus by the procedure outlined above.

Host DNA was prepared from isolated nuclei of Vero cells according to the method of *Berns* and *Thomas* [1] as modified by *McCarthy* and *Hoyer* [12] and *Martin* [10].

Rates of reassociation of DNA were measured as described by *Britten* and *Kohne* [3]. The salt concentration of the 400–500 nucleotide fragments was adjusted by dialysis to 0.12 M phosphate buffer (PB), pH 6.8. DNA was denatured by heating at 100 °C for 6 minutes, then was brought to 60 °C and incubated at this temperature for the duration of the experiment. Samples were removed periodically and passed over a column of hydroxyapatite (24 mm diameter by 10 mm height at 60 °C). Nonspecific binding to the column, determined from a sample of DNA removed at the beginning of the incubation, ranged between 0.9 and 1.9 %. Single-stranded DNA was removed by washing with 0.12 M PB with 0.4 % SDS at 60 °C, and the double-stranded DNA was then eluted with 0.4 M PB with 0.4 % SDS at 60 °C. The percentage of the total trichloroacetic acid (TCA) precipitable radioactivity recovered in the 0.4 M PB washes was taken as a measure of the amount of DNA reassociated. Since the variables of temperature, salt concentration, and DNA size were kept constant, the reassociation of the DNA was a function of DNA concentration and incubation time. The data are presented as the fraction of DNA reassociated as a function of time incubated or as a function of Cot (DNA conc. × time, expressed as moles of nucleotide per liter × seconds). A Cot of 1 mole sec/liter is equivalent to incubating 83 μg/ml of DNA for 1 hour [3].

Linear 5 to 20 (w/v) sucrose density gradients containing 0.7 M NaCl, 0.3 M NaOH, 0.01 M Tris-HCl (pH 7.5), and 0.001 M EDTA were used for sedimentation analysis of DNA extracted from virus particles. After the addition of marker [^{14}C] DNA-I, virus samples were lysed in 0.1 N NaOH–0.005 M EDTA for 10 minutes at 25 °C, cooled to 0 °C, layered on gradients, then centrifuged in the SW40 rotor for 4.75 hours at 40,000 rpm and 6 °C. Tubes were pierced at the bottom and fractions collected were precipitated with 5 % TCA.

Virus infectivity was assayed by plaquing on primary AGMK cells [22], and the infectivity of DNA-I from upper and lower band virus was assayed by plaquing in the presence of 300 μg/ml DEAE-dextran [13]. Concentration of purified virus was determined spectrophotometrically from the assumption that an A_{260} of 3.85 was equal to 1.0 mg of total virus per ml.

DNA-DNA hybridization on nitrocellulose filters was performed basically as described by *Denhardt* [5] but on a microscale as described previously [10]. For radioactivity measurements, samples were precipitated with TCA at a concentration of 5 % w/v, collected on membrane filters, dried, and counted in a scintillation counter.

Immunofluorescent identification of SV40-T in mouse cells

SE polyoma grown on 10-day-old mouse kidney cells was used to infect confluent monolayers of an SV40-transformed mouse cell line, SV3T3-clone 6, passage 8 (obtained from Dr. *G. Todaro* of the National Institutes of Health, Bethesda, Maryland). Because previous experiments had demonstrated that the percentage of pseudovirions was increased during a high multiplicity infection, a multiplicity of 100 was used. Infected cells were harvested at 5 days, and virus was extracted as described by *Crawford* [4]. Purification of the virus was carried out as previously outlined for SV40. The purified polyoma virus PV (3T3) was used to infect coverslip monolayers of primary AGMK cells in Leighton tubes. At 7 and 14 days, infected and control coverslips were fixed and stained for the SV40 specific T antigen. The indirect method with hamster ascitic fluid containing SV40-T antibody [18] and fluorescein-conjugated rabbit antihamster gamma globulin was used. To determine the presence of any infectious SV40 in the polyoma population, plaque assays of the polyoma virus were also carried out on primary AGMK cells.

Results

Characteristics of virus

After equilibrium density gradient centrifugation in CsCl, the small plaque SV40 was distributed into two discrete bands. The upper band virus had a density of 1.322 g/cm^3 and comprised on the average 58 % of the virus. The lower band virus, containing the remaining 42 % of the virus, had a density of 1.332 g/cm^3. The specific infectivity, plaque-forming units (PFU) per mg of total virus, of the lower band virus was 2–4 times greater than that of the upper band virus (Fig. 1).

Band	g/cm^3	Absorbance ratios 260/242	Absorbance ratios 260/250	Specific activities cpm/mg virus, Labeled, Postinfection 0.25 μCi/ml	Postinfection 1.65 μCi/ml	Preinfection 1.0 μCi/ml	Specific infectivities PFU/mg total virus	PFU/μg DNA-I	Virus (%) Pseudo-	Mature	Defective
Upper	1.322	1.09	1.40	165.000	848.000	446.000	1.44×10^{11}	1.5×10^5	38.2	28.1	33.7
Lower	1.332	1.08	1.37	150.000	836.000	219.000	3.43×10^{11}	3.3×10^5	17.3	82.7	–

Fig. 1. Characteristics of upper and lower band (AGMK-1°) SV40

When virus was labeled by the addition of [^{3}H] Tdr after infection, the virus in the two bands had similar specific activities (cpm per mg of virus). Alkaline sucrose density gradients demonstrated that both bands consisted primarily of fast-sedimenting DNA. Although the fast-sedimenting DNA of the lower band moved with the marker [^{14}C] DNA-I, the fast-sedimenting DNA of the upper band sedimented 10 % slower than the 53S DNA-I marker of mature SV40. The slower component, 48S, had a greater heterogeneity as shown by the broader distribution. The circular DNA molecules were extracted from the virus, purified in CsCl-ethidium bromide gradients, and assayed for infectivity in the presence of DEAE-dextran. The 53S DNA-I from the lower band had more than twice the specific infectivity of the 48S circular DNA from the upper band. This difference in specific infectivity can only be explained by the presence of noninfectious, defective DNA in the upper band.

In addition to the defective viral DNA, a second minor component sedimenting slower than 18S was extracted from the upper band virus. This fraction represented host DNA derived from pseudovirus [23]. The [^{14}C] DNA in the 18S region of the gradient was nicked, circular DNA (DNA-11) which was produced from the [^{14}C] DNA-I by repeated cycles of freezing and thawing.

When nonconfluent cells were labeled with [^{3}H] Tdr and infected with SV40 after reaching confluence, the specific activity of the upper band virus was twice that of the lower band virus. Only trace amounts of [^{3}H] DNA from either the upper or lower band virus sedimented in the 53S part of the gradient. The [^{3}H] DNA sedimented slower than 18S and was present in the upper viral band in twice the quantity of the lower band (Fig. 1). This component was the host DNA of the pseudovirus and has been previously shown to sediment at 15S [7,23].

Results obtained from DNA-DNA hybridization experiments confirmed the sedimentation analysis. When label was added to cells after infection, the [^{3}H] DNA extracted from the virus reacted primarily with the SV40 DNA. When the cells were labeled prior to infection, the [^{3}H] DNA extracted from virus reacted primarily with the host Vero cell DNA (Fig. 2).

	1	2	3	4	5	6	7	8
Expt.	Sample	Input (μg/ml) SV40 [^{14}C]DNA	Total DNA	Half-time reassoc. $t_{1/2}$ (hr)	Factor increase SV40 DNA $t_{1/2}$ / (SV40 DNA + sample $t_{1/2}$)	SV40 DNA μg/ml (2) X (5)	Host DNA μg/ml (3)–(6)	% Host DNA relative to input (7) : (1) X 100
A	–	0.20	0.20	2.62	1.00	0.20	–	–
B	0.78 non-labeled SV40 DNA	0.20	0.98	0.54	4.84	0.97	–	–
C	0.80 Vero DNA	0.20	1.00	2.62	1.00	0.20	0.80	100
D	1.80 Vero DNA	0.20	2.00	2.62	1.00	0.20	1.80	100
E	0.84 Upper band DNA	0.20	1.04	0.71	3.59	0.72	0.32	38.2
F	0.81 Lower band DNA	0.20	1.01	0.60	4.36	0.87	0.14	17.3

Fig. 2. Determination of SV40 and host DNA concentrations by reassociation rates

Reassociation rate measurements

Since this rate of reassociation of SV40 DNA follows second-order kinetics in a free solution reaction [7,3], the rate of reassociation of the DNA is dependent on its concentration. Thus, the amount of SV40 DNA in a heterogeneous population of DNAs can be determined [6] from its rate of reassociation relative to a standard concentration of SV40 DNA, if the different species of DNA molecules do not cross-react.

A curve for the reassociation of SV40 DNA at 0.20 μg/ml was determined. The results were reproducible and fix the equation for a second-order reaction C/Co = 1/(1 + K Cot) where C is the concentration of single-stranded DNA and Co is total concentration [3]. The $Cot_{1/2}$, the Cot (DNA conc. and time) required for one-half of the DNA to reassociate, was determined to be 6.3×10^{-3} mol. sec/liter. The rate of reassociation of the labeled DNA was then determined in the presence of added nonlabeled SV40. A 4.90-fold increase in the concentration of SV40 corresponded to a 4.84-fold increase in the role of reassociation or measured by $t_{1/2}$, the time required for one-half of the SV40 DNA to reassociate. This determination was made by plotting fraction reassociated on a linear scale as a function of time on a logarithmic scale. The $t_{1/2}$ of 0.20 μg/ml SV40 DNA was then divided by the $t_{1/2}$ of 0.98 μg/ml SV40 DNA to give the increased rate of reassociation (Fig. 3). In the presence of excess host Vero cell DNA, there was no change in the rate of reannealing of SV40 [^{14}C] DNA because there was no reaction between the host repeated sequences and SV40 DNA (Figs. 3 and 4A).

	DNA in solution	Amounts incubated Virus or DNA		Specifically bound (% of input)	
		μg	cpm	SV40 filter	Vero filter
1	Combined bands labeled postinfection	2.3	3.638	24.1	2.2
2	Upper band labeled preinfection	5.4	2.428	3.1	12.7
3	Lower band labeled preinfection	7.0	1.533	3.1	12.3
4	Combined bands labeled preinfection	5.7	1.894	1.5	14.5
5	Vero [^{14}C]DNA	1.1	24.724	<0.1	N.P.
6	Vero [^{14}C]DNA	1.1	24.724	N.P.	17.1
7	SV40 [^{14}C]DNA	0.2	5.242	48.9	N.P.
8	SV40 [^{14}C]DNA	0.2	5.242	N.P.	0.3

N.P. indicates filters not present in reaction mixture; otherwise, both filters present in same reaction.

Fig. 3. Hybridization analysis of DNAs from virus labeled postinfection and from virus replicated in prelabeled cells

This kind of analysis was applied to determine the concentration of SV40 DNA in the upper and lower band virus. By subtraction from the total DNA, the concentrations of host DNA were determined. The upper band contained 38.2 % host DNA, and the lower band, 17.3 %. This twofold difference in host DNA content of the upper and lower bands is consistent with the twofold difference in specific activity between the upper and lower band virus obtained from cells labeled prior to infection.

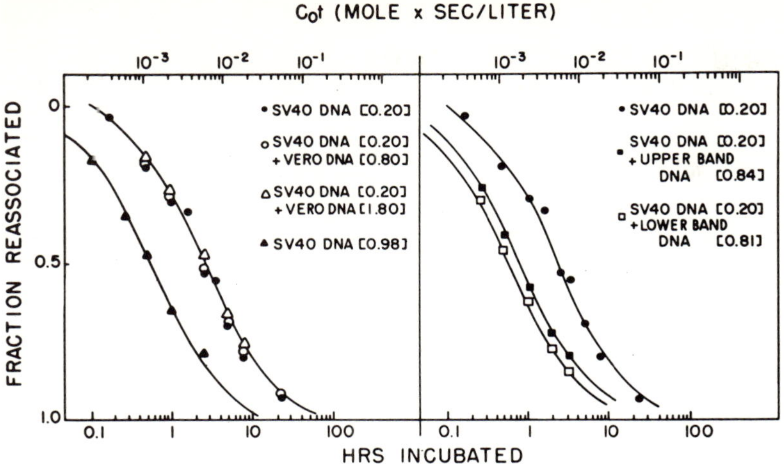

Fig. 4. Reassociation of [^{14}C] SV40 DNA at differing concentrations and in the presence of DNAs from the host cell and from the upper and lower band virus.

(A) Measurements of the reassociation of SV40 DNA at 0.20 μg/ml, at 0.98 μg/ml, and at 0.20 μg/ml with 0.80 μg/ml Vero DNA or 1.80 μg/ml Vero DNA.

(B) Measurements of reassociation or SV40 DNA at 0.20 μg/ml and at 0.20 μg/ml with either 0.84 μg/ml DNA from upper band virus or 0.81 μg/ml DNA from lower band virus.

All results were plotted by fraction of DNA reassociated on a linear scale as a function of time (hr) incubated on a logarithmic scale. The results for the reassociation of SV40 DNA at 0.20 μg/ml were also plotted by fraction reassociated as a function of Cot (DNA conc. x time) on a logarithmic scale corresponding to the time scale. Concentrations of DNA in μg/ml are in parentheses.

Hyperchromicity of total DNA

Since SV40 DNA-I is a double-stranded, closed, circular molecule, its strands fail to separate until a thermal scission is introduced [24]. The host component, however, is a double-stranded linear molecule, and its strands separate at relatively low temperatures (depending on counter ion concentrations) with an accompanying increase in chromicity. As illustrated in Fig. 5, authentic viral DNA yielded only 5.5 % of its maximum hyperchromicity by 81 °C, whereas the total DNA from the upper band yielded 36 % of its maximum hyperchromicity at this temperature. If the total DNA was first sheared to 400,000 daltons by sonication, it yielded 95 % of its maximum hyperchromicity by 81 °C with a T_m of 70 °C. When the samples were held at 100 °C for 8 minutes, the authentic DNA-I yielded 46 % of its maximum hyperchromicity, and the total DNA of the upper band virus yielded 57 %. At 100 °C for 48 minutes, maximum hyperchromic effect was achieved in all samples. To calculate the amount

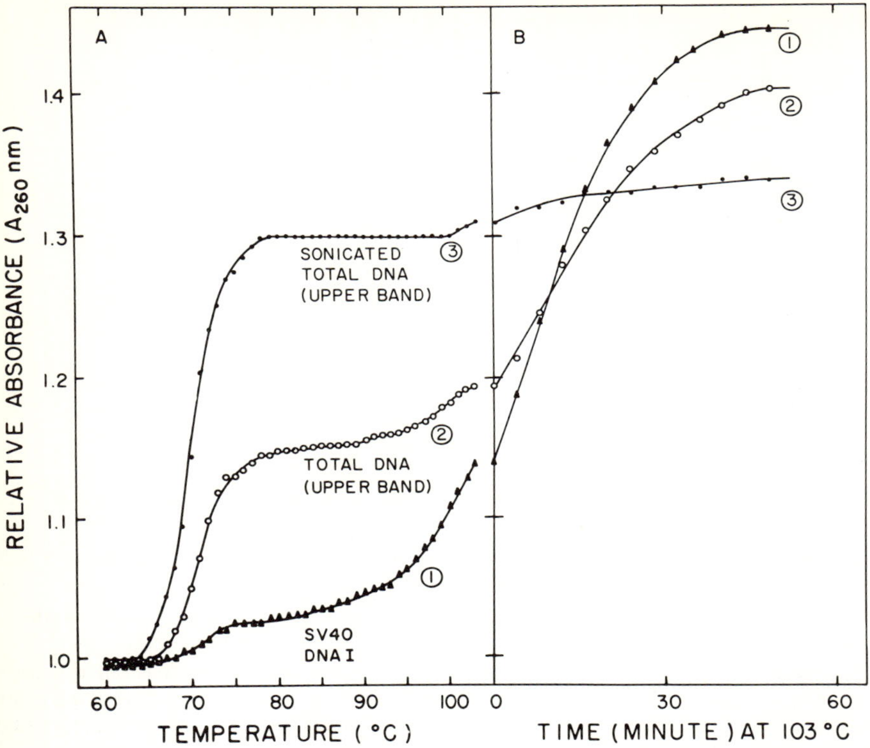

Fig. 5. Thermal transition curves. Thermal transitions of SV40 DNA-I (25.4 μg/ml), total DNA from upper band virus (21.9 μg/ml), and sonicated total DNA from upper band virus (21.9 μg/ml) in 0.1 x SSC. The rate of temperature increase was 1 °C per minute up to 100 °C; then the temperature was held at approximately 103 °C. The host cell DNA curve superimposed the sonicated total DNA curve and has been omitted.

of host DNA, the data obtained for thermal denaturation of nonsheared monkey DNA were used to convert the hyperchromicity data of the total DNA isolated from upper band virus into its linear host DNA component. At 85 °C in 0.1 X SSC, monkey DNA yielded a hyperchromic effect equal to 38 % of its initial chromicity or an increase of 0.076 A_{260} per 10 μg/ml. Thus, the hyperchromic effect of 0.0675 A_{260} at 85 °C in the total DNA sample from the upper band virus was equivalent to 8.9 μg/ml which was 40.5 % of the DNA (21.9 μg/ml) in the thermal transition experiment. This figure for percent host DNA in the upper band virus agreed with the data independently derived from the association analysis (38.2 %) (Fig. 5).

The PV (SV3T3) obtained had a mean density of 1.31 %, somewhat lower than that of the original SE-PV used to infect the SV3T3 cells. The hemagglutinin (HA) titer of the PV (SV3T3) was 1.3 × 10^4/ml. In the first experiment, a single cell with characteristic nuclear flecks was observed with immunofluorescence. Assays for the presence of infectious SV40 failed to yield any evidence for SV40 virions. In a second experiment, carried out in an identical manner, the primary AGMK cells on coverslips were infected at a particle-cell multiplicity of 2 × 10^6 particles/cell. No specific fluorescence was observed in this second experiment.

Discussion

The quantitative relationship of the component particles of SV40 progeny virions has not previously been determined because of limitations in the methods employed. Hybridization is insufficient in itself because of the rapid reassociation of the viral DNA in solution and the inability of unique host sequences to hybridize [7]. As with viral DNA, the repeated host sequences may also reassociate in solution and an incomplete reaction result.

The methods outlined provide a convenient procedure for the accurate determination of the quantities of the three components. The percent pseudovirus in each band was determined from the parent host DNA in each preparation based on the reassociation analysis (Fig. 2). The remainder in each band, 61.8 % of the upper band and 82.7 % of the lower band, was virus with viral DNA. The data on the infectivity of the DNA-I molecules from upper and lower bands could be used to determine the fraction of defective SV40 virus in the upper band. This calculation was based on the assumption that all DNA-I molecules in the lower band were derived from mature virus. Since the DNA-I molecules from the upper band had 45.5 % the specific infectivity of the DNA-I molecules from the lower band, 28.1 % of the upper band virus was mature SV40 (45.4 % of 61.8 %). The noninfectious fraction of the DNA-I molecules, 54.6 %, was equivalent to the defective virus, which thus represents 33.7 % of the total virus in the upper band (54.6 % of 61.8 %). From these data, the ratio of the proportion of mature virus in the upper band (28.1 %) to mature virus (82.7 %) in the lower band was calculated to be 1 : 2.9. This figure agreed well with the ratio of specific infectivity of total virus (PFU/μg total virus) as measured in the upper and lower bands (1 : 2.4).

The data from the thermal denaturation experiments indicated that 40.5 % of the DNA extracted from the upper band virus was in the linear configuration of host DNA as opposed to the circular configuration of viral DNA. This determination for host DNA agreed with the estimate for host DNA derived independently by the analysis of reassociation rates of the DNA molecules extracted from the upper band virus (38.2 %). The slightly higher value obtained from the thermal denaturation data was probably the result of the hyperchromic effect at 85 °C of a small percentage of nicked, circular DNA molecules (DNA-II) in the preparation. These broken circles also accounted for the initial increase in chromicity of the authentic DNA-I preparation.

The results obtained with the reassociation rate analysis were in good agreement with those obtained with the thermal transition experiment.

Based on our previous determinations of the relationship between HA units and plaque-forming units at different multiplicities, 1 HA unit was equivalent to approximately 10^7 particles. Since each of the coverslips contained 10^5 cells, the number of particles per cell was of the order of 5×10^5. If the pseudovirions represented 10 % of the population, a very conservative estimate, the number of pseudovirions per cell was 5×10^4. With the assumption that the encapsidated DNA is a random representation of the host DNA [7] and has a molecular weight of $2\text{–}3 \times 10^6$ daltons, the quantity of mouse DNA per green monkey cell represented approximately 10 % of the mouse genome or 1×10^{11} daltons. If the gene responsible for the expression of the SV40-T antigen has a molecular weight of 5×10^5 daltons, one would have anticipated that 10 % of the cells would have received the genetic information capable of coding for the antigen. The various calculations, however, of the quantity of pseudovirus DNA associated with the nucleus of an infected cell [7, 19] suggest that only about 10 % of the pseudovirus DNA can be accounted for. This would decrease the number of cells which received the SV40-T gene to only 1 %, but even this would nevertheless represent a significant number of the cells on the coverslip. Our ability to find SV40-T antigen fluorescence in only a single cell makes us very cautious in assessing the value of these pseudovirions as vehicles for gene therapy.

Conclusion

Techniques have been described by which the encapsidated DNA in a population of virions may be characterized and readily quantitated. Because populations of papova viruses have variable quantities of mature, infectious virions, defective virions, and pseudovirions, the interpretation of results obtained with these viruses may be difficult. Despite our awareness of the formidable difficulties in demonstrating phenotypic alterations produced by pseudoviruses in infected host cells, we have attempted several methods of detecting evidence for virus-mediated transduction. To date these efforts have been uniformly unsuccessful.

While our statistical analysis of the likelihood of the interaction between DNA encoding a specific gene and an infected host cell is replete with unwarranted assumptions, we feel that our studies provide additional evidence for a sober view of the use of pseudovirions for genetic engineering. As we have previously pointed out [7], the similarities between the general transducing phages and the pseudovirions are intriguing, but the importance of these similarities remains to be demonstrated by the replication and expression of the transduced genetic information. Our failures will most certainly not dampen the enthusiasm for the use of pseudovirions to insert functional genes into mammalian cells. However, we would suggest that other mechanisms of gene transfer in mammalian cells may be more profitably studied [2, 14, 16, 21]. Much remains to be learned concerning the initiation, the integration, and a myriad of other problems involving the interaction between a randomly selected segment of foreign DNA and the nuclear machinery of a eukaryotic cell. The apparent spectacular success of *Merril* and coworkers [14], as well as that of *Munyon* et al. [16], may be a result of the fact that the genes transmitted had well defined points of initiation.

References

[1] *Berns, K. I., and Thomas, C. A., Jr.* 1965. Isolation of high molecular weight DNA from Hemophilus influenzae. J. Mol. Biol. **11:** 476–90.

[2] *Borenfreund, E., Honda, Y., Steinglass, M., and Bendich, A.* 1970. Studies of DNA-induced heritable alteration of mammalian cells. J. Exp. Med. **132:** 1071–89.

[3] *Britten, R. J., and Kohne, D. E.* 1968. Repeated sequences in DNA. Science **161:** 529–40.

[4] *Crawford, L. V.* 1969. Purification of polyoma virus. In: *Habel and Salzman* (ed.): Fundamental techniques in virology. Academic Press, New York.

[5] *Denhardt, D. T.* 1966. A membrane-filter technique for the detection of complementary DNA. Biochem. Biophys. Res. Commun. **23:** 641–46.

[6] *Gelb, L., Kohne, D. E., and Martin, M. A.* 1971. Quantitation of simian virus 40 sequences in African green monkey, mouse and virus-transformed cell genomes. J. Mol. Biol. **57:** 129–45.

[7] *Grady, L., Axelrod, D., and Trilling, D.* 1970. The SV40 pseudovirus: its potential for general transduction in animal cells. Proc. Nat. Acad. Sci. U.S. **67:** 1886–93.

[8] *Koch, M. A., Eggers, H. J., Anderer, F. A., Schlumberger, H. D., and Frank, H.* 1967. Structure of simian virus 40. I. Purification and physical characterization of the virus particle. Virology **32:** 503–10.

[9] *Levine, A. J., and Teresky, A. K.* 1970. Deoxyribonucleic acid replication in simian virus 40-infected cells. II. Detection and characterization of simian virus 40 pseudovirions. J. Virol. **5:** 451–57.

[10] *Martin, M. A.* 1969. Characteristics of the syrian hamster ribonucleic acid present in cells transformed by polyoma, simian virus 40, or adenovirus 12. J. Virol. **3:** 119–25.

[11] *Martin, M. A., and Axelrod, D.* 1969. SV40 gene activity during lytic infection and in a series of SV40 transformed mouse cells. Proc. Nat. Acad. Sci., U.S. **64:** 1203–10.

[12] *McCarthy, B. J., and Hoyer, B. H.* 1964. Identity of DNA and diversity of messenger RNA molecules in normal mouse tissues. Proc. Nat. Acad. Sci., U.S. **52:** 915–22.

[13] *McCutchan, J. H., and Pagano, J. S.* 1968. Enhancement of the infectivity of simian virus 40 deoxyribonucleic acid with diethylaminoethyl-dextran. J. Nat. Cancer Inst. **41:** 351–57.

[14] *Merril, C. R., Geier, M. R., and Petricciani, J. C.* 1971. Bacterial virus gene expression in human cells. Nature **233:** 398–400.

[15] *Michel, M. R., Hirt, B., and Weil, R.* 1967. Mouse cellular DNA enclosed in polyoma viral capsids (Pseudovirions). Proc. Nat. Acad. Sci., U.S. **58:** 1381–88.

[16] *Munyon, W., Kraiselbrud, E., Davis, D., and Mann, J.* 1971. Transfer of thymidine kinase to thymidine kinaseless L cells by infection with ultraviolet-irradiated Herpes simplex virus. J. Virol. **7:** 813–20.

[17] *Osterman, J. V., Waddell, A., and Aposhian, H. V.* 1970. DNA and gene therapy: uncoating of polyoma pseudovirus in mouse embryo cells. Proc. Nat. Acad. Sci., U.S. **67:** 37–40.

[18] *Pope, J. H., and Rowe, W. P.* 1964. Detection of specific antigen in SV40-transformed cells by immunofluorescence. J. Exp. Med. **120:** 121–28.

[19] *Qasba, P. K., and Aposhian, H. V.* 1971. DNA and gene therapy: transfer of mouse DNA to human and mouse embryonic cells by polyoma pseudovirions. Proc. Nat. Acad. Sci., U.S. **68:** 2345–49.

[20] *Radloff, R., Bauer, W., and Vinograd, J.* 1967. A dye-buoyant-density method for the detection and isolation of closed circular duplex DNA: The closed circular DNA in HeLa cells. Proc. Nat. Acad. Sci., U.S. **57:** 1514–21.

[21] *Schwartz, A. G., Cook, P. R., and Harris, H.* 1971. Correction of a genetic defect in a mammalian cell. Nature New Biology **230:** 5–8.

[22] *Takemoto, K. K., Kirschstein, R. L., and Habel, K.* 1966. Mutants of simian virus 40 differing in plaque size, oncogenicity and heat sensitivity. J. Bact., **92:** 990–94.

[23] *Trilling, D. M., and Axelrod, D.* 1970. Encapsidation of free host DNA by simian virus 40: A simian virus 40 pseudovirus. Science **168:** 268–71.

[24] *Vinograd, J., Lebowitz, J., and Watson, R.* 1968. Early and late helix-coil transitions in closed circular DNA: the number of superhelical turns in polyoma DNA. J. Mol. Biol. **33:** 173–97.

[25] *Weil, R., and Vinograd, J.* 1963. The cyclic helix and cyclic coil forms of polyoma viral DNA. Proc. Nat. Acad. Sci., U.S. **50:** 730–38.

[26] *Winocour, E.* 1968. Further studies on the incorporation of cell DNA into polyoma-related particles. Virology **34:** 571–82.

[27] *Yoshiike, K.* 1968. Studies on DNA from low-density particles of SV40. I. Heterogeneous defective virions produced by successive undiluted passages. II. Noninfectious virions associated with a large-plaque variant. Virology **34:** 391–409.

Discussion

Sinsheimer: Why are there intact virions in the upper band at all? Or, is that just overlap?

Axelrod: Yes, it's overlap. If you do Poisson distributions on the two populations, you will find that there is considerable overlap. The slide of the gradient indicates how close the bands are.

Bodmer: I was interested in your data on the amino-acylating enzymes. I wondered whether you have tried this in hybrids. It is a less exciting situation, but one where we do know we can get some genetic crosses. It seems to me you should be able to verify that you can use this as a marker in interspecific hybrids.

Axelrod: Exactly. Possibly these may be among the most useful markers there are. Since we have obtained evidence for considerably more t-RNA genes than has Attardi, we are able to look for differences in the entire pool of approximately 700 t-RNA genes. In most other instances, you are faced with the identification of a single phenotypic marker, a single enzyme or single gene. Here you have a whole group. Hopefully, the t-RNA genes would permit the identification of marked differences for different species.

Lysolecithin-induced Fusion of Rabbit Spermatozoa with Hamster Somatic Cells

C. M. Croce, B. L. Gledhill*, B. Gabara, W. Sawicki, and H. Koprowski

The Wistar Institute of Anatomy and Biology, Philadelphia, Pennsylvania 19104, USA

*) School of Veterinary Medicine, New Bolton Center, University of Pennsylvania, Kennett Square, Pennsylvania 19348, USA

Summary: Rabbit spermatozoa have been successfully fused with hamster somatic cells by the use of lysolecithin. The primary effect of lysolecithin on spermatozoa seems to be related to an alteration of cell membranes. Many of the spermatozoal nuclei were modified in appearance, and observations of the ultrastructure show that a small percentage proceed to a state of partial disaggregation.

Incorporation of tritiated thymidine administered in vitro indicated the initiation of DNA synthesis by disaggregated spermatozoal chromatin in the cytoplasm of cultured heterokaryocytes.

Introduction

The efficiency of fusion between somatic cells can be enhanced by Sendai virus inactivated by ultraviolet irradiation or β-propiolactone [7, 13]. Viable homokaryocytes, heterokaryocytes, and hybrid cells also can be obtained after fusion of somatic cells by lysolecithin, a phospholipid which seems to act on the cell membrane, perhaps changing its structure from a bimolecular leaflet to a globular micellar arrangement [2, 9]. When spermatozoa were fused with somatic cells by the use of inactivated Sendai virus, their morphology did not appear to be modified, and they persisted in the cytoplasm of the somatic cells [12]. However, after fusion of trypsinized bovine spermatozoa with mitotic HeLa cells using UV-inactivated Sendai virus, *Johnson* et al. obtained a frequency of 0.01 % of premature chromosome condensation in sperm nuclei in the heterokaryocytes [8].

Manuscript received: 15 December 1971

In view of this, we have used lysolecithin to fuse rabbit spermatozoa with somatic cells. Our purpose was to modify the sperm membranous envelope with the hope of allowing activation of the sperm nuclei and subsequent induction of DNA synthesis. Ultrastructural studies also were made to determine the modifications of spermatozoal membranes after pretreatment with lysolecithin to visualize the sequences of the fusion process, and to follow the structural modifications of the sperm nuclei in the heterokaryocytes.

Materials and methods

Somatic cells

Hamster simian virus 40 (SV40)-transformed cells (F5–1 cells) [3] were trypsinized (0.25 % trypsin in phosphate-buffered saline (PBS)), centrifuged at 180 g for 5 minutes at 4 °C, and resuspended in Hanks' solution. The F5–1 cells were checked for mycoplasma contamination by several methods (colony-formation, autoradiography, and electron microscopy [14]), and they were at all times found uninfected.

Spermatozoa

Ejaculated spermatozoa were obtained from New Zealand bucks by an artificial vagina. Spermatozoa were also obtained from the cauda epididymides of the same or similar rabbits immediately after killing [4]. The spermatozoa were washed three times by centrifugation at 45 g with calcium-free Krebs-Ringer solution. The spermatozoa were resuspended in Hanks' solution, counted in a hemocytometric chamber, divided into aliquots, and recentrifuged at 180 g at room temperature to obtain pellets containing 5×10^7 spermatozoa.

To obtain labeled spermatozoa, an adult New Zealand buck was given intravenous injections[1]) daily for four consecutive days with ^{3}H-methyl thymidine (0.25 mCi/kg body weight). Ejaculated spermatozoa were obtained from day 36 to day 52 following the first injection of the isotope. After precipitation with trichloroacetic acid, the radioactivity of the spermatozoa was evaluated by counting in a Beckman LS 100 scintillation counter and by autoradiography.

Pretreatment of spermatozoa with lysolecithin

Three to six mg of lysolecithin (LL) (Supelco, Bellefonte, Pa.) were dissolved in absolute ethanol (0.05 ml). To this solution 5 ml of PBS, containing 5 mg/ml of albumin (Pentex, fraction V), were added to make a final concentration of LL of 600–1,200 μg/ml. LL (0.1 ml) at a concentration of 1,200 μg/ml was added to the sperm pellets containing 5×10^7 spermatozoa. After 0.5–30 minutes, the action

---

[1]) We are grateful to Dr. *R. P. Amann* of Pennsylvania State University for his cooperation in this aspect of the study.

of LL was neutralized with 0.2–0.4 ml of inactivated (56 °C for 30 minutes) fetal bovine serum. The tubes containing the pretreated spermatozoa were immediately centrifuged at room temperature (25 °C) at 180 g and the pellets were resuspended in Hanks' solution. For comparison, samples of 5 X 10^7 spermatozoa were treated in 1 % sodium dodecyl sulfate (SDS) in 0.05 M sodium borate (pH 9.0) in the presence or absence of 10^{-2} M dithiothreitol (DTT) for 15 minutes. The samples were fixed for electron microscopy as described below.

Fusion procedure

Pretreated spermatozoa (5 X 10^7) were mixed with 5 X 10^6 somatic cells in Hanks' solution. The mixtures were spun down at room temperature at 180 g and the pellets containing spermatozoa and F5–1 somatic cells were treated for 0.5–30 minutes with 0.1 ml of LL at a concentration of 600 µg/ml in PBS with albumin. At appropriate intervals (Fig. 1), the action of LL was neutralized by the addition of inactivated fetal bovine serum. The samples were again centrifuged at 180 g at room temperature. The supernatants were then discarded, and 1 ml of Eagle's basal medium (BME) was added to each tube without disturbing the pellets. The tubes were incubated at 37 °C for 15–20 minutes, and the pellets were then resuspended. The cells were seeded over glass coverslips in 60-mm Petri dishes containing 5 ml of BME, with 2 mM glutamine, 100 IU/ml penicillin, and 50 µg/ml streptomycin added.

The cells on the coverslips were fixed in a mixture of ice-cold ethanol:glacial acetic acid (3 : 1) for light microscopy at 18, 24, 48, 72, and 96 hours after fusion. In some experiments, 0.25 µCi/ml ^{3}H-methyl thymidine (^{3}H-TdR; specific activity, 12.5 Ci/mM, New England Nuclear Corp., Boston, Mass.) was added immediately after the cells were seeded in the Petri dishes. These cells were fixed 4.5 or 16–20 hours after fusion, and were processed for stripping-film autoradiography in the conventional manner.

Feulgen reaction

Some preparations of heterokaryocytes were stained by the Feulgen method as described by *Gledhill* et al. [5].

Acridine orange staining

The binding of acridine orange (AO) to the nuclear chromatin was evaluated by microfluorimetry, measuring the green (double-stranded) and red (single-stranded) AO fluorescence at 530 and 590 nm, respectively, as described by *Ringertz* et al. [10]. Fluorescence measurements were made on AO-stained heterokaryocyte nuclei that had been previously heated to 22, 85, and 100 °C in a 0.15 M NaCl, 0.015 M Na citrate (SSC) solution containing 4 % (w/v) formaldehyde (pH 7.0) for 20 minutes. After heating, the cells attached to the slides were rapidly cooled in ice-cold SSC and dehydrated in a series of ethanol baths, the concentrations of which were progressively increased until absolute ethanol was reached. The fluorescence values ($F_{590}/F_{530} = \alpha$) of sperm heads, F5–1 nuclei, and disaggregated sperm nuclei in heterokaryocytes were compared [10] for an estimate of relative resistance to heat denaturation.

Autoradiographic techniques

Preparations of smeared, in vivo-labeled spermatozoa, of heterokaryocytes resulting from use of in vivo-labeled spermatozoa, and of heterokaryocytes exposed to tritium during culture, were fixed in ice-cold glacial acetic acid:ethanol (1 : 3) for 15 minutes and washed in 70 % ethanol for 15 minutes at 0 °C. These preparations were treated with 5 % trichloroacetic acid for 15 minutes at 0 °C. The coverslips were dried, mounted cell-side up in Permount on gelatinized glass slides, dried again, and covered with Kodak AR stripping film [11] for 4–52 days. After routine autoradiographic development and fixing, the slides were stained with Feulgen light green SF.

Electron microscopic techniques

The spermatozoa, the F5–1 somatic cells, and the F5–1 somatic cells fused with spermatozoa were fixed in a cold (4 °C) solution of 3 % glutaraldehyde in cacodylate buffer for 2 hours, followed by postosmication in 2 % OsO_4 in the same buffer for 4 hours. After dehydration in a series of alcohol steps, the samples were embedded in Epon 812. Thin sections were stained with uranyl acetate followed by lead citrate, and examined in a Hitachi HS-8 microscope at 50 kV.

Quantitation procedures

Mono- and multinucleated F5–1 cells were counted, and the percentage of F5–1 multinucleated homokaryocytes was calculated after counting at least 1,000 cells from at least 4 coverslips. Likewise, to evaluate the percentage of heterokaryocytes, a minimum of 1,000 cells from at least 4 coverslips was counted. Nuclei in heterokaryocytes were recognized on the basis of their morphology, isotope labeling, or cytochemical reactions.

Results

After treatment of 5×10^7 spermatozoa with 0.1 ml of LL (1,200 μg/ml) for 1–3 minutes, more than 95 % of the spermatozoa became immotile.

The spontaneous fusion efficiency between rabbit spermatozoa and F5–1 somatic cells was 3 %, as detected 16–20 hours after the untreated cells were brought together. After pretreatment of spermatozoa and fusion with LL, 10–30 % of the F5–1 somatic cells contained, in addition to one or several somatic cell nuclei, 1 to 6 structures in their cytoplasm (Figs. 1 and 2). These structures were either sperm heads or ovoid to irregularly-shaped clumps of Feulgen-positive material now assumed to be chromatin. The LL treatment also induced fusion between F5–1 cells with the production of 20–30 % of F5–1 multinucleated homokaryocytes.

Experiment No.	Time of pretreatment of spermatozoa with lyso-lecithin (minutes)	Time of exposure to lysolecithin in the fusion pellet (minutes)	% of hetero-karyocytes**)
1	0.5	1	9.4
2	1	1	16.5
3	3	0.5	20.8
4	3	1	28.8
5	3	3	26.1
6	3	30	25.7
7	30	0.5	15.1
8	30	1	15.7
9	30	3	16.7
10	30	30	19.0

Fig. 1. Fusion efficiency between rabbit spermatozoa and F5–1 somatic cells*)

*) 18 hours after fusion.

**) Percentage of F5–1 somatic cells containing 1 or more sperm heads.

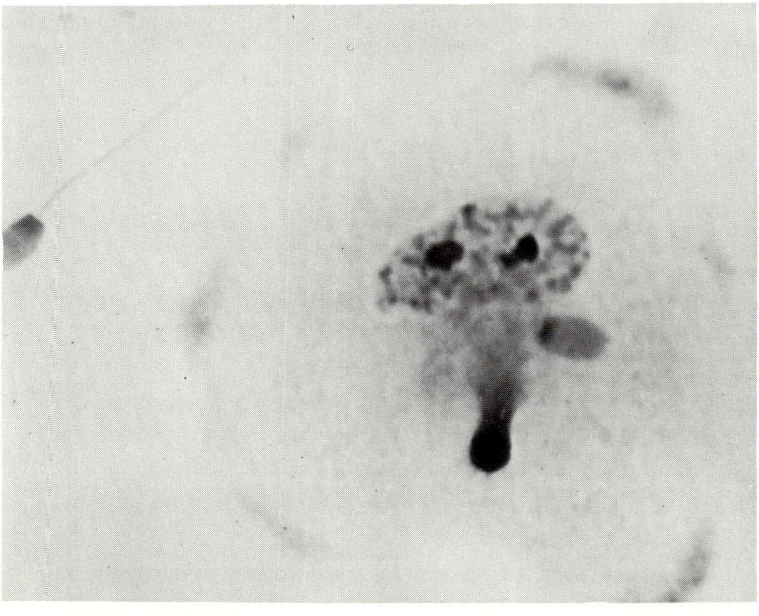

Fig. 2. Photomicrograph of an F5–1 somatic cell-rabbit spermatozoa heterokaryocyte. The preparation was fixed 16 hours after fusion and stained by the Feulgen method. Beginning of disaggregation of chromatin is seen in one spermatozoon. x 1.000 (from [6] with permission).

To remove spermatozoa which were adhering to the surface of the somatic cells, the procedure described by *Sawicki* and *Koprowski* [12] was followed. Mild treatment for 1 minute with 0.25 % of trypsin in PBS at room temperature removed most, if not all, of these spermatozoa. The fused spermatozoa mostly lacked flagella, and some of them caused deformation of the somatic nuclei.

In addition, the spermatozoa and the chromatin clumps were studied by phase contrast microscopy. The chromatin clumps were not removed by trypsinization. No nucleolus was seen in these clumps.

After treatment at 22 °C and staining with AO, the chromatin clumps fluoresced green in the same way as the somatic cell nuclei and the unfused sperm heads. After heating at 85 °C and staining with AO, they fluoresced predominantly green as did the spermatozoa. At this temperature the somatic cell nuclei fluoresced a red green. After melting at 100 °C, the clumps fluoresced red green. This experiment showed a similarity in resistance to heat denaturation of the DNA of these clumps of chromatin to that found in spermatozoa, and a dissimilarity to that found in F5–1 nuclei. It seems valid, therefore, to call these clumps of chromatin "sperm bodies." The alpha values, i.e., F_{590}/F_{530}, as evaluated by microfluorimetry for rabbit spermatozoa, sperm bodies, and F5–1 somatic cell nuclei are presented in Fig. 3.

Temperature (°C)	Spermatozoa	Sperm Bodies	F5–1 Nuclei
22	0.18	0.16	0.11
85	0.19	0.21	0.32
100	0.28	0.28	0.31

Fig. 3. Binding of acridine orange to spermatozoa, sperm bodies, and F5–1 nuclei 16 hours after fusion: α values*) [6].

*) $\alpha = F_{590}/F_{530}$ and is independent of the amount of DNA present. (n = 10).

Furthermore, when rabbit spermatozoa labeled in vivo with tritiated thymidine (Fig. 4) were fused with F5–1 somatic cells, the silver grains in autoradiograms could be recognized only over spermatozoal heads and sperm bodies.

This series of results clearly indicates the spermatozoal origin of these clumps of chromatin [6].

To determine if there was incorporation of ^{3}H-methyl thymidine in the sperm heads or in the sperm bodies, cultures of heterokaryocytes were grown in the continuous presence of ^{3}H-TdR for 4.5 and 16–20 hours after fusion. The auto-

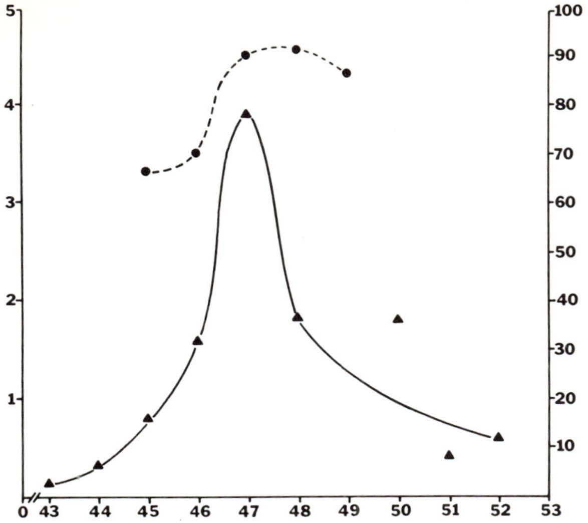

Fig. 4.

Radioactivity curve of rabbit spermatozoa labeled in vivo. *Ordinate* (left): cpm/10^3/10^7 spermatozoa (▲——▲) (background= 20–30 cpm); *ordinate* (right): percentage of labeled spermatozoa, as determined by autoradiography (●– – – – –●). *Abscissa:* days after injection of ^{3}H-TdR into a mature New Zealand buck (from [6] with permission).

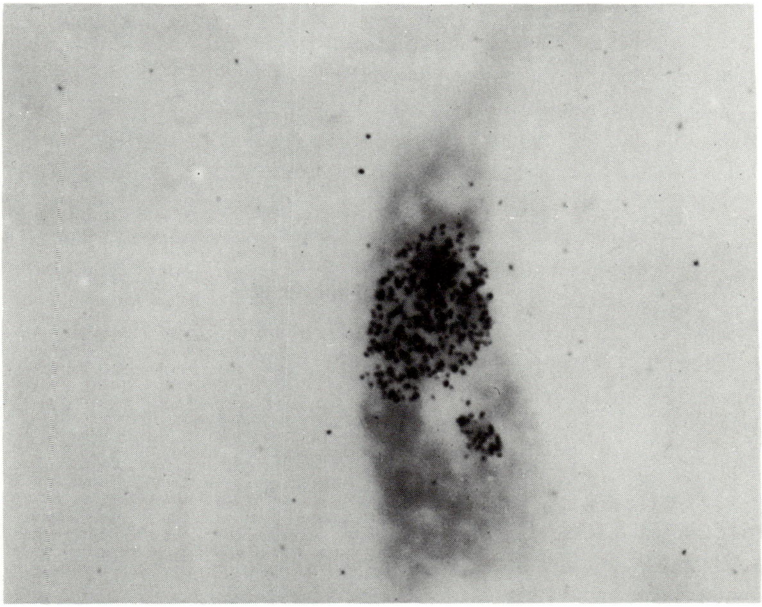

Fig. 5. Photomicrograph of an F5–1 somatic cell-rabbit spermatozoa heterokaryocyte grown in the continuous presence of ^{3}H-TdR and fixed 20 hours after fusion. The silver grains are over the F5–1 nucleus and over the sperm body, indicating incorporation of the DNA precursor. × 750 (from [6] with permission).

radiograms did not show incorporation of ^{3}H-TdR in the unfused and fused, but morphologically-intact, spermatozoal heads. In contrast, the Feulgen-positive clumps of disaggregated sperm chromatin were labeled (Fig. 5) [6]. The incorporated labeled precursor was not removed after treatment with 5 % trichloroacetic acid. In control cultures of F5–1 cells, no incorporation of ^{3}H-TdR was seen in the cytoplasm under the same experimental conditions. Nuclear incorporation was evident. These facts imply synthesis of DNA by the sperm bodies.

Ultrastructural studies

As seen with the electron microscope, the action of LL appeared to affect primarily the acrosomal portion of the spermatozoa. After treatment with LL, the plasma membrane and the external membrane of the acrosome were disrupted and formed vesicles. The internal membrane of the acrosome was changed, but still present after LL treatment (Fig. 6A–D).

After treatment with SDS, loss of the plasma membrane and external acrosomal membrane of the spermatozoa were seen, and the acrosomal contents disappeared, but no vesiculation was seen. After treatment with SDS and DTT, complete disruption of the spermatozoa was observed.

Twenty minutes after treatment with LL, the spermatozoa were surrounded by pseuopods of the F5–1 cytoplasm (Fig. 7A–C). After this step, fusion took place and the spermatozoa could be seen in the cytoplasm of the somatic cells. At 18–24 hours after fusion, one to six spermatozoa, mostly without tails, were seen in many of the heterokaryocytes. Some spermatozoa were inside vacuoles in the cytoplasm of the somatic cells (Fig. 8) and some were not (Fig. 9). Only the sperm nuclear membrane and, in some cases, the partially degraded inner acrosomal membrane remained around the spermatozoal nucleus. As is shown in Fig. 9, the spermatozoal nuclei occasionally maintained their spatial relationship with the mid-piece of flagellum. At 48, 72, and 96 hours after fusion, many of the sperm nuclei persisted in cytoplasmic vacuoles in the somatic cells. However, some of them appeared to be disaggregated, as shown in Fig. 10.

Discussion

Disaggregation of the nuclear chromatin appears to be a sine qua non for synthesis of DNA by the activated spermatozoon [1]. No synthesis of DNA by spermatozoa fused with somatic cells using either UV- or β-propiolactone inactivated Sendai virus was obtained [12]. The attachment and the successful fusion of the spermatozoal plasma membrane and the cytoplasmic membrane of a somatic cell do not seem to induce the disaggregation of the spermatozoal chromatin necessary for the initiation of DNA synthesis.

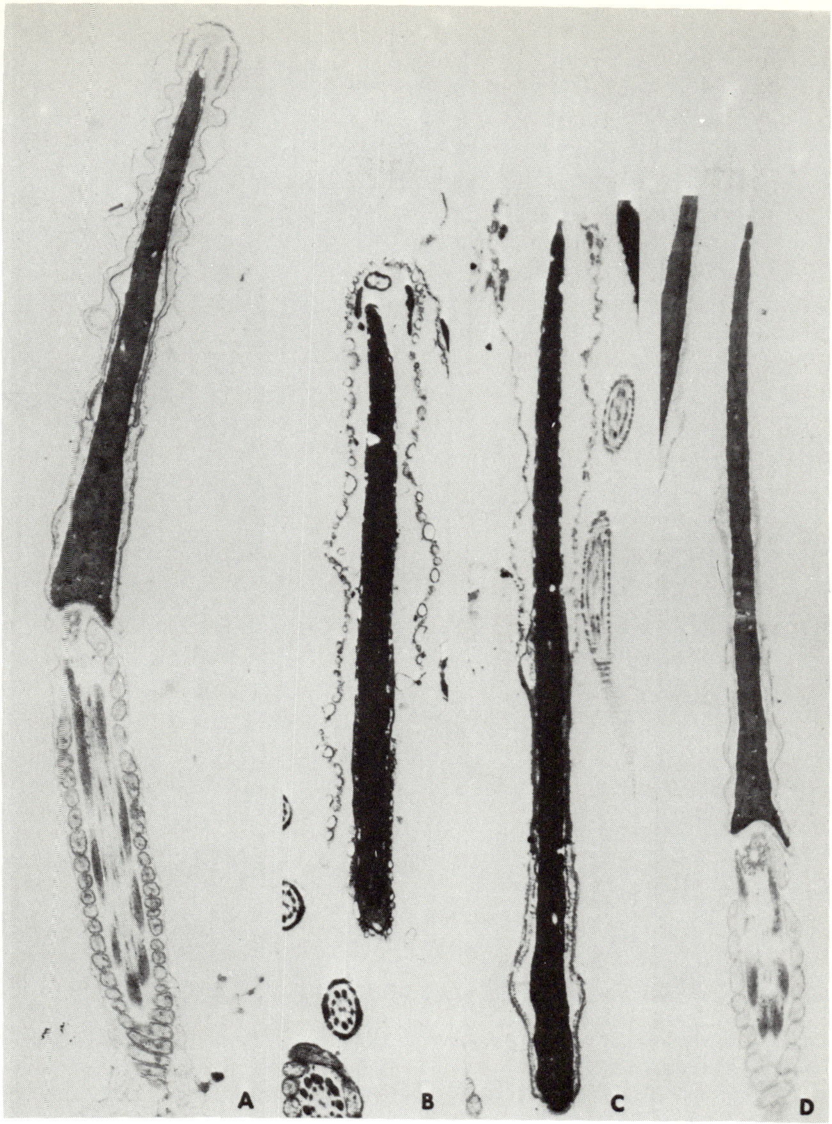

Fig. 6. Effect of treatment with LL on rabbit spermatozoa.

(A) Control spermatozoon: Note the presence of the acrosome and of the contents which appear to have mimicked a false acrosome reaction. Perforation and a typical wrinkled cell membrane are evident. x 17.400.

(B) The plasma and outer acrosomal membranes are disrupted. A row of vesicles is visible instead of these membranes. x 23.200.

(C) The acrosome is lacking. Note the breakage of the cell membrane on the top of the sperm head. x 19.250.

(D) The plasma and outer acrosomal membrane are completely dissolved. The inner acrosomal membrane is only partially intact. x 17.400.

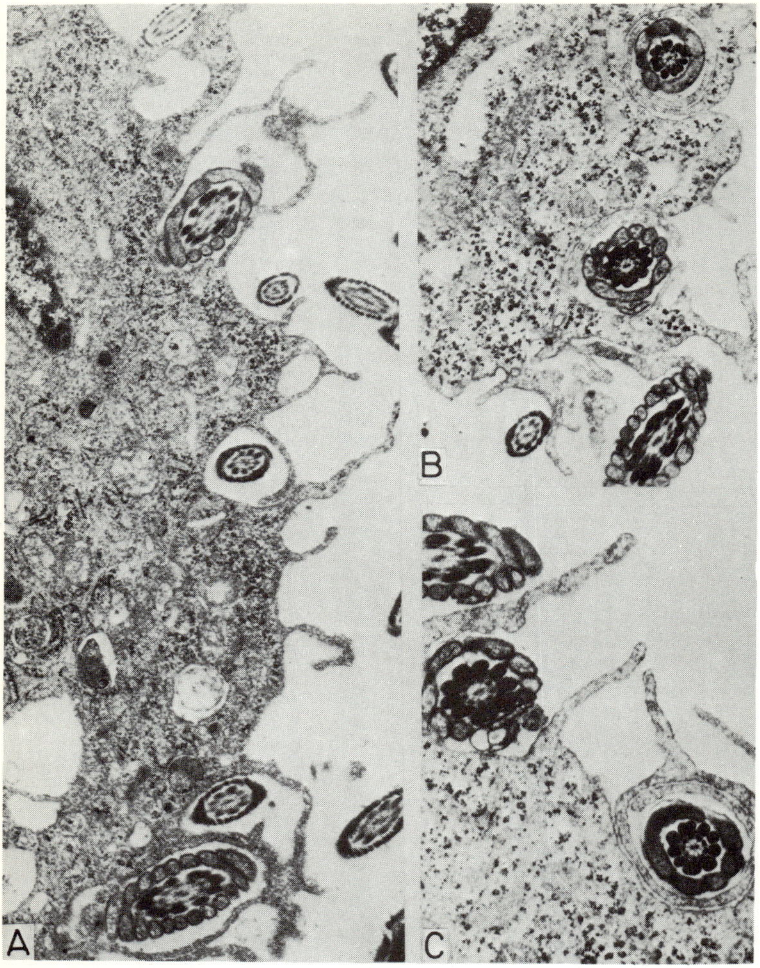

Fig. 7. Electron micrographs of preparations fixed 20 minutes after fusion between rabbit spermatozoa and F5–1 cells.

(A) Cross section of the spermatozoon at the beginning of the cell fusion. The spermatozoa are close to the cell membrane, usually surrounded by pseudopods of F5–1 cells. x 23.200.

(B) and (C) The spermatozoa are closely surrounded by pseudopods. (B) x 23.200. (C) x 30.800.

With the use of LL to assist fusion in the present study, 10–30 % of the F5–1 somatic cells contained spermatozoal heads, 10–15 % of which underwent nuclear disaggregation. That these structures of disaggregated chromatin (sperm bodies) were derived from spermatozoa is shown by several facts: (1) only spermatozoal heads and sperm

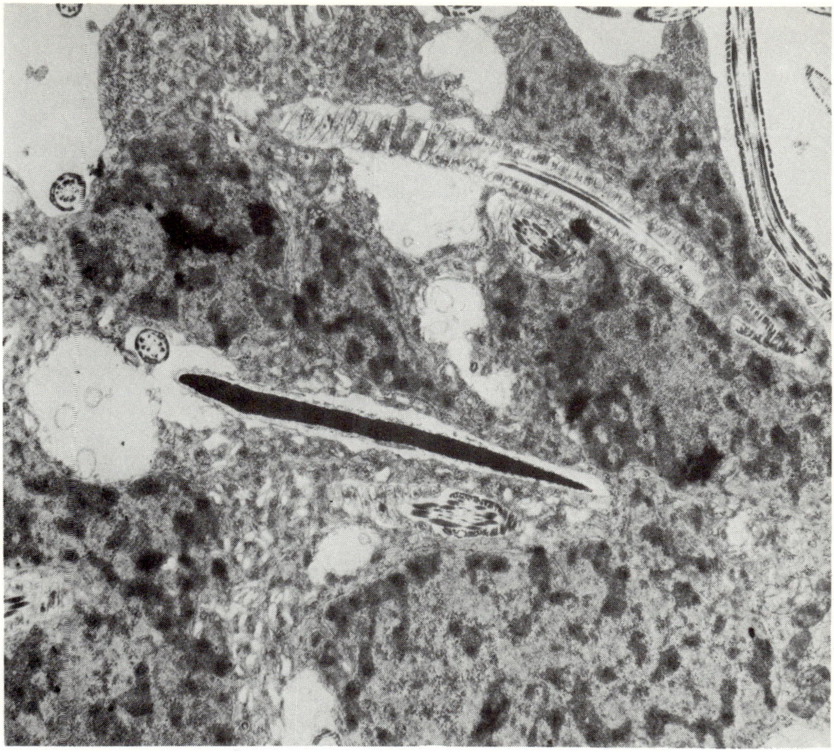

Fig. 8. A somatic F5–1 cell 48 hours after fusion with rabbit spermatozoa. Note the presence of the LL-pretreated spermatozoon in a vacuolelike space in the F5–1 cell. × 38.500.

bodies showed labeling after fusion of spermatozoa labeled with tritium in vivo with F5–1 somatic cells; (2) the number of somatic cells in the ejaculates used was extremely low (< 0.1 %) and was not sufficient to account for the percentage of the sperm bodies seen; (3) the concentration of LL used for pretreatment of the sperm pellets was sufficient to disrupt more than 95 % of the somatic cells in the pellet (*Croce* and *Koprowski,* unpublished results); and (4) autoradiograms of smears of the ejaculate and epididymal preparations did not show incorporation of the ^{3}H-TdR in those few somatic cells remaining. Additional support for the origin of the sperm bodies can be found in the fact that after staining with acridine orange, they showed a metachromatic response to thermal denaturation which was similar, but not identical, to that obtained with spermatozoa.

Ultrastructural studies showed the presence of spermatozoal nuclei in the cytoplasm of the somatic cells after fusion by means of LL. A percentage of these nuclei

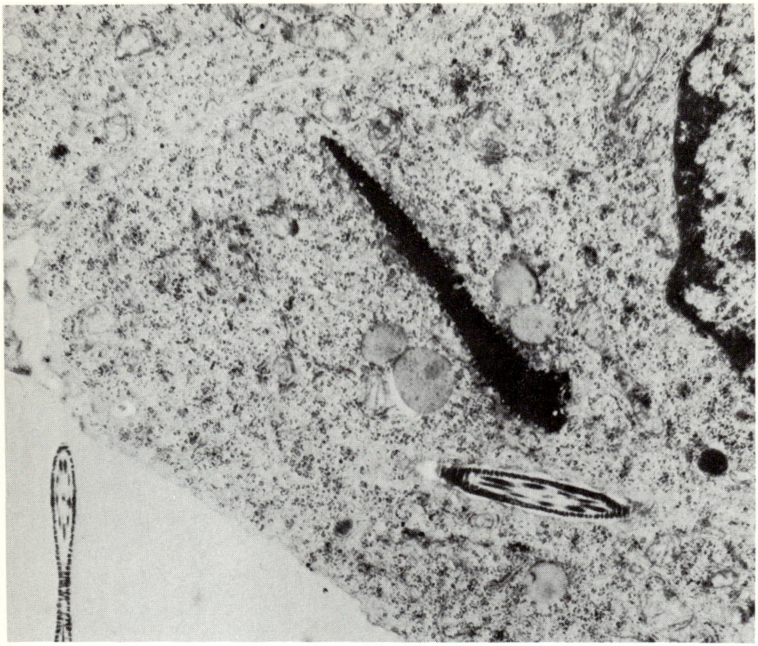

Fig. 9. A somatic F5–1 cell 48 hours after fusion with rabbit spermatozoa. Note the presence of the sperm chromatin in the cytoplasm of the somatic cell. × 15.080.

appeared to undergo profound chromatin disaggregation. Presumably these nuclei were the type of sperm bodies which would incorporate ^{3}H-TdR.

Incorporation of ^{3}H-TdR by the nuclei of somatic cells and by sperm bodies, but not by the morphologically intact sperm heads, was obtained after culture of heterokaryocytes in the continuous presence of the isotopically labeled precursor. This fact and the stability of the incorporated isotope to cold treatment with trichloroacetic acid strongly indicates DNA synthesis by the sperm bodies.

LL has also been used to fuse rabbit spermatozoa with cell lines from several different species. Studies on the fusion of rabbit spermatozoa with either human or mouse mutant cell lines are in progress in an attempt to see if there is expression of the rabbit genome in the heterokaryocytes and if it is possible to obtain hybridization between spermatozoa and somatic cells.

The major immediate effect of LL on the rabbit sperm head seen was vesiculation of the plasma and external acrosomal membranes. These alterations of the sperm envelope seem to play an important role in the disaggregation of sperm nuclei prior to initiation of DNA synthesis.

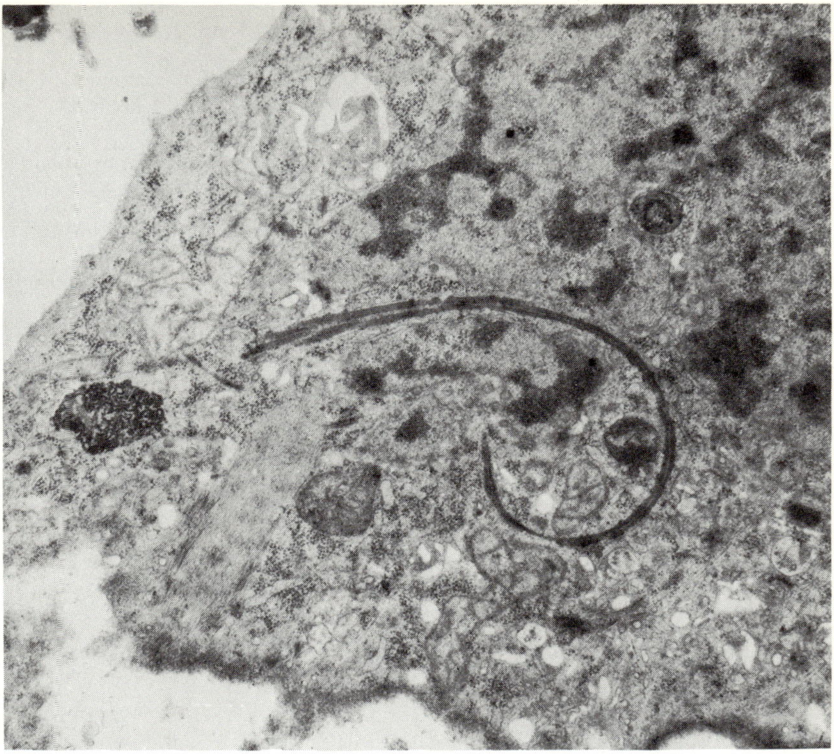

Fig. 10. A somatic F5–1 cell 48 hours after the fusion process. The chromatin of the spermatozoon is visible in the cytoplasm of the F5–1 cell. Note the disaggregation of the chromatin of the sperm head × 47.000.

Acknowledgements: This investigation was supported by U.S. Public Health Service Research grants CA-10815 and CA-04534 from the National Cancer Institute, and HD-03577 from the National Institute of Child Health and Human Development (to BLG), and by funds from the Labor Foundation. *C. M. C.* is a Fellow of the Italian Society of Electron Microscopy, at the Wistar Institute. *B. G.* is a Fellow of the W. H. O. at the Wistar Institute from the Laboratory of Cytochemistry, University of Łodz, Łodz, Poland, and *W. S.* was a Fellow of the Wistar Institute from the Department of Histology and Embryology, Academy of Medicine, Warsaw, Poland.

References

[1] *Colwin, L. H., and Colwin, A. L.* 1967. Membrane fusion in relation to sperm-egg association. *C. B. Metz,* and *A. Monroy. In* Fertilization, eds. New York: Academic Press.

[2] *Croce, C. M., Sawicki, W., Kritchevsky, D., and Koprowski, H.* 1971. Induction of homokaryocyte, heterokaryocyte, and hybrid formation by lysolecithin. Exptl. Cell Res. **67:** 427–35.

[3] *Girardi, A. J., Sweet, B. H., and Hilleman, M. R.* 1963. Factors influencing tumor induction in hamsters by vacuolating virus SV40. Proc. Soc. exptl. Biol. Med. **112:** 662–67.

[4] *Gledhill, B. L.* 1966. Studies on the DNA content, dry mass and optical area of bull spermatozoa heads during epididymal maturation. Acta Vet. Scand. **7:** 131–42.

[5] *Gledhill, B. L., Gledhill, M. P., Rigler, R., Jr., and Ringertz, N. R.* 1966. Changes in deoxyribonucleoprotein during spermiogenesis in the bull. Exptl. cell. Res. **41:** 652–65.

[6] *Gledhill, B. L., Sawicki, W., Croce, C. M., and Koprowski, H.* In press. DNA synthesis in rabbit spermatozoa after treatment with lysolecithin and fusion with somatic cells. Exptl. Cell Res.

[7] *Harris, H., and Watkins, J. F.* 1965. Hybrid cells derived from mouse and man: Artificial heterokaryons of mammalian cells from different species. Nature **205:** 640–46.

[8] *Johnson, R. T., Rao, P. N., and Hughes, H. D.* 1970. Mammalian cell fusion. III. A Hela cell inducer of premature chromosome condensation active in cells from a variety of animal species. J. cell. Physiol. **76:** 151–57.

[9] *Poole, A. R., Howell, J. I., and Lucy, J. A.* 1970. Lysolecithin and cell fusion. Nature **227:** 810–14.

[10] *Ringertz, N. R., Gledhill, B. L., and Darzynkiewicz, Z.* 1970. Changes in deoxyribonucleoprotein during spermiogenesis in the bull. Sensitivity of DNA to heat denaturation. Exptl. Cell Res. **62:** 204–18.

[11] *Rogers, R. W.* ed. 1967. Techniques of autoradiography, 1st ed. Amsterdam: Elsevier Publ.

[12] *Sawicki, W., and Koprowski, H.* 1971. Fusion of rabbit spermatozoa with somatic cells cultivated in vitro. Exptl. Cell Res. **66:** 145–51.

[13] *Steplewski, Z., and Koprowski, H.* 1970. Somatic cell fusion and hybridization. Methods in Cancer Res. **5:** 155.

[14] *Zgorniak-Nowosielska, I., Sedwick, D. W., Hummeler, K., and Koprowski, H.* 1967. New assay procedure for separation of mycoplasmas from virus pools and tissue culture systems. J. Virol. **1:** 1227–37.

Discussion

follows the paper of *Ronald J. Ericsson,* p. 205 f.

Logic in Germinal Cell Membranes

Ronald J. Ericsson

Department of Endocrinpharmacology, Schering AG, Berlin, Germany

Summary: Sendai virus effectively monitors a relationship between sperm membrane differences and sperm fertility. Head membranes of mature sperm, specifically the acrosome region, adsorb with Sendai virus. Whereas immature (infertile) sperm do not associate with the virus. The virus is also active in initiating fertility in vitro. It is thought to induce an artificial capacitation – acrosome reaction – which changes sperm from the potential to the actual fertile state. Another, although unrelated, factor in germinal cell membranes is their premature exfoliation. Certain chemicals cause these cells to break the intercellular bonds and migrate in mass through the excurrent ducts. Immature germinal cells unfettered and freely available await the unfettered scientist.

Introduction

Why not logic in germinal cell membranes? Scientific dogma does not deny membrane fusion of gametes for genetic recombination. The confusion is over the status of membranes and their function in sexual reproduction. Admittedly the word logic in the title expresses more hope than knowledge. Nonetheless, my aim is to give some explanations on the commitment of germinal cell membranes to male fertility. The reviews cited provide background information on sperm structure-function relationship [4, 12, 14]. Results are not necessarily in the order acquired, but rearranged to make some biological sense. Some may think placing parainfluenza virus with mammalian sperm so as to fertilize ova in vitro borders on the nonbiological – in any sequence.

Manuscript received: 26 January 1972

Rabbit sperm and Sendai virus

Sperm from the caput epididymidis are infertile [3, 14]. A gradual change from the infertile state to the fertile one occurs as sperm move toward the cauda epididymidis. Our experiments [6] give a fit between infertile and fertile epididymal sperm and their ability to adsorb with Sendai virus. Since sperm from the caput do not associate with the virus, one can tentatively respect this condition of the plasma membrane as an indicator of sperm immaturity. The opposite response occurs in the cauda – sperm adsorb with Sendai virus – which is taken as a positive indicator of sperm maturity. The sperm – virus interaction is not simply adsorption or nonadsorption because the virus adsorbs on a selective membrane basis. An immediate check on sperm response when mixed with virus is head-to-head agglutination [11]. When observing such sperm through a light microscope, one concludes the virus is associated on the sperm head. Work with transmission and scanning electron microscopes [6] confirms virus like particles adsorb almost exclusively over the acrosome (anterior head region) (Fig. 1). And this is important because the acrosome is the sperm structure directly involved in attachment to and penetration of the ovum. These sperm-membrane conditions, specifically the moiety of the anterior head, then can and do serve as a monitor of sperm maturity, or as also stated, sperm fertility. Reasons for differences in virus-membrane adsorption remain unknown. However, one might research the possibilities of coating substances [8], changes in electrostatic charge [2], or the factor of underlying acrosomal membranes.

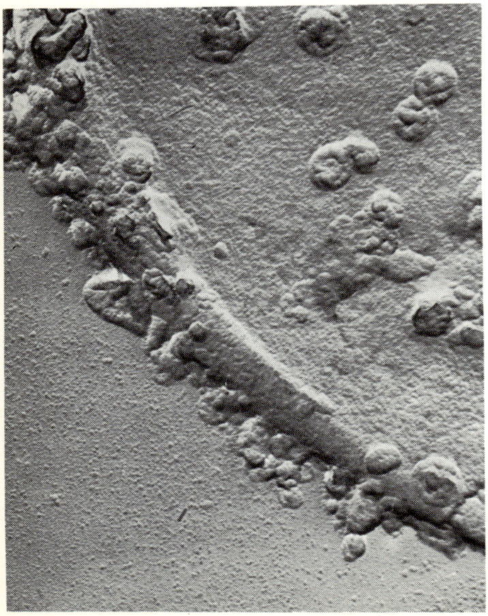

Fig. 1

Acrosome segment from a washed ejaculated rabbit sperm exposed to Sendai virus. × 19.000. Viruslike particles adsorb almost exclusively over this region.

In preparing sperm with virus for in vitro fertilization experiments, we took unquantitative notes on motile versus nonmotile sperm agglutination. Motile sperm agglutinate, in general terms, whereas nonmotile sperm do not agglutinate. Or restated, motile sperm adsorb with virus whereas nonmotile sperm most likely do not; the latter agrees with membrane changes known to occur postmortem [10]. This observation, although indirect, is more evidence for the use of Sendai virus as a monitor of sperm fertility. Sendai virus also take an active part in initiating sperm fertility in vitro. It is possible to fertilize rabbit ova in vitro provided sperm first undergo capacitation (the phenomenon of sperm acquiring the ability to penetrate the ovum) in vitro [5]. Our experiments use basically the same procedures [11] except sperm are not first capacitated. The idea is to induce an artificial capacitation – acrosome reaction [13] – through the sperm-virus interaction. The acrosome reaction [1] follows capacitation and includes vesiculation of acrosomal membranes, which allows release of lytic enzymes from the acrosome [15]. Fertilization of approximately 50 percent of over 200 ova signifies an active role of the virus in altering sperm fertility in vitro. Ova incubated with sperm but without virus remain unfertilized. An obvious and simple explanation for the change in sperm fertility is a virus induced vesiculation of sperm membranes. No doubt the final explanation is more complex, but absence of knowledge does not distract from the fact that it works. More important, nothing else is available to induce an artificial capacitation – acrosome reaction.

Immature germinal cell release and capture

U–29,409 [2,3-dihydro-2-(1-naphthy)-4(1H)-quinazolinone], U–32,104 [2-benzimidazolecarbamic acid, methyl ester], and U–32,422 [urea, 1,1-dimethyl-3-(5-methyl-2-benzimidazolyl)–] have a common biological property. They cause exfoliation of immature germinal cells (Fig. 2). The scientific importance of this discovery is still ruminating in the minds of a few "gonad-directed" researchers. What I have to say about immature germinal cells is both direct and limited. All work relates to the male.

Awareness, in my mind, is tantamount to experimental success. I, therefore, advocate research in areas which intimately and ultimately include germinal cells. Since the means are at hand to acquire many ($1–40 \times 10^6$) cells [9], it requires only that ideas expand to meet the supply. My point is not to outline ways in which immature germinal cells can be used in fruitful research. I wish only to expose a segment of the scientific community to their existence in an unfettered form. Perhaps I am still scientifically naive, but a microscopic view of thousands upon thousands of immature germinal cells free of intercellular bonds [7, 9] never fails to cause wonderment. Readers must decide if cells voided from the testis, before their time, and captured in the epididymis, before their time, are of value to biological research.

Chemical	Species	Single effective dose per kg	Recovered from	
			Epididymis	Semen
U–29,409	rat	75 mg ip	+	
	mouse	200 mg ip	+	
U–32,104	rat	150 mg oral	+	
	guinea pig	500 mg oral*		
U–32,422	rat	75 mg oral	+	
	rhesus monkey	50 mg oral		+
	bovine	12 mg oral**		+

Fig. 2. Exfoliated immature germinal cells.

* seven daily doses.

** four daily doses.

Acknowledgements: Dr. *D. A. Buthala* is credited for the electron photomicrograph. Laboratory work was at the Upjohn Company, Kalamazoo, Michigan, USA. And the "logic" originated on both sides of the Atlantic.

References

[1] *Barros, C., Bedford, J. M., Franklin, L. E., and Austin, C. R.* 1967. Membrane vesiculation as a feature of the mammalian acrosome reaction. J. Cell. Biol. **34**: C1–C5.

[2] *Bedford, J. M.* 1963. Changes in the electrophoretic properties of rabbit spermatozoa during passage through the epididymis. Nature, Lond. **200:** 1178–1180.

[3] *Bedford, J. M.* 1966. Development of the fertilizing ability of spermatozoa in the epididymis of the rabbit. J. exp. Zool. **163**: 319–30.

[4] *Bedford, J. M.* 1970. Sperm capacitation and fertilization in mammals. Biol. Reprod. Suppl. **2**: 128–58.

[5] *Brackett, B. G.* 1970. In vitro fertilization of mammalian ova. *In* Schering Symposium on Mechanisms Involved in Conception, Advances in the Biosciences 4, ed. G. Raspé, pp. 73–94. Braunschweig: Pergamon Press/Vieweg.

[6] *Buthala, D. A., Ericsson, R. J., and Chubb, G. T.* 1971. Interaction of Sendai virus and rabbit sperm: transmission and scanning electron microscopy. Biol. Reprod. **5:** 325–329.

[7] *Dym, M., and Fawcett, D. W.* 1971. Further observations on the numbers of spermatogonia, spermatocytes, and spermatids connected by intercellular bridges in the mammalian testis. Biol. Reprod. **4**: 195–215.

[8] *Ericsson, R. J.* 1970. Methods for measurement of sperm capacitation in different species. *In* Schering Symposium on Mechanisms Involved in Conception, Advances in the Biosciences 4, ed. G. Raspé, pp 51–59. Braunschweig: Pergamon Press/Vieweg.

[9] *Ericsson, R. J.* 1971. Antispermatogenic properties of 2,3-dihydro-2-(1-naphthyl)-4 (1H) -quinazolinone (U–29,409). Proc. Soc. Exp. Biol. Med. **137:** 532–35.

[10] *Ericsson, R. J., and Buthala, D. A.* 1970. A fluorescence staining method for studying sperm membranes. Biol. Reprod. **3**: 8–12.

[11] *Ericsson, R. J., Buthala, D. A., and Norland, J. F.* 1971. Fertilization of rabbit ova in vitro by sperm with adsorbed Sendai virus. Science **173**: 54–55.

[12] *Fawcett, D. W.* 1970. A comparative view of sperm ultrastructure. Biol. Reprod. Suppl. **2**: 90–127.

[13] *Lucy, J. A.* 1970. The fusion of biological membranes. Nature, Lond. **227**: 814–17.

[14] *Orgebin-Crist, M.-C.* 1969. Studies on the function of the epididymis. Biol. Reprod. Suppl. **1**: 155–75.

[15] *Stambaugh, R., and Buckley, J.* 1969. Identification and subcellular localization of the enzymes effecting penetration of the zona pellucida by rabbit spermatozoa. J. Reprod. Fertil. **19**: 423–32.

Discussion

Koprowski: Does the general silence indicate that there are some problems that have been presented that you do not understand? Perhaps the terminology that has been used to describe the fusing of mammalian cells should be clarified. Feel free to ask any questions you like concerning technical terms.

Graham: Is it correct to assume that only sperm which will agglutinate with Sendai are able to fuse into somatic cells?

Croce: After the Sendai virus treatment of spermatozoa, we noticed agglutination of the spermatozoa. It seems that many of these sperm can fuse with the somatic cells.

Graham: Therefore, these sperm from the head of the epididymis, which do not agglutinate with Sendai, are also unable to fuse with somatic cells?

Croce: No. They can fuse, too.

Graham: What is the relationship between agglutination of sperm and their ability to fuse?

Croce: For example, with the lysolecithin we see little or no agglutination of spermatozoa, but observe fusion.

Ericsson: I think the fusion of sperm is only significant in that it allows the artifical activation and the acrosome reaction. If you don't have the fusion taking place, then probably you don't have this process, which I would simply relate as a kind of poking holes in the membranes of the sperm.

Croce: In fact, it is not a problem of putting a hole in the membrane. May I have Fig. 4 again which shows that after lysolecithin treatment, the spermatozoa show changes very similar to the acrosomal reaction, with vesiculation of the plasma membrane and of the outer acrosomal membrane. We see that when we fuse these treated spermatozoa which are nonmotile, they fuse well with somatic cells. In Fig. 2, we can see vesiculation of the plasma membrane and of the outer acrosomal membrane. Only about 10 to 15 % of the spermatozoa which are fused with the somatic cell undergo chromatin disaggregation, and only a fraction of them synthesized DNA. When we used Sendai virus, we did not get chromatin disaggregation even when we got about 50–60 % of spermatozoa inside the somatic cell. That seems to confirm the finding that *Johnson* and coworkers reported that after trypsinization, the bovine spermatozoa can fuse with human mitotic cells and that a small fraction of them, less than 0.01 %, presented the phenomemon of premature chromatin condensation.

Milgrom: Dr. *Ericsson,* did you try interspecies fertilization by using Sendai virus?

Ericsson: We did try to fuse species of sperm other than rabbit and we were not successful. Specific examples are bull, human, and rhesus monkey.

Fischer: Dr. *Croce,* lysolecithin is a highly surface agent and usually it lyses cells, so you must have used a technical trick because the efficiency of fusing was rather high. What is this trick? Furthermore, is lysolecithin good enough to fuse other types of cells?

Croce: Yes. In this kind of experiment we also got a fusion percentage of between 20 and 30 % of homokaryocyte F5-1 cells.

Koprowski: Excuse me for one second. Does the audience understand just what are homokaryocytes and heterokaryocytes? These terms have been used so many times. Please clear these up.

Croce: The homokaryocyte is a cell which has two or more nuclei which derive from the same parental cell line. Whereas a heterokaryocyte is a cell with two or more nuclei which come from two different parental cell lines. In this case, we call our fused cells heterokaryocytes because we have the F5-1 somatic cell nuclei and the spermatozoa nuclei inside the same somatic cell. One is of hamster origin, the other of rabbit origin. Let me now answer Dr. *Fischer's* question. For the lysolecithin treatment we resuspend the pure lysolecithin phosphate-buffered solution at various pHs, containg 5 mg per ml of albumin. The presence of albumin seems to decrease the toxicity of lysolecithin for the somatic cell.

Duesberg: Dr. *Ericsson,* I was just wondering, why do you need Sendai virus to achieve fertilization? Is not a sperm supposed to be capable of doing that itself, without any virus?

Ericsson: In most species which have been studied, primarily domestic animals and the smaller normal laboratory mammals, and perhaps man, sperm are not capable of fertility until they have resided in the reproductive tract of an estrus female. If the female is under the influence of a progestogen, either artificially by experimentation or naturally through pseudopregnancy, the sperm do not undergo capacitation, which is defined as the physiological and biochemical changes which sperm must undergo to bring about the acrosome reaction. The latter is the vesiculation of plasma and outer acrosomal membranes. This physical change in integrity of the membranes allows for lytic enzymes, which are stored within the acrosome, to be liberated and thereby bring about penetration of the zona pellucida. Sendai virus does nothing more than artificially activate the sperm just as if one had placed the sperm in an estrous female reproductive tract and allowed this process to occur naturally.

Hofschneider: It was reported that following fusion of sperm and cells, one can observe DNA synthesis within sperm bodies inside of cells. Is it possible to use, instead of intact cells, cell extracts to study DNA synthesis in such sperm bodies?

Koprowski: Drs. *Croce* and *Ericsson,* you have been challenged. Now please reply.

Croce: No. To obtain formation of sperm bodies and DNA synthesis in these structures it is necessary to fuse intact cells with LL pretreated spermatozoa.

Studies on Genetic Regulation Utilizing Microinjection of Nuclei and DNA into Living Eggs and Oocytes

H. R. Woodland, C. C. Ford, and J. B. Gurdon

Department of Zoology, Oxford University, South Parks Road, Oxford, England

Summary: The microinjection of nuclei and other substances into eggs and oocytes is a powerful tool in analyzing phenomena occurring in living cells. As an example of the way in which the technique may be used, a study of the role of deoxynucleotide triphosphates in regulating DNA synthesis is described. The injection into oocytes of brain nuclei together with the four triphosphates fails to cause the nuclei to synthesize DNA, although they synthesize DNA in the usual way when the same mixture is injected into eggs. The four common deoxynucleotide triphosphates also fail to cause purified native DNA to be replicated in oocytes, although it is replicated in eggs under similar conditions. The fact that denatured DNA is replicated in oocytes enable us to calculate that, in an oocyte, at least sufficient precursor is available to enable the replication of 100–150 nuclei in a 6 hour period.

Introduction

The experiments reported in this paper utilize the technique of microinjection into two types of living cell, the egg and oocyte of the toad *Xenopus laevis.* One use of this technique is the manipulation of development. This end is to some extent achieved in nuclear transplant experiments, where nuclei of differentiated somatic cells are inserted into eggs and support the development of tadpoles [5, 12, 19] and in some cases normal adults [14]. As well as permitting the preparation of clones of genetically identical individuals [6], experiments of this sort show that differentiated cells contain all the genes necessary to direct the formation of a normal individual. In the future, nuclear transplantation should permit the Mendelian analysis of the genes in single somatic cells [2].

Manuscript received: 13 December 1971

A second way in which microinjection into embryonic cells may be used is the analysis of the way in which the cytoplasm controls nuclear activity. In the two cells used, it has been shown that the presence or absence of DNA and RNA synthesis, and the condensation of chromosomes, are under cytoplasmic control [15]. Microinjection permits the analysis of this control by enabling the experimenter to alter the composition of the cytoplasm. Experiments of this type have so far involved studies of the translation of exogenous mRNA [11, 18, 20], and of RNA and DNA synthesis on injected DNA templates, both in a purified form [10, 17] and as injected nuclei [3,7]. In order to illustrate the way in which investigations of this sort may be conducted using micromanipulation methods, we describe here some experiments designed to show whether the regulation of DNA synthesis is mediated by the availability of deoxynucleotide triphosphates.

Large oocytes of *X. laevis* are inactive in DNA synthesis [7]. By hormone stimulation in vivo or in vitro these cells may be induced to complete meiosis and form eggs, which on activation enter a phase of rapid and frequent DNA synthesis in which cell divisions follow on one another at intervals of as little as 10 minutes [4]. This change in nuclear activity derives from a change in the cytoplasm, rather than the nucleus itself, since nuclei from nondividing tissues replicate their DNA when introduced into eggs, but not when introduced into oocytes. A simple explanation of this difference is that the precursors of DNA synthesis are absent from oocytes, but are present in eggs. This hypothesis is tested by the experiments described in this paper.

Materials and Methods

Eggs and oocytes were obtained from *X. laevis* by methods previously described [8, 9]. Nuclei were prepared, injected, and autoradiographed as described by *Graham* et al. [3]. The conversion of radioactive nucleosides to nucleotides was analyzed by perchloric acid extraction and Dowex-1-formate chromatography [22]. The pure DNA used was a commercial preparation from calf thymus (Sigma Chemical Co., Ltd., 12, Lettice Street, London, S. W. 6). It was denatured by heating for 10 minutes at 100 °C in microinjection saline (88 mM-NaCl, 1 mM-KCl, 15 mM-Tris-HCl, pH 7.6), followed by rapid cooling in iced water. DNA was extracted by a method which involved extensive pronase and ribonuclease digestion, phenol extraction, Sephadex G-50 chromatography, ethanol precipitation, and finally CsCl isopycnic gradients. The full details will be described elsewhere [1]. Recovery of DNA was estimated by addition of nonradioactive DNA to the homogenate.

Radiochemicals were obtained from the Radiochemical Centre (Amersham, Bucks., England), with the exception of [^{3}H] guanosine, which was obtained from Schwartz/Mann (Orangeburg, New York 10962). Nucleotide triphosphates were obtained from the Sigma Chemical Co., Ltd. All other chemicals were obtained from British Drug Houses (Poole, Dorset, England.).

Results and Discussion

Direct estimation of precursor pools

Woodland and *Pestell* [24] have attempted to identify the deoxynucleotides present in the eggs and oocytes of *X. laevis.* The main components of the pool were found to be the deoxynucleotide triphosphates, dCTP, dTTP, dATP, and dGTP [1]). These were present in similar amounts, approximately 15 p moles of each per egg. These amounts are sufficient to synthesize the DNA of about 2,500 nuclei, indicating that the egg is well prepared for the rapid phase of nuclear replication which follows fertilization. In oocytes, there appear to be less pyrimidine deoxynucleotide triphosphates than in eggs, and purine deoxynucleotide triphosphates could not be detected. This observation would seem to indicate that the apparent deficiency of deoxynucleotide triphosphates prevents DNA synthesis from occurring in oocytes. However, the methods used have a lower level of sensitivity of 1–2 p moles of nucleotide per oocyte, an amount still sufficient to enable the replication of 100–200 nuclei. Thus conventional biochemical techniques do not, and formally *cannot,* show that nuclei are not replicated in oocytes because they contain too little deoxynucleotide triphosphate. A direct approach to this problem is to raise the size of the pools artificially.

It is appropriate to state at this point that we assume that deoxynucleotide triphosphates are direct DNA precursors in higher organisms, since there is no direct evidence to the contrary. There is, however, a report that other substances may perform this role in bacteria [21]. If this is also so in *Xenopus* cells, these other substances must at least be in equilibrium with the triphosphate pool, for the following reason. *Ford* and *Woodland* (unpublished) injected [^{3}H] thymidine into unfertilized eggs and estimated the radioactive DNA formed. Calculating the intracellular [^{3}H] dTTP specific activity from the amount of [^{3}H] dTTP formed from the injected [^{3}H] thymidine and the estimate of the total pool size obtained by *Woodland* and *Pestell,* they estimated that 7 pg of DNA were synthesized after injection. At this time the expected value from two replications of the egg nucleus is 9 pg. Allowing for variation in pool sizes in different batches of eggs, and the low amounts of radioactivity involved, the estimate is well within the limits expected. The experiment therefore shows that the dTTP pool actually estimated is all immediately relevant to DNA synthesis, and that if it is not the direct precursor pool, it is at least in equilibrium with it. Unfortunately it is not possible to do experiments with the very short incubation necessary to demonstrate differential entry of [^{3}H]thymidine into dTTP and a hypothetical precursor pool with which it is in moderately rapid equilibrium.

---

1) Abbreviations used: dCTP, deoxycytidine-5′-triphosphate; dTTP, deoxythymidine-5′-triphosphate; dATP, deoxyadenosine-5′-triphosphate; dGTP, deoxyguanosine-5′-triphosphate.

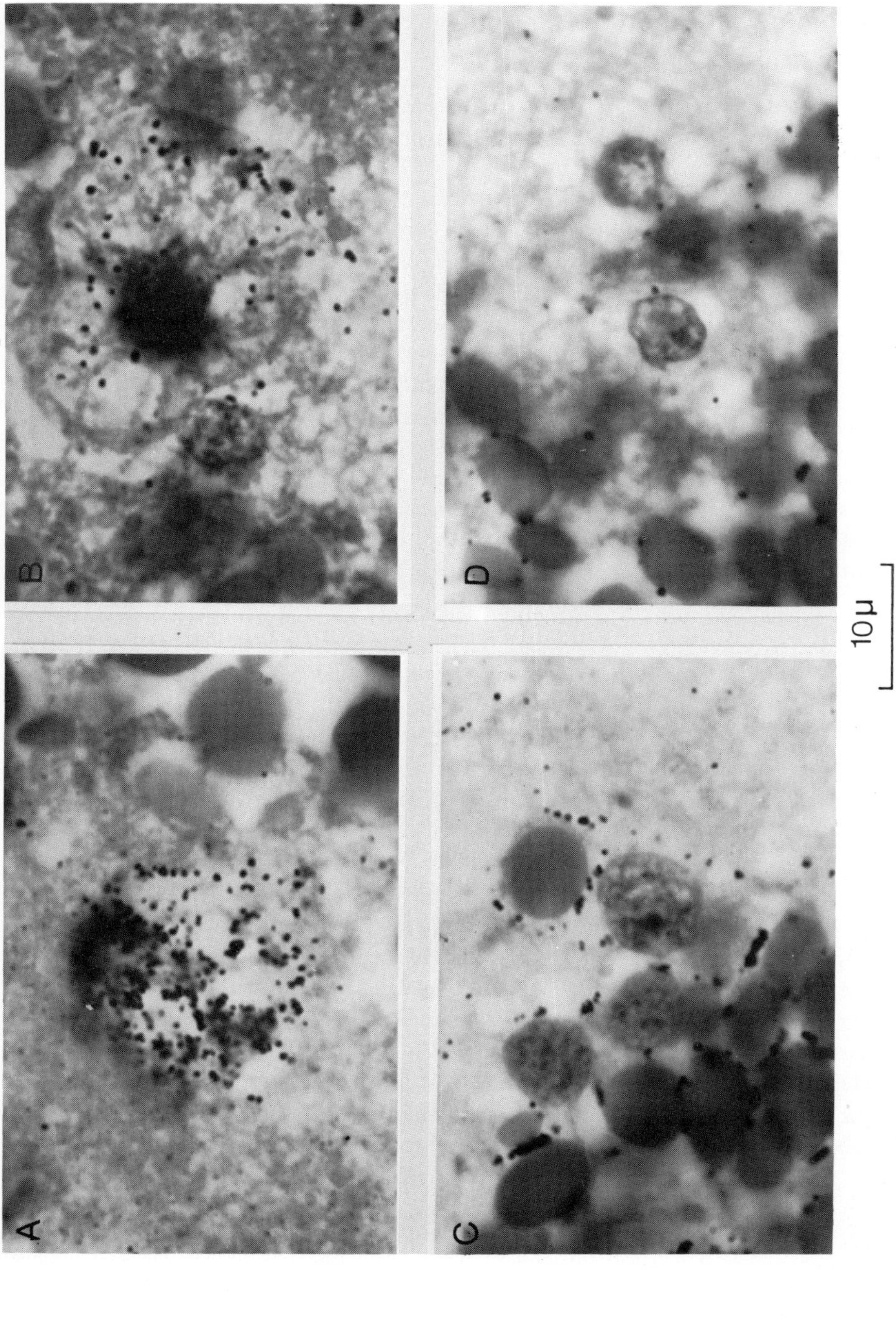
A
B
C
D
10μ

Effects of deoxynucleotide triphosphate injection on nuclear DNA synthesis

It has previously been shown that brain nuclei of frogs synthesize DNA when inserted into eggs [3] but not when inserted into oocytes [7]. The design of the experiments reported in this section is that brain nuclei are injected into oocytes together with deoxynucleotide triphosphates and a radioactive DNA precursor. The presence or absence of DNA synthesis is then detected by autoradiography of fixed, sectioned material.

A trivial explantation of any failure to observe labeling of nuclei in experiments of this design is that introducing large amounts of deoxynucleotide triphosphates might inhibit DNA synthesis. In addition, the injection of dTTP dilutes the [^{3}H] dTTP formed from injected [^{3}H] thymidine, thus preventing any DNA synthesis which is taking place from being visible. These possibilities were tested by first conducting the experiments with eggs. Fig. 1 (A, B) shows that the DNA of brain nuclei is labeled when the three triphosphates other than dTTP are injected, and also when all four are injected. The triphosphates other than dTTP would have expanded the pools in eggs by a factor of about 20–30, and the dTTP would have expanded its pools by a factor of eight. A smaller amount of dTTP was used or else the dilution of [^{3}H] dTTP would have been too great to have allowed DNA synthesis to have been detected readily. The experiments with eggs, Fig. 1, show that injecting deoxynucleotide triphosphates into eggs does not inhibit DNA synthesis, nor render it indetectable by the methods we use.

The experiment described for eggs was repeated using oocytes, No autoradiographic labeling was detected (Fig. 1, C and D). Two possible explanations of this result are that the nucleotides might have been broken down in cells which do not synthesize DNA, and that the radioactive [^{3}H] thymidine might not have been converted to the direct DNA precursor (which we assume to be [^{3}H] dTTP). The first possibility was tested by adding radioactive nucleosides to the brain nuclei and triphosphates injected into oocytes. All four deoxynucleosides were converted to triphosphates, although the extent of this conversion was low for both [^{3}H] guanosine and [^{3}H] thymidine (Fig. 2). The former is converted to its triphosphate to only a small

Fig. 1. Autoradiographic analysis of DNA synthesis by brain nuclei in eggs and oocytes. (A): Brain nuclei injected into eggs together with (^{3}H) thymidine and approximately 400 p-moles dCTP, dATP, and dGTP per egg and fixed after 45 minutes. (B): Similar to (A), but with the addition of 80 p-moles dTTP per egg, and fixed after 90 minutes. (C) and (D): Similar to (B), but injections into oocytes not eggs and fixed after 2 hours. It may be observed that the nuclei in eggs have synthesized DNA, but not those in oocytes.

extent, even when injected alone into eggs. The phosphorylation of the latter is known to be inhibited by injected dTTP in embryonic frog cells [22]. The phosphorylation of all four nucleosides is, however, sufficient to show that there could have been no substantial degradation of the deoxynucleotide triphosphates which accompanied them. Fig. 2 also provides evidence against the second explanation; even though the conversion of [^{3}H] thymidine to [^{3}H]dTTP was small, it was not much less than in eggs. In other experiments, the [^{3}H] thymidine was injected into oocytes before the nuclei and triophosphates. At the end of the 2 hour incubation period, 50 % of the injected [^{3}H] TdR was converted to [^{3}H] dTTP, but the nuclei were unlabeled. On the other hand, eggs injected first with brain nuclei, then with [^{3}H] thymidine and deoxynucleotide triphosphates had only 13 % of the injected [^{3}H] thymidine converted to [^{3}H] dTTP, but still produced nuclei which were well labeled in autoradiographs.

We may conclude from these experiments that little, if any, DNA synthesis is induced in nuclei by the introduction of deoxynucleotide triphosphates into the cytoplasm. One possible explanation of this result is that the nucleotides were unable to enter the nuclei. Evidence against this explanation is presented in the next section.

1		2	3	4	5	6
Type of Cell		Radioactive nucleoside	Other substances injected	% phosphorylation	% converted to triphosphate	Autoradiographic labeling
	1	[^{3}H] thymidine	None	51	49	Not tested
Eggs	2	[^{3}H] thymidine	Brain nuclei, dCTP, dATP, dGTP	21	19	+
	3	[^{3}H] thymidine	Brain nuclei, dTTP, dCTP, dATP, dGTP	18	16	+
	4	[^{3}H] thymidine		31	8	-
Oocytes	5	[^{3}H] deoxycytidine	Brain nuclei, dTTP, dCTP, dATP, dGTP	88	75	-
	6	[^{3}H] deoxyadenosine		24	17	-
	7	[^{3}H] deoxyguanosine		8	7	-

Fig. 2. Conversion of radioactive nucleosides to nucleotides when injected into oocytes or eggs together with brain nuclei.

400 p-moles dCTP, dATP, and dGTP were injected into each egg or oocyte. 80 p-moles of dTTP were injected. Experiments 1–3 utilized a single batch of eggs, and the autoradiographic results are shown in Fig. 1, (A) and (B). Experiments 4–7 utilized oocytes from one ovary; the autoradiographic results of the [^{3}H] thymidine experiments being shown in Fig. 1, (C) and (D). Columns 4 and 5 show phosphorylation of radioactive nucleosides indicated in column 2.

Effect of deoxynucleotide injection on the replication of injected DNA

It has been shown by *Gurdon* et al. [10] that both single- and double-stranded DNA injected into eggs stimulate the synthesis of DNA of similar buoyant density. Evidence that true replication is involved has been obtained for native polyoma DNA by *Laskey* and *Gurdon* (unpublished). In oocytes, on the other hand, native DNA is not replicated [1, 13], though denatured DNA is replicated [1]. The behaviour of native DNA thus reflects that of intact nuclei in this respect, as it does in others [13, 23]. It is therefore possible to inject native DNA and the deoxynucleotide triphosphates into oocytes, in order to test whether it is replicated under these conditions. Fig. 3 shows that it is not replicated when 20 p-moles of dCTP, dATP, and dGTP, together with [^{3}H] thymidine which forms 0.7 p-moles of dTTP (about 7 p-moles are already present in the cell [24]) are injected into the oocyte (i.e., expanding the pools to a little more than their size in eggs).

The fact that pure native DNA behaves in the same way as nuclei, enables us to conclude that nuclei do not fail to synthesize DNA in oocytes injected with triphosphates because these substances fail to pass through the nuclear membrane.

Type of injected template	ng of template injected per cell	Precursors injected	pg DNA synthesized
Saline control	0	[^{3}H] thymidine, dCTP, dATP, dGTP	171
Native DNA	30	[^{3}H] thymidine	157
Native DNA	30	[^{3}H] thymidine, dCTP, dATP, dGTP	237
Denatured DNA	30	[^{3}H] thymidine	987
Denatured DNA	30	[^{3}H] thymidine, dCTP, dATP, dGTP	1017

Fig. 3. Effect of exogenous deoxynucleotide triphosphates on the replication of injected DNA in oocytes.

Large oocytes were taken from a female that had ovulated three days previously, injected in groups of 30, and incubated for 6 hours. Substances were injected at the following concentrations: dATP, 2.5 mM; dCTP, 2.5 mM; dGTP, 2.3 mM; [^{3}H] thymidine (25 Ci/mmole) 8 mCi/ml; calf thymus DNA, 600 μg/ml. The DNA synthesized was calculated from a knowledge of the recovery of DNA, the amount of [^{3}H] thymidine converted to [^{3}H] dTTP and the size of the dTTP pool [24]. The radioactivity in DNA was taken as acid insoluble counts/min banding within the limits of calf thymus DNA marker on CsCl gradients. The G + C content was assumed to be 42.9 %, and the average molecular weight for deoxynucleotide monomers in DNA to be 335 daltons. The high value for DNA synthesis in the controls, produced by follicle cells, should be noticed when assessing the results with injected native DNA.

Measurement of the available deoxynucleotide triphosphate using replication of denatured DNA in oocytes

It was mentioned in the previous section that denatured DNA is replicated in oocytes. This result indicates that all four deoxynucleotide triphosphates must be present in the oocyte. Quite large amounts of DNA were injected to give a large stimulation of DNA synthesis over the rather high background from the follicle cells which invest the oocyte. The percentage replication is therefore low, but it enables us to set a minimum value on the amount of precursor available during the incubation period. This calculation is shown in Fig. 4, which indicates that during the six-hour incubation period sufficient triphosphate was available to synthesize between 100 and 150 nuclei. This level of purine triphosphate would not have been detected by *Woodland* and *Pestell* [24].

The pools of deoxynucleotide might conceivably be lower at any one time than the estimates obtained here, because they are continually replenished by de novo synthesis. Even if this were so, the total amount of deoxynucleotide does not seem to be limiting the rate of DNA synthesis, since the injection of additional deoxynucleotide triphosphates does not modify the rate of DNA synthesis on a denatured DNA template (Fig. 3).

Type of cell	Labeled precursor		ng template injected per cell	pg DNA synthesized	Nuclei equivalents of DNA synthesized	p-moles nucleotide incorporated			
	Type	p-moles triphosphate per cell				dTTP	dAMP	dCMP	dGMP
Oocytes Type 1*)	[^{3}H] thymidine	0.73	30	814	135	0.69	0.69	0.52	0.52
Oocytes Type 2*)	[^{3}H] dATP	2.52	30	728	122	0.62	0.62	0.47	0.47
Eggs	[^{3}H] thymidine	1.62	30	1505	251	1.28	1.28	0.97	0.97

Fig. 4. Estimation of minimum functional deoxynucleotide triphosphate pool sizes by replication of injected DNA.

*) Type 1 oocytes were large oocytes from a female that had laid three days previously; Type 2 oocytes were large oocytes from a female that had not laid for many months. These oocytes differ in their ability to convert [^{3}H] thymidine to [^{3}H] dTTP.

Denatured calf thymus DNA (600 μg/ml) was injected together with either [^{3}H] thymidine (25 Ci/mmole) or [^{3}H] dATP (4.1 Ci/mmole), and the cells incubated for 90 minutes (eggs) or 6 hours (oocytes). Details of extraction, analysis, and calculation are similar to those indicated in Fig. 3. The dATP pool of an oocyte was assumed to be zero, and the dTTP pool to be 7 p-moles [24]. The dTTP pool of an egg was taken as 12 p-moles [24]. The amount of DNA synthesized in control samples injected with [^{3}H] thymidine or [^{3}H] dATP has been subtracted.

Conclusions

The data presented in this paper show that the initiation of DNA synthesis at the start of amphibian development is not mediated through the supply of deoxynucleotide triphosphates. This study provides an example of the way in which the control of gene activity may be studied in these embryonic cells by microinjection techniques. In particular, it shows how the alteration of the composition of the cell may be achieved, and the sort of way in which the availability of substances for reactions in living cytoplasm may be measured.

Acknowledgements: The authors gratefully acknowledge the technical assistance of Miss *S. E. Ayers* and Mrs. *V. A. Moar.* The work was financially supported by the Medical Research Council of Great Britain.

References

[1] *Ford, C. C., and Woodland, H. R.* Manuscript in preparation.

[2] *Fischberg, M.* 1960. Will nuclear transplantation lead to a genetics of somatic cells? Symposium on Germ Cells and Development. Institut Intern. d'Embryologie and Fondazione A. Baselli. 478.

[3] *Graham, C. F., Arms, K., and Gurdon, J. B.* 1966. The induction of DNA synthesis by frog egg cytoplasm. Devl. Biol. **14:** 349.

[4] *Graham, C. F., and Morgan, R. W.* 1966. Changes in the cell cycle during early amphibian development. Devl. Biol. **14:** 439.

[5] *Gurdon, J. B.* 1962a. The developmental capacity of nuclei taken from intestinal epithelium cells of feeding tadpoles. J. Embryol. exp. Morph. **10:** 622.

[6] *Gurdon, J. B.* 1962b. Multiple genetically identical frogs. J. Heredity **53:** 5.

[7] *Gurdon, J. B.* 1967a. On the origin and persistance of a cytoplasmic state inducing nuclear DNA synthesis in frogs eggs. Proc. natn. Acad. Sci., U.S.A. **58:** 545.

[8] *Gurdon, J. B.* 1967b. African clawed toads. In: *Wilt, F. H. and Wessels, N. K.* (ed.): Methods in developmental biology. Thomas Y. Cromwell Co., New York: 75.

[9] *Gurdon, J. B.* 1968. Changes in somatic cell nuclei inserted into growing and maturing amphibian oocytes. J. Embryol. exp. Morph. **20:** 401.

[10] *Gurdon, J. B., Birnstiel, M. L., and Speight, V. A.* 1969. The replication of purified DNA introduced into living egg cytoplasm. Biochim. Biophys. Acta. **74:** 614.

[11] *Gurdon, J. B., Lane, C. D., Woodland, H. R., and Marbaix, G.* 1971. Use of frog eggs and oocytes for the study of messenger RNA and its translation in living cells. Nature **233:** 177.

[12] *Gurdon, J. B., and Laskey, R. A.* 1970. Methods of transplanting nuclei from single cultured cells to unfertilized frogs eggs. J. Embryol. exp. Morph. **24:** 249.

[13] *Gurdon, J. B., and Speight, V. A.* 1969. The appearance of cytoplasmic DNA polymerase activity during the maturation of amphibian oocytes into eggs. Exptl. Cell Res. **55:** 253.

[14] *Gurdon, J. B., and Uehlinger, V.* 1966. "Fertile" intestine nuclei. Nature **210:** 1240.

[15] *Gurdon, J. B., and Woodland, H. R.* 1968. The cytoplasmic control of nuclear activity in animal development. Biol. Rev. **43:** 233.

[16] *Gurdon, J. B., and Woodland, H. R.* 1969. The influence of the cytoplasm on the nucleus during cell differentiation, with special reference to RNA synthesis during amphibian cleavage. Proc. Roy. Soc. B. **173:** 99.

[17] *Gurdon, J. B., and Woodland, H. R.* 1970. On the long term control of nuclear activity during cell differentiation. Devl. Biol. **5:** 39.

[18] *Lane, C. D., Marbaix, G., and Gurdon, J. B.* 1970. Rabbit haemoglobin synthesis in frog cells: the translation of reticulocyte 9s RNA in frog oocytes. J. Mol. Biol. **61:** 73.

[19] *Laskey, R. A., and Gurdon, J. B.* 1970. Genetic content of adult somatic cells tested by nuclear transplantation from cultured cells. Nature **228:** 1332.

[20] *Moar, V. A., Gurdon, J. B., Lane, C. D., and Marbaix, G.* 1971. Translational capacity of living frog eggs and oocytes, as judged by messenger RNA injection. J. Mol. Biol. **61:** 93.

[21] *Werner, R.* 1971. Nature of DNA precursors. Nature New Biology **233:** 99.

[22] *Woodland, H. R.* 1969. The phosphorylation of thymidine by oocytes and eggs of Xenopus laevis Daudin. Biochim. Biophys. Acta. **186:** 1.

[23] *Woodland, H. R., Ford, C. C., Gurdon, J. B., and Lane, C. D.* 1972. In press.

[24] *Woodland, H. R., and Pestell, R. Q. W.* 1972. Determination of the nucleoside triphosphate contents of eggs and oocytes of Xenopus laevis. Biochem. J. **127:** 597.

Discussion

Fischberg: Dr. *Woodland* stated many experiments and mentioned them in the context of genetic engineering. Let us dive right into this field about which we have so far not spoken much – genetic engineering. Let us pose the question: Could transplantation of nuclei from somatic cells be used as an improvement of livestock or other organisms? Dr. *Woodland* cites that epithelial cells of the gut can, after transplantation, lead to normal development; the problem has frequently been raised whether one could transfer somatic nuclei from a superior donor into the eggs of a genetically inferior female and thereby improve these inferior strains. A few things must however be pointed out. Firstly, somatic nuclei of advanced developmental stages give rise only rarely to normal development. Usually, abortive cleavage results, and many monstrous embryos are created. Secondly, one could think of improving a strain by transplanting nuclei from early embryos, let's say from a blastula stage rather than either from adult tissue or advanced embryonic hosts. By such so-called cloning experiments, one can very well transmit the genotype of a chosen individual, at least in frogs. These cloning experiments, by serial transplantation or by single transplantation of blastula nuclei, can also be used for genetical analysis to compare the effects of genotype and environment in development.

Woodland: I think there is not very much that I would add, except that I'd emphasize that the main use of this sort of work is in genetic analysis, rather than engineering for it's own sake. In other sorts of animals, it might be of some value in stock breeding, but nobody has yet thought of a commercial use for the sort of frog we use!

Bodmer: I should like to come down to more prosaic things, since I gather we will discuss these aspects of engineering tomorrow afternoon. I would like to ask a question concerning the injection of the DNA. Do you know whether native DNA gets into the nucleus of the oocyte?

Woodland: We don't know. In the egg there is only a very small nucleus, and one can insert quite a large amount of DNA, so it is unlikely that all of it enters the nucleus. Entry into the nucleus is not necessary for replication to occur.

Bodmer: Do you have a guess as to the source of the enzymes which are doing the DNA synthesis? One thought that occurred to me was that it is occurring in the cytoplasm, since there is DNA synthesis in the cytoplasm of the oocyte. What about mitochondrial enzymes, for example? What spectrum of enzymes is found in the oocyte?

Woodland: I would think that the enzymes are more or less free in the cell, and they can be found in the ribosomal supernatant of homogenized cells (*C. C. Ford,* personal communication). There is a reason for this because the fertilized egg has just nine hours to make 30.000 nuclei, and it's unlikely that it will be able to synthesize all the relevant substances *de novo* in such a short interval. We should, therefore, expect a cytoplasmic store to exist.

Bodmer Then, why the difference between the oocyte and the egg? Is this a matter of the availability of the enzyme?

Woodland: We don't know. In the oocyte, in contrast to the egg, there is very inefficient, if any, replication of native DNA, both when the DNA is injected into the cell and in extracts of the cell (23), but a nonenzymic factor could be responsible for the difference between eggs and occytes.

Ringertz: Do you know what chemical difference might be responsible for triggering the DNA synthesis in one case and not in the other?

Woodland: No, at the moment we have no ideas at all.

Ringertz: You have no way of manipulating noninducing cytoplasm so that it supports DNA synthesis?

Woodland: No, as yet that hasn't been done. One thing that one might do is inject some of the egg cytoplasm into the oocyte in order to see if one is able to alter its DNA metabolism. This has been done with *Rana pipiens* by *Masui* and *Markert* [1]), but unfortunately the oocyte alters and becomes an egg.

Hirt: I would like to ask a question about the work you referred to by *Laskey* and *Gurdon.* So far as is known the host range of polyoma virus is restricted absolutely to mammals.

Woodland: Yes.

Hirt: And there we have a lag phase after infection of about 12 hours before you see DNA synthesis. How long is this lag in the frog?

Woodland: Well, we don't know if there is a lag, but there is certainly synthesis within two hours. We would expect a lag of 20 minutes because in this cell that's when the nuclear DNA synthesis normally happens. Injected nonviral native DNA certainly shows this effect.

Glick: Have you injected nuclei from different species into the egg? Secondly, when you inject DNA into the eggs, do you get any stimulation of RNA synthesis?

Woodland: In answer to the first question, it depends on the source of the nuclei, but normally one gets development as far as the late blastula with nuclei from all sorts of frogs, and then they die. If you inject DNA into the egg, then there is no stimulation of RNA synthesis, but in the oocyte there is. This fits with the observations that when one inserts nuclei into eggs and oocytes, one gets exactly the same effects, and that the resident nuclei also behave in this way.

Lewandowski: Would you care to comment further on the insertion of hemoglobin message?

Woodland: The evidence makes it quite certain that hemoglobin was being made (gel electrophoresis, peptide maps, and so on). The amount that's made is more than double what the cell makes of its own sorts of proteins. There seems to be little competition between the message you insert and the message that is working in the egg. The message is translated with an efficiency comparable to that in the reticulocyte. I think that they are the main points.

---

1) *Y. Masui,* and *C. L. Markert,* (1971): J. Exp. Zool. **177,** 129.

v. Ehrenstein: When you inject denatured DNA into the oocyte, what is the product? Is it native DNA?

Woodland: It's a mixture of native and denatured. The proportions depend on how much you insert. We normally insert large amounts of denatured DNA to get the maximum possible amount of DNA synthesis, but the product is normally a 50/50 mixture of native and denatured DNA. This is interesting because if one injects large amounts of DNA, there is only a small amount of replication, yet, there is a singificant amount of native DNA made. So it appears that when the enzyme begins on one DNA molecule, it finishes replicating that fragment completely.

Hurwitz: How much replication did you get in the experiments you described?

Woodland: About 0.5–1 % for native DNA and 3 % for denatured DNA. This is after injecting a large amount of DNA. If as little as 200 pg are injected, there is about 10 % replication.

Hurwitz: Well, then the following questions come up: (1) What is the extent of repair in a system such as you study, and (2) What happens if you use heterologous DNA in egg; what is the type of DNA which is made?

Woodland: If you use heterologous DNA, then the DNA that's made, seems to be the same as that which was injected. For instance, if you use *Xenopus* cistrons for ribosomal RNA, then it's ribosomal DNA that's synthesized. If you use DNA from microorganisms, then that is the sort of DNA that is synthesized, although this sort of DNA does not seem to be replicat when injected in a native form. I wouldn't like to say very much about replication versus repair, because it is very hard to sort out these phenomena. The only real evidence that we have at the moment is using the circular viral DNA (*Laskey* and *Gurdon,* unpublished), since it's very hard to explain these results on a basis of repair. Here you begin with DNA which is in the form of super coils, and then you get new helices formed which are also in the form of super coils, indicating that the DNA is not nicked, there is no detectable degradation of the DNA.

Hurwitz: Maybe I misunderstood that experiment. I thought that you were using a density label and radioactivity. Was the density label throughout the entire molecule or only in one strand?

Woodland: The density label was in one strand; the radiation was in both strands. Radioactive DNA was formed, both strands of which lacked the density label.

Hurwitz: What happens to the BUdR if there is no degradation?

Woodland: It remains as high molecular weight material, but we don't know if it is replicated. If one uses DNA which is "heavy/heavy", then there is no replication at all. So for some reason or other the enzyme doesn't seem to function with "heavy" strands.

Hurwitz: What you are saying here is that you have one strand which is labelled with BU while the entire molecule is labelled with radioactivity. You may then have a selective removal of the density label, but this might be achieved by degradation.

Woodland: Yes, but if degraded it would presumably form acid-soluble material, and none was detected by *Laskey* and *Gurdon.*

Hurwitz: But it may not necessarily be acidsoluble, it might also be nucleotides which are acid-insoluble.

Duesberg: Can your system be used for DNA-dependent RNA synthesis?

Woodland: Yes, in the oocyte DNA will stimulate the synthesis of RNA.

Nucleocytoplasmic Interactions and the Control of Nuclear Activity

Nils R. Ringertz, Sten-Anders Carlsson, and Robert E. Savage *)

Institute for Medical Cell Research and Genetics, Medical Nobel Institute, Karolinska Institutet, 104 01 Stockholm 60, Sweden

*) Present address: Dept of Biology, Swarthmore College, Swarthmore, Penn. 19081, USA

Summary: The nuclei of mature chick erythrocytes are genetically inactive and do not synthesize RNA, DNA, or protein. In heterokaryons formed by fusing chick red cells with human HeLa cells, the erythrocyte nucleus is reactivated, grows in size, and resumes RNA synthesis. The reactivation process occupies several days after cell fusion and results in the synthesis of new chick specific protein. The signals activating the chick genome appear to operate at several different levels. Early changes in the physicochemical properties of chick deoxyribonucleoprotein appear to be triggered by changes in the ionic environment. Nuclear growth during the first two days after fusion is due mainly to an accumulation in the chick nucleus of human nucleospecific molecules, probably proteins. These proteins are taken up from the surrounding HeLa cytoplasm or directly from HeLa nuclei lying in close contact with the chick nuclei. With increasing numbers of chick erythrocyte nuclei in the cytoplasm, less human nucleospecific proteins accumulate in each chick nucleus and the rate at which RNA synthesis accelerates is slowed down.

By reactivating chick erythrocyte nuclei in the cytoplasms of rat and mouse cells representing different lines of cell differentiation, attempts have been made to study how the cytoplasm influences the phenotype expressed by the reactivated chick erythrocyte nucleus. Chick erythrocyte nuclei have been reactivated in rat myoblasts and myotubes at a time when rat myosin synthesis starts. No significant chick myosin synthesis has been demonstrated in spite of the fact that the chick erythrocyte nuclei developed a nucleolus, produced RNA, and expressed other chick genes in the form of chick antigens. Hybrids made by fusing chick myoblasts with rat myoblasts, however, synthesized both chick and rat myosin.

Introduction

The role of different forms of nucleocytoplasmic interactions in the control of nuclear activity has been studied in cell fusion experiments in which a completely dormant cell nucleus, that of the chick erythrocyte, is reactivated in the cytoplasm

Manuscript received: 10 December 1971

of human tumor cells. The nuclei of mature chick erythrocytes are inactive with respect to DNA, RNA, and protein synthesis [3]. Their chromatin is tightly condensed and nucleoli are missing. *Harris* and coworkers [12, 13] have demonstrated that in heterokaryons made by virus induced fusion of chick erythrocytes with human HeLa cells, the chick erythrocyte nucleus undergoes an activation process and resumes RNA synthesis. We have subsequently examined:

a) changes in nuclear composition during this activation process
b) the nature of the cytoplasmic signals which activate the chick genome
c) the phenotype expressed by the chick nucleus when it is reactivated in the cytoplasm of different types of cells.

Materials and methods

Cell fusion

The heterokaryons and synkaryons studied have been produced by mixing different types of cells with Sendai virus killed by UV-irradiation [2, 12]. Hybrids between rat myoblasts and chick myoblasts form spontaneously [30] without the addition of virus in cultures maintained on Eagles MEM, horse serum, and chick embryo extract (89 : 10 : 1).

Cytochemical methods

The cytochemical methods used in the analysis of cell nuclei in cell hybrids or nuclei isolated from such cells are summarized in Fig. 1.

Molecule	Property	Parameter	Measuring technique
DNA	quantity	Feulgen absorbancy	microspectrophotometry
		UV absorbancy	UV-microspectrophotometry
	synthesis	^{3}H-thymidine incorp	autoradiography
RNA	synthesis	^{3}H-uridine incorp	autoradiography
total dry mass	quantity	optical path difference	microinterferometry
DNA-protein complex	"free" PO_4	acridine orange binding [19]	microfluorimetry
	thermal stability	acridine orange binding of formaldehyde heated nuclei [20, 25]	microfluorimetry
	acid stability	hydrolysis curve [16a, 22]	microspectrophotometry

Fig. 1. Survey of cytochemical methods used in characterizing cell hybrids

Immunological methods

Antibodies against human nucleolar, nucleoplasmic, and cytoplasmic antigens were obtained from patients suffering from lupus erythematosus, rheumathoid arthritis, and other "autoimmune" diseases [24]. Antibodies against human membrane antigens were obtained by injecting rabbits with HeLa cells. In a similar manner, antibodies were obtained against chick membrane antigens, chick nucleoli, and chick myosin. Binding of antibodies to cells fixed in ethanol acetone (1 : 1) was demonstrated with the indirect immune fluorescence method. After "staining" with patient or rabbit antibodies, the binding of these antibodies was demonstrated by a second staining step in which the cells were exposed to fluorescein conjugated sheep antibodies, to human immunoglobulins, and rabbit immunoglobulins respectively.

Results

Reactivation of chick erythrocyte nuclei

Reactivation of the chick erythrocyte nucleus in heterokaryons is a process which occupies several days following cell fusion. The rate at which reactivation occurs varies considerably depending on: (a) what type of chick erythrocytes are used, (b) the state of the cytoplasm in which the erythrocyte nucleus is inserted, (c) the ratio of erythrocyte nuclei to nuclei derived from the other parental cells, and (d) general culture conditions. The fastest reactivation takes place when embryonic chick erythrocytes are fused with rapidly growing cells (e.g., tumor cells, fibroblasts, etc.), and the heterokaryons are maintained under optimal culture conditions. Among the different heterokaryons formed, reactivation will be most rapid in those containing a high ratio of tumor cell nuclei to chick erythrocyte nuclei. During reactivation, the chick erythrocyte nucleus undergoes a series of morphological and chemical changes. The following is a list of the main events taking place when thirteen-day embryonic chick erythrocyte nuclei are reactivated in the cytoplasm of human tumor cells (HeLa):

a) early changes in the physicochemical properties of deoxyribonucleoprotein ("chromatin activation") [2, 21, 22],
b) increase in nuclear volume and dry mass, and a dispersion of the condensed chromatin [2, 12],
c) migration of *human* nucleospecific proteins into the chick nucleus [7, 24],
d) initiation of RNA synthesis (hight-molecular, polydisperse RNA) [12,15],
e) migration of human nucleolospecific proteins into the chick nucleus [7, 24], and the formation of nucleolus [12, 28],

f) initiation of ribosomal RNA (rRNA) synthesis [15],

g) initiation of chick specific protein synthesis including chick nucleolar antigens, surface antigens, enzymes, etc. [6, 14, 15, 24].

h) migration of chick nucleolar components into the chick nucleolus [24],

i) DNA synthesis [2, 12],

j) synchronous entry of HeLa and chick nuclei into mitosis and the formation of mononucleate HeLa-chick synkaryons capable of continued proliferation [24, 27].

The approximate timing for the main morphological changes is indicated in Fig. 2. This presentation will be restricted to some steps in which nucleocytoplasmic interactions appear to control nuclear reactivation.

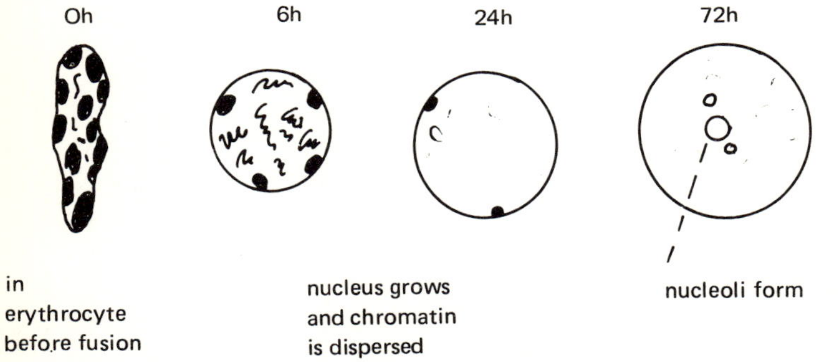

Fig. 2. Morphology of the chick erythrocyte nucleus in mature erythrocytes and in heterokaryons at different times after cell fusion.

"Chromatin activation"

At an early stage of the reactivation of the chick erythrocyte nucleus in HeLa cells, the DNA protein complex (DNP) undergoes drastic physicochemical changes which manifest themselves in an increased binding of basic intercalating dyes (acridine orange, ethidium bromide, Fig. 3), and an increased susceptibility to thermal denaturation (Fig. 4) and to acid hydrolysis [2].

Similar observations have been made in other situations where inactive cell nuclei suddenly become active and show an accelerated RNA synthesis [23, 16b]. We have also found the opposite type of cytochemical changes in situations where cells with an active RNA synthesis differentiate into a more inactive form [26, 32]. The nuclear condensation during spermiogenesis [8,25] and erythropoiesis [16a] is accompanied by a drastic decrease in acridine orange binding, increased stability of DNA to thermal denaturation, and acid hydrolysis. The cytochemical changes

Nuclei in	Acridine orange fluorescence arb. units ± S.E.M. (n)	Ethidium fluorescence arb. units ± S.E.M. (n)
chick erythrocytes	0.59 ± 0.02 (10)	0.65 ± 0.01 (10)
chick erythrocytes mixed with HeLa cells	0.77 ± 0.01 (15)	–
chick erythrocytes mixed with inactivated Sendai virus	0.96 ± 0.03 (20)	–
chick erythrocyte nuclei in heterokaryons		
14.5 hours after fusion	2.29 ± 0.05 (42)	–
24 hours after fusion	3.04 ± 0.13 (15)	–
46 hours after fusion	–	1.75 ± 0.09 (30)

Fig. 3. Binding of intercalating fluorochromes to chick nuclear deoxyribonucleoprotein.

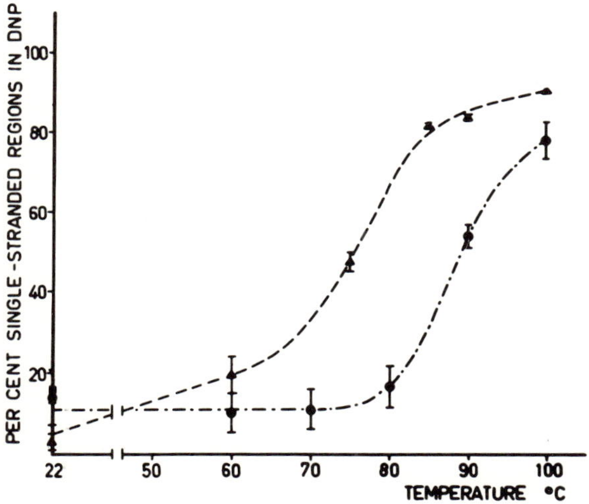

Fig. 4
Cytochemical heat denaturation curves for chick erythrocyte DNA in the nuclei of mature erythrocytes (●) and in heterokaryons 41 hours after cell fusion (▲).

in DNP during nuclear activation suggest alterations in the interactions between DNA-phosphate groups and their cationic counterions. This change may be due to a removal of proteins and/or divalent metal ions in the DNP complex as well as to more subtle conformational changes. Model experiments with chick erythrocyte ghosts and isolated nuclei [22] show that DNP changes similar to those in heterokaryons can be induced by manipulating the ionic environment. The first signal in the reactivation of the chick erythrocyte nucleus may therefore be the equilibration of ions bound to the chick chromatin with the ionic milieu of the

HeLa cells cytoplasm. This may "unmask" DNA phosphate groups, causing increased dye binding and greater sensitivity of DNA to thermal denaturation. A DNP complex thus "activated" can be expected to attract more water and show an increased affinity for proteins. The "activated" DNP molecules would have to stretch out due to repulsion between charged groups. Such a phenomenon, together with the increased water and protein content of the chick nucleus could explain the transition of the chromatin from a tightly condensed to a more dispersed state.

RNA synthesis

Shortly after fusion of chick erythrocyte nuclei with HeLa cells, the erythrocyte nucleus resumes RNA synthesis [12]. As the erythrocyte nucleus grows in size and the chromatin becomes more dispersed, the rate of labeling with ^{3}H-uridine increases. Autoradiographic data obtained by Harris, and confirmed in our work, suggest that the rate of RNA synthesis increases in direct relation to nuclear [12] growth. The RNA produced in the erythrocyte nucleus during the first days after fusion has a high molecular weight and is polydisperse [15]. This RNA is not transported to the cytoplasm and appears to be similar to the HnRNA found in other cell systems. Ribosomal RNA fractions (rRNA) cannot be detected until about the third day after fusion. The appearance of these RNA fractions coincides with the appearance of nucleoli visible under the light microscope. Transport to the cytoplasm of RNA synthesized in the chick nucleus and new synthesis of chick proteins (surface antigens) can only be detected from the stage where nucleoli develop [28] and rRNA fractions are synthesized. For further details concerning RNA synthesis in chick erythrocyte – HeLa cell hybrid cells – a recent review by *Harris* may be consulted [13].

Nuclear growth

During the first 24 hours after cell fusion, the chick erythrocyte nucleus grows markedly in size and the chromatin assumes a more dispersed form. The increase in nuclear volume is paralled by an increase in nuclear dry mass (Fig. 5). Microinterferometric and microspectrophotometric measurements demonstrate that the increase in dry mass is due to a drastic increase in nuclear protein content as well as to some increase in nucleic acid content [1,2]. The increase in nuclear protein could be due to the synthesis of chick proteins specified by the RNA molecules produced in the chick erythrocyte nucleus or to a redistribution of protein from the cytoplasmic compartment into the chick nucleus. In the latter case, the proteins accumulating in a chick nucleus could be expected to be of human origin. The chick erythrocytes lyse and lose their cytoplasm before fusion, and virtually all the cytoplasm of the hybrid cell therefore originates from the HeLa cell. In order to examine how the chick erythrocyte nuclei grew, erythrocytes were irradiated with

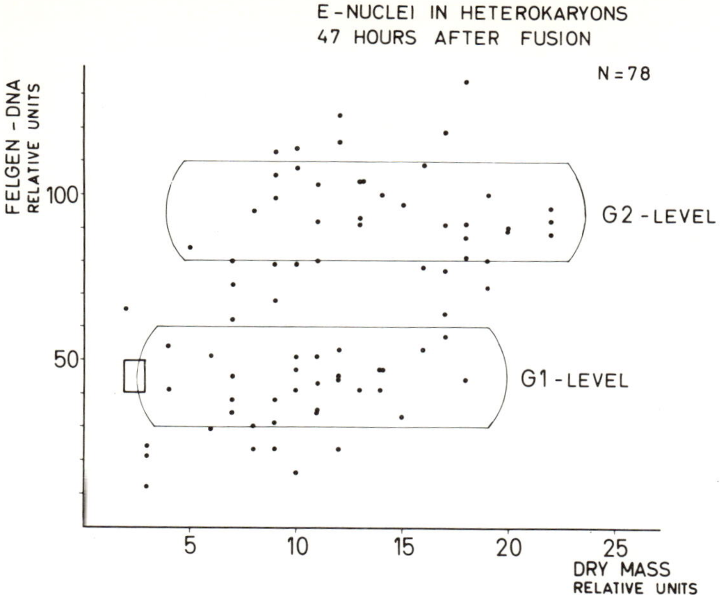

Fig. 5. DNA and dry mass of individual chick erythrocyte nuclei in heterokaryons 47 hours after fusion. The small rectangle represents the range of values found in nuclei isolated from chick erythrocytes. A large number of nuclei have completely replicated their DNA and virtually all nuclei have increased in dry mass. (Modified from *Bolund* et al. [2].)

UV-light prior to fusion. This effectively prevented the chick erythrocyte nuclei from starting RNA synthesis after fusion with HeLa cells. In spite of this, the chick erythrocyte nuclei were found to grow markedly in both volume and dry mass [1]. Thus it appeared that the proteins accumulating in the chick nucleus were in fact of human origin. In order to examine this question further, and in order to obtain more direct evidence, the chick nuclei in HeLa cell cytoplasms were examined for human nuclear antigens [7, 24]. For this analysis antisera were collected from patients with various "autoimmune" diseases. Using the indirect immune fluorescence reaction, the autoantibodies in these sera were found to bind to antigens in human cells including fixed HeLa cells but not to antigens in chick cells. At early stages of reactivation, these sera did not react with chick erythrocyte nuclei in heterokaryons. However, as the nuclei enlarged, human antigens could be detected in the nucleoplasm, nuclear membrane, and nucleolus of the chick nucleus (Fig. 6). This provided direct evidence for a migration of human macromolecules (antigens) into chick nuclei. More interesting, however, was the observation that only antigens characteristic of human cell nuclei accumulated in the growing chick erythrocyte nuclei. Antigens characteristic of human cytoplasm did not enter the chick nuclei. Therefore, the growth of the chick nucleus is not due to a random uptake of material from the surrounding human cytoplasm but to a *selective concentration of human nucleospecific macromolecules.* Moreover, the individual human antigens showed exactly the same distribution in the fully reactivated chick nucleus as they did in the HeLa nucleus. Human nuclear membrane, nucleoplasmic, and nucleolar antigens were thus found in equivalent structures in the chick nuclei and not randomly distributed in the chick nucleus [24]. This phenomenon may be due to selective penetration of antigenic macromolecules through the nuclear membrane and selective binding in the nuclear compartment. Altered properties of the DNP, such as those observed in this system, may be the reason for trapping of certain proteins (antigens) in the chick nucleus. The "activated" DNP molecules may function as an ion exchange resin and retain proteins in the nuclear compartment. The characteristic intranuclear distribution of human nucleolar and nucleoplasmic antigens in the chick nucleus is probably due to a specific affinity for specialized DNP regions. The fact that human membrane antigens are concentrated in the chick nuclear membrane may be due to an affinity of membrane protomers for preexisting membrane structures irrespective of species differences.

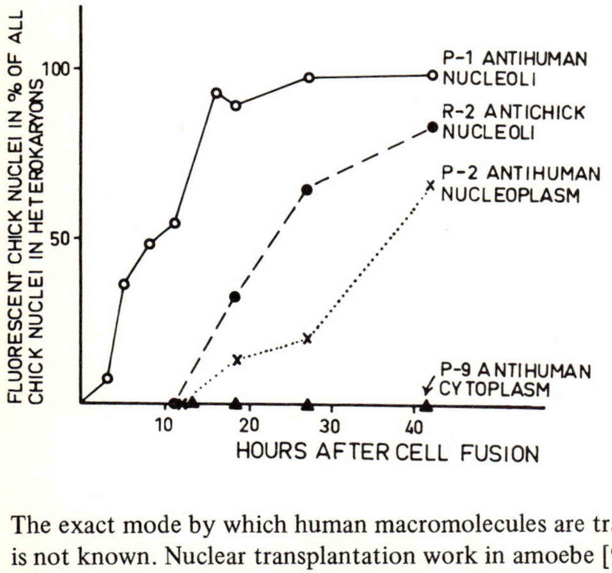

Fig. 6
Appearance of human and chick antigens in chick erythrocyte nuclei undergoing reactivation in HeLa x chick erythrocyte heterokaryons. (Modified from *Ringertz* et al. [24].)

The exact mode by which human macromolecules are transferred to the chick nucleus is not known. Nuclear transplantation work in amoebe [9, 10] and cytochemical studies on mouse fibroblasts [31] suggest that virtually all nuclear proteins are synthesized in the cytoplasm and migrate into the nucleus where some tend to remain, whereas others constantly shuttle back and forth between nucleus and

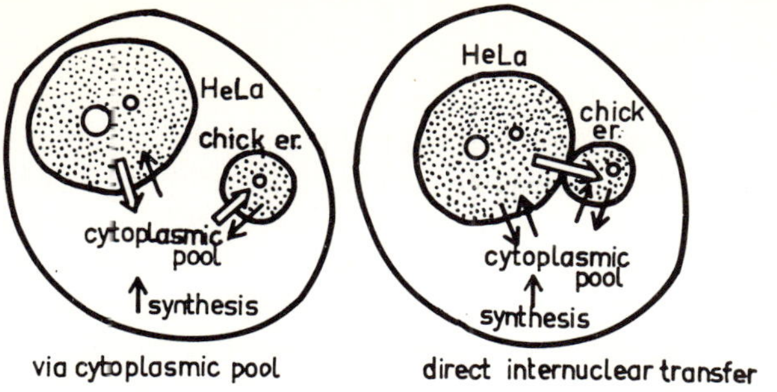

Fig. 7
Schematic illustration of the two way exchange of nucleospecific proteins between nuclear and cytoplasmic compartments. In HeLa x chick erythrocyte heterokaryons, the chick nucleus grows mainly by concentrating human nucleospecific proteins. These enter the chick via a cytoplasmic pool or by direct internuclear transfer.

cytoplasm [9, 10]. In the heterokaryons (Fig. 7), human nucleospecific proteins may move into the chick nucleus from a cytoplasmic pool or by direct internuclear transfer.

When the chick erythrocyte nucleus had been reactivated, chick nucleolar antigens were synthesized [24]. These antigens were then found in nucleoli or nucleoluslike bodies of both chick and HeLa nuclei. Similarly we have found that chick nuclear membrane antigens may appear in HeLa nuclear membranes (unpublished observations). It seems possible in this specific case, that chick nuclear membrane components formed before fusion, simply migrate over and become integrated into the HeLa nuclear membrane.

The analysis of human and chick nuclear antigens in heterokaryons at different time intervals after fusion show that both human and chick nuclei become molecular mosaics containing macromolecules specified by both species. Fig. 8 summarizes in a schematic way our present information about the composition of chick erythrocyte nuclei at different stages of reactivation. In our work with other types of heterokaryons, human-chick, rat-chick, mouse-chick, we have also noted that the rate at which nucleospecific antigens are exchanged varies with the type of cells fused and the antigens studied.

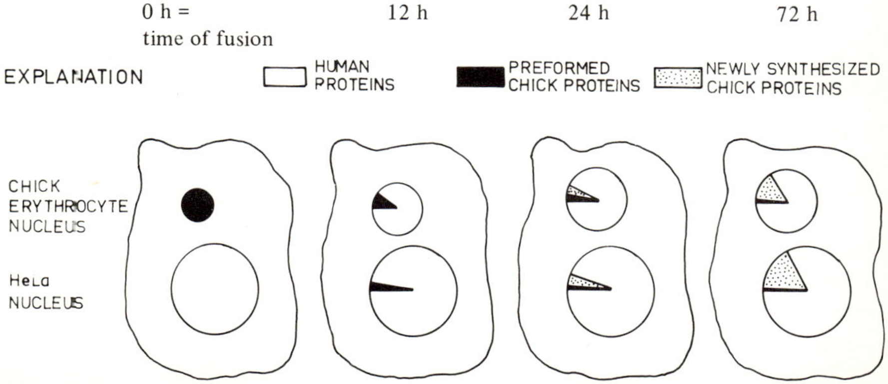

Fig. 8. Schematic illustration of the distribution of human and chick nuclear proteins in HeLa x chick erythrocyte heterokaryons at different times after cell fusion.

"Competition" between nuclei in heterokaryons

The growth of chick erythrocyte nuclei during the first 48 hours after the formation of a heterokaryon between chick erythrocytes and HeLa cells is a fairly rapid process. The average increase in nuclear dry mass during this time increases on the average by a factor of 5; a few individual nuclei increase by as much as 9–10 times [2]. Since this growth appeared mainly to be due to an uptake of human nucleospecific proteins, it seemed that the supply of human nucleospecific proteins could be a rate limiting factor in the reactivation of the chick genome. In order to explore this possibility, we examined the distribution of human nucleoplasmic antigens in heterokaryons containing varying proportions of HeLa and chick erythrocyte nuclei [7]. Fixed heterokaryons on slides were stained with human autoantibodies to nucleoplasmic antigens and then stained with fluorescein conjugated sheep antibodies to human immuno-globulins. The immune fluorescence was subsequently measured by point measurements in a special microfluorimeter where the fluorescence of each individual nucleus in the heterokaryons could be quantified. The results are given in Fig. 9. The "corrected" values were obtained by subtracting the unspecific cross reaction of chick fibroblast nuclei rather than chick erythrocyte nuclei. Although neither of these corrections is altogether satisfactory, the data clearly show that with an increase in the ratio of chick/HeLa nuclei, the chick erythrocyte nuclei pick up less human nucleospecific antigens. This suggests that the new synthesis of at least some human nucleospecific molecules may be so slow as to be a rate limiting factor when many chick erythrocyte nuclei are reactivated in the same cytoplasm [7].

Type of nucleus and cell	Nucleoplasmic fluorescence: Fluorescence ± S.E.M. arbitrary units	n	"Corrected" values
HeLa nuclei			
in HeLa cells	59.8 ± 4.7	17	
in heterokaryons HeLa + chick nuclei 1 + 1, 2 + 2, or 3 + 3	53.4 ± 4.0	15	
Chick nuclei			
in erythrocytes	1.0 ± 0.2	8	–
in fibroblasts	11.2 ± 0.7	10	0
in heterokaryons HeLa + chick erythrocyte nuclei			
(1 + 1, 2 + 2, 3 + 3) mean = 1 : 1	27.7 ± 2.8	15	16.5
(1 + 2, 4 + 6, 1 + 3, 2 + 3) mean = 1 : 2	18.6 ± 2.1	11	7.4
(1 + 7) mean = 1 : 7	14.0 ± 1.6	14	2.8

Fig. 9. Microfluorimetric analysis of the distribution of human nucleoplasmic antigens in HeLa-chick erythrocyte heterokaryons, four days after cell fusion [7].

In order to test if the "competition" for nucleoplasmic human antigens represented an exceptional situation without any significance for nuclear function, we examined heterokaryons made by fusing normal rat epithelial cells with chick erythrocytes. Heterokaryons containing varying proportions of rat and chick nuclei were examined for nuclear size and for rate of ^{3}H-uridine incorporation into RNA during a 30 minute pulse. It was found that at 32 hours after fusion, chick erythrocyte nuclei in heterokaryons containing rat and chick nuclei in combination 1 + 1, 1 + 2, and 1 + 3 had projected nuclear areas of 38±2, 29±1, and 25±1 μ^2 respectively, suggesting that nuclear enlargement was slowed down when many chick nuclei were being reactivated in the same cytoplasm. The ^{3}H-uridine autoradiographs were counted with respect to silver grains over each individual nucleus present in the hybrid cell. As shown in Fig. 10, the number of silver grains over each erythrocyte nucleus decreases with increasing ratios of chick/rat nuclei in the heterokaryon. Furthermore, the rate of RNA synthesis in the rat nucleus is reduced if many chick erythrocyte nuclei are present in the cytoplasm. The total number of silver grains over all nuclei in the heterokaryons (rat + chick), however, does not decrease with increasing numbers of erythrocyte nuclei, suggesting that the total RNA synthesis in the heterokaryons remains fairly constant. The decreased labeling of individual nuclei therefore is not due to reduced vitality of heterokaryons containing many erythrocyte nuclei. Rather it appears that the different nuclei in the heterokaryons compete with each other for factors which are limiting for the rate of RNA synthesis.

Nuclear ratio in heterokaryons rat : chick	Grains counted ± S.E.M.				
	Over the rat nucleus		Over the chick erythrocyte nucleus or nuclei		Over rat + all chick nuclei
1 : 1	105 ± 3	(n = 52)	24 ± 1	(n = 52)	129
1 : 2	97 ± 5	(n = 23)	15 ± 1	(n = 46)	127
1 : 3	86 ± 5	(n = 23)	12 ± 1	(n = 69)	122
1 : >3 (mean : 1 : 4.3)	76 ± 7	(n = 20)	13 ± 1	(n = 87)	133

Fig. 10. ^{3}H-uridine incorporation into RNA in rat epithelial cells + chick erythrocyte heterokaryons, 32 hrs after cell fusion.

Phenotype of the reactivated chick erythrocyte nuclei

The observation that the chick erythrocyte nucleus in HeLa heterokaryons grows by selectively concentrating human nucleospecific macromolecules and the indication that competition between chick erythrocyte nuclei for these molecules was a rate limiting factor for nuclear growth and also for the rate at which RNA synthesis accelerates poses a number of questions. One of these is whether the pattern of human macromolecules accumulating in the chick nucleus also determine the gene expression pattern. Greatly simplified, the question is whether the reactivated chick erythrocyte nuclei express an *erythroid* phenotype or an *undifferentiated* phenotype, of if the chick nucleus is *reprogramed* to express a phenotype similar to that of the other parental nuclei.

In order to study these questions, cell hybrids have been made by fusing chick erythrocytes with various highly differentiated cells. In particular, we have studied heterokaryons obtained after fusion of rat myogenic cells with embryonic chick erythrocytes [4,5]. Rat myoblasts have a characteristic morphology and will under certain conditions undergo spontaneous fusion and form very large multinucleated myotubes. At about the same time as the myoblasts fuse, the pattern of protein synthesis changes. New proteins characteristic of a myogenic phenotype begin to accumulate. Among these are myosin, actin, creatine phosphokinase, myoglobin, and glycogen phosphorylase [30]. Our primary aim in producing heterokaryons from chick erythrocytes and rat myoblasts has been to examine (a) if the chick erythrocyte nucleus can be reactivated in the cytoplasm of a differentiating cell, and (b) to examine if a reactivated chick erythrocyte nucleus in a rat myoblast or myotube will be reprogramed to specify the synthesis of *chick* myosin.

We found that in the presence of inactivated Sendai virus, chick erythrocytes readily fused with rat myogenic cells (Fig. 11). Furthermore, the chick nuclei were reactivated in the usual way in both rat myoblasts and in multinucleated myotubes. In agreement with our observations on HeLa-chick erythrocyte heterokaryons, we found that rat nuclear antigens migrated into the chick nucleus, and chick nucleolar and nuclear membrane antigens were also detected in the rat nuclei some time after fusion [4]. The observation that new chick nucleolar antigens are formed constitutes evidence that at least some chick genes have been fully reactivated and are being both transcribed and translated.

The search for chick myosin accumulation was carried out by immunological methods. Different rabbits and guinea pigs were immunized with rat and chick myosin respectively. After absorption of the anti-rat-myosin sera with chick myosin and vice versa, we obtained antibodies which reacted only with one type of myosin and which could be used in the indirect immune fluorescence reaction for the detection of rat or chick myosin. Although *rat* myosin synthesis is easily detected in some mononucleated rat myoblasts and in all myotubes except some of the smallest, we have so far been unable to obtain any convincing evidence that chick myosin accumulates in equivalent heterokaryons. Even relatively large myotubes containing 15–20 nuclei, of which one fourth were chick erythrocyte nuclei, failed to give any significant reaction with the rabbit antichick myosin antibodies. In order to test if this could be due to lack of sensitivity or specificity of our methods, we examined myotubes formed spontaneously from mixtures of chick myoblasts and rat myoblasts. The ratio of rat and chick myoblasts was varied so as to produce myotubes containing different proportions of rat and chick nuclei. Using the same immunological technique as described above, these hybrid myotubes were examined for the accumulation of rat and chick myosin by quantitative microfluorimetric measurements. The results clearly showed that both rat and chick myosin accumulated and that the relative proportions of the two myosins was directly related to the proportion of rat and chick nuclei in the individual myotubes [4].

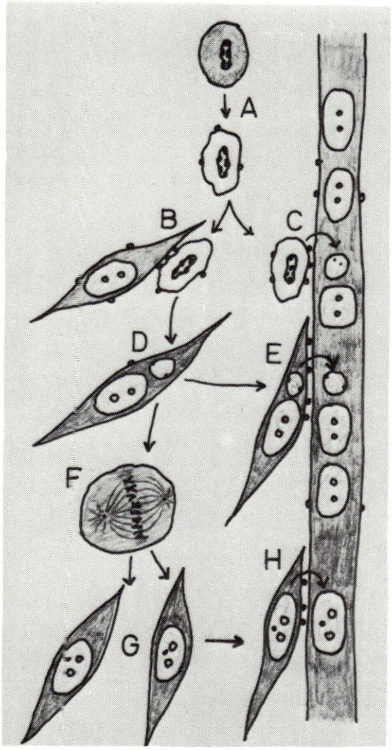

Fig. 11

Introduction of chick erythrocyte genomes into rat myogenic cells.

A. Lysis of chick erythrocytes exposed to inactivated Sendai virus.
B. Formation of heterokaryon by fusion of chick erythrocyte ghost with rat myoblast.
C. Direct fusion of erythrocyte ghost with myotube.
D. Chick erythrocyte nucleus undergoing reactivation in heterokaryon.
E. Fusion of heterokaryon with rat myotube.
F. Mitosis in heterokaryon resulting in the formation of –
G. two rat-chick synkaryons.
H. Introduction of a chick genome into a myotube by fusion of a rat-chick synkaryon with a rat myotube.

There may be many reasons why the chick erythrocyte nuclei reactivated in rat myogenic cells have failed to specify any detectable chick myosin synthesis. There are indications in the literature that myosin is synthesized with the aid of a long lived messenger formed some considerable time before myotube formation [30]. It may be therefore that we have not yet introduced erythrocyte nuclei at the right time point for reprograming to occur. Secondly, there are indications that a critical round of DNA synthesis may be required for a change in cell differentiation. These aspects are presently being explored by fusing chick erythrocytes with rat myoblasts long before myotubes form and by screening myotubes containing rat-chick synkaryotic nuclei for chick myosin accumulation.

Conclusions

The chick erythrocyte nucleus offers a useful model for studies on gene regulation since it can be reactivated both in cell hybrids and in cell-free extracts [17b, 29]. The evidence obtained with cell hybrids suggests that reactivation is a highly ordered self-assembly process involving different forms of nucleocytoplasmic interactions. Early changes ("chromatin activation") in the deoxyribonucleoprotein (DNP) may be induced by the equilibration of ions bound to the chick chromatin with the ionic milieu of the HeLa cytoplasm. This interpretation is supported by the model experiments [22] on erythrocyte ghosts and by observations on isolated polytene chromosomes, where changes in the ionic composition of the medium can mimic hormone induced puffing patterns [18].

Early growth of the chick erythrocyte nucleus in HeLa cells is primarily due to an uptake and selective concentration of *human* nucleospecific proteins. This phenomenon may be related to the altered physical properties of the DNP. Probably the "activated" DNP not only binds nucleospecific proteins but also determines the intranuclear localization of each individual protein molecule. The observation that several antigens found only in the nucleus of interphase HeLa cells leave the chromatin and become evenly distributed in the cytoplasm when the DNP condenses into metaphase chromosomes further stresses the role of the physical state of DNP in determining protein (antigen) binding in the nucleus (to be published). In the chick erythrocyte nucleus undergoing reactivation in a heterokaryon, the binding of a first generation of proteins to the DNP may create specific sites for the binding of later proteins by means of protein-protein interaction. The erythrocyte nucleus may grow through a self-assembly process quite similar to the assembly of ribosome subunits [17a].

During the reactivation process in cell hybrids, the chick DNA becomes surrounded by nucleospecific proteins typical of the "host" cell nucleus. This clearly may affect the phenotype expressed by the reactivated erythrocyte nucleus. *Gurdon* [11] has demonstrated that many somatic cell nuclei in frogs are totipotent with respect to their ability to undergo cell differentiation. In view of this, we have produced cell hybrids between chick erythrocytes and rat myogenic cells where it should be possible to detect if the chick erythrocyte nucleus becomes reprogramed to express a phenotype equivalent to that of the host cell, if it "remembers" its previous erythroid phenotype, or if it looses all markers characteristic of histodifferentiation.

Acknowledgement: Parts of this presentation are based on work performed in collaboration with Drs *H. Harris, L. Bolund, Z. Darzynkiewicz, O. Luger,* and *T. Ege* (see references).

References

[1] *Bolund, L., Darzynkiewicz, Z., and Ringertz, N. R.* 1969. Growth of hen erythrocyte nuclei undergoing reactivation in heterokaryons. Exptl. Cell. Res. **56:** 406–10.

[2] *Bolund, L., Ringertz, N. R., and Harris, H.* 1969. Changes in the cytochemical properties of erythrocyte nuclei reactivated by cell fusion. J. Cell. Sci. **4:** 71–83.

[3] *Cameron, I. L., and Prescott, D. M.* 1963. RNA and protein metabolism in the maturation of the nucleated chicken erythrocyte. Exptl. Cell. Res. **30:** 609.

[4] *Carlsson, S.-A., Luger, O., Ringertz, N. R., and Savage, R. E.* To be published.

[5] *Carlsson, S.-A., Savage, R. E., and Ringertz, N. R.* 1970. Behaviour of differentiated hen nuclei in the cytoplasm of rat myoblasts and myotubes. Nature **228:** 869–71.

[6] *Cook, P. R.* 1970. Species specificity of an enzyme determined by an erythrocyte nucleus in an interspecific hybrid cell. J. Cell. Sci. **7:** 1–3.

[7] *Ege, T., Carlsson, S.-A., and Ringertz, N. R.* 1971. Immune microfluorimetric analysis of the distribution of species specific nuclear antigens in the HeLa-chick erythrocyte heterokaryons. Exptl. Cell. Res. **69:** 472.

[8] *Gledhill, G. L., Gledhill, M. P., Rigler, R., and Ringertz, N. R.* 1966. Changes in deoxyribonucleoprotein during spermiogenesis in the bull. Exptl. Cell. Res. **41:** 652–65.

[9] *Goldstein, L., and Prescott, D. M.* 1967. Proteins in nucleocytoplasmic interactions. I. The fundamental characteristics of the rapidly migrating proteins and the slow turnover proteins of the amoeba proteus nucleus. J. Cell. Biol. **33:** 637–44.

[10] *Goldstein, L. and Prescott, D. M.* 1968. Proteins in nucleocytoplasmic interactions. II. Turnover and changes in nuclear protein distribution with time and growth. J. Cell. Biol. **36:** 53.

[11] *Gurdon, J. B.* 1970. Nuclear transplantation and the control of gene activity in animal development. Proc. R. Soc. Lond. (Biol.) **176:** 303–14.

[12] *Harris, H.* 1966. The reactivation of the red cell nucleus. J. Cell. Sci. **2:** 358–68.

[13] *Harris, H.* 1970. Nucleus and cytoplasm. 2nd edition, Clarendon Press, Oxford.

[14] *Harris, H., and Cook, P. R.* 1969. Synthesis of an enzyme determined by an erythrocyte nucleus in a hybrid cell. J. Cell. Sci. **5:** 121–34.

[15] *Harris, H., Sidebottom, E., Grace, D. M., and Bramwell, M. E.* 1969. The expression of genetic information. A study with hybrid animal cells. J. Cell. Sci. **4:** 499.

[16a] *Kernell, A.-M., Bolund, L., and Ringertz, N. R.* 1971. Chromatin changes during erythropoiesis. Exptl. Cell. Res. **65:** 1–6.

[16b] *Killander, D., and Rigler, R.* 1969. Activation of deoxyribonucleoprotein in human leucocytes stimulated by phytohemagglutinin. Exptl. Cell. Res. **54:** 163–70.

[17a] *Kurland, C. G.* 1970. Ribosome structure and function emergent. Science **169:** 1171–77.

[17b] *Leake, R. E., and Barry, J. M.* 1969. Stimulation of hen erythrocyte chromatin by extracts of rat liver or uterus. Biochem. J. **114:** 61.

[18] *Lezzi, M.* 1970. Differential gene activation in isolated chromosomes. Int. Rev. Cytol. **29:** 127.

[19] *Rigler, R.* 1966. Microfluorimetric characterization of intracellular nucleic acids and nucleoproteins by acridine orange. Acta Phys. Scand. **67:** suppl. 267.

[20] *Rigler, R., Killander, D., Bolund, L., and Ringertz, N. R.* 1969. Cytochemical characterization of deoxyribonucleoprotein in individual cell nuclei. Techniques for obtaining heat denaturation curves with the aid of acridine orange microfluorimetry and ultraviolet microspectrophotometry. Exptl. Cell. Res. **55:** 215–24.

[21] *Ringertz, N. R.* 1969. Cytochemical properties of nuclear proteins and deoxyribonucleoprotein complexes in relation to nuclear function. In: Lima-de-Faria, Handbook of molecular cytology, North Holland Publ. Co., Amsterdam.

[22] *Ringertz, N. R., and Bolund, L.* 1969. "Activation" of hen erythrocyte deoxyribonucleoprotein. Exptl. Cell. Res. **55:** 205–14.

[23] *Ringertz, N. R., Bolund, L., and Darzynkiewicz, Z.* 1970. AO binding of intracellular nucleic acids in fixed cells in relation to cell growth. Exptl. Cell. Res. **63:** 233–38.

[24] *Ringertz, N. R., Carlsson, S.-A., Ege, T., and Bolund, L.* 1971. Appearance of human and chick nuclear antigens in chick erythrocyte nuclei during reactivation in heterokaryons with HeLa cells. Proc. Natl. Acad. Sci. **68:** 3228–3232.

[25] *Ringertz, N. R., Gledhill, B. L., and Darzynkiewicz, Z.* 1970. Changes in deoxyribonucleoprotein during spermiogenesis in the bull. Sensitivity of DNA to heat denaturation. Exptl. Cell. Res. **62:** 204–18.

[26] *Ritzén, M., Carlsson, S.-A., and Darzynkiewicz, Z.* 1972. Cytochemical evidence of nucleoprotein changes in rat adrenal cortex following hypophysectomy or dexamethasone suppression. Exptl. Cell. Res. **70,** 462–465.

[27] *Schwartz, A. G., Cook, P. R., and Harris, H.* 1971. Correction of a genetic defect in a mammalian cell. Nature New Biol. **230:** 5–8.

[28] *Sidebottom, E., and Harris, H.* 1969. The role of the nucleolus in the transfer of RNA from nucleus to cytoplasm. J. Cell. Sci. **5:** 361–64.

[29] *Thompson, L. R., and McCarthy, B. J.* 1968. Stimulation of nuclear DNA and RNA synthesis by cytoplasmic extracts in vitro. Biochem. Biophys. Res. Comm. **30:** 166.

[30] *Yaffe, D.* 1969. Cellular aspects of muscle differentiation in vitro. In: *Moscona, A. A.* and *Monroy, A.* (eds.). Current topics in Dev. Biol. Chapter 2, Academic Press.

[31] *Zetterberg, A.* 1966. Protein migration between cytoplasm and cell nucleus during interphase in mouse fibroblasts in vitro. Exptl. Cell. Res. **43:** 526.

[32] *Zetterberg, A., and Auer, G.* 1970. Proliferative activity and cytochemical properties of nuclear chromatin related to local cell density of epithelial cells. Exptl. Cell. Res. **62:** 262.

Discussion

Hirt: I would like to ask Dr. *Ringertz* a question. He stated that the dry mass of the nuclei of erythrocytes increases. How is this determined?

Ringertz: By microinterferometry.

Milgrom: I am afraid that I did not follow the origin of Dr. *Ringertz*'s reagents. To the best of my knowledge, the sera from lupus patients are not species specific, and they would react with nuclear material from all species. I know very little about cytoplasmic antibodies in rheumatoid arthritis. I do not think that rabbits readily form nucleus-specific antibodies. Could you please define more precisely the origin of your sera?

Ringertz: As to the species specificity of the patient sera human nucleospecific antibodies react with human, mouse, and rat cells but not with chick cells. There is a trick in the method we use for antibody "staining". By increasing the salt concentration in the solutions, we increase the species specificity of our antisera. The antibodies to chick nucleoli were obtained inadvertently during immunization of rabbits with crude actomyosin. These very useful antibodies are species specific, reacting with chick but not with rat, mouse, or human cells. The rheumatoid arthritis antiserum, we think, is directed toward some ribosomal component which occurs mainly in the cytoplasm, but to some extent around the nucleolus too. We are presently characterizing these sera, and I can't say much more about it.

Zipser: Our speaker has just said that specific proteins that were chosen for examination, apparently at random, from the chick were made, whereas another speaker tells us that certain other proteins that were specifically looked for were *not* made, for example myosin. Did I understand this correctly?

Ringertz: The phenomena I was looking at was early nuclear growth during the first 48 hours after cell fusion when the chick erythrocyte nucleus is not yet specifying chick-specific protein synthesis. The only other comment I made was about chick erythrocyte nuclei four days after cell fusion. These nuclei had developed a nucleolus and were expressing chick genes. At this stage, there were nucleo-specific chick proteins which moved into the chick erythrocyte nucleolus and into the human nucleus as well, but this is a late phenomenon. Now what *Deák* has been looking at is a fully reactivated chick erythrocyte nucleus which has a nucleolus which is functioning and expressing chick genes. When he knocks out the nucleolus, the chick markers disappear, showing that the nucleolus is necessary for maintaining the expression of chick genes. The previous work by Dr. *Harris'* group in Oxford indicates that this effect occurs at the level of RNA transport from the nucleus to the cytoplasm.

Zipser: O. K., but I am still confused because you said that later on in your attempt to find out whether the nuclei remembered what they were, they did not make the myosin or whatever it was you were looking for. Is that true? Those must have been functioning nuclei because otherwise the experiment wouldn't make any sense.

Koprowski: Dr. *Ringertz* will answer your question about myosin.

Deák: I think I should make it clear that these are two separate experiments. My work and the earlier part of Dr. *Ringertz'* talk were concerned with the fusion of chick erythrocytes onto tissue culture cells. The chick erythrocyte nucleus then directs the synthesis of proteins which are typical of tissue culture cells, e. g., hypoxanthineguaninephosphoribosyltransferase and surface antigens.

Zipser: That is an incredibly important point; I am not trying to say that it is wrong, it's just got to be clarified.

Deák: In the experiment which Dr. *Ringertz* described toward the end of his talk, he introduced a chick erythrocyte nucleus into a differentiated cell and asked a different question: Can this nucleus direct the synthesis of proteins, such as myosin, which are typical of this differentiated cell? The answer is that although the chick nucleus is reactivated in such cells, no chick myosin synthesis can be detected.

Zipser: If that is true, then this apparently arbitrary distinction between differentiated and nondifferentiated makes some sense. Thank you.

Konings: Dr. *Ringertz* can you tell me whether chicken hemoglobin is made in this hybrid?

Ringertz: We don't know yet but are trying to find out by Fe^{59} autoradiography, microspectrophotometry at 410 nm, and by immunological methods.

Konings: But if you fuse, for example, Hela cells with chicken erythrocytes, and at the same time you try to introduce cytoplasm from the chicken erythrocyte into the hybrid, probably you then can find an expression of the globin genes.

Ringertz: For all we know the cytoplasm of the mature chicken erythrocyte consists of little but hemoglobin. All cytoplasmic material except possibly some membrane material is lost due to lysis before fusion occurs, so the experiments may be equivalent to nuclear transplantation experiments.

Konings: But, can you exclude definitely the effect of the cytoplasm of the erythrocyte on the expression of the globin genes?

Ringertz: No, we cannot.

Short Term Interactions in Animal Cell Heterokaryons Leading to Induction of DNA Synthesis and Premature Chromosome Condensation

Robert T. Johnson and Potu N. Rao*)

Department of Zoology, University of Cambridge, Cambridge, U.K.

*) Department of Developmental Therapeutics, The University of Texas, M.D. Anderson Hospital and Tumor Institute, Houston, Texas, USA

Summary: Fusion between cells in different phases of the cell cycle can result in the induction of DNA synthesis and premature chromosome condensation. Of the fusions between mitotic and interphase complementing auxotrophic mutants of Chinese hamster cells, the mitotic and S phase fusion gave the lowest rate of hybrid survival and the greatest loss of chromosomes. It is proposed that use can be made of the internal asynchrony of multinucleate cells immediately after fusion for the directional loss of chromosomes from the hybrid progeny.

Introduction

When somatic cells are fused by means of inactivated Sendai virus, the initial products are multinucleate, and only after the first few mitotic events do mononucleate hybrid cells predominate. The multinucleate heterokaryon and the mononucleate hybrid differ in at least two important respects. First, unless synchronization procedures are employed before cell fusion, the multinucleate cell will frequently contain nuclei and cytoplasms which occupy different positions in the cell cycle. The monunucleate hybrid does not, however, suffer from "internal asynchrony" in this manner. Second, the total chromosomal complement of both partners is present in the multinucleate cell, while in the hybrid there is very commonly a reduction in the number of chromosomes, particularly from one parent.

Our studies have been concerned with exploiting the internal asynchrony of multinucleate cells during the first cell cycle after fusion in order to elucidate the control mechanisms which regulate cellular reproduction.

Manuscript received: 8 March 1972

Some consequences of internal asynchrony in heterophasic multinucleate cells

DNA synthesis

When a cell in the phase of DNA synthesis is fused with another in the preceding (G1) phase, the G1 nucleus is commonly induced to synthesize DNA [9, 13]. Evidence has been provided that the induced G1 nucleus continued to replicate DNA after the S phase component had finished [13]. In addition, *Rao* and *Johnson* [9] found that the arrival of G1/S HeLa cells at the first mitosis after fusion was delayed compared with the time of mitotic accumulation of binucleate S/S cells, indicating that the G1 component had slowed down the more advanced nucleus. No data are presently available which confirm that a full round of replication occurs in the induced G1 nucleus before the onset of the first mitosis. When G2 and S phase HeLa cells are fused, up to 10 percent of the multinucleate cells arrive at mitosis with a prematurely condensed S phase nucleus. In these cells, the G2 component enters mitosis ahead of the S nucleus and induces the chromatin of the latter to condense. The exact pattern of condensation depends on the amount of replication which has preceded induction. Thus the further the replication has progressed, the more the chromatin will condense to resemble bivalent metaphase chromosomes (Fig. 1). Since the induction of premature chromosome condensation in S phase nuclei inhibits DNA synthesis, it is clear that some of the S/G2 heterophasic cells will arrive at mitosis bearing less than the 8c amount of DNA [5].

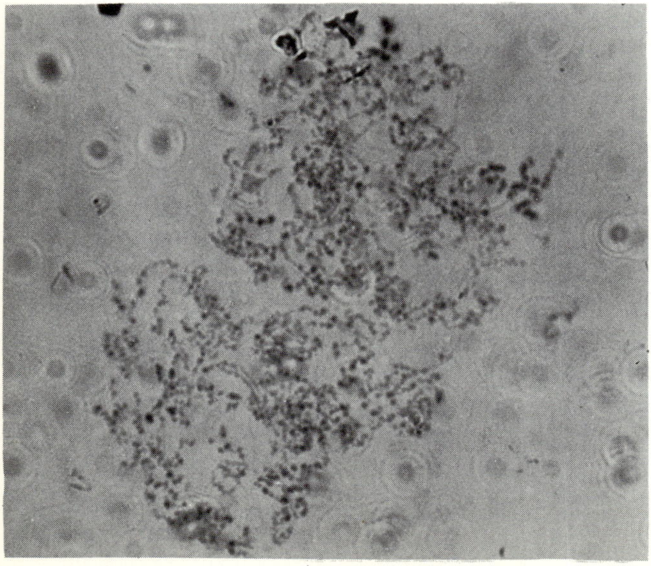

A

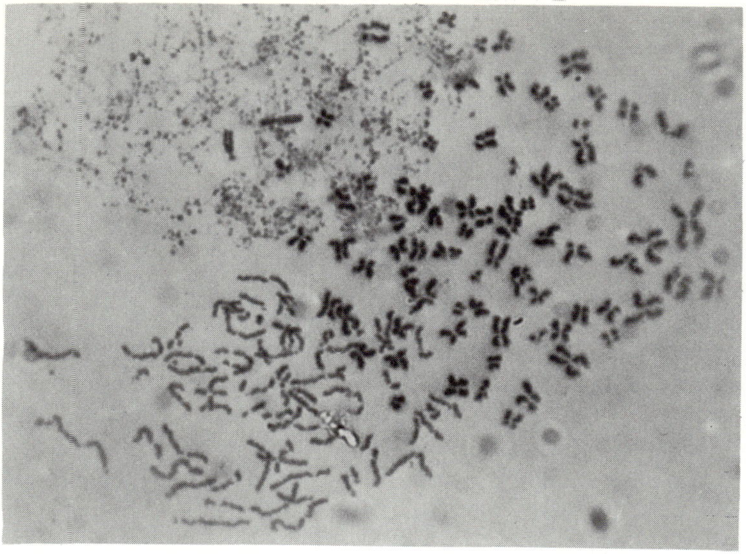

B

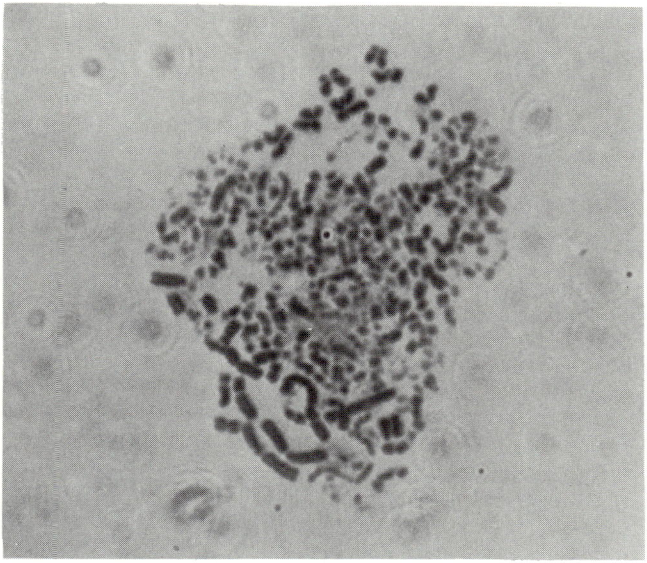

C

Fig. 1. Premature chromosome condensation of S and G1 HeLa nuclei.

A. Early S nucleus. Note the beaded appearance and fragmentary nature of the strands of chromatin.

B. Trinucleate cell containing an early S PCC, a G1 PCC showing the single chromatids, and the mitotic chromosomes of the inducing component.

C. PCC of a late S phase nucleus showing a few single G1 chromatids, much fragmentary S material, and a few elongated G2 bivalent chromosomes.

By cell fusion, it is possible to induce DNA synthesis in nuclei which are in a non-cycling Go state. For example, embryonic and adult fowl erythrocytes are inducible when fused with HeLa cells [3, 4] and, at 40 hours after fusion, many of the erythrocyte nuclei have reached the 4c level of DNA [1]. Similarly, macrophage nuclei are induced in macrophage-melanocyte heterokaryons [2].

Mitosis

When mitotic and interphase cells are fused, the latter are commonly induced to undergo a premature chromosome condensation (PCC). The morphology of the prematurely induced chromosomes depends on the stage in the cell cycle which the interphase nucleus occupied at the time of fusion [5, 11]. Thus, G1 nuclei are induced to form single chromatids, S nuclei appear as more or less partially condensed material depending on how much replication has occurred, and G2 chromatin condenses into morphologically normal but elongated chromosomes. Noncycling differentiated cells such as the avian erythrocyte and the unstimulated lymphocyte are also induced to form PCC by fusion with mitotic cells [6].

We have recently investigated the fate of prematurely induced chromosomes in the hope of gaining insight into the mechanisms of chromosome loss in hybrid cells [10]. It might be expected that PCC would be lost from fused cells for two reasons:

1. The condensation phenomenon is not accompanied by the induction of a spindle in the interphase cell. The only spindle present is, therefore, that of the mitotic component.

2. The structure of G1, and particularly of S, PCC differs greatly from the typical metaphase chromosome and may not be capable of ordered segregation into the daughter cells.

In order to investigate the fate of PCC, synchronized populations of G1, S, and G2 HeLa cells were each separately fused with reversibly blocked mitotic HeLa, and the patterns of chromosome and PCC segregation were examined. In most cases, the first mitosis after fusion was completed in seven hours. Immediately after fusion, PCC of G1, S, and G2 types formed discrete bodies adjacent to the metaphase plate of the mitotic component (Fig. 2). There was, however, no evidence to suggest that they were integrated into the infrastructure of the spindle. With the approach of anaphase, the S PCC became scattered and for each type of PCC there was random segregation into the daughter cells. In general PCC of G1 and G2 types segregated as single entities (Fig. 3) while PCC of the S type was often dispersed throughout the daughter cells. In radioautographs of the succeeding interphase cells, the PCC from nuclei which were prelabelled with ^{3}H-thymidine frequently appeared as local concentrations of radioactivity in the reconstituted nuclei of the hybrid cells [10].

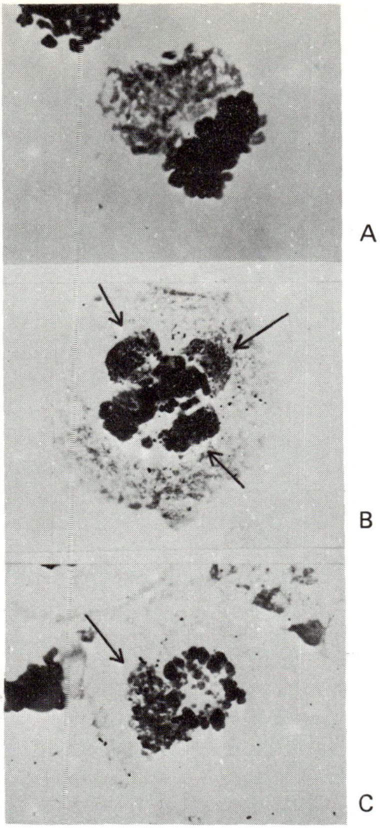

Fig. 2

PCC in M/G1, M/S, and M/G2 fusions of HeLa cells at 30 minutes after fusion.

A. M/G1. PCC of the G1 type which is lightly stained lying alongside the metaphase plate (darkly stained) of the mitotic cell.

B. M/S. A radioautograph showing labelled S type PCC (indicated by arrows) lying around the metaphase plate.

C. M/G2. A radioautograph showing labelled PCC of the G2 type (indicated by the arrow) lying adjacent to the metaphase chromosomes of the mitotic cell. (From *Rao* and *Johnson* [10]).

To examine whether PCC became genetically integrated into the progeny of hybrid cells, fusions were carried out between synchronized populations of complementing auxotrophic mutants of Chinese hamster ovary origin [7]. The products of fusion were grown in the absence of glycine, and it was found [10] that the rate of recovery of hybrids was related to the initial internal asynchrony of the fused cells (Fig. 4). Thus, homophasic fusions, i.e., where both parent cells were in the same phase of the cycle, produced almost twice the number of hybrid colonies as did heterophasic interphase-interphase fusions, e.g., S x G1, G1 x G2, S x G2. In addition, the lowest rate of survival occurred in the mitotic x interphase crosses, and of these the M x S was the lowest. These data suggest that the imposition of internal asynchrony in multinucleate cells leads to a lower survival, and that the greater the asynchrony, e. g., mitotic x interphase, the greater the cell death.

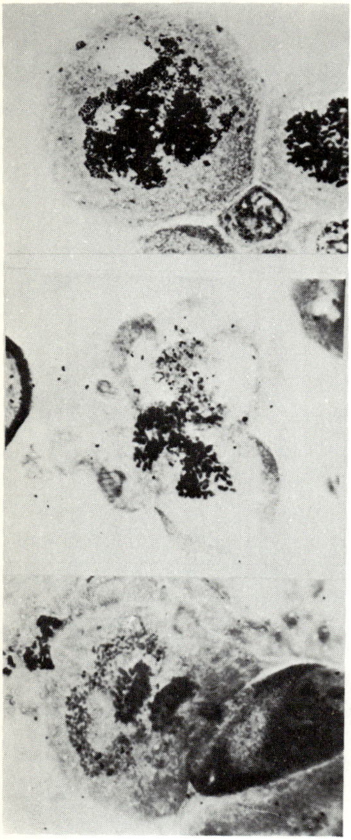

A

B

C

Fig. 3

Scattering of PCC in M/S fused HeLa cells at 4 hours after fusion. S phase cells prelabelled with ^{3}H-TdR were fused with mitotic cells reversibly blocked with N_2O.

A. The mitotic chromosomes are completing anaphase while the PCC are scattered around them. PCC will probably be distributed randomly between the daughter nuclei.

B. Radioautograph of an M/S fused cell during late anaphase. The labelled PCC are segregating separately into one of the daughtei cells.

C. The labelled S PCC form a crescent around one of the anaphase chromosome groups and in all probability will become part of that daughter cell. (From *Rao* and *Johnson* [10].

At approximately 20 generations after fusion, the chromosomes of the various hybrids were examined (Fig. 5). Apart from the M/S hybrids with a mode of 37 chromosomes, both the homophasic and heterophasic cells had modes of 39, 40, and 41. This suggests that the continued selection imposed on these cells leads to the survival of those hybrids which retain almost all the chromosomes from each parent. If the selection pressure were to be relaxed at an early stage after fusion when the parental populations had been removed, it is possible that hybrids containing much smaller numbers of chromosomes might be found.

These data show that fusion between mitotic and S phase cells leads to the greatest loss of chromosomes from the hybrids, and that this occurs relatively early. In this connection, it is of interest that *Kao, Johnson,* and *Puck* [8], working with other complementing auxotrophic mutants of the CHO line, found that the modal chro-

Fig. 4. (Adapted from *Rao* and *Johnson* [10].)

Rate of survival of hybrids in homo- and heterophasic fusions between synchronized populations of complementing auxotrophic Chinese hamster cell mutants.

	Type of fusion	Number of colonies per 1000 cells of each parental type
1.	R^a/R^b	15.6
2.	R^a_c/R^b_c	5.0
3.	$R^a + R^b$(w/o virus)	0
	Homophasic fusions)*	
4.	M^a/M^b	4.0
5.	$G1^a/G1^b$	5.5
6.	S^a/S^b	4.6
7.	$G2^a/G2^b$	4.4
	Heterophasic fusions)*	
	(i) Mitosis/Interphase	
8.	$M^a/G1^b$	2.0
9.	M^a/S^b	0.62
10.	$M^a/G2^b$	1.56
	(ii) Interphase/Interphase	
11.	$S^a/G1^b$	3.2
12.	$S^a/G2^b$	2.7
13.	$G1^a/G2^b$	2.5

R = Random population
a = Parent glycine-requiring mutant a
b = Parent glycine-requiring mutant b
c = Exposed to colcemide (0.05 μg/ml) for 2 hours prior to fusion
*) Synchronization of the cells achieved by exposure to colcemide (0.05 μg/ml) for 2 hours followed by release and collection of the various phases.

Fig. 5. (Adapted from *Rao* and *Johnson* [10].)

Modal chromosome numbers among the progenies of the various types of fusions between the complementing auxotrophic mutants.

Fusion	Mode
R/R	41
M/M	40
M/G1	40
M/S	37
G1/G1	39
G1/S	39
S/S	40

Note: Because of the difficulties in obtaining even a reasonable degree of synchrony among the G2 population, the M/G2 fusion for chromosome studies was not attempted.

R = Random population.

mosomal value of the hybrids was reduced to 37 after 116 generations of culture. We believe that the induction of PCC in the S phase component of the cell and the scattered nature of the S chromosomes are a contributing factor to this loss in M/S hybrids.

Discussion

The internal asynchrony imposed by fusing cells in different phases of the cell cycle leads in many cases to death of the hybrids. In those that survive, however, there may frequently be reduction in the amount of genetic material from one parent. This may be brought about by incomplete replication of DNA in the lagging nucleus, or by elimination of prematurely induced chromosomes.

In certain heterologous systems, the imbalance may be increased by combining a multiplying with a noncycling Go cell as, for example, when A9 mouse fibroblasts and chick erythrocytes are fused. In these heterokaryons, it is unlikely that a complete round of DNA synthesis occurs in the erythrocyte nucleus before the active nucleus enters division. At division, the erythrocyte nucleus may be induced to form PCC (Fig. 6) and, because of the many avian microchromosomes, small pieces of chick DNA may be segregated with and integrated into the genetic material of the active component (see *Schwartz, Cook,* and *Harris* [12]).

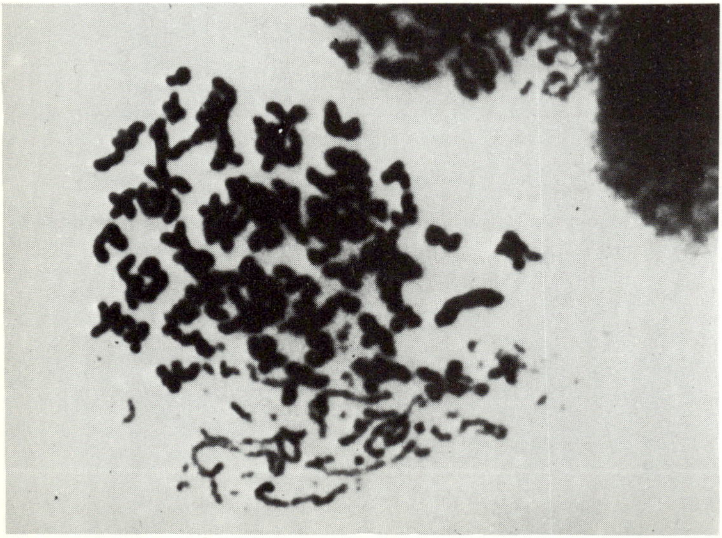

A

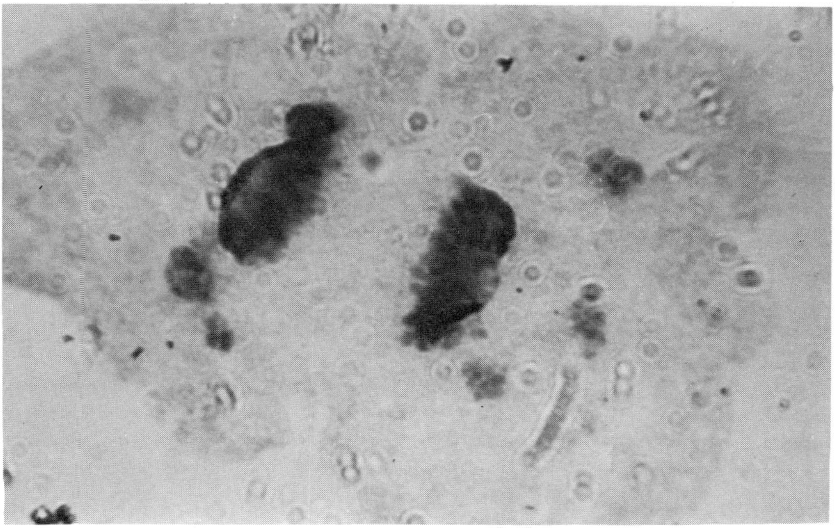

B

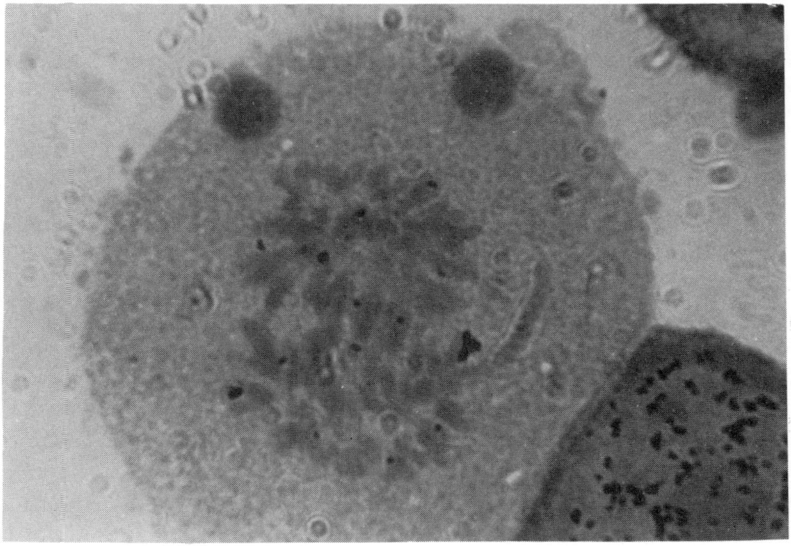

C

Fig. 6. HeLa-embryonic chick erythrocyte heterokaryons.

A. G1 type of PCC of an embryonic chick erythrocyte nucleus in a HeLa/chick RBC heterokaryon. Note the many dotlike chick chromosomes. (From *Johnson* et al. [6].

B. HeLa/chick RBC heterokaryon in telophase. Six erythrocyte nuclei, whose irregular outlines are probably a consequence of PCC, are segregating equally into the daughter cells.

C. PCC is not an inevitable consequence of fusion between mitotic HeLa cells and chick erythrocytes. The two erythrocyte nuclei in this cell will probably pass unaffected into the next cell cycle of the hybrid.

We believe that by fusing mitotic and interphase cells it may become possible to direct the loss of chromosomes from one component of the hybrid cell. This would clearly be useful in genetic studies with hybrids which are formed either by fusing cells of the same or different species. Hybrids derived by fusing cells of one species are frequently stable with respect to their chromosomal constitution, and inter-species crosses are often characterized by large-scale loss of the chromosomes from one species. In the latter case, it might be of interest to direct the chromosome loss from the other species.

Conclusion

The internal asynchrony in heterokaryons during the first cycle after cell fusion constitutes a way of generating genetic diversity in the hybrid progeny.

Acknowledgements: Much of this investigation was carried out in the Department of Biophysics and Genetics and the Eleanor Roosevelt Institute for Cancer Research, University of Colorado Medical Center, Denver, and we wish to thank Dr. *T.T. Puck* for his interest in this work. This investigation was aided by research grants from the Medical Research Council of Great Britain, and from the Damon Runyon Memorial Fund for Cancer Research, DRG-1110.

References

[1] *Bolund, L., Ringertz, N. R.,* and *Harris, H.* 1969. Changes in the cytochemical properties of erythrocyte nuclei reactivated by cell fusion. J. Cell Sci. **4:** 71–87.

[2] *Gordon, S.,* and *Cohn, Z.* 1970. Macrophage-melanocyte heterokaryons. I. Preparation and properties. J. Exptl. Med. **131**: 981-1003.

[3] *Harris, H.* 1965. Behaviour of differentiated nuclei in heterokaryons of animal cells from different species. Nature **206**: 583-88.

[4] *Johnson, R.T.,* and *Harris, H.* 1969. DNA synthesis and mitosis in fused cells. II. HeLa-chick erythrocyte heterokaryons. J. Cell Sci. **5**: 625-44.

[5] *Johnson, R.T.,* and *Rao, P.N.* 1970. Mammalian cell fusion: Induction of premature chromosome condensation in interphase nuclei. Nature **226**: 717-22.

[6] *Johnson, R.T., Rao, P.N.,* and *Hughes, S.D.* 1970. Mammalian cell fusion. III. A HeLa cell inducer of premature chromosome condensation active in cells from a variety of species. J. Cell. Physiol. **77**: 151-58.

[7] *Kao, F.T., Chasin, L.,* and *Puck, T.T.* 1969. Genetics of mammalian cells. X. Complementation analysis of glycine-requiring mutants. Proc. Natn. Acad. Sci., USA **64**: 1284-91.

[8] *Kao, F.T., Johnson, R.T.,* and *Puck, T.T.* 1969. Complementation analysis on virus-fused Chinese hamster cells with nutritional markers. Science **164**: 312-14.

[9] *Rao, P.N.,* and *Johnson, R.T.* 1970. Mammalian cell fusion: Studies on the regulation of DNA synthesis and mitosis. Nature **225**: 159-64.

[10] *Rao, P. N.,* and *Johnson, R. T.* 1972. Premature chromosome condensation: a mechanism for the elimination of chromosomes in virus-fused cells. J. Cell Sci. **10:** 495–513.

[11] *Sandberg, A. A., Aya, T., Ikeuchi, T.,* and *Weinfield, H.* 1970. Definition and morphologic features of chromosome pulverization: a hypothesis to explain the phenomenon. J. Natn. Cancer Inst. **45**: 615-23.

[12] *Schwartz, A., Cook, P.R.,* and *Harris, H.* 1971. Correction of a genetic defect in a mammalian cell. Nature New Biol. **230**: 5-8.

[13] *Westerveld, A.,* and *Freeke, M.A.* 1971. Cell cycle of multinucleate cells after fusion. Exptl. Cell Res. **65**: 140-44.

Source of Genetic Information of Chloroplast Proteins in Acetabularia

Hans-Georg Schweiger, Klaus Apel, and Klaus Kloppstech

Max-Planck-Institut für Zellbiologie, 2940 Wilhelmshaven, Germany

Summary: Nuclear exchange experiments enabled the investigation of the origin of genetic information for a number of chloroplast proteins in *Acetabularia*. It could be shown that chloroplast proteins like malic dehydrogenase, lactic dehydrogenase, membrane proteins, and ribosomal proteins are coded by the nuclear genome.

Introduction

It is generally accepted that besides the cell nucleus the cytoplasm is a site of genetic information. Supporting evidence comes from genetic experiments and from the fact that all the components involved in protein synthesis are present in chloroplasts and mitochondria. One of the major contributions to the problem of organelle autonomy was the detection of DNA in chloroplasts. The presence of DNA in chloroplasts has been demonstrated by biochemical and electronmicroscopic methods. The amount of DNA in the cell nucleus exceeds that in a chloroplast by several orders of magnitude.

Due to the fact that the number of chloroplasts per cell is in the order of magnitude of 10^7 in *Acetabularia* cells, the amount of chloroplast DNA per cell surpasses the amount of nuclear DNA by a factor of about 10^4. The DNA content of the chloroplast of *Acetabularia* has been determined by biochemical and electronmicroscopic methods to be about 10^{-16} g (*Baltus* and *Brachet* [2], *Gibor* and *Izawa* [3], *Werz* and *Kellner* [8]) which corresponds to about 6×10^7 Daltons or 1×10^5 nucleotide pairs corresponding to 3×10^4 amino acids. It is obvious that this quantity is not sufficient to code for all the protein components of the chloroplasts besides different types of

Manuscript received: 20 December 1971

RNA in these organelles. At least part of the genetic information should originate in other sources than in the chloroplast genome. One of the favourite candidates for such sources might be the cell nucleus. The question that is raised is which of the chloroplast proteins are coded by the nuclear genome and which by the chloroplast genome.

Recently two techniques have proved suitable for approaching this problem. The first one represents a rather indirect method since it aims at the localization of protein synthesis. All the components which are involved in protein synthesis including ribosomes are present in the chloroplasts. The chloroplast ribosomes are different from those in the cytosol which sediment with 80 S while the chloroplast ribosomes have a sedimentation coefficient of 70 S. Similar to bacterial ribosomes, the chloroplast ribosomes are sensitive to chloramphenicol while the 80 S ribosomes are sensitive to cycloheximide. A number of experiments have been performed in order to study the action of the two inhibitors on the synthesis of different chloroplast proteins. Suppose the nuclear information is translated exclusively on the 80 S ribosomes and the chloroplast information only on the 70 S, the inhibition experiments might answer the question of where the information for chloroplast proteins comes from. However, the final proof for such assumptions is still missing.

The second approach seems to be a much more promising one. The technique is to look for species-specific chloroplast proteins in closely related species and to investigate the influence of a heterologous nucleus on these proteins.

For this type of experiment, the unicellular and uninuclear green alga *Acetabularia* has proved to be suitable (*Schweiger* [6]) since it is possible:

1. to prepare anucleate cells by cutting off the basal part of the cell (rhizoid) which contains the nucleus,
2. to produce heterologous combinations by grafting a rhizoid of one species to an anucleate cell of another species,
3. to isolate the nucleus from a rhizoid of one species and to transplant it to an anucleate cell of another species.

Malic dehydrogenase

One of the proteins which are to be talked about is the malic dehydrogenase, an enzyme which exhibits a high activity in *Acetabularia* cells. Species-specific properties were sought or, in other words, differences between two species of *Acetabularia* with respect to the protein under consideration. Then the nuclei of two cells belonging to different species were exchanged in a cross experiment. Finally after several weeks, it was determined whether the properties of the protein under consideration had changed into that of the species of the nucleus or had remained that of the cytoplasm (*Schweiger, Master,* and *Werz* [7]).

Species specificity of malic dehydrogenase can easily be shown by gel electrophoresis. After electrophoresis on a polyacrylamide gel, reproducible and species-specific isozyme patterns have been established. The species-specificity is determined by observing the number and the position of the bands after specific staining. The enzyme activity of a single cell of *Acetabularia* is sufficient for at least ten separations. When a mixture of the cytoplasms from cells of different species are subjected to electrophoresis, the resulting isozyme pattern contains all the bands of both patterns.

The nuclear dependency or nondependency was proved by the following experiment. The nuclei of two cells belonging to different species of Acetabularia (*A. mediterranea* and *Polyphysa cliftonii*) were exchanged, and after four weeks the cytoplasms of both heterologous combinations were subjected to gel electrophoresis. The result (Fig. 1) can be summarized as follows:

1. The isozyme pattern of that species from which the cell nucleus originates has been established in the cytoplasm.
2. The isozyme pattern of the host cytoplasm has disappeared.

It is shown by this result that the cell nucleus releases genetic information into the cytoplasm and that this information is realized in the cytoplasm. Furthermore, it is shown that under the influence of the implanted nucleus the host isozyme pattern of the cytoplasm disappears. This is a striking fact since in the anucleate cell the activity of the enzyme is increased and the isozyme pattern is retained. Therefore, it must be

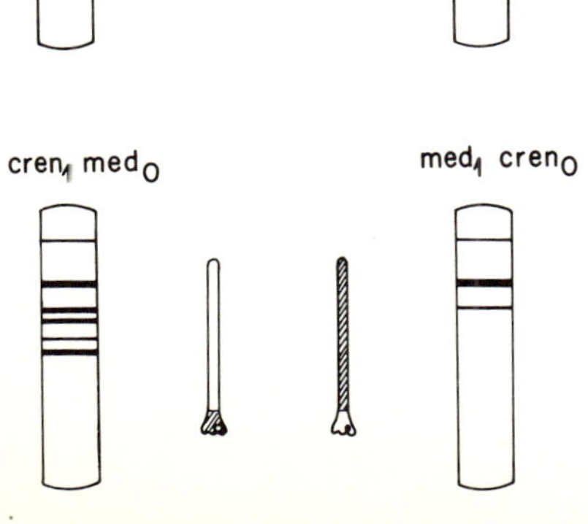

Fig. 1
Isozyme pattern of malic dehydrogenase in *A. mediterranea* and *P. cliftonii* before and four weeks after nucleus exchange experiment [7].

concluded that it is the influence of the implanted nucleus which makes the host isozymes disappear. Until now such a role could be demonstrated only for the heterologous nucleus. However, it seems at least possible that the homologous nucleus also exerts an accelerating action on the intracellular turnover of enzyme proteins.

In this connection, the question is raised whether the increase in enzyme activity just described is due to synthesis of new enzyme proteins or only represents an activation of protein molecules already present. This question has been answered by the following type of experiment.

Nucleate cells were grown for four weeks in an Erd-Schreiber medium in which 50 % of the water had been replaced by deuterium oxide. The density of the enzyme protein (enzyme activity) was determined before and after the growth period in deuterium oxide by caesium chloride density centrifugation. During the incubation in deuterium oxide, the density of the enzyme protein is increased (Fig. 2a), that means, deuterium atoms have replaced hydrogen atoms in the protein molecule. That this increase in density is not due to a physicochemical exchange but rather reflects de novo synthesis follows from another observation: If the deuterium content of the medium is increased to more than 50 %, the growth of the cells stops. In this case, no increase in the density of the enzyme protein can be observed.

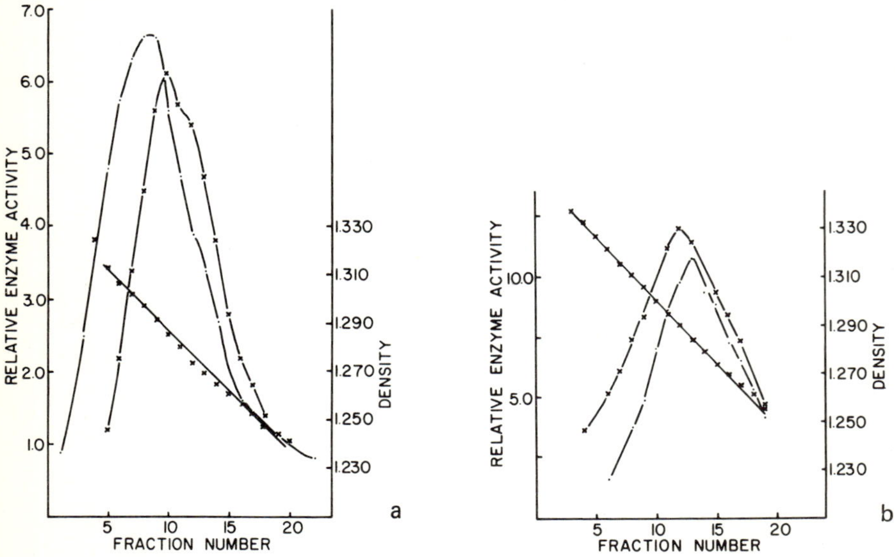

Fig. 2. CsCl-density gradient centrifugation of MDH before (x–x–x) and four weeks after (·–·–·) growing *Acetabularia* cells in a medium containing 50 % D_2O.
(a) Nucleate cells, (b) anucleate cells.
(*Schael* and *Schweiger*, in preparation.)

The situation in the anucleate cells is comparable to that in nucleate cells. During the four week treatment with deuterium oxide, a shift in the density of the enzyme protein takes place (Fig. 2b). So it is obvious that the anucleate cell is indeed able to synthesize specific protein molecules.

The fact that the MDH is coded by the nuclear gene raises the question as to where the enzyme is located in the cell. The answer to this question comes from histochemical and density gradient centrifugation experiments. Both techniques show that at least part of the enzyme activity is associated with the chloroplasts. This means that part of the chloroplast proteins are coded by the nuclear genomes.

Lactic dehydrogenase

The results on malic dehydrogenase have been supported by similar results on lactic dehydrogenase (*Reuter* and *Schweiger* [5]). Under the influence of the implanted heterologous nucleus, the enzyme pattern of the species of the nucleus donor in the host cytoplasm is established (Fig. 3). At the same time the enzyme pattern of the host cytoplasm disappears. Moreover, it was shown that at least part of the enzyme activity is located in the chloroplasts. Both the results on MDH and on LDH raise the interesting question as to which role the chloroplast genes have with respect to the total amount of information which is necessary for the synthesis of the chloroplast proteins. In order to approach this problem, two more groups of chloroplast proteins were investigated with methods which are similar to those which had been used for the malic dehydrogenase and for the lactic dehydrogenase.

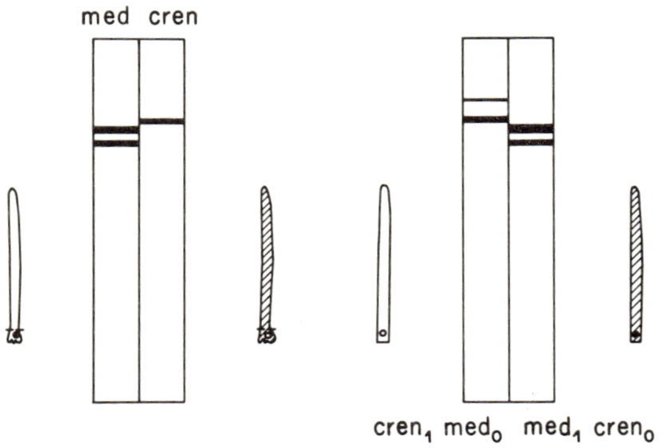

Fig. 3
Isozyme pattern of lactic dehydrogenase in *A. mediterranea* and *A. crenulata* before and four weeks after nucleus exchange experiment [5].

Membrane proteins

The first group was the membrane proteins of the chloroplasts. There is no doubt that the membrane proteins represent integral constituents of the chloroplasts, and one may expect that if the chloroplast genes have anything to do with coding of chloroplast proteins, then this should also be true for the membrane proteins.

As already mentioned, in order to perform experiments like those on malic dehydrogenase and lactic dehydrogenase, it is necessary to find differences between the membrane proteins of different species. Such differences have been found between *Acetabularia mediterranea* and *Acetabularia calyculus* after subjecting isolated membrane proteins to gel electrophoresis *(Apel* and *Schweiger* [1]). The membrane proteins were dissolved in a formic acid-phenol mixture and run in a polyacrylamide gel which was saturated with the same mixture. In this way, a separation of at least six sections representing twelve bands could be obtained which were traced in a photometer after staining with amido black.

Although the electrophoretic pattern of the chloroplast proteins of *Acetabularia calyculus* resembled those from *Acetabularia mediterranea,* one cannot overlook some differences. In contrast to *Acetabularia mediterranea* (Fig. 4a) whose section d appears as a single peak, section d of *Acetabularia calyculus* (Fig. 4b) is characterized by the distinct resolution into the components d_1 and d_2. If a mixture of chloroplast proteins of *Acetabularia mediterranea* and *Acetabularia calyculus* was subjected to electrophoresis, the two *Acetabularia calyculus* peaks disappeared in an unresolved broad peak d. Probably this broad peak was due to the fact that the resolving capacity of the method was not sufficient to separate the components d, d_1, and d_2. Other differences between the two species are exhibited by peaks e_4 and e_3 and the single peak f in *Acetabularia calyculus* which corresponds to the double peak f_1 and f_2 in *Acetabularia mediterranea.*

Since the protein patterns of the two species were distinguished by several bands, the cross experiment as already described could be performed. Six weeks after the implantation of the heterologous nucleus chloroplasts were isolated, the membrane proteins were prepared, and the protein pattern was established.

It is evident that the species-specific membrane proteins behave similar to the malic dehydrogenase and lactic dehydrogenase. After the nucleus exchange experiment, the protein pattern had been changed in a nucleus-specific way. This holds for the combination of *Acetabularia mediterranea* nucleus with the *Acetabularia calyculus* cytoplasm and vice versa (Fig. 4c, Fig. 4d).

In this case also the determining influence of the nucleus on the chloroplast proteins becomes obvious. Although the direct evidence is limited to the species-specific bands, one might assume that those bands which are common in both species are also coded by the cell nucleus.

Effect of chloramphenicol and cycloheximide

Although there is no direct answer to the question as to whether there are any membrane proteins which are coded by the chloroplast DNA, it is indicated that part of the membrane proteins are synthesized within the chloroplast (*Apel* and

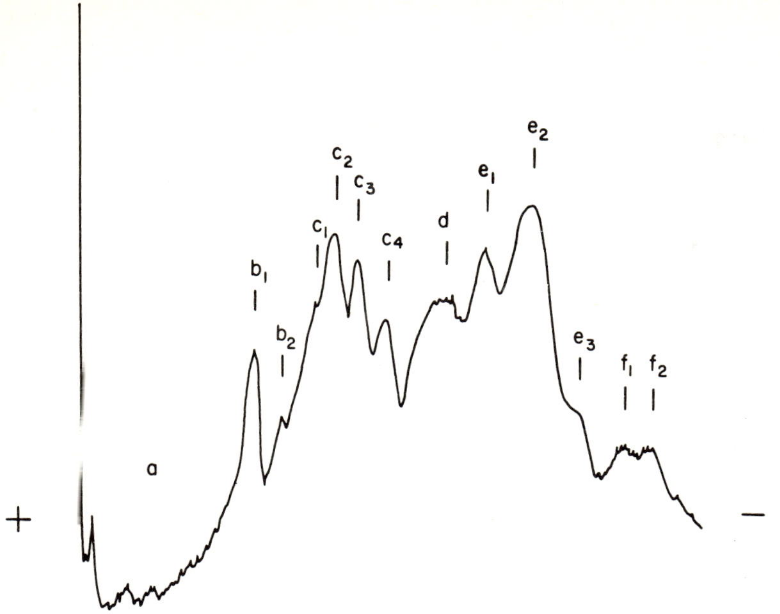

a

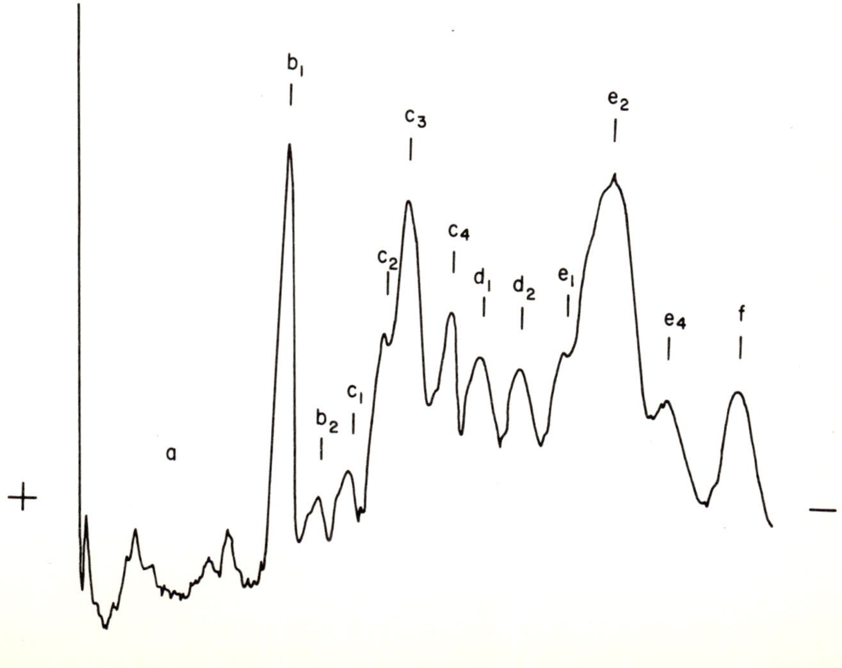

b

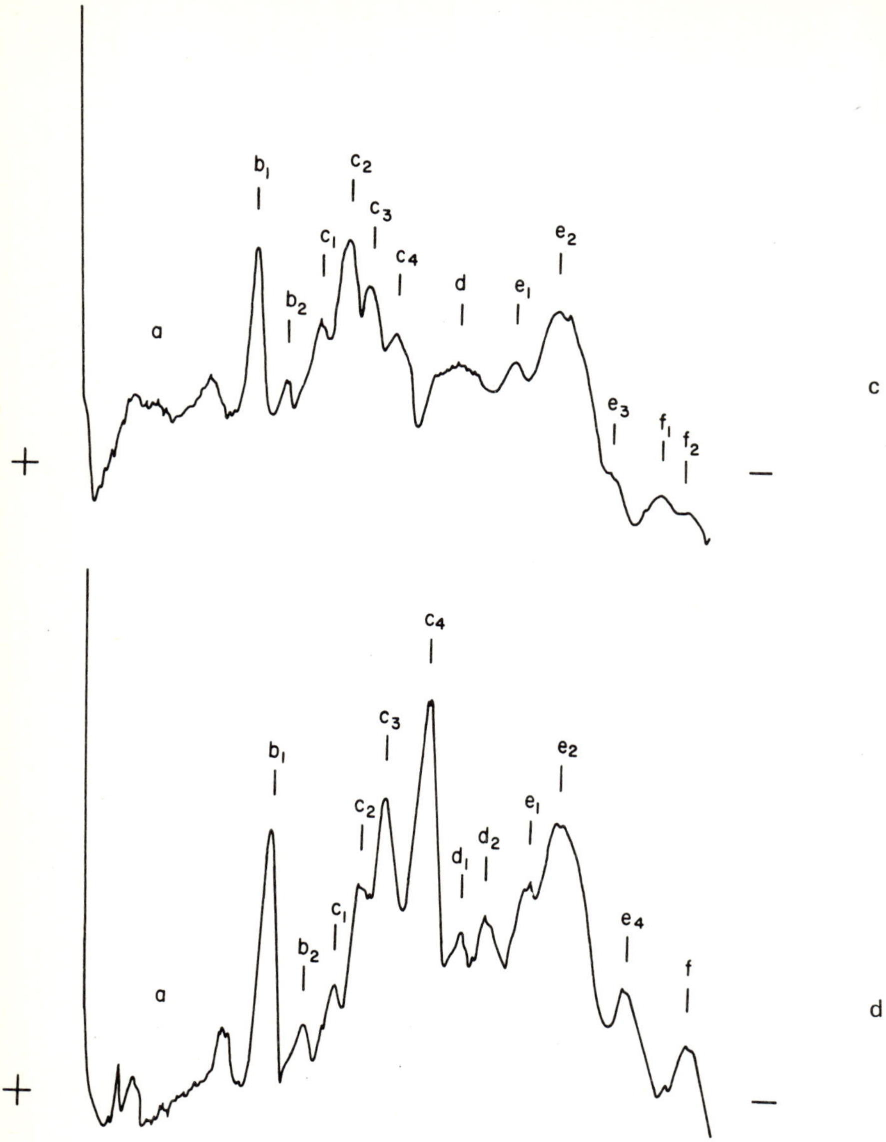

Fig. 4. Electrophoresis of chloroplast membrane proteins prepared from *Acetabularia:* (a) *A. mediterranea,* (b) *A. calyculus,* (c) med_1cal_0-transplant, (d) cal_1med_0-transplant. The proteins isolated from ten plants were subjected to electrophoresis. After staining with amido black, the gel was scanned at 578 nm. The peaks of the different sections (a–f) were numbered sequentially according to their migration distance from the origin of the gel (+) [1].

Schweiger [1]). The experimental evidence is based on the well known fact that the chloroplast ribosomes belong to the 70 S type whereas the cytosol ribosomes belong to the 80 S type. Evidence for the involvement of these types in the protein synthesizing mechanism under consideration comes from the inhibitory action of chloramphenicol on the 70 S ribosomes and of cycloheximide on the 80 S ribosomes. This technique has been used in a number of laboratories to locate the site of synthesis of mitochondrial enzymes.

With respect to the membrane proteins of *Acetabularia* chloroplasts, it has been shown that chloramphenicol (10 μg/ml) exerts an inhibitory action on the incorporation in vivo into peaks 1 and 2 but not on peak 3 (Fig. 5a). On the other hand, it is obvious that it is predominantly the incorporation into peak 3 which is inhibited by cycloheximide (0.05 μg/ml) while the inhibition of the incorporation into peak 1 and 2 is less pronounced (Fig. 5b). From these results, it can be concluded that the membrane proteins of the chloroplasts are synthesized inside as well as outside the chloroplasts. This conclusion is supported by the finding that the incorporation of radioactive amino acids in vitro, that is in isolated chloroplasts, reveals a pattern which is similar to that of the in vivo incorporation in the presence of cycloheximide (Fig. 5c). In both cases, the 80 S ribosomes could not contribute to protein synthesis.

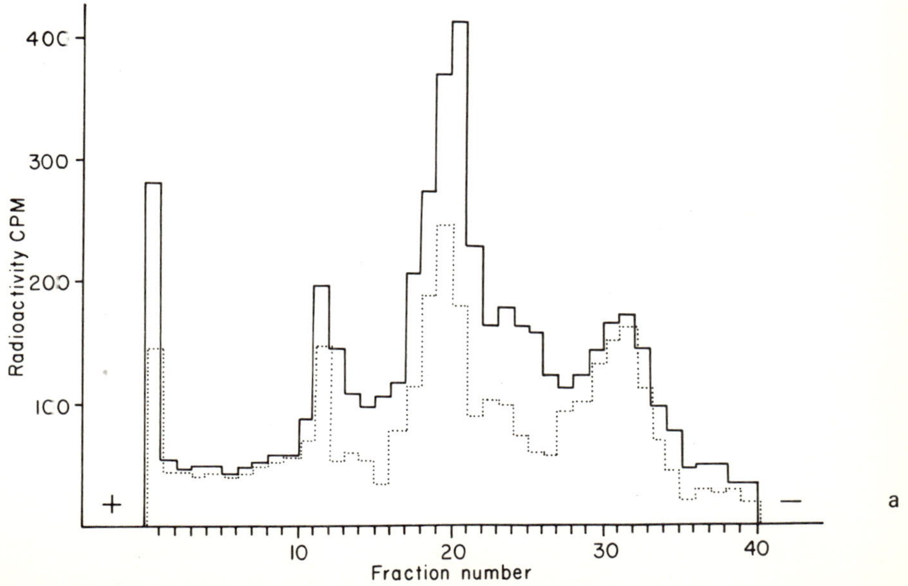

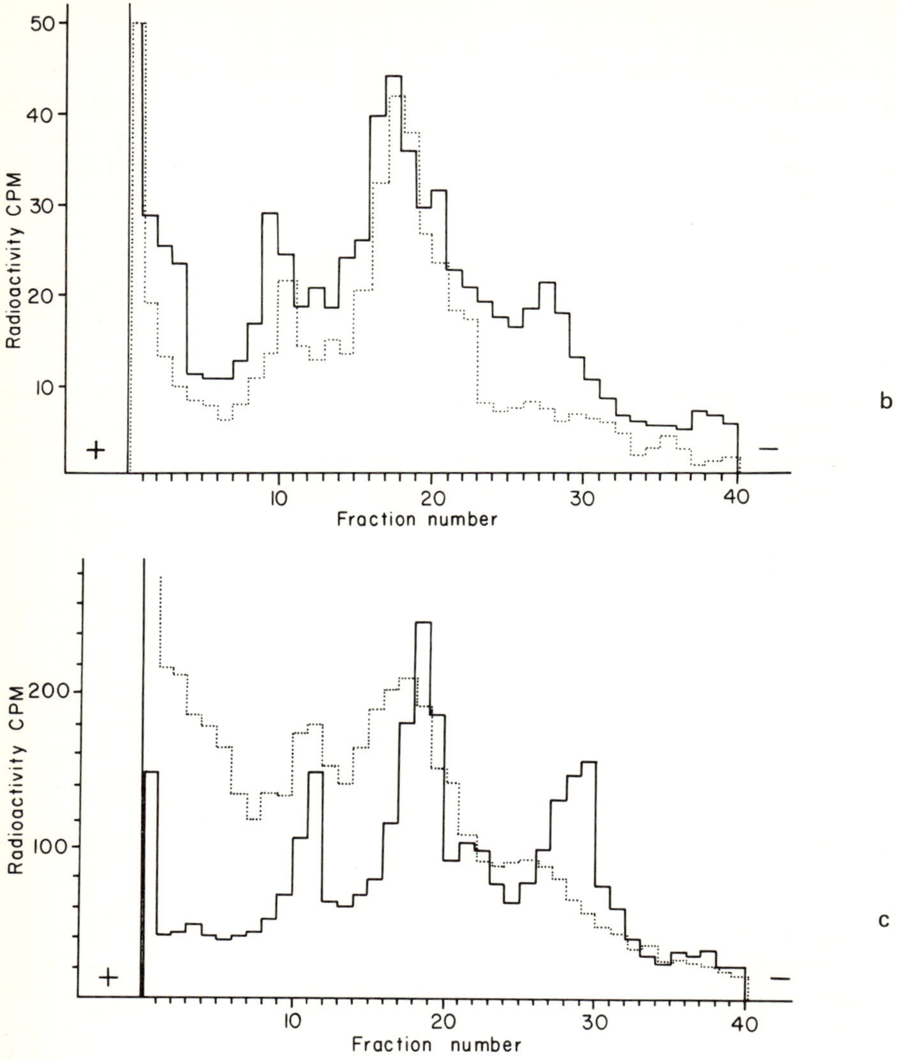

Fig. 5. Electrophoretic pattern of C^{14}-labelled chloroplast membrane proteins of *A. mediterranea*. Ten plants were incubated

a) in the presence (·········) or in the absence (————) of chloramphenicol (10 μg/ml) with 5 μ Ci ^{14}C-glycine,

b) in the presence (········) or in the absence (————) of cycloheximide (0.05 μg/ml) with 1 μ Ci ^{14}C-glycine.

c) Isolated chloroplasts from 100 plants were incubated with 6 μ Ci ^{14}C-labelled amino acid mixture (········). Ten plants were incubated with 5 μ Ci ^{14}C-glycine (————) [1].

Ribosomal proteins

In order to further elucidate the possible role of the chloroplast DNA in coding for chloroplast proteins, the synthesis of another group of proteins has been investigated, namely the proteins of the chloroplast ribosomes or 70 S particles (*Kloppstech* and *Schweiger* [4]). From isolated chloroplasts 44 S ribosomal particles were prepared, the proteins of these ribosomes were isolated, subjected to gel electrophoresis, stained, and the pattern was scanned. In this way, the proteins of the 44 S particles from *Polyphysa cliftonii, Acetabularia mediterranea,* and *Acetabularia crenulata* were compared. A reference run was performed with proteins from *E. coli* 70 S ribosomes. By means of the gel electrophoresis technique, it was possible to demonstrate differences between the protein patterns of these three species.

The experimental approach was similar to that which had been described for malic dehydrogenase, lactic dehydrogenase, and for the chloroplast membrane proteins. However, a serious difficulty arose because of the shortage of material which was available from a limited number of cells, especially grafts. This difficulty was overcome in the following way.

The cell disposes of two types of genetic information namely in the cell nucleus and in the organelles, e.g., in the chloroplasts. If the nucleus A is replaced by a nucleus B, a heterologous combination results which contains the nucleus B and the chloroplasts A. It is known, mainly from genetic experiments, that during the reproductive step the combination nucleus B and chloroplast A is retained. Since by reproduction thousands of cells can be produced, it is much more convenient to use the F_1-generation of the grafts than the limited number of grafts themselves.

From such cells of the F_1-generation chloroplasts, chloroplast ribosomes and ribosomal proteins from 44 S subparticles were prepared. The proteins were subjected to gel electrophoresis and the protein patterns of *Acetabularia crenulata* (Fig. 6 K), the F_1-generation of $cren_1$ $clift_0$ (Fig. 6H), that is a combination of a nucleus of *Acetabularia crenulata* and cytoplasm of *Polyphysa cliftonii,* and the pattern of *Polyphysa cliftonii* were compared (Fig. 6F). In order to facilitate the exact localization of the bands, the ribosomal proteins were run together with a reference preparation from *E. coli.*

In the region "7" there is a prominent peak in *Acetabularia crenulata* but not in *Polyphysa cliftonii.* In the pattern of the F_1-generation of the combination $cren_1$ $clift_0$, the same group of peaks appears. This again means that the components under consideration are determined by the nucleus, or, in other words, the genetic information for these proteins originates in the nuclear DNA. Similar results have been obtained with *Acetabularia mediterranea* (Fig. 6B) and *Polyphysa clifonii.* In the case of the combination med_1 $clift_0$ (Fig. 6D), it is not the appearance but rather the disappearance of a band which is caused by the heterologous nucleus.

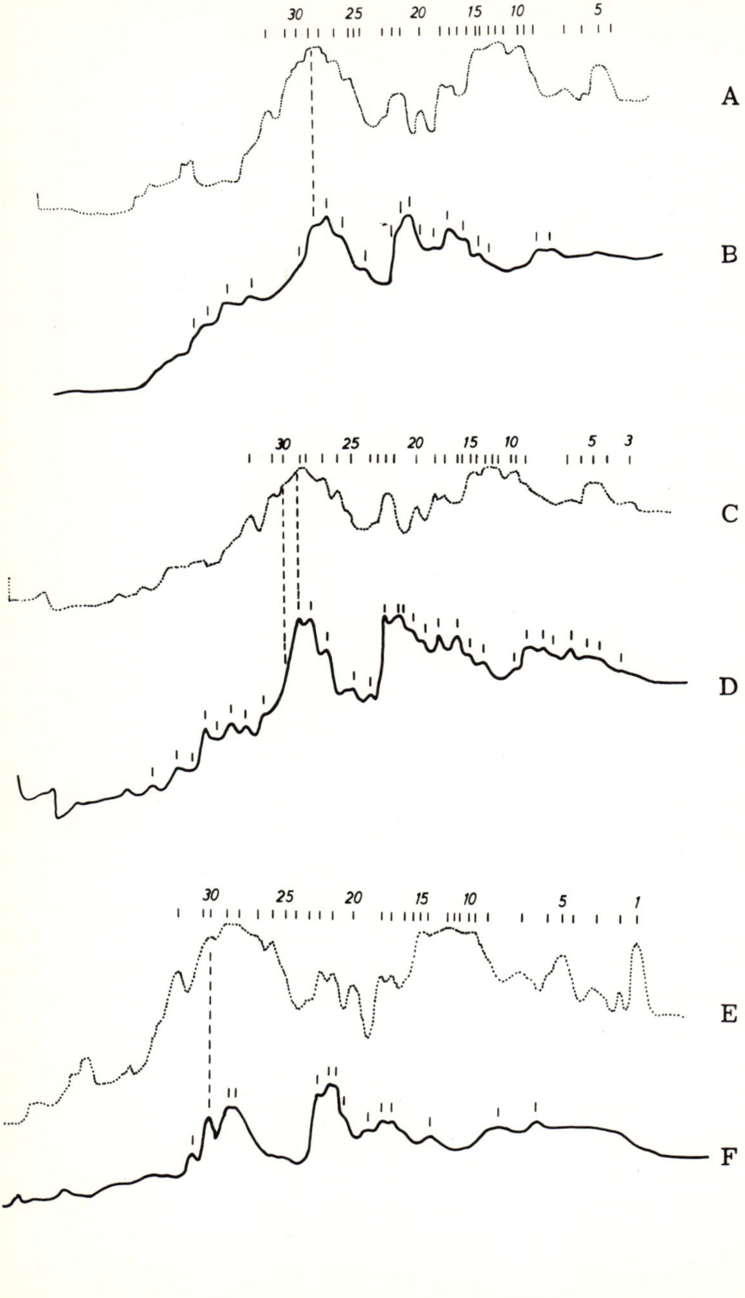
30
25
20
15
10
5
A
B
30
25
20
15
10
5
3
C
D
30
25
20
15
10
5
1
E
F

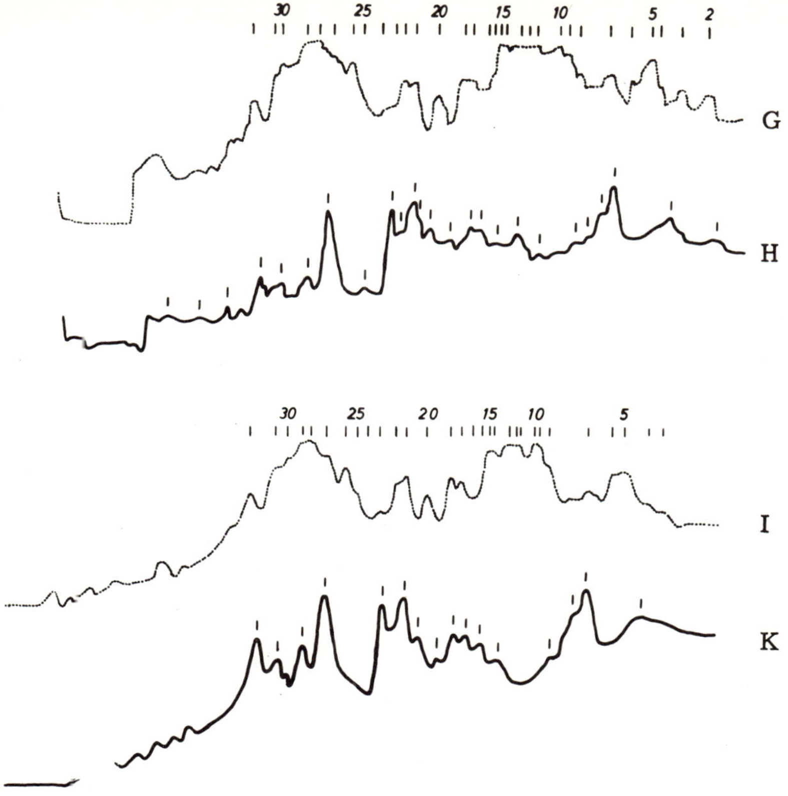

Fig. 6. Proteins of the 44 S ribosomal subunit isolated from the chloroplasts of a F_1-generation from heterologous combinations: med (B), med_1 $clift_0$ (D), clift (F), $cren_1$ $clift_0$ (H), cren (K), reference proteins from *E. coli* 70 S ribosomes (A, C, E, G, I) [4].

More recently we have succeeded in preparing and growing one thousand grafts of each of the types $clift_1$ med_0 and $cren_1$ $clift_0$. In this case, the proteins of the 44 S particles were compared between the two species and the grafts. In accordance with the results obtained with the F_1-generation, the species-specific bands were unequivocally determined by the species of the nucleus (*Kloppstech* and *Schweiger* [4]).

It is manifest that what has been shown for two enzymes and for chloroplast membrane proteins holds also for chloroplast ribosomal proteins: at least some of them are coded by nuclear DNA. This is surprising since the combination of all the different ribosomal proteins with the ribosomal ribonucleic acids presumes a high degree of adequacy in order to result in functional units. It is well known that even minor errors in the primary structure of the polypeptide chains may result in substantial changes in the tertiary and quarternary structure and may seriously impair the function of the ribosomes.

In the case under consideration, the genetic information for the protein moiety originates at least in part in the nuclear genes while the ribosomal RNA is coded by the chloroplast DNA. Nevertheless, this combination results in units whose function is shown by the participation of chloroplast ribosomes in protein synthesis and by the viability of such combinations. A complex cooperation between the cell nucleus and the organelles must exist, and it seems to be obvious that the cell nucleus plays an important part in this cooperation. The problem which remains to be solved is which chloroplast proteins are coded by the chloroplast genome.

References

[1] *Apel, K.*, and *Schweiger, H. G.* 1972. Nuclear dependency of chloroplast proteins in Acetabularia. Europ. J. Biochem. **25:** 229–238.

[2] *Baltus, E., and Brachet, J.* 1963. Presence of deoxyribonucleic acid in the chloroplasts of Acetabularia mediterranea. Biochim. biophys. Acta **76:** 490–92.

[3] *Gibor, A., and Izawa, M.* 1963. The DNA content of the chloroplasts of Acetabularia. Proc. Nat. Acad. Sci. **50:** 1164–69.

[4] *Kloppstech, K., and Schweiger, H. G.* 1971. Nuclear coded ribosomal proteins in chloroplasts of Acetabularia. Proc. 7. Meet. Fed. Europ. Biochem. Soc., Varna.

[5] *Reuter, W., and Schweiger, H. G.* 1969. Kernkontrollierte Lactatdehydrogenase in Acetabularia. Protoplasma **68:** 357–68.

[6] *Schweiger, H. G.* 1969. Cell Biology of Acetabularia. Current Topics in Microbiol. and Immunol. **50:** 1–36.

[7] *Schweiger, H. G., Master, R. W. P., and Werz, G.* 1967. Nuclear control of a cytoplasmic enzyme in Acetabularia. Nature **216:** 554–57.

[8] *Werz, G., and Kellner, G.* 1968. Molecular characteristics of chloroplast DNA of Acetabularia cells. J. Ultrastructure Res. **24;** 109–15.

Discussion

Graham: I would like you to comment on the stability of the host malic dehydrogenase following nuclear transfer in comparison with the stability of that enzyme following enucleation. Presumably the enzyme is being preferentially inactivated following nuclear transfer.

Schweiger: That is a question which needs further attention. We do not know whether the disappearance of the host isozyme pattern is due to degradation of the enzyme proteins involved or to inactivation. In case of the other proteins, for example the species-specific membrane proteins, it is apparently due to degradation.

Genetic Manipulation of Mouse Embryos

Christopher F. Graham

Zoology Department, Oxford, U.K.

Summary: This paper reviews the most recent attempts to introduce genetic material into the developing mouse embryo.

First, attempts have been made to suppress second polar body formation to obtain parthenogenetic diploid embryos. This manipulation has been performed successfully but the embryos do not develop beyond the seventh day of pregnancy.

Secondly, attempts have been made to fuse adult cell nuclei into fertilized and unfertilized eggs. These additional nuclei undergo chromosome condensation, and chromosome fragmentation. However, unlike somatic cell hybrids, the fragmented genome does not appear to be incorporated into the developing embryos. However, embryos exposed to these fusion conditions develop into live young, showing that fusion itself does not block the development of mammalian embryos.

Introduction

This is a review of the attempts which have been made to introduce genetic material into the mammalian embryo.

At present, it is only possible to work with preimplantation stages of development; postimplantation stages can be cultured in vitro for prolonged periods, but they do not develop into live young (reviewed [12]). The mouse and rabbit embryo can be grown from the fertilized one-cell egg to the blastocyst in culture and will subsequently develop into healthy young when transferred to a foster mother at the correct stage of pseudopregnancy [18]. Since the cells of the embryo are known to be viable in culture up to the blastocyst stage, attempts to introduce extraneous genetic material have been confined to preimplantation development.

Manuscript received: 10 December 1971

The problems

The problems of maintaining the second polar body nucleus in egg cytoplasm by suppressing second polar body formation

Many attempts have been made to produce parthenogentic mammals (reviewed [15]). The failure of these efforts appears to be due to the inability to reproduce those conditions which have lead to successful parthenogenesis in amphibians [6].

If second polar body formation is suppressed in amphibians, then gynogenetic diploid embryos may develop into adult frogs. About one-quarter of the blastulae can complete development into an adult. Second polar body formation can be suppressed by pressure applied at the time that the second polar body nucleus is being extruded [17]. The frogs will be heterozygous if crossing-over occurs at meiosis, and the frequency of heterozygosity for a particular gene can be used to map its distance from the centromere [17].

Parthenogenetic activation of mouse eggs in vitro normally results in the egg developing in three ways [6]:

1) Second polar body formation is surpressed and the activated egg contains two pronuclei. These eggs might give rise to heterozygous diploid parthenogenones,
2) Second polar body formation occurs, but the polar body contains half the cytoplasm of the egg. These parthenogenones are described as undergoing "immediate cleavage." Both the blastomere containing the egg pronucleus and the blastomere containing the second polar body nucleus continue to divide. The parthenogenone is a presumptive haploid chimera.
3) Second polar body formation is surpressed and the activated egg contains two pronuclei. These eggs might give rise to heterozygous diploid parthenogenones, but it has been observed that the pronuclei rarely fuse and usually segregate into separate blastomeres near the time that the fertilized egg would normally reach the two-cell stage.

It was decided to look for the conditions which would suppress second polar body formation and encourage the second polar body nucleus to fuse with the egg pronucleus.

Methods

F_1 hybrid eggs were obtained from superovulated virgin females produced by the crosses C57BL × CBA. In some of the transfer experiments, eggs from PO strain females were used. These eggs were activated in culture as described previously [5]. Following hyaluronidase treatment, the eggs were placed for 2–3 hours in White's culture medium at different dilutions [9]. After this treatment, they were cultured in *Whitten*'s medium [18] and treated in the following ways:

1) Cultured overnight in *Whitten's* medium containing 1 µg/ml of colcemid. Chromosome preparations were made from these embryos by the air drying method [14].
2) Transferred to the oviduct of pseudopregnant recipients on the day of plug, by Dr. *A. K. Tarkowski.* The development of these embryos was followed on different days of pregnancy.

Results

As the osmolarity of the culture medium is decreased, so the frequency of suppression of second polar body formation is increased (Fig. 1). When the culture medium is diluted to 3/5, a new class of activated eggs is observed. These contain one pronucleus and no second polar body. It must be presumed that the whole spindle of the second meiotic metaphase sinks to the centre of the egg and forms a single nucleus. At the first cleavage division, these parthenogenones are diploid (Fig. 2).

Activation medium	Culture medium	2^0 Polar body one pronucleus	Immediate cleavage (two-cells)	No polar body two pronuclei	No polar body one pronuclei
White's	Whitten's	38 (78)	5 (10)	6 (12)	0
4/5 White's	Whitten's	3 (2)	25 (19)	103 (79)	0
3/5 White's	Whitten's	23 (37)	11 (18)	16 (25)	13 (20)

Fig. 1. Activation of mouse eggs in media differing in osmolarity

All eggs were from virgin females which were F_1 hybrids from crosses of C57BL and CBA or C57BL and CBAT6T6. The figures in brackets refer to the percentage of eggs developing along difference pathways after particular activation treatments.

Observed cytology	Chromosome counts
One pronucleus, 2^0 polar body	3 × 20, 22, 19
Immediate cleavage, two cells	6 × (20 + 20), 3 × (21 + 19), 22 + 18, 21 + 18, 20 + 19
Two pronuclei, no second polar body	4 × 20 + 20, 20 + 19, 20 + 19
One pronucleus, no second polar body	6 × 40, 37

Fig. 2. Chromosome counts from eggs activated in different media

The eggs were separated into different categories at twelve hours after activation. They were cultured in separate drops of colcemid over night and fixed the following morning.

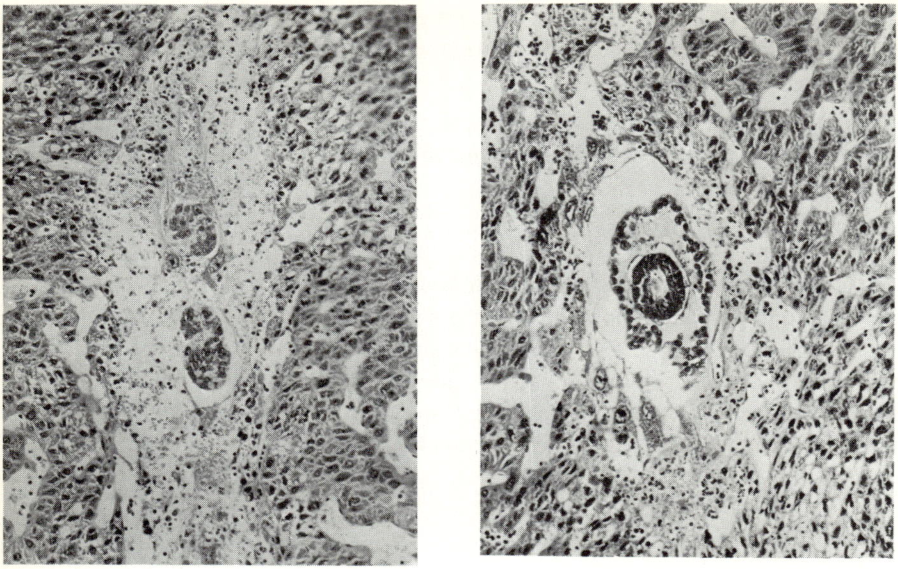

Fig. 3. Embryos on the seventh day of pregnancy. These embryos developed from activated eggs in which only one pronucleus was found and which did not form a second polar body. They were presumed to be diploid heterozygotes formed by the incorporation of the chromosomes of the second polar body and the egg pronucleus.

On the seventh day of pregnancy, disorganized egg cylinders can be seen inside the extraembryonic membranes and trophoblast giant cells.

Tarkowski has transferred many of these eggs to the oviducts of pseudopregnant recipients. None have come to term. These eggs form morphologically normal blastocysts but only develop into disorganized day 7 embryos (Fig. 3).

If these embryos are diploid heterozygotes, it is difficult to understand why they should not develop to term.

The problem of fusing genetic material into the developing embryo and obtaining subsequent development

Many different cell types have been fused into the fertilized and unfertilized one-cell mouse egg using the Sendai virus [1, 4, 7]. The majority of these experiments have been conducted in a culture medium which does not support the development of normal untreated eggs, and it is therefore not surprising that little subsequent development was achieved.

It was also necessary to modify the normal method of fusion which killed the egg. The UV-inactivated Sendai virus is cytopathic to the one-cell egg, and it was only when beta propio-lactone- (BPL) inactivated Sendai virus was used for fusion that the egg remained viable [1].

Using the BPL-inactivated Sendai virus and *Whitten*'s culture medium, it has proved possible to fuse one-cell eggs and the blastomeres of two-cell eggs together and obtain subsequent development [7]. The fused cells developed into morphologically normal blastocysts which contained a similar number of cells to untreated control blastocysts developing under identical culture conditions. The DNA content of the cells of these blastocysts was consistent with a tetraploid chromosome number.

However, on transfer to a suitable pseudopregnant recipient, these blastocysts only gave rise to large trophoblast outgrowths and did not appear to form any fetal or extraembryonic derivatives.

The successful development of these fused blastomeres up to the blastocyst stage makes it unlikely that fusion had damaged the cells. It appeared that their subsequent failure to develop into a foetus was the consequence of tetraploidy. To demonstrate that this was the case, diploid: tetraploid chimeras were made by fusing one blastomere of a four-cell embryo to one blastomere of a two-cell stage which was still in contact with its normal partner (Fig. 4). Six of these combinations developed into blastocysts and were transferred into a pseudopregnant recipient. Three young were born, but these could be shown by coat color markers only to contain cells from the unfused blastomere (Fig. 5).

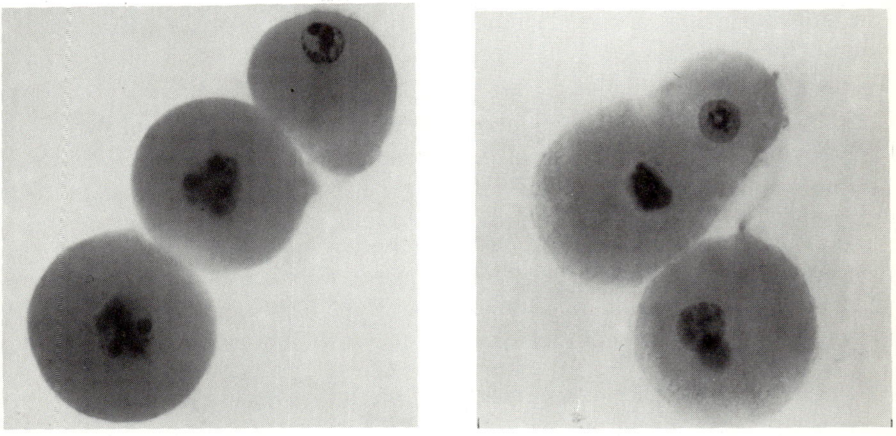

Fig. 4. Fusion of one blastomere of a four-cell stage with one blastomere of a two-cell embryo to make a diploid: tetraploid chimera. (a) Whole mount of the blastomeres placed together. (b) Whole mount of the blastomeres 15 minutes after the start of fusion.

[Development of two-cell PO embryos with one fused blastomere]

Donor Stage	Donor Genotype	Recipient Stage	Recipient Genotype	Transfer	Pregnant	Result
four-cell embryo	CBAT6T6	one-cell of a two-cell embryo	PO	5/100	0/1	–
four-cell embryo	F_1 CBAT6T6/PO	one-cell of a two-cell embryo	PO	6/7	1/1	three white mice

Development of fused and unfused blastomeres

Female	Male	Stage	Combination	Transfer	Result
F_1 C57BL/CBAT6T6	PO	two-cell	fused in pairs	36	tenth day of pseudopregnancy – 11 growths
PO	C57BL	two-cell	unfused in chimeric association with PO two-cell	9	3 born – 1 brown, 1 white, 1 mixed

Fig. 5. Development of embryos with fused cells.

The fused blastomere in this combination was inevitably a larger cell than its unfused partner. It appeared from microscopic observation of the developing combination that the fused blastomere only gave rise to cells on the outside of the blastocysts which developed. It has also been found that in normally developing embryos, the largest blastomeres tend to form the outside of the morula [2]. It was therefore not surprising that derivations of the tetraploid blastomere were not seen in the fetus whcih developed.

The problem of fusing cells with the fetal precursor cells

The next difficulty is that of introducing genetic material into those cells of the preimplantation embryo which will give rise to the fetus. The one-cell egg and the blastocyst form the trophoblast and the extraembryonic membranes as well as the fetus. And there are several reasons for thinking that only a few cells of the blastocyst form the fetus:

1) Histological studies suggest that the cells of the inner cell mass (icm) of the blastocyst form the extraembryonic membranes as well as the fetus [13]. Since there are on average thirteen cells in the icm of a sixty-four-cell blastocyst, there can not be many cells from which to form the fetus (*Barlow* et al. [2]).

2) One quarter of chimera formed by combining two eight-cell embryos of different genotype develop into a fetus of only one genotype. This suggests that only a few cells are selected from the chimeric blastocyst to form part of the fetus [11].

3) A single cell from the icm of one blastocyst, when transferred into another blastocyst, can form half the coat color of the mouse which is born [3]. Although it is not known that a single cell left in an intact blastocyst does form half the coat color of a mouse, this experiment shows that a single cell has the potentiality to do so.

The problem would be overcome if it were possible to identify those cells which were determined to form the fetus and also if it were possible to ensure that the genetic material was introduced into them. Unfortunately there are no methods for achieving this at present.

Instead, attempts were made to force cells which contained extraneous genetic material to form the fetus rather than the extraembryonic membranes or the trophoblast. To achieve this, it was necessary to discover the conditions which determined cell fate during early development. It had been suggested by *Tarkowski* and *Wroblewska* [16] that cell position determined cell fate during cleavage in the mouse. They proposed that as a consequence of their relative positions, cells on the outside of the morula formed the trophoblast and cells on the inside of the morula formed the icm of the blastocyst. The consequence of this theory is that if a cell is placed on the inside of the morula, it should have an increased chance of contributing to the icm of the blastocyst and therefore an increased chance of forming part of the fetus. This prediction was shown to be correct by combining embryos or parts of embryos together in such a way that the cells of one genotype were either in an outside or inside position at the time of combination [8].

Hillman and her associates found that if all the cells of a four-cell mouse embryo were placed outside another embryo, then 90 % of the division products of these blastomeres were found on the outside of the blastocyst which developed. Conversely, if all the cells of four-cell embryo were placed between the blastomeres of other embryos, then over 40 % of their division products were found in the icm of the blastocysts which developed.

Experiments were therefore designed to encourage the fused blastomeres to form the icm of the blastocyst.

Methods

Embryonic fibroblasts and adult bone marrow cells were fused to 8–16 cell embryos of different genotype. The embryonic fibroblasts and adult bone marrow cells were dissociated and exposed to BPL-inactivated Sendai virus as described previously [4]. Following fusion, the recipient morulae were cultured in *Whitten*'s medium up to the blastocyst stage and then transferred to the uterus of pseudopregnant recipients [10]. The recipient morulae were either grown in pairs or surrounded by five unfused morulae (Fig. 6c). The development of these composites is described in Fig. 7.

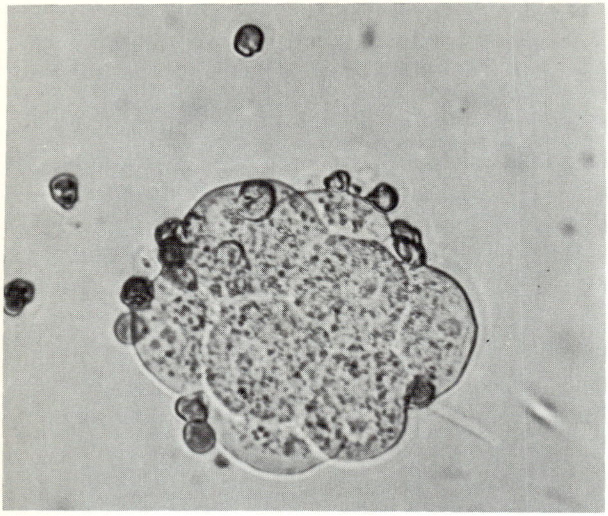

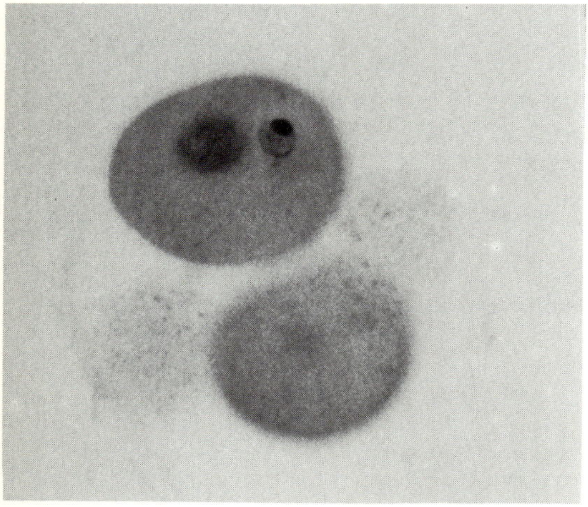

Fig. 6. Fusion of adult bone marrow cells to 8–16 cell morulae. (a) The morula with many adult cells attached to its surface 5 minutes after the start of fusion. (b) Section of a fused blastomere of the morula. The bone marrow nucleus (arrow) is at the periphery of the recipient blastomere. (c) A composite consisting of one central morula fused to adult cells, surrounded by five unfused morulae.

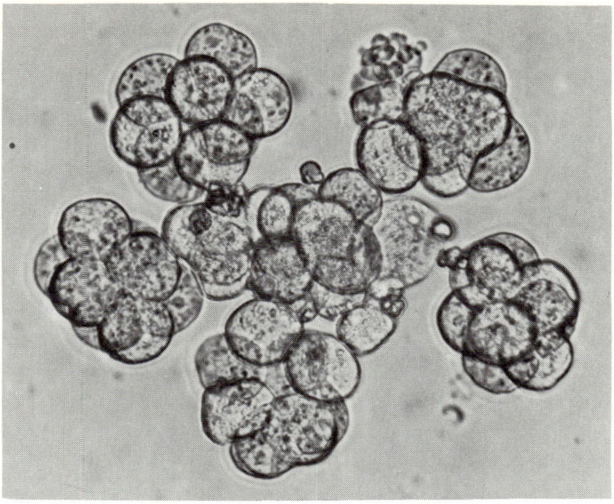

Donor		Recipient					
Tissue	Genotype	Stage	Genotype	Combination	Transfer	Pregnant	Result
Embryonic fibroblasts	F_1 PO/CBAT6T6	8–16 cell	PO	fused pairs and singles	152/180	6/10	25 white mice
Adult bone marrow	C57BL	8–16 cell	PO	1 fused embryo + 5 unfused embryos	18/18	1/1	2 white mice
Adult bone marrow	CBAT6T6	8–16 cell	PO	1 fused embryo + 5 unfused embryos	20/20	1/1	d. 11–20 embryos
Adult bone marrow	CBAT6T6	8–16 cell	PO	1 fused embryo + 5 unfused embryos	207/210	9/18	70 resorbtion sites 3 white mice

Fig. 7. Development of PO morulae fused with embryonic fibroblasts and adult bone marrow

Results

Culture in pairs

After fusion, the donor cell nuclei are mainly found in the peripheral cytoplasm of the outside cells of the 8–16 cell recipient embryos (Fig. 6a, b). The recipient morulae were therefore cultured in pairs to increase the probability of the outside fused cells contributing to the icm of the blastocyst.

152 blastocysts were transferred to pseudopregnant recipients and only 25 mice were born. None of these showed the coat color of the donor cells (Fig. 7). It appeared that recipient cells containing the donor nuclei did not form fetal tissue.

Culture in groups

To increase the probability of the fused cells contributing to the fetus, the fetal morulae were surrounded by five unfused morulae (Fig. 6, c). It is known that in these circumstances the inside morulae always contributes to the icm of the blastocyst [8].

These composites did not develop well in the uterus (Fig. 7). 245 blastocysts were transferred and only 5 mice were born. These had the coat color of the recipient morulae.

Discussion

The failure of the fused genotype of the cells to contribute to the fetus in these experiments could be the consequence of (1) tetraploidy, (2) slower growth rate, (3) chromosome condensation of the donor nucleus. This is known to occur when adult cells are fused to two-cell stages [4].

Conclusion

So far it has been impossible to introduce extraneous genetic material into the mammalian embryo and observe a contribution of this material to the fetus at term. (1) The gynogenetic parthenogenones develop up to the seventh day of pregnancy. (2) Tetraploid blastocysts only form large trophoblast growths in the uterus. (3) Morulae containing fused adult cells develop to term, but they show no evidence of genetic contributions from the donor nuclei.

Acknowledgements: This work was generously supported by the MRC. The experiments were performed in the Sir William Dunn School of Pathology and Prof. *H. Harris,* and the staff provided excellent facilities. Dr. *A. K. Tarkowski* performed many of the experiments on parthenogenetic embryo. Mrs. *Z. A. Deussen* assisted in all the experiments.

References

[1] *Baranska, W., and Koprowski, H.* 1970. Fusion of unfertilized mouse eggs with somatic cells. J. exp. Zool. **174:** 1.

[2] *Barlow, P. B., Owen, D., and Graham, C. F.* DNA synthesis in the preimplantation embryo. J. Emb. exp. Morph. **27:** 431.

[3] *Gardner, R. L.* 1971. Manipulations on the blastocyst. *In* Schering Symposium on . . . Mammalian Development, Advances in the Biosciences 6, ed. G. Raspé, p. 279. Braunschweig: Pergamon Press/Vieweg.

[4] *Graham, C. F.* 1969. The fusion of cells with one- and two-cell mouse embryos. *In* Heterospecific genome interactions, Wistar Institute Symposium Monograph 9, ed. V. Defendi, p. 19.

[5] *Graham, C. F.* 1970. Parthenogenetic mouse blastocysts. Nature, Lond. **226:** 165.

[6] *Graham, C. F.* 1971. Experimental early parthenogenesis in mammals. *In* Schering Symposium on . . . Mammalian Development, Advances in the Bioscience 6, ed. G. Raspé, p. 87. Braunschweig: Pergamon Press/Vieweg.

[7] *Graham, C. F.* 1971. Virus assisted fusion of embryonic cells. *In* In vitro methods in Reproductive Biology, Karolinska Symposium on Research Methods in Reproductive Cell Biology 3, ed. E. Diczfalusy, p. 154.

[8] *Hillman, N., Sherman, M. S., and Graham, C. F.* In press. The effect of spatial arrangement on cell determination during mouse development. J. Emb. exp. Morph.

[9] *Knowland, J. S., and Graham, C. F.* RNA synthesis at the two-cell stage of mouse development. J. Emb. exp. Morph. **27:** 167.

[10] *McLaren, A., and Michie, D.* 1956. Studies on the transfer of fertilized mouse eggs to uterine foster mothers. 1. Factors affecting the implantation and survival of native and transferred eggs. J. exp. Biol. **33:** 394.

[11] *Mintz, B.* 1971. Clonal basis of mammalian differentiation. *In* Control mechanisms in growth and differentiation, ed. D. D. Davis and M. Balls. S. E. B. Symposium 25, p. 345.

[12] *New, D. A. T.* 1971. Culture of fetuses in vitro. *In* Schering Symposium on . . . Mammalian Development, Advances in the Biosciences 6, ed. G. Raspé, p. 367. Braunschweig: Pergamon Press/Vieweg.

[13] *Snell, G. D., and Stevens, L. C.* 1966. *In* Biology of the laboratory mouse, ed. E. L. Green, p. 205. New York: McGraw-Hill.

[14] *Tarkowski, A. K.* 1966. An air drying method for chromosome preparations from mouse eggs. Cytogenetics **5:** 394.

[15] *Tarkowski, A. K.* 1971. Recent studies on parthenogenesis in the mouse. J. Reprod. Fert. Suppl. **14:** 31.

[16] *Tarkowski, A. K., and Wroblewska, J.* 1967. Development of blastomeres of mouse eggs isolated at the four- and eight-cell stage. J. Emb. exp. Morph. **18:** 155.

[17] *Volpe, E. P.* 1970. Chromosome mapping in the leopard frog. Genetics **64:** 11.

[18] *Whitten, W. K.* 1971. Nutrient requirements for culture of preimplantation embryo in vitro. *In* Schering Symposium on . . . Mammalian Development, Advances in the Biosciences 6, ed. G. Raspe, p. 129. Braunschweig: Pergamon Press/Vieweg.

Discussion

follows the paper of *Andrzeij K. Tarkowski,* p. 275.

Comments on the Application of Cell Fusion Techniques to the Study of Mammalian Embryology

Andrzej K. Tarkowski

University of Warsaw, Department of Embryology, Warsaw 64, Poland

I would like to discuss briefly two applications of cell fusion techniques to mammalian embryology. The first one is to investigate the nature of nuclear changes underlying cell differentiation by putting the nucleus from embryonic or differentiated cells back into the egg cytoplasm. There already have been two such attempts [2, 5], but they were not very successful. The second application is to obtain diploid hybrid embryos.

These goals are still very remote and the real progress is not to be expected until a number of technical problems are solved. First, it must be known that a cell produced by the fusion of two eggs or blastomeres can develop at least as far as a blastocyst. Such an experiment has been performed by Dr. *Graham* [6] and, subsequently, by our group. It consisted of the fusion of two blastomeres from the two-cell mouse embryo. This experimental system fulfills two essential requirements: first, contrary to one-cell eggs, blastomeres from a two-cell stage can be easily cultured to the blastocyst stage and, second, both partners are in the same phase of the cell cycle. In this experiment, the "parental" cells were diploid and the resulting embryo was consequently tetraploid. Since fusion reduces the number of cells by two, the tetraploid embryos should have half the number of cells of comparable diploids. According to our observations – and this is the point at which our results disagree with those obtained by Dr. *Graham* – the tetraploid embryos contain less than half the number of cells in diploid controls. This effect may be due to the treatment or to the longer cell cycle in tetraploid blastomeres as compared to diploid ones. In view of the fact that polyploidy appears to slow down preimplantation development of mouse and rabbit embryos [3, 4], the latter explanation is likely.

Manuscript received: 28 February 1972

The fusion of diploid blastomeres is just to practice the technique; tetraploid embryonic cells may be useful for special studies in cell biology, but at the embryonic level tetraploidy is most probably lethal soon after implantation. In order to obtain diploid blastocysts and hopefully diploid individuals in this way, it is necessary to use haploid eggs or blastomeres. As Dr. *Graham* said, we can easily obtain haploid eggs by artificial activation. However, unless it is known that the viability of artificially activated eggs is not inferior to the viability of sperm activated eggs, it seems more safe (although more troublesome) to use the latter type of eggs. During the last two years, my collaborators have been trying to obtain haploid androgenetic mouse embryos using various methods: for example, by irradiating the egg nucleus with a UV microbeam, or by submitting the egg at the time of second meiotic division to colcemid (in some eggs, this treatment causes the extrusion of the whole nuclear apparatus). Unfortunately, all these methods are very erratic and do not produce consistent results. The method which looks most encouraging at the moment consists of the microsurgical removal of one of the two pronuclei. I hope that this technique will prove very useful, not only for providing haploid fusion partners but also for studying the development of androgenetic embryos, especially those which do not carry X chromosome.

Obtaining full development of diploid embryos produced by fusion of haploid eggs or blastomeres is, in a sense, a trivial result as nature achieves the same effect simply by fertilization. However, I consider this experiment a necessary step for further experimentation, especially for obtaining diploid hybrid embryos.

Hybrids, especially lethal hybrids, present a very useful material for studying nucleo-cytoplasmic interactions and the time of onset of gene action in development. This approach proved very fruitful in echinoderms and amphibia but, so far, has not been extensively explored in mammals. The reason for this is that it is very difficult to obtain cross-hybrydization between laboratory animals. Experiments by Drs. *Austin, Braden, Edwards,* and myself carried out several years ago showed that in many combinations of species foreign sperms achieved the site of fertilization but could not penetrate the zona pellucida [1]. It appears, therefore, that fertilization in vitro of zona-free eggs or virus-assisted fusion of haploid eggs (or blastomeres) is the only solution to the problem. Hybrids obtained by the latter method could have even better developmental prospects than "normal" hybrids because fusion, unlike fertilization, endows the hybrid cell with cytoplasm from both parental species thus creating a more favourable environment for the paternal genome. So far, we have been looking for a species (1) whose eggs could be cultured in vitro in the same way as mouse eggs, and (2) whose rate of development is similar to that of mouse embryos. The bank vole (Clethrionomys glareolus) appears to fulfill these requirements, and experiments on fusing blastomeres from these two species are just getting underway.

References

[1] *Austin, C. R.* 1961. The mammalian egg. pp. 95–96. Oxford: Blackwell Scientific Publications.

[2] *Baranska, W.*, and *Koprowski, H.* 1970. J. exp. Zool. **174**.

[3] *Beatty, R. A.* 1957. Parthenogenesis and polyploidy in mammalian development. Cambridge: Cambridge University Press.

[4] *Bomsel-Helmreich, O.* 1965. Preimplantation stages of pregnancy. A Ciba Foundation Symposium, eds. *G. E. W. Wolstenholme* and *M. O'Connor*, pp. 246–67. London: J. and A. Churchill.

[5] *Graham, C. F.* 1969. Heterospecific genome interaction. The Wistar Institute Symposium Monograph, ed. *V. Defendi*, p. 19.

[6] *Graham, C. F.* 1971. Karolinska symposia on research methods in reproductive physiology, 3rd Symposium, p. 154.

Discussion

Hofschneider: I have one technical question. Apparently the term "fusion" is now being used in two different senses. Sometimes cells are really fused, other times they are only mixed. Will one of the speakers clarify which technique is used for which experiments?

Koprowski: May I answer that question? I think that applies to only one experiment presented here, the last experiment that Dr. *Graham* summarized where he actually mixed cells in the absence of Sendai or other fusion factors. Everything else that he had discussed prior to this applied to fused cells. Dr. *Tarkowski,* were you talking about fused cells or mixed cells?

Graham: All my experiments involved cell fusion. The last experiments I discussed concerned fusion of 8 to 16 cell stages. In addition to those cells which fused, however, there were also cells present which mixed with the embryonic cells. These mixed cells and these fused cells appeared not to contribute to the mouse which was born.

Tarkowski: I have also been talking about cell fusion with Sendai virus and not about association of embryonic cells.

Fischberg: I want to say a word about the cells number in polyploids and haploids in mammals. Drs. *Graham* and *Tarkowski* are probably aware that *Beatty* and I have analyzed the cell number of blastocysts in haploid, diploid, triploid, and tetraploid mouse embryos, way back in 1949–51. We actually found that triploids and tetraploids had only about two-thirds or half the cell number of diploids, whereas haploids possessed a higher cell number. This agrees with your results. Another point is that we all know of hybridization without the use of viruses. I mean natural or artificial insemination between different species. If one crosses a lion and a tiger, one obtains viable hybrids. During the last year, *Blackler* in America and my collaborators and I have hybridized different species of *Xenopus,* and found that initially in the hybrid only one of the two nucleolar organizers becomes active. Even more astonishing is, as *D. D. Brown* showed, that in the hybrid females resulting from the cross between *Xenopus laevis* and *Xenopus mulleri,* amplification of the *laevis* r(ribosomal)DNA region only and not of that from *mulleri* occurs. Such examples of natural or artificial insemination in organisms which survive as hybrids form an excellent source for studies of genetic regulation.

Tarkowski: I did not mean to say that it is impossible to obtain interspecific hybrids in mammals. Obviously, there is a long list of viable hybrids, some of which are fertile, some infertile. There is a real difficulty, however, in obtaining hybrids between species which can be used for routine laboratory work. We have tested several combinations of rodent species without success. One solution may be to remove the zona pellucida, thereby fertilizing naked eggs; another solution would be to use the technique of virus-mediated cell fusion to eggs or blastomeres. Hybrids obtained by these two methods may not be comparable to each other because in one the paternal partner supplies only the genome, while in the other the two partners contribute on equal footing both to the nucleus and the cytoplasm of the hybrid cell.

Duesberg: I have what is probably a naive question. I want to find out what is known about the viability of hybrids in general. In tissue culture where they are usually polyploid these hybrid cells do not do very well.

Koprowski: The viability of which hybrid? Are you inquiring about the somatic cell hybrid?

Duesberg: Of any hybrid, I mean of any fused cell. How long can it divide? In tissue culture, if you fuse a chicken cell with a chicken cell or a chicken cell with a mouse cell, will they divide forever?

Koprowski: Oh yes, they are immortal. I think that Dr. *Ruddle* will probably speak about this tomorrow. A hybrid between a human cell, which has a finite lifetime, and a mouse cell is immortal. Hence, one way for us to live forever is to hybridize with mice!

Duesberg: I will try this as soon as I return to Berkeley.

Expression of New Antigens

Felix Milgrom and Kyoichi Kano

Department of Microbiology, School of Medicine, State University of New York at Buffalo, Buffalo, New York 14214, USA

Summary: Denaturation of proteins results in the emergence of new antigenic specificities. In the case of IgG, the new antigens resemble native configurations occuring in IgG of members of other species and some individuals belonging to the same species. The antigenic structure of cultured cells may be transformed by "implanting" foreign antigens through the process of hybridization. By using this procedure, human HL-A antigens of peripheral blood leukocytes were "added" to the cell surface of murine fibroblast cultures. Antigenic transformation of cultured human lymphoid cells may occur without any intentionally induced modification. This transformation may include acquistition of antigens belonging to HL-A system.

Antibodies detected in some diseases may be formed in response to new antigens that emerged during the morbid process. Evidence was presented that formation of the rheumatoid factor may be stimulated by the patient's own IgG antibodies that underwent molecular transformation in reacting with their corresponding antigens. Heterophile antibodies might be elicited by stimulus exerted by a new antigen, the heterophile antigen, which was created on the patient's cells during the disease. This hypothesis is made plausible by the observation that the cultured human lymphoid cells carry the heterophile antigen.

New antigens on isolated proteins and cell cultures

Expression of new antigenic determinants on isolated proteins and whole cells has attracted considerable attention among immunologists. *Rothen* and *Landsteiner* [29] demonstrated that heat-coagulated crystalline ovalbumin and equine serum albumin differed considerably from native proteins in antigenic structure. Antisera to native proteins reacted very weakly with denatured proteins and conversely antisera to denatured proteins gave weak reactions with native proteins.

Manuscript received: 7 November 1971

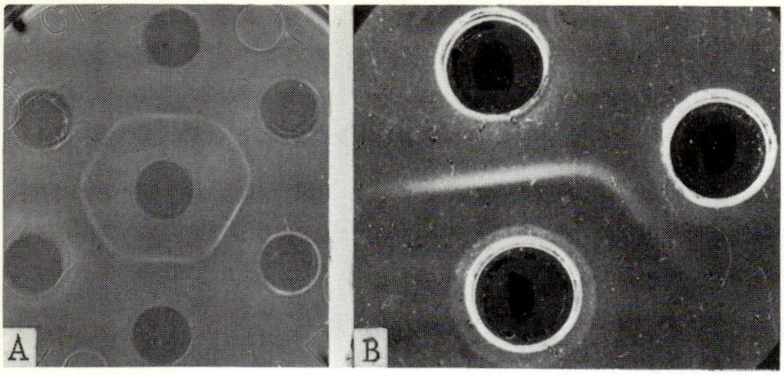

Fig. 1. (A) Central well: Fraction II of pooled normal human serum, 0.1 mg/ml. Peripheral wells: Rabbit antisera to gamma globulin; clockwise, starting from the uppermost well: Antiautologous, antihuman, antiautologous, antihuman, antiautologous, antihuman.
(B) Lower well: Rabbit antiserum to denatured autologous gamma globulin. Upper left well: Fraction II of pooled human serum, 0.1 mg/ml. Upper right well: Fraction II of pooled rabbit serum, 200 mg/ml.

We studied antibody responses in rabbits injected with autologous denatured gamma globulin [24] and found that the animals formed antibodies which combined in a variety of serological reactions with IgG of foreign species origin. Reactions with human IgG are presented in fig. 1A. It may be seen that precipitation lines produced by rabbit antisera to denatured autologous gamma globulin merged into identity reactions with lines formed by rabbit antisera to human IgG. In contrast, antisera to denatured autologous gamma globulin reacted very poorly with the rabbit IgG. As a matter of fact, rabbit IgG had to be used at a concentration over 1000 times higher to give a reaction comparable to that obtained with human IgG (Fig. 1B). These data have been confirmed and extended by *McCluskey* et al. [17] and by *Williams* and *Kunkel* [33]. In addition to antibodies combining with IgG of foreign species origin, antibodies acting upon IgG of some rabbits were detected. These experiments supplied evidence that the denaturation of gamma globulin molecules reveals new antigenic determinants which stimulate antibody formation in the autologus host. Interestingly, similar antigenic sites might be present as native configuration on the IgG of some other species and of other individuals of the same species.

Acquisition of new antigens on cultured cells may be readily produced by the procedure of hybridization. In our recent studies [14] conducted in collaboration with Drs. *Koprowski* and *Knowles* of the Wistar Institute, we used murine fibroblast cell cultures as the "vehicle" for antigens of human peripheral blood leukocytes. Heterokaryons were produced with the aid of the Sendai virus and hybrid cells were isolated through biochemical pressures. As shown in fig. 2, hybrid cells contained both human and murine species-specific antigens which could be conveniently

Serum dilution 1 to:		Parents: Murine cell line IT	Parents: Human leukocytes PL	KK	BK	Hybrids: IT X PL	IT X KK	IT X BK
Antiman serum	300	–	+++	+++	+++	+++	+++	+++
	900	–	+++	++	++	+++	+++	+++
	2700	–	++	+	–	++	+	++
	8100	–	–	–	–	–	–	–
Antimouse serum	300	+++	–	–	–	+++	+++	+++
	900	+++	–	–	–	+++	+++	+++
	2700	+++	–	–	–	+++	+++	+++
	8100	++	–	–	–	++	+	+

Fig. 2. Reactions of parental and hybrid cells with rabbit antisera to human and murine species-specific antigen*).

*) Tests with human parental cells were performed by means of lymphocytotoxicity, all other te tests were performed by means of mixed agglutination.

	IT X PL							IT X KK						
	Typing sera to HL-A							Typing sera to HL-A						
Serum dilution	1	2	3	9	5	7	8	1	2	3	9	5	7	8
1 to: 100	+++	–	+	+++	–	++	+++	+	+++	–	+++	+++	+	++
300	++	–	–	+++	–	–	+++	–	++	–	+++	++	–	–
900	+	–	–	++	–	–	+++	–	+	–	++	++	–	–
2700	–	–	–	++	–	–	–	–	–	–	–	–	–	–
LCT*)	+++	–	–	+++	–	–	+++	–	+++	–	+++	+++	–	–

Fig. 3. Mixed agglutination with hybrid cells

*) Standard cytotoxicity tests with lymphocytes of PL and KK donors.

detected by means of mixed agglutination tests with cell cultures and rabbit antisera to human and murine antigens. It was also shown that mice injected with man-mouse hybrid cells would form antibodies to human species-specific antigens.

All hybrid cells contained murine alloantigens detectable by means of proper alloantisera with transplantation antibodies. No further efforts were made to identify these antigens. The hybrid cells also contained human alloantigens. In two hybrids, HL-A specificities were defined. It may be seen in fig. 3 that hybrid cells originating from human parent PL gave definite positive reactions with antisera to HL-A specificities 1, 9, and 8, which corresponded very well with the phenotype of the leukocyte donor. Similarly on the second hybrid, antigens 2, 9, and 5 were detected which were also present on leukocytes of the donor KK. Results of tests with other anti-HL-A sera were either negative or very weakly positive. These tests were repeated using four sera for each specificity and the results were very similar.

Further studies were performed on clones obtained from the hybrid IT × KK (Fig. 4). It may be seen that all clones but clone 3 retained human species-specific antigens as evidenced by the reactions with antiman serum. Of three HL-A antigens present on the original hybrid, HL-A 1 and 5 were lost from all clones, whereas HL-A 9 was retained by all clones except clone 3. Our interpretation of these findings is the following: Upon cloning, subpopulations of hybrid cells are selected which may contain one, both or neither of the HL-A chromosomes. The fact that clones 2, 4, 6, and 7 retained HL-A 9 antigen whereas they lost HL-A 2 and 5 antigens may be explained by selection of cells with one of the two HL-A chromosomes. This would infer that the KK donor has one chromosome carrying the information for HL-A 2 and 5 and another chromosome controlling HL-A 9, which is consistent with the most likely genotype of this individual, established on the basis of the haplotype frequency in the population. Clone 3 must have lost all chromosomes controlling the appearance of human species-specific antigens. It may be said that clone 3 lost all the "manness" returning to pure "mouseness"; it obviously retained murine

		Original hybrid	Clones				
			2	4	6	7	3
Antiman serum	100	+++	+++	+++	+++	+++	±
	300	+++	++	+++	+++	+++	–
	900	+++	–	++	+++	+++	–
	2700	++	–	–	+	–	–
Anti-HL-A 9 serum	100	+++	+++	+++	+++	+++	–
	300	+++	++	++	+++	+++	–
	900	++	–	+	+	±	–
Anti-HL-A 2 serum	100	+++	–	–	–	–	–
	300	+++	–	–	–	–	–
	900	+	–	–	–	–	–
Anti-HL-A 5 serum	100	+++	–	–	–	–	–
	300	+	–	–	–	–	–
	900	–	–	–	–	–	–

Fig. 4. Mixed agglutination with hybrid IT × KK and its clones

species-specific antigens but, for brevity's sake, this has not been shown in the figure. The fact that the loss of human species-specific antigens was accompanied by the loss of HL-A antigens is consistent with our recent experiments which demonstrated that the HL-A specificity is located in close proximity to the species specificity [13]. This could be construed as an assumption that the HL-A chromosome also controls the appearance of some species-specific antigen on the cell surface.

The acquisition of new antigens by cell cultures may occur in a more "natural" way than the highly artificial hybridization procedure. Several investigators have recently reported that cultured lymphoid cells acquire new antigenic specificities which can be demonstrated in mixed leukocyte culture by stimulation of peripheral blood lymphocytes originating from the same donor [9]. The acquisition of novel antigenic specificities by cultured human lymphoid cells has been recently studied in our laboratory using a different approach [15]. Five patients of the Roswell Park Memorial Institute, Buffalo, New York, received injections of autologous cultured lymphoid cells for therapeutical trials. It was anticipated that sera of these patients may contain antibodies to the antigenic determinants developed by the cells during propagation on artificial media. This prediction proved to be correct. Sera of four patients contained antibodies to human alloantigens, which could be conveniently detected by means of mixed agglutination with cultures of established human cell lines (Fig. 5). One of these sera gave also definite cytotoxicity tests with 21 of 40 human lymphocyte samples. The antibodies responsible for this reaction were of HL-A 2 specificity as demonstrated by comparing the reactions produced by the patient's serum with those given by five standard typing sera with HL-A 2 specificity (Fig. 6). In addition, lymphocytotoxins could be absorbed from the patient's serum with HL-A 2(+) but not with HL-A 2(–) human leukocytes. Evidence was also presented that the lymphoid cell cultures of this patient contained HL-A 2 antigen in addition to antigens 3, 9, and 5. The appearance of antigens 2, 3, and 9 is governed by the first HL-A sublocus. According to the current gentic theory, only two antigens governed by one sublocus appear on the cells. Therefore, it appears plausible that the antigen HL-A 2 against which the patient responded immunologically was developed during in vitro propagation.

	Antibody titers of sera of patients									
	BH		MC		CC		DG		RC	
Cell lines	B	A	B	A	B	A	B	A	B	A
HEp-2	< 100	24,300	< 100	8100	< 100	2700	100	2700	< 100	< 100
HeLa	< 100	900	< 100	2700	< 100	8100	100	900	< 100	< 100
AV-3	< 100	2700	< 100	2700	< 100	2700	100	2700	< 100	< 100

Fig. 5. Mixed agglutination with sera of patients before (B) and after (A) injections of autologous lymphoid cell cultures.

Lymphocyte panel	BH	Anti-HL-A 2 sera JO	PN	PQ	DG	RV
1	+	+	+	+	+	+
2	+	+	+	+	+	+
3	+	+	+	+	+	+
4	+	+	+	+	+	+
5	+	+	+	+	+	+
6	+	+	+	+	+	+
7	+	+	+	+	+	+
8	+	+	+	+	+	+
9	+	+	+	+	+	+
10	+	+	+	+	+	+
11	+	+	+	+	+	+
12	+	+	+	+	+	+
13	+	+	+	+	+	+
14	+	+	+	+	+	+
15	+	+	+	+	+	+
16	+	+	+	+	+	+
17	+	+	+	+	+	+
18	+	+	+	+	+	+
19	+	+	+	+	–	+
20	+	+	–	–	–	–
21	+	+	–	–	–	–
22	–	–	+	+	+	+
23	–	–	–	+	+	+
24	–	–	–	–	–	+
25	–	–	–	–	–	+
26	–	–	–	–	–	–
27	–	–	–	–	–	–
28	–	–	–	–	–	–
29	–	–	–	–	–	–
30	–	–	–	–	–	–
31	–	–	–	–	–	–
32	–	–	–	–	–	–
33	–	–	–	–	–	–
34	–	–	–	–	–	–
35	–	–	–	–	–	–
36	–	–	–	–	–	–
37	–	–	–	–	–	–
38	–	–	–	–	–	–
39	–	–	–	–	–	–
40	–	–	–	–	–	–

Fig. 6. Lymphocytotoxicity tests with serum of patient BH and anti-HL-A 2 typing sera

The most plausible explanation for the observed phenomenon would be that the lymphoid cells cultured in vitro might synthesize new antigens. This phenomenon may be related to infections by viruses since lymphoid cell cultures are known to harbor EB and other viruses. On the other hand, the possibility also exists that the development of new antigens is an attribute of lymphoid cells grown on an artificial medium and deprived of the control which exists under physiological conditions in the body.

New antigens as immunizing stimuli in diseases

The principle of most serodiagnostic tests is based on the detection of serum antibodies which combine in test tube reactions with the antigens of the microbial agent eliciting the particular disease. There are, however, several valuable serodiagnostic tests which deviate from this principle in that the antibodies detected in the patient's serum give completely unexpected reactions with an antigen different from the causative agent of the disease from which the patient suffers [20]. In this category of tests, one may include the Wassermann test for syphilis, the Weil-Felix test for typhus fever, heterophile antibody tests for infectious mononucleosis, cold hemagglutination and MG streptococcus agglutination tests for atypical pneumonia, and antigamma globulin tests for rheumatoid arthritis. The occurrence of these unusual antibodies in the course of listed diseases has been frequently attributed to an antigenic relationship between the causative agent of the given disease and the antigen used as a test tube reagent. This explanation may well be correct for some of these tests but it certainly does not appear to be plausible for all of them. An alternative explanation proposes that new antigenic determinants emerge in the course of the disease as a result of transformation of autologous antigenic constituents, and that these new determinants happen to be shared by the particular test tube antigen. Studies supporting the second hypothesis in relation to antibodies in rheumatoid arthritis and infectious mononucleosis will be reviewed.

The vast majority of patients suffering from rheumatoid arthritis form a serum factor called rheumatoid factor which is IgM in nature and which has serological properties of antibodies to IgG of various species origin [4]. Most serodiagnostic tests have been performed using rabbit [30] and human [31] IgG. Convincing evidence has been presented that separate antibodies combine with IgG of these two species [26, 34]. The factor combining with rabbit IgG should be called heteroantibody whereas evidence has been presented that the factor combining with human IgG is predominantly an isoantibody [11] but may also be an autoantibody [16]. We anticipated that the rheumatoid factor is formed in response to the patient's own IgG which had been altered in the course of the disease [18, 25]. This hypothesis was supported by the results of experiments on immunization of rabbits with denatured autologous gamma globulin quoted previously [17, 24, 33].

It has been suggested that in the course of antigen-antibody reactions taking place in vivo, there are structural alterations of antibody. Experimental evidence for such alterations was brought by the work on serum factors which specifically combine with the IgG antibody portion of the immune complex but which do not react with unaltered IgG [19, 22]. It was speculated that rheumatoid factor may be produced as a result of in vivo antigen-antibody reactions [18, 25]. If this hypothesis were correct, one would expect that the rheumatoid factor would not be restricted to the rheumatoid arthritis but would be a stereotype response to various antigen-antibody reactions occurring in vivo for prolonged periods of time. This indeed proved to be correct since rheumatoid factor was described in chronic infectious diseases first of all subacute bacterial endocarditis [35]. Rheumatoid factor was also demonstrated in many human recipients of renal allografts [5, 11, 12, 32]. The appearance of the rheumatoid factor after transplantation could be related to IgG transplantation antibodies which had combined with the graft and apparently underwent molecular transformation. Furthermore, it was demonstrated that the rheumatoid factor is present in the graft, apparently attached to the transplantation antibodies that had combined with their corresponding antigens.

Experimental support for the hypothesis on the role of antigen-antibody complexes in the stimulation of the rheumatoid factor formation called for prolonged immunization of animals with the same antigens with anticipation that antigen-antibody complexes formed in vivo will stimulate production of anti-IgG antibodies resembling the rheumatoid factor. In the first studies along this line, guinea pig leukocytes were selected for immunization of rabbits [21]. Thereafter several other antigens were employed [8, 23, 33]. In all these studies, formation of anti-IgG antibodies was noted after several months of immunization. Very convincing experiments along these lines were presented by *Abruzzo* and *Christian* [1] who found antibodies resembling the rheumatoid factor in rabbits injected with bacteria grown on protein-free media.

On the basis of the experiments described, it seems quite plausible that rheumatoid factor is formed as a result of protracted antigen-antibody reaction in vivo. Whether the original antigen is a microorganism or some autologous tissue constituent is a matter of conjecture.

Another disease with unusual antibodies is infectious mononucleosis. Patients suffering from this disease form heterophile antibodies which are conveniently detected by means of agglutination of sheep erythrocytes or lysis of bovine erythrocytes [7]. In addition, some proportion of patients with infectious mononucleosis produce antibodies which agglutinate human erythrocytes modified by Newcastle Disease Virus (NDV) [6]. In previous studies in our laboratory, these two reactions of infectious mononucleosis sera were studied by means of the technique of double diffusion in agar gel [3]. Finely dispersed sheep erythrocyte stromata and suspension of NDV particles were used as antigens. As seen in fig. 7, the lines formed by an infectious mononucleosis serum with these two antigens crossed in a reaction of nonidentity. This experiment provided direct evidence that the two types of antibodies under study are unrelated. It indicated also that the anti-NDV antibodies most likely react with the virion itself.

The mode of formation of heterophile antibodies remains an enigma in spite of the fact that they have served for the serodiagnosis of infectious mononucleosis for 40 years. Here again, we postulated that heterophile antibody is unrelated to the causative agent of infectious mononucleosis and that it represents a response to antigenic determinants created on the host's tissues in the course of the disease [3, 20]. Some experimental evidence supporting this hypothesis was obtained from recent studies performed in our laboratory in collaboration with Dr. *Ines Malave.*

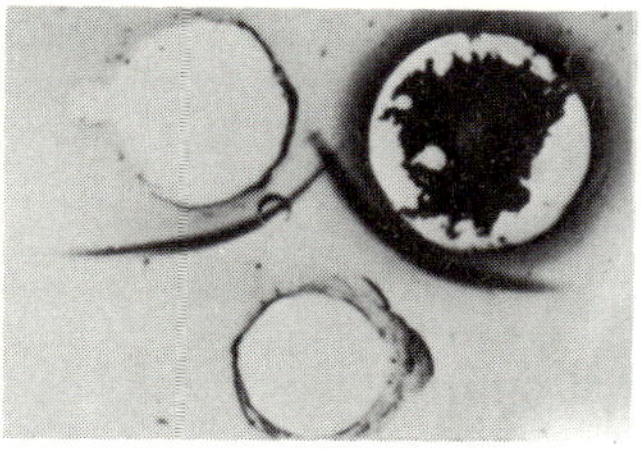

Fig. 7
Lower well: Infectious mononucleosis serum. Upper left well: Bovine erythrocyte stromata. Upper right well: Newcastle disease virus.

We found that human lymphoid cell cultures would remove heterophile antibodies from infectious mononucleosis sera whereas buffy coat from normal human blood or cultures of nonlymphoid cells such as HeLa or HEp-2 would have very little effect on the heterophile antibody titer. The results of a graded absorption experiment are presented in fig. 8. We also found that about 30 % of infectious mononucleosis sera would agglutinate cultured lymphoid cells. This agglutination was caused by IgM antibodies and could be obtained in a broad range of temperatures from 4 to 37 °C. We believe that the observed phenomenon was completely independent from reactions observed by *Mottironi* and *Terasaki* [27] in which infectious mononucleosis sera would induce complement-mediated cytotoxicity of normal human lymphocytes at a temperature of 15 °C. Strangely enough, the agglutination of cultured lymphoid cells could not be abolished by absorption of the sera with sheep or bovine erythrocytes.

Our interpretation of these findings is as follows. Cultured human lymphoid cells develop new antigenic determinants, which is consistent with the above described results of a study on immune responses to cultured autologous cells. The experimental data presented seem to indicate that at least two specificities might have been developed by the cultured cells. One antigen would be identical with or very similar

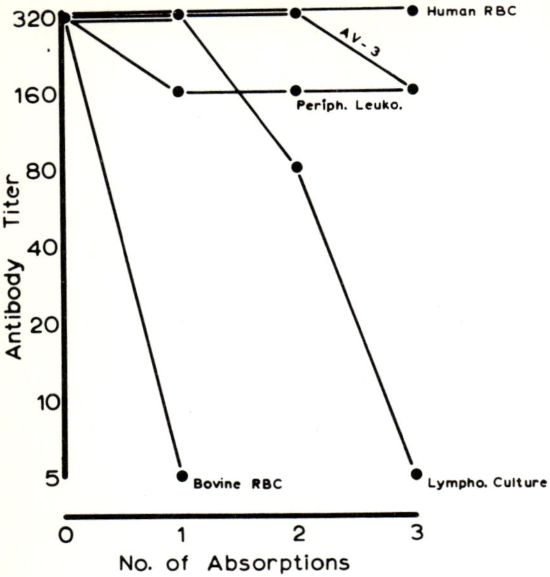

Fig. 8
Lysis of bovine red blood cells by an infectious mononucleosis serum. Effect of absorption of the serum.

to the heterophile antigen and it would account for the absorption of heterophile antibodies. The other antigen would be primarily or exclusively responsible for the agglutination of lymphoid cells. This antigen would be unrelated to the heterophile antigen as evidenced by the fact that the removal of the heterophile antibody did not influence agglutination of cultured cells. This would imply that infectious mononucleosis sera contained two different types of antibodies participating in these reactions – the heterophile antibody and the antibody predominantly or exclusively responsible for agglutination of cultured cells. No correlation of this second antibody to the antibody against NDV could be demonstrated in our preliminary experiments. It has been convincingly demonstrated that antibodies to Epstein-Barr virus (EBV) [10] are unrelated to either heterophile [28] or NDV antibodies [2]. It also appears highly unlikely that the EBV antibodies are related to the agglutinin for cultured lymphoid cells since this agglutinin was never found in normal sera containing EBV antibodies.

The appearance of a heterophilelike antigen on human lymphoid cell cultures makes it more plausible that similar antigen may be developed on the cells of infectious mononucleosis patients and that it may engender formation of heterophile antibodies.

Acknowledgement: Supported by USPHS Research Grant No. AI-06754 from the National Institute of Allergy and Infectious Diseases.

Conclusion

Denaturation of proteins results in the emergence of new antigenic specificities. In the case of IgG, the new antigens resemble native configurations occurring in IgG of members of other species and some individuals belonging to the same species. The antigenic structure of cultured cells may be transformed by "implanting" foreign antigens through the process of hybridization. By using this procedure, human HL-A antigens of peripheral blood leukocytes were "added" to the cell surface of murine fibroblast cultures. Antigenic transformation of cultured human lymphoid cells may occur without any intentionally induced modification. This transformation may include acquistition of antigens belonging to HL-A system.

Antibodies detected in some diseases may be formed in response to new antigens that emerged during the morbid process. Evidence was presented that formation of the rheumatoid factor may be stimulated by the patient's own IgG antibodies that underwent molecular transformation in reacting with their corresponding antigens. Heterophile antibodies might be elicited by stimulus exerted by a new antigen, the heterophile antigen, which was created on the patient's cells during the disease. This hypothesis is made plausible by the observation that the cultured human lymphoid cells carry the heterophile antigen.

References

[1] *Abruzzo, J. L., and Christian, C. L.* 1961. The induction of a rheumatoid factor-like substance in rabbits. J. Exp. Med. **114:** 791–806,

[2] *Barron, A. L.* Personal communication.

[3] *Barron, A. L., Friedman, A., and Milgrom, F.* 1967. Reactions of Newcastle Disease Virus with infectious mononucleosis sera in agar gel. J. Immunol. **99:** 778–84.

[4] *Bartfeld, H., and Epstein, W. V.*, eds. 1969. Rheumatioid factors and their biological significance. Ann. N. Y. Acad. Sci. **168:** 1–207.

[5] *Bravo, J. F., Herman, J. H., and Smyth, C. J.* 1967. Musculoskeletal disorders after renal homotransplantation. A clinical and laboratory analysis of 60 cases. Ann Int. Med. **66:** 87–104.

[6] *Burnet, F. M., and Anderson, S. G.* 1946. Modification of human red cells by virus action. II. Agglutination of modified human red cells by sera from cases of infectious mononucleosis. Brit. J. Exp. Path. **27:** 236–44.

[7] *Eyquem, A.* 1961. Sérologie de la mononucléose infectieuse. Nouv. Rev. Franc. Hemat. **1:** 312–23.

[8] *Eyquem, A., Guyot-Jeannin, N., and Podliachouk, L.* 1959. Présence dans les immunsérums anti-bactériens de facteurs anti-globuliniques analogues a ceux de la polyarthrite chronique évolutive. Ann. Inst. Pasteur **96:** 295–302.

[9] *Green, S. S., and Sell, K. W.* 1970. Mixed leukocyte stimulation of normal peripheral leukocytes by autologous lymphoblastoid cells. Science **170:** 989–90.

[10] *Henle, G., Henle, W., and Diehl, V.* 1968. Relation of Burkitt's tumor-associated herpes-type virus to infectious mononucleosis. Proc. Nat. Acad. Sci. **59:** 94–101.

[11] *Kano, K., and Milgrom, F.* 1968. Anti-γ-Globulin Factors in human allograft recipients. Transplantation **6:** 111–20.

[12] *Kano, K., and Milgrom, F.* 1969. Relation of anti-γ-globulin antibodies to transplantation antibodies in human renal allograft recipients. Transplantation **7:** 281–89.

[13] *Kano, K., Andres, G. A., and Milgrom, F.* To be published. Relation of HL-A antigens to species-specific antigens. Intern. Arch. Allg. Appl. Immunol.

[14] *Kano, K., Knowles, B. B., Koprowski, H., and Milgrom, F.* To be published. HL-A antigens on man-mouse hybrid cells. Eur. J. Immunol.

[15] *Kano, K., Minowada, J., Moore, G. E., and Milgrom, F.* To be published. Humoral antibody response to autologous lymphoid cell cultures. Cell. Immunol.

[16] *Kunkel, H. G., and Tan, E. M.* 1964. Autoantibodies and disease. Adv. Immunol. **4:** 351–95.

[17] *McCluskey, R. T., Miller, F., and Benacerraf, B.* 1962. Sensitization to denatured autologous gamma globulin. J. Exp. Med. **115:** 253–73.

[18] *Milgrom, F.* 1959. Antigenicity of antibodies. *In* Proc. 7th Cong. Intern. Soc. Blood Transf., Rome 1958. New York/Basel: Karger.

[19] *Milgrom, F.* 1962. Rabbit sera with "anti-antibody." Vox Sang. **7:** 545–58.

[20] *Milgrom, F.* 1964. The unusual serology of syphilis, infectious mononucleosis and rheumatoid arthritis. Transf. **4:** 407–13.

[21] *Milgrom, F., and Dubiski, S.* 1957. Antigenicity of antibodies of the same species. Nature **179:** 1351–52.

[22] *Milgrom, F., Dubiski, S., and Wozniczko, G.* 1956. Human sera with "anti-antibody." Vox Sang. **1:** 172–83.

[23] *Milgrom, F., and Witebsky, E.* 1960. Rabbit antibodies against γ-globulins resembling the rheumatoid factor. Fed. Proc. **19:** 197.

[24] *Milgrom, F., and Witebsky, E.* 1960. Studies on the rheumatoid and related serum factors. I. Autoimmunization of rabbits with gamma globulin. J.A.M.A. **174:** 56–63.

[25] *Milgrom, F., and Witebsky, E.* 1962. The rheumatoid factor and its possible pathogenic role. *In* Mechanism of cell and tissue damage produced by immune reactions, Proc. Second Intern. Symp. on Immunopathology, 1961. Basei: Benno Schwabe.

[26] *Milgrom, F., Witebsky, E., Goldstein, R., and Loza, U.* 1962. Studies on the rheumatoid and related serum factors. II. Relation of anti-human and anti-rabbit gamma globulin factors in rheumatoid arthritis serums. J.A.M.A. **181:** 476–84.

[27] *Mottironi, V. D., and Terasaki, P. I.* 1970. Lymphocytotoxins in disease. I. Infectious mononucleosis, rubella, and measles. *In* Histocompatibility Testing 1970, ed. P. I. Teraski. Copenhangen: Munksgaard.

[28] *Niederman, J. C., McCollum, R. W., Henle, G., and Henle, W.* 1968. Infectious mononucleosis. Clinical manifestations in relation to EB virus antibodies. J.A.M.A. **203:** 139–43.

[29] *Rothen, A., and Landsteiner, K.* 1942. Serological reactions of protein films and denatured proteins. J. Exp. Med. **76:** 437–50.

[30] *Waaler, E.* 1940. On the occurrence of a factor in human serum activating the specific agglutination of sheep blood corpuscles. Acta. path. microbiol. scand. **17:** 198–200.

[31] *Waller, M. V., and Vaughan, J. H.* 1956. Use of anti-Rh sera for demonstrating agglutination activating factor in rheumatoid arthritis. Proc. Soc. Exp. Biol. Med. **92:** 198–200.

[32] *Waller, M., Irby, R., Mullinax, F., and Toone, E. C., Jr.* 1965. Connective-tissue disease and rheumatoid factors in patients with renal transplants. New Eng. J. Med. **273:** 12–18.

[33] *Williams, R. C., Jr., and Kunkel, H. G.* 1963. Antibodies to rabbit γ-globulin. Proc. Soc. Exp. Biol. Med. **112:** 554–61.

[34] *Williams, R. C., Jr., and Kunkel, H. G.* 1963. Separation of rheumatoid factors of different specificities using columns conjugated with gamma globulin. Arch. & Rheum. **6:** 665–75.

[35] *Witebsky, E., and Milgrom, F.* 1960. Immunological research in rheumatic disease. Presented at the Annual Meeting of the Amer. Rheum. Assoc., Hollywood-by-the-Sea, Florida, June 1960.

Discussion

Bodmer: Results analogous to, and in agreement with Dr. *Milgrom*'s demonstration of new antigens on a lymphocyte line have been obtained by direct cytotoxicity typing. *Ceppellini* and coworkers, Drs. *Dick, Steel,* and *Evans* of Glasgow and Edinburgh, and ourselves in Oxford have, independently, found that after longer periods of time in culture, the lines seem to acquire reactions to sera whose specificity does not correspond to antigens that were originally present on the cells. We think it is possible that changes in the cell-surface architecture of the cells as they grow, allow the detection of cross-reacting antigens.

Siniscalco: I wonder whether the possibility has been considered that the appearance of a new antigen could be the result of somatic recombination. This is admittedly an unlikely event, but perhaps it would be worthwhile to bear it in mind. Would it be compatible with the genotypes of the parent cells?

Regarding the possibility of contamination between different cell lines, this is unfortunately a risk that can hardly be excluded with certainty, considering the number of people involved in the handling of the cell lines from their establishment to the time they are used for a particular experiment. We had such an experience ourselves with a lymphoblastic line that was given to us as being derived from a Negro male. It turned out that this line was indeed composed of cells with a 46XY chromosome complement as expected, but showed a G6PD phenotype suggesting the presence of both Gd^B and Gd^B alleles. An obvious case of contamination of which nobody was aware in the lab where the line had been started. It turned out that this was a line among the first ones ever set up in that lab at a stage when every device was utilized to improve the chance of establishing these permanent cultures. One of the devices was apparently that of mixing two different lymphocyte cultures.

Bodmer: I can answer that question not so much from our own work but from the work done in Glasgow and Edinburgh, where they have established some 30 to 40 lymphocyte lines in people of known genotype. Professor *Harry Harris* of London has been doing the enzyme phenotypes on these, and therefore has been able to control, I think, very well that these lines are not contaminated. As far as recombination goes, I think the only thing that one can say is that the results that we have found could not readily be accounted for by somatic recombination, or, for that matter, by a simple genetic change.

Milgrom: The data which I presented describe the first demonstration of HL-A antigens on hybrid cells. There have been other attempts, but the mixed agglutination procedure which we employed is much more sensitive than procedures which other groups used, and I think this is why we have been more successful. Cloning experiments offer a possibility of identifying chromosomes which carry genes controlling the appearance of HL-A and/or species-specific antigens. I hope that these chromosomes will be identified eventually.

Contamination of a lymphoid cell culture by other human or animal cells can hardly ever be excluded. However, I do not think that contamination accounts for the results of our studies with these cells. Otherwise one would have to assume that practically all human lymphoid cell cultures had been contaminated. Our results showed that the antigens acquired by cultured lymphoid cells are immunogenic for the original donor. I believe that this is stronger evidence for acquisition of new antigens than any in vitro studies.

Studies on Malignant Behavior and Antigen Expression in Cell Hybrids

George Klein
Department of Tumor Biology, Karolinska Institute, Stockholm, Sweden

This abstract is a brief summary of the collaborative studies performed by Professor *Henry Harris* at the Sir William Dunn School of Pathology, University of Oxford, England, and our laboratory. The purpose of the work is to assess the malignant behavior of hybrid cells produced by the fusion of various normal and highly malignant mouse tumor cells, slightly malignant and highly malignant lines of mouse cells, and two different kinds of highly malignant mouse cells.

At the same time, the expression of certain virally or genetically determined antigens contributed by the parental lines is also studied in the hybrids. The third aspect of the work concerns the release of C-type virus in parental hybrid cells. The karyological work involved has been done mainly by Dr. *Francis Wiener.* The work on antigen expression and C-type virus release was performed in collaboration with *Eva Maria Fenyö, Gertrud Grundner, Ulla Gars,* and *Sten Friberg.*

Abstract

It was found [1, 4, 5, 6, 7, 10] that when normal cells or cells of low malignancy were fused with several highly malignant, established tumor lines, *malignancy was suppressed* as a rule. The A9 cell, an HGPRT-deficient L-cell derivative of low malignancy, suppressed the malignant behavior of all of the following highly malignant ascites tumors: Ehrlich carcinoma, methylcholanthrene-induced MSWBS sarcoma, polyoma-induced SEWA sarcoma, and the two Moloney virus-induced lymphoma sublines, YAC and YACIR. By continued animal passage, it was possible to select highly malignant segregant sublines of the original nonmalignant hybrid cell population. Compared to the original population the malignant sublines

Manuscript received: 14 February 1972

were always characterized by extensive chromosome losses, often with preferential disappearance of the A9-derived metacentric elements. In contrast, the lines propagated in vitro maintained a complete or nearly complete set of the A9-derived metacentric markers, although they lost telocentric and acrocentric elements. The malignancy of these lines maintained in vitro did not increase on continued cultivation, but tended rather to decrease even further.

Three hybrid cell series, each consisting of several clonal populations, were obtained by fusing normal diploid fibroblasts with the highly malignant Ehrlich, SEWA, and TA3 ascites tumors. The three series differed with regard to their chromosomal stability and malignant behavior, and individual clones within a given series also differed. The Ehrlich fibroblast hybrids had very unstable chromosomal constitutions, and all clones showed substantial chromosome losses as soon as they were isolated. All were highly malignant.

In the SEWA/fibroblast series, eight clones with substantially reduced chromosome numbers were found to be highly malignant, but three with essentially complete chromosome constitutions gave low take incidences. On serial propagation as monolayers in vitro, the clones showing low take incidences lost chromosomes and became highly malignant. The TA3-fibroblast hybrid clones initially had high chromosome numbers and a low take incidence, but, after a long period of propagation in vitro chromosome losses occurred and were paralleled by shifts to a high level of malignancy.

The antigenic analysis of the parental and the hybrid cells has proceeded along a number of different lines. Two new phenomena were encountered. They can be referred to as antigen suppression of the Ehrlich type [9] and of the TA3/Ha type [8].

The first was found in all hybrids where the Ehrlich ascites tumor was one of the parents, namely Ehrlich/A9, Ehrlich/A9RI, Ehrlich/B82 hybrids, and a series of Ehrlich/fibroblast clones. Whereas all other hybrids studied showed full expression of both parental H-2 antigens and virus-induced surface antigens, all Ehrlich hybrids showed an unexpected suppression of the antigen(s) contributed by the partner cell. This could be demonstrated by a number of independent serological techniques and could not be accounted for by the increase of cell size.

In some of the highly malignant segregants derived from the hybrid cell population and showing various degrees of chromosome loss, the suppressed antigens reappeared; in others they remained suppressed [3].

The second case of antigen suppression was encountered in two sublines of the TA3 tumor, a strain A mammary adenocarcinoma, now adapted to the ascitic form of growth. The TA3/Ha (Hauschka) subline has a low $H\text{-}2^a$ antigen expression, and its growth is restricted to the A strain. The difference in the absorptive capacity of the two sublines for the appropriate antiserum is approximately 60-fold. In syngeneic mice, the two sublines have similar growth characteristics. Although it transgresses histocompatibility barriers like the Ehrlich ascites tumor, the TA3/Ha line has a very different history for it has never been subjected to selection in genetically foreign (allogeneic) mice.

When the TA3/Ha cell was hybridized with a normal ACA fibroblast, it did not suppress the ACA-derived H-2^f complex. On the contrary, it reacquired the high level of H-2^a antigen expression characteristic for the TA3/St line [8]. This is the opposite of what has been observed for the Ehrlich cell and must reflect a fundamentally different mechanism of antigen suppression. Some step involved in antigen synthesis or incorporation in the cell membrane must be defective in the TA3/Ha line.

Studies on the *release of C-type virus* were carried out in two different series of hybrids. L-cell sublines, known to release a C-type virus, were hybridized with a series of normal and neoplastic cells that did not release C-type virus, as judged by infectivity and immunogenicity tests. In some of these hybrids, virus production continued, whereas in others it was switched off. When chromosomes were lost from hybrids in which virus production had been switched off, some of the segregant sublines again began to produce virus.

In another series of hybrids, one of the parent cells was the immunoresistant YACIR subline of the Moloney virus-induced YAC lymphoma. This subline has been selected for low FMR-surface antigen concentration. It releases only very small amounts of virus, in contrast to the original YAC line. When YACIR was hybridized with the A9 cell, which itself releases a C-type virus, the expression of the FMR-antigen of YACIR was switched back to the high level characteristic of YAC. Antigenic suppression in both cases behaved as a "recessive" character in the hybrid cell. Surprisingly, however, FMR virus production was not switched on. On the contrary, the normal C-type virus production of the A9 cell was switched off [2].

This confirms a previous suggestion, made on more indirect grounds, that the production of C-type virus and the expression of the virus-induced membrane antigen are under independent control, at least in part.

Details of these experiments can be found in the following papers.

References

[1] *Bregula, U., Klein, G.,* and *Harris H.* 1971. The analysis of malignancy by cell fusion. II. Hybrids between Ehrlich cells and normal diploid cells. J. Cell Sci. **8**: 673–80.

[2] *Fenyo, E.M., Grundner, G., Klein, G., Klein, E.* and *Harris, H.* 1971. Surface antigens and release of virus in hybrid cells produced by the fusion of A9 fibroblasts with Moloney lymphoma cells. Exp. Cell Res. **68**: 323-31.

[3] *Grundner, G., Fenyo, E.M., Klein, G., Klein, El, Bregula, U.,* and *Harris, H.* 1971 Surface antigen expression in malignant sublines derived from low-malignant cell hybrids. Exp. Cell Res. **68**: 315-22.

[4] *Harris, H.* 1971. Croonian lecture. Proc. Roy. Soc. **179:** 1–20.

[5] *Harris, H.,* and *Klein, G.* 1969. Malignancy of somatic cell hybrids. Nature **224**: 1314-16.

[6] *Harris, H., Miller, O.J., Klein, G., Worst, P.,* and *Tachibana, T.* 1969. Suppression of malignancy by cell fusion. Nature **223**: 363-68.

[7] *Klein, G., Bregula, U., Wiener, F.,* and *Harris, H.* 1971. The analysis of malignancy by cell fusion. I. Hybrids between tumour cells and L cell derivatives. J. Cell Sci. **8:** 659–72.

[8] *Klein, G., Friberg, S.,* and *Harris, H.* 1972. Two kinds of antigen suppression in tumour cells revealed by cell fusion. J. Exp. Med. **135:** 839–49.

[9] *Klein, G., Gars, U.,* and *Harris, H.* 1970. Isoantigen expression in hybrid mouse cells. Exp. Cell Res. **62:** 149–60.

[10] *Wiener, F., Klein, G.,* and *Harris, H.* 1971. The analysis of malignancy by cell fusion. III. Hybrids between diploid fibroblasts and other tumour cells. J. Cell Sci. **8:** 681–92.

Discussion

Ruddle: I would like to ask a very general question about the first part of your talk. It would seem that in cell lines, grown for long periods of time in animals in serial passage, there would be the selection of mutants and also gene balances which would be highly adaptive with regard to that mode of growth in the animals. Similiarly, there would be a selection of gene – balance situations in cells which are adaptive over a long period of time – to grow in tissue culture. In the hybrids which form between the tissue culture adapted cell lines and the tumor adapted cell line, one might accept coexpression of these particular phenotypes, and, perhaps under these, the situation that the hybrid line would behave neither as a good tumor or as a good tissue culture cell line. Now, if one takes the cells and passes them through animals and selects for those which grow as tumors, one might well select for cells which have lost chromosomes or genes of the tissue culture adaptive-type. And, if one also grows the cells in tissue culture, one might also lose, in this kind of selection system, genes and chromosomes which are adaptive to the tumor type. Now, I think that actually what I've said is fairly consistant with what you have reported today and, of course, the selection does offer a chance to earmark those factors, genetic factors or otherwise, which contribute either to tumor growth or to tissue culture growth. But the question I would like to ask you is, how many genetic factors are involved? It may be incredibly large, many of them may have only a trivial individual effect. There is a second difference, and that is in the adaptation of a particular cell line to a highly adaptive mode of growth either as tumor or a tissue culture cell line. Different genes or combinations of genes may be selected for, so that the kind of genes or the genetic factors which are involved in the selection process may be different in different cell lines and, therefore, you may obtain a different result when you carry out the analysis of the segregation of these factors for the individual cases.

Klein: I have no quarrel at all with this argument. These lines are adapted to the conditions under which they grow. The hybrids are all fully adapted to tissue culture, incidentally, even in the cases where two lines that have not been adapted were fused with each other and a hybrid line was selected subsequently in vitro. The number of controlling factors may be quite large, as you say. This is why we would like to make a start at one point. The reason for choosing the A9 system lies in the fact that it is relatively easily analyzed, and there is a good possibility to link the reversion of the suppression of malignancy to certain specific chromosomes.

Jaenisch: I would like to mention the in vitro studies of *R. Pollack* (*Pollack, Wolman,* and *Vogel,* 1970, Nature **228,** 967) using SV 40 and polyoma transformed cell lines which have a high saturation density in tissue culture. He selected for "flat" revertants of these transformed cell lines by selective killing with BUDR. These flat revertants had a low saturation density. When he examined their karyotype, he found that these flat cells had about double the chromosome number of the original cell lines.When he subsequently selected for rerevertants of these flat cells, which again displayed high saturation density, he observed that these cells had lost their excess of chromosomes. I think these studies somehow parallel the observations you have reported.

Klein: The revertants are an interesting case in point, with the qualification that in vitro reversion of "transformed" morphology is not the same as a reversion of tumorigenic behavior in vivo.

Siniscalco: I would like to elaborate further what *Frank Ruddle* said about the possibility of interpreting the suppression of an antigen or the expression of a new one in terms of co-expression of different genomes in a hybrid cell.

The usual situation for the multimeric enzyme markers in somatic cell hybrids is a pattern that includes the parental types as well one or more heteropolymeric hybrid molecules, depending upon the number of the subunits present in the functional molecule. However, in certain cases the pattern of the hybrid cell turns out to be different altogether from those of the parental cells. For instance in the man-rat hybrid, one sees practically only the heteropolymeric G6PD as if the affinity between unlike subunits were greater than that between like subunits.

Accordingly, the disappearance of parental antigens and the appearance of new antigenic determinants do not necessarily have to be explained in terms of specific suppressors or regulatory phenomena. What I just said, by the way, could also apply to the case reported by Dr. *Milgrom* earlier. I have a question regarding the formation of hybrids between the injected tumor cells and those of the recipient animal. This is an interesting thought, but how can one establish whether the fusion event has occurred in vivo or in vitro after the removal of the tumor cells which very likely may have been contamined with the cells of the recipient animal

Bauer: I would like to ask a question concerning the antigenic makeup of your hybrid cells. You have used at least a few tumor cell lines in your fusion experiments which have well established tumor specific surface antigens and which, after fusion, became a nontumor cell or a cell with a low yield of tumors. I am wondering whether in those cases the tumor-specific surface antigen had disappeared or persisted.

Rajewsky: If the Ehrlich tumor cell indeed possesses a suppressor which suppresses the expression of the H-2 antigens on the cell surface, then one would expect that loss of the suppressor from the cell hybrid, which you described, might also result in the expression of the original H-2 antigens of the Ehrlich cell itself. Have you tested for this?

Klein: Tumor associated antigens that we have studied in the hybrids include the FMR antigens of the Moloney cells. These are maintained in the low malignant hybrids and show no change in the high malignant segregants. The same is true for the polyoma transplantation antigen. We have not tested for the expression of "Ehrlich-H-2".

New Developments in the Establishment of Gene Linkage Relationships in Man

Frank H. Ruddle

Department of Biology, Yale University, Kline Biology Tower, New Haven, Connecticut 06520, USA

Summary: Mouse/man somatic cell hybrids segregate human chromosomes. It is possible to recover independently derived hybrids which possess different numbers and combinations of human chromosomes plus an intact mouse genome. The human chromosomes can be identified by a variety of cytological techniques. The presence or absence of specific human chromosomes can be correlated with the presence or absence of specific human enzyme phenotypes. Concordant segregation of phenotypes and chromosomes provides evidence for human gene assignment to particular chromosomes. Using this approach, we have assigned LDH-A to C-11, Pep B to C-12, IDH to F-20, and TK to E-17. Evidence has been obtained for the restriction of TK to the long arm of E-17, making use of spontaneous chromosome rearrangements. Evidence also has been obtained which supports the translocation between man and mouse chromosomes. These and related matters will be discussed.

Introduction

Gene mapping in man by kindred analysis is limited by long generation time, small progeny size, and limited numbers of genetic polymorphisms. It is desirable to augment kindred analysis with other approaches to linkage which are free from these limitations. Somatic cell genetic systems offer this possibility. Parasexual processes such as cell fusion and chromosome segregation occur in somatic cell populations. Therefore, somatic cell systems can be manipulated for the purpose of extracting linkage information. Such systems permit the reduction of effective generation time by several orders of magnitude, allow substantially larger progeny sizes, and provide catalogs of genetic markers of unlimited extent [6].

Manuscript received: 28 February 1972

The somatic cell genetic system

Cell fusion allows the combination of diverse genetic information within a single cell. It is therefore an analog to fertilization. Combinations of evolutionally related or unrelated parental cells can be formed. For example, mouse/mouse, mouse/Chinese hamster, man/mouse, mouse/chick, and man/mosquito hybrids have been reported. The fusion of cells in vitro can be enhanced by exposure to certain viruses, such as the parainfluenza virus-Sendai; or to chemical reagents, such as lysolecithin. Such agents produce plasma membrane lesions. In some instances, where lesions exist cojointly between adjacent cells, the repair process extends between the cells giving rise to intercellular bridges. Such connections broaden, and the parental nuclei subsequently reside in a common cytoplasm bounded by a single plasma membrane. Cells containing two nuclei of diverse genetic origins are termed *heterokaryons.* Following the first mitotic division, the chromosomes of both parental types are dispersed into daughter cells and are enclosed within a single common nuclear envelope. Such a cell is termed a *hybrid*, and may continue to proliferate indefinitely [6].

Only a small proportion of cells undergo fusion. Even in instances where fusion is enhanced, the frequency of viable hybrids may be relatively low (10^{-3} to 10^{-6}). It is therefore necessary to employ methods for the enrichment of hybrids. Enrichment procedures currently employed commonly make use of genetic complementation. In a system devised by *Littlefield*, complementation of the purine and pyrimidine salvage pathway enzymes, hypoxanthine-guanine phosphoribosyl transferase (HGPRT) and thymidine kinase (TK), is used [2]. Endogenous synthesis of purines and pyrimidines can be blocked by the drug aminopterin. Thus, normal cells containing the salvage enzymes survive on a medium containing hypoxanthine, aminopterin, and thymidine (HAT). However, mutant cells which are deficient in either HGPRT or TK cannot survive on HAT medium. It is possible to fuse cell populations which are TK and HGPRT deficient. In this case, the enzyme deficiencies are complemented and, in HAT medium, the hybrid cells survive whereas the parental cells die. It is possible to extend this system of selection based on deficiency complementation to a broad spectrum of conditional mutants of the nutritional auxotrophic and temperature sensitive types.

Chromosome segregation is a property shared by all somatic cell hybrids. In general, chromosome segregation is minimal in intraspecific hybrids, but can be pronounced in certain interspecific combinations. In man/mouse and man/Chinese hamster hybrids, the human chromosomes are invariably lost. The unilateral reduction of human chromosomes in these hybrids has not yet been satisfactorily explained; however, the phenomenon is well established. The loss of human chromosomes in man/mouse hybrids is not necessarily continously progressive. Frequently, hybrid populations continue through numerous generations with relatively stable partial human chromosome constitutions [6].

Selection systems based on complementation play a direct role in linkage analysis. If, for example, man/mouse hybrids are isolated in HAT and the mouse parent is HGPRT deficient, then in all viable cells retention of human HGPRT is required. Thus, in this case, one selects against any hybrid which has segregated, most probably by random nondisjunction, the HGPRT chromosome. Therefore, specific deficiencies in the mouse genome should result in the retention of specific human chromosomes. It is possible to conceive of 23 conditionally mutant mouse or Chinese hamster parental cell populations which under appropriate selective conditions would select for retention of each of the 23 human chromosomes. Four complementation systems of this general type have been described to date for TK, HGPRT, and Adenine Phosphoribosyl Transferase (APRT) [1].

In order to make gene assignments to chromosomes using somatic cell linkage systems, it is necessary to accurately identify the chromosomes. One must be able to discriminate between the parental chromosome sets; also one must be able to identify each individual chromosome in the segregating genome. New cytological staining techniques make these requirements possible. All normal human and murine homolog pairs can now be identified unambiguously by means of Giemsa banding and quinacrine mustard fluorescence. Moreover, staining procedures that reveal centromeric heterochromatin and *in situ* annealing procedures that permit detection of redundant DNA localized in different parts of the chromosomes are useful for the purpose of identifying specific chromosomes or parts of chromosomes [6].

Linkage strategy

Linkage analysis in man/mouse hybrids is based on the following properties: (1) the human chromosomes segregate, (2) the homologous enzymes between man and mouse differ in their physical properties and can, as a consequence, be distinguished, and (3) the human chromosome constitution of hybrid clones can be determined. It is therefore possible to determine linkage relationship by determining the correlated loss or retention of markers. If concordant segregation of markers occurs, then linkage may be presumed; if discordant segregation occurs, then linkage can be denied. Linkage in somatic cell systems is not limited to 50 map units but includes entire chromosomes. For this reason, it is more appropriate to use the term *synteny* when referring to linkage relationships ascertained by means of cell hybrid analysis. It is unlikely that false-positive *syntenic* relationships will be detected if adequate numbers of hybrids of independent origin are examined. More likely is the identification of false-negative syntenic associations. False negative associations can result from misreading of enzyme phenotypes and the occurrence of chromosome rearrangements which disrupt genuine syntenic associations [6].

Results

Definition of linkage groups

Kindred studies have shown that the enzymes glucose-6-phosphate dehydrogenase (G6PD), phosphoglycerate kinase (PGK), and HGPRT are X-linked [8]. Cell hybrid analysis has similarly shown synteny between HGPRT, G6PD, and PGK. Studies on rearranged X-chromosomes in cell hybrids have also provided information on the regional location of X-linked markers (see *Siniscalco's* contribution to this Symposium p. 307 ff.).

Synteny between human peptidase B and LDH-B has been established by cell genetics [7, 11]. This finding has subsequently been confirmed by other laboratories.

Man/Chinese hamster hybrids have been used to demonstrate synteny between 6 phosphogluconate dehydrogenase (6PGD) and phosphoglucomutase 1 (PGM-1) [14]. Man/mouse hybrids have provided evidence for synteny between PGM-1 and Pep C [13]. *Ruddle* et al. reported the absence of synteny between PGM-1 and Pep C, but this was due to the unrecognized heterozygosity for PGM-1 in the human parent (WI-38) [8]. In this instance, one allelic form can not be distinguished from the murine form of PGM-1. Thus, possibly spurious PGM-1 minus/Pep C plus clones were reported. Subsequently, hybrids made with human cells which are homozygous at the PGM-1 locus and which differ from the murine enzyme have been established in order to reexamine this possible syntenic association. If one accepts the PGM-1/Pep C synteny, then it is likely that the loci for PGM-1/Pep C/ and 6PDG are all syntenic. Kindred data also provide evidence that the rhesus *(Rh)* locus is closely linked to the locus controlling the red cell trait elliptocytosis (*El*). Moreover, family data show linkage between the *Rh* and 6PGD loci. It is thus likely that the following loci are autosomally linked: *PGM-1/6PGD/Pep C/ Rh/El* [5]. This constitutes the largest currently established autosomal human linkage group.

Puck et al. have recently reported an association between the sensitivity of human/ Chinese hamster hybrid cells to antisera directed against human cell surface antigen(s) and the expression of human lactate dehydrogenase A (LDH-A) [4]. This result suggests the syntenic association between the LDH-A locus and a locus or loci which control(s) the expression of a human cell surface antigen(s). *Bodmer* and his associates previously have reported a similar correlation [3].

T. B. Shows has recently reported a possible syntenic relationship between the human markers isocitrate dehydrogenase (IDH) and malate dehydrogenase (NAD dependent soluble form, MOR) in man/mouse somatic cell hybrids [12]. We have seen a similar association between these two markers in three out of four separate man/mouse somatic hybridization experiments. Work is in progress to obtain more extensive data bearing on this possible syntenic relationship in new primary and secondary hybrid clones.

Gene assignments to chromosomes

The availability of new cytological procedures for the identification of chromosomes contributes significantly to the assignment of genes to chromosomes in cell genetic systems. It has been possible to assign the structural loci for the enzymes HGPRT, PGK, and G6PD to the X-chromosome and its subparts (see *Siniscalco's* contribution to this Symposium p. 307 ff.). *Ruddle* et al., using man/mouse hybrids, have assigned the LDH-A linkage unit to the C-11 chromosome, the IDH linkage unit to F-20, and confirmed earlier reports that TK was located on chromosome E-17 [10]. Moreover, *Ruddle* et al. made use of spontaneous mouse/man chromosome translocations to restrict the location of TK to the long arm of the E-17 chromosome [10]. *Ruddle* et al. have demonstrated the localization of the LDH-B linkage unit to the C-12 chromosome [9].

Conclusions and discussion

It is apparent that somatic cell genetics can contribute significantly to linkage analysis in man. The current rate of progress suggests that linkage units will be defined for and assigned to each of the human chromosomes within the next half decade. However currently available methodologies are still too crude and slow for the rapid development of high resolution human maps. The assignment of genes to parts of chromosomes based on chromosome rearrangement is extremely coarse. Possibly, approaches based on somatic recombination or gene transfer can be developed to achieve high resolution mapping. Perhaps these approaches will also serve to speed up the acquisition of linkage data. It can be calculated that gene mapping can be accomplished 10–100 times faster in microorganisms than in man by existing somatic genetic techniques. This means, at the present rate, it may require 50–100 years to develop a human map as comparably complete as that which currently exists for *E. coli.* This is a somewhat frustrating and discouraging estimate especially since a complex human gene map can be expected to contribute meaningfully to human medical genetics, give insight into developmental mechanisms in man, and provide detailed knowledge of human evolution. The solution to this dilemma can be obtained only through the future development of new, more powerful, somatic cell approaches to linkage analysis.

References

[1] *Kusano, T., Long, C.,* and *Green, H.* 1972. A new reduced human an-mouse somatic cell hybrid containing the human gene for adenine phosphoribosyl transferase. Proc. Natl. Acad. Sci., U.S. **68**: 82.

[2] *Littlefield, J. W.* 1964. Selection of hybrids from matings of fibroblasts in vitro and their presumed recombinants. Science **145**: 709.

[3] *Miggiano, V., Nabholz, M.,* and *Bodmer, W.* 1969. Hybrids between human leucocytes and a mouse cell line: Production and characterization. *In* Heterospecific genome interaction. The Wistar Institute Symposium Monograph No. 9, p. 61.

[4] *Puck, T., Wuthier, R., Jones, C.,* and *Kao, F.* 1971. Genetics of somatic mammalian cells: Lethal antigens as genetic markers for study of human linkage groups. Proc. Natl. Acad. Sci., U.S. **68**: 3102.

[5] *Renwick, J. H.* 1971. The mapping of human chromosomes. *In* Ann. revs. of genetics, ed. *H. L. Roman,* Palo Alto: Annual Reviews, **5**: 81.

[6] *Ruddle, F. H.* in press. Linkage analysis using somatic cell hybrids. *In* Advances in human genetics, eds. *Harris* and *Hirschhorn,* Vol. 3. New York: Plenum Press.

[7] *Ruddle, F. H., Chapman, V. M., Chen, T. R.,* and *Klebe, R. J.* 1970. Linkage relationships between human lactate dehydrogenase genes A and B and the peptidase B gene in man/mouse somatic cell hybrids. Nature **227**: 251.

[8] *Ruddle, F. H., Chapman, V. M., Ricciuti, F., Murnane, M., Klebe, R.,* and *Meera Khan, P.* 1971. Linkage relationships of seventeen human gene loci as determined by man/mouse somatic cell hybrids. Nature **232**: 69.

[9] *Ruddle, F. H.,* and *Chen, T. R.,* in press. Gene assignment in man by somatic cell genetics. *In* 21st annual symposium on the biology of the skin: Genetics and the skin, eds. *W. Montagna* and *L. Pastor.*

[10] *Ruddle, F. H., Chen, T. R.,* and *Boone, C.* 1971. Assignment of genes to chromosomes using somatic cell hybrids (TK: E-17; LDH-A: C11; IDH: F-20). *In* Proc. 4th intl. symp. human genet, ed. *J. De Grouchy.*

[11] *Santachiara, A. S., Nabholz, M., Miggiano, V., Darlington, A. J.,* and *Bodmer, W.* 1970. Genetic analysis with man/mouse somatic cell hybrids: Linkage between human lactate dehydrogenase B and peptidase B genes. Nature **227**: 248.

[12] *Shows, T. B.* Personal communication.

[13] *Van Cong, N., Billardon, C., Picard, J.-Y., Feingold, J.,* and *Frèzal, J.*, 1971. Laison probable (linkage) entre les locus PGM-1 et Pep–C chez l'homme. C. R. Acad. Sci. **272**: 485.

[14] *Westerveld, A.,* and *Meera Khan, P.,* in press. Evidence for linkage between human loci for 6-phosphogluconate dehydrogenase and phosphogluconate-1 in man/Chinese hamster somatic cell hybrids. Nature.

Discussion

Hausen: Since you do have a progressive loss of human chromosomes, I would like to know how stable these clones are and how much variation you have in the cells within a clone.

Ruddle: The loss of chromosomes in not actually progressive. After the clones have been collected, after the cells have gone through maybe thirty generations or so, the human chromosome consitution of the clones is relatively stable, so there appears to be a kind of coadaptation between the remaining human chromosomes and the mouse chromosomes. If one grows the cells under difficult conditions, in other words, if one starves the cells or mistreats them, it is possible to stimulate further human chromosome loss, although not in all cases, and in some instances it is difficult to obtain further segregation of the human chromosomes which in some cases is desirable. There is variation between cells within a clone, as you pointed out. For example, where we see a linkage relationship between a gene and a particular chromosome, even though we see the enzyme expressed in the clone at all times in the sample, the chromosome may be present or identifiable in only, say, 50 to 80 percent of the cells. This is due in part to our failure to recognize a chromosome even when it is present. In other cases, I think it probably represents a segregation of the chromosomes by nondisjunction. Now, where the chromosome is retained for a long period of time, this probably indicates that there is selection against those cells which have lost that chromosome.

Krone: Is the frequency of rearrangements related to the titre of Sendai virus used for the fusion, and do you also observe rearrangements in cells fused with lysolecithin?

Ruddle: We have always used beta propriolactone and inaetivaled Sendai, and we have used comparable titres of Sendai virus, usually 1000 HAU for 4 million cells. So I really cannot answer your question. We have never used lysolecithin in our experiments.

Klingmüller: I think it is important in your method that the hybrid cells you are using are segregating out the human chromosomes. Why don't they segregate out mouse chromosomes, too? Have you a special trick to avoid this?

Ruddle: I think this is really a philosophical question, or at least I can answer it only in a philosophical way. We just don't know what is causing the unilateral segregation of the human chromosomes. I can speculate on it. I think it might have something to do with the possible repression of cell division in those clones which retain certain human chromosomes but this is really restating the observations in a way. We just don't know what the mechanism is for the unilateral segragation.

Klingmüller: So it was just a fortunate finding?

Ruddle: A very fortunate finding.

Correction of Genetic Defects in Cultured Mammalian Cells

Marcello Siniscalco

Department of Genetics, Faculty of Science, University of Naples, Italy, and Albert Einstein College of Medicine, Bronx, New York, USA

Summary: The talk reviews the current knowledge on the attempts made for the correction of genetic defects in cultured somatic cells. Apart from the special case of "metabolic cooperation", the correction of genetic defects has thus far been unequivocally achieved only by somatic cell hybridization (intergenomic, intergenic or interallelic complementation). Though these studies have provided critical data to the students of human genetics, they seem unlikely to be of practical value for the so-called "gene therapy" (which perhaps would be better labelled as "phene therapy").

Of close relevance to this problem are the attempts made, by different authors and with different approaches, for a more direct repair of the abnormal function in the cultured diseased cells by means of DNA-transformation or virus-mediated transformation. Some preliminary results from experiments bearing on this subject and performed in the author's laboratory are presented.

The need for basic data on the mapping of mammalian chromosomes is emphasized for a better understanding of the transfer of genetic information in mammalian cells. Strategies and perspectives of somatic cell genetics are discussed with particular reference to the mapping of the human genome. A rudimental map of the human X-chromosome is presented.

I would like to begin by stating my personal biased views on the question that may occur spontaneously to the casual reader of the proceedings of this symposium: will therapy of genetic defects soon be feasible in man?

In my opinion, the answer to this question differs depending on what and whom we are referring to. I suppose that everybody will share my skepticism about the possibility of reverting an "abnormal gene" to its "normal state" in a given living organism and in all of its cells, including those of the germinal line, for the "change" to become inheritable. Obviously, for this to be possible one should be able to detect the "genetic defect" at a very early stage of embryogenesis, almost soon

Manuscript received: 17 March 1972

after zygote formation, and at the same time be able to induce the minute change desired and only that one. This is not only beyond reach, but perhaps is also beyond our scope since we may have very fallacious opinions about what a "normal gene" really means. All we should aim to achieve is the correction of a genetic defect within the lifetime of the individual and his ecological environment, since we have learned from population genetics that a gene "deleterious" in a given environment may even be "favourable" in a different one.

Such an aim could be achieved, for instance, by correcting the "genetic defect" only within the diseased tissue which happens to be most directly connected with the manifestation of the disease. This is, by the way, what medicine has been trying to do since prehistoric times, using different means such as dietary measures, avoidance of toxic drugs, replacement of missing gene products or defective tissues, enzyme inductions, etc., all being examples of "phene" rather than "gene" therapy since the change brought about involves the "expression" and not the structure of the genes.

The novelty today lies in the possibility of culturing in vitro at least some of the "diseased tissues" so that the correction of genetic defects can be sought at the level of the single cells and much nearer to the primary gene defect. In principle, a patient affected by sickle cell anemia could be cured even if the genetic defect were corrected only in his hematopoietic stem cells. Such a goal would, perhaps, have been already achieved if these cells could be grown in vitro and the rare "normal cells", arisen by spontaneous back mutation or induced transformation or otherwise, could be ' selected for" and transferred back into the diseased donor.

The purpose of the present report is to summarize the different ways by which the correction of genetic defects has been achieved so far in cultured cells.

Alternatives to sex in mammalian somatic cells

During the last ten years, it has been realized that some of the so-called "alternatives to sex", such as the formation of heterokaryons and true somatic hybrids, once thought to be the prerogative of fungi and microorganisms at large, can in fact be brought about in mammalian cells cultured in vitro.

We have heard from Dr. *Ruddle* [26] and Dr. *Bodmer* [2] how precious these tools are for the genetic analysis of those animal species which do not lend themselves to the classical approach of experimental genetics.

I want to discuss here the usefulness of the same tools for studies on genetic complementation which represents one way for the correction of genetic defects in cultured mammalian cells.

We also heard, from Dr. *Merril* [16] and Dr. *Munyon* [19], that there are other ways for reaching the same goal, through phenomena which resemble very closely phage-transduction and DNA-transformation, two additional examples of "alternatives to sex" in microorganisms, whose occurrence in mammalian cells cannot be considered unequivocally proven, as yet.

In relation to the latter topic, I am going to mention some attempts that Dr. *Caneva*, Dr. *Shin*, Dr. *Schildkraut*, Dr. *Klinger*, and myself have been doing recently, at the Albert Einstein, to induce the correction of HGPRT-deficiency in mammalian cells with the addition of isologous or heterologous DNA derived from HGPRT-positive mammalian cells. These studies are far from being finalized but some preliminary though negative, data seem fit for discussion in an occasion like this.

I am grateful to my colleagues for their consent to the mentioning of these data before publication.

Complementation in somatic cell hybrids

Intraspecific hybrids

Numerous examples of genetic complementation have been reported in intraspecific and interspecific somatic cell hybrids of different kinds. The list was opened by the classical experiment of *Littlefield* [14] who found evidence of intergenomic complementation for the enzymes hypoxanthine-guanine-phospho-rybosil-transferase (HGPRT) and thymidine-kinase (TK) in a somatic cell hybrid between two mouse lines which were deficient in one or the other of these enzymes.

You may recall that these deficient parental lines had been obtained through selection of the "forward mutants" (which, however, have not yet been proven to be structural mutants at specific chromosomal genes) in media containing respectively critical concentrations of 8-azaguanine or 5-bromo-de-oxy-uridine (5-BUDR). The isolation of the hybrid was in turn fulfilled by growing the cocultivation mixture in the so-called HAT-selective medium which kills both types of deficient parental cells.

This experiment was repeated by Dr. *Harris* and his colleagues [6] in Oxford, after the introduction of the inactivated Sendai virus as a better inducer of cell fusion, to investigate the evolution of the hybrid karyotype during a long period of propagation in the HAT-medium. It was thus realized that the ability of the hybrid of the type described to grow in the selective medium remained unaltered after extensive chromosome loss that usually brought the hybrid karyotype back to the standard chromosome number and to the morphology of the parental genomes. This observation was in some ways reminiscent of the "haploidization" in *Aspergillus* and proved for the first time the occurrence of chromosomal segregation in mammalian somatic cell hybrids. As work went on along these lines, it became apparent that not all somatic cell hybrids behave the same way in this respect.

In general, the karyotype of somatic hybrids between cells of the same species seems to remain remarkably stable when at least one or both the parental cells are diploid or quasi-diploid.

Numerous examples of genetic complementation have been reported in these types of stable intraspecific hybrids. *Puck* and colleagues [10] have shown intergenomic and perhaps interallelic complementation in somatic hybrids derived from the fusion of different auxotrophic mutants of Chinese hamster. *Silagi* et al. [29] observed restoration of normal activity for HGPRT, orotidine-5′-monophosphate (OMP)-decarboxylase and OMP-pyrophosphorilase in a stable hybrid between a human strain deficient in both the latter two enzymes (from a patient with orotic-aciduria) and a human heteroploid line deficient in HGPRT.

Our group [31] reported the occurrence of intergenic complementation in a stable diploid human hybrid derived from the fusion of two male fibroblastic strains respectively deficient in HGPRT or glucose-6-phosphate-dehydrogenase (G6PD), two well-known examples of X-linked inborn errors of metabolism in man.

Nadler et al. [21] produced several human somatic hybrids of the same type but using, as parental strains, fibroblastic cells derived from seven unrelated galactosemic patients and therefore deficient in the enzyme α-D-galactose-1-phosphate-uridyl (GPU)-transferase. All of the twenty-one hybrids produced between these galactosemic strains had again a stable karyotype, but only three of them exhibited GPU-transferase activity indicating that there must exist "different" mutants for galactosemia.

A similar observation has been made recently by *Bootsma* and his colleagues (personal communication) who have reported complementation for the mechanism of DNA-repair in a somatic hybrid between two human diploid strains derived from unrelated patients affected by xeroderma pigmentosum, a recessive condition that may be diagnosed in cultured cells for their typical inability to DNA-repair.

The last two examples quoted could also represent instances of interallelic complementation if the parental cells were known for certain to carry mutations allelic at the same locus. Unfortunately, this is not known.

All these observations illustrate the use of the stable intraspecific somatic cell hybrids for carrying out complementation analysis in mammalian somatic cells, even in the absence of selective techniques capable of selecting directly in favour of the "complemented phenotype".

The potentials of this tool for analysing the "fine structure" of the human genome, especially for what concerns X-linked mutants, has been emphasized elsewhere [32]. It is worth stressing here its value in preventive medicine when, for instance, the children of unrelated galactosemic patients, who have decided to marry, would want to know for eugenical purpose whether or not they are carriers of the same mutation.

Interspecific hybrids

All interspecific somatic cell hybrids thus far described have shown a clear-cut tendency toward a preferential chromosome loss at the expense of one or the other of the parental genomes.

We have already heard a lot about the man-mouse hybrid [2, 26, 37] which undergoes loss of the human chromosomes. The man-Chinese hamster [11, 38] and the man-Syrian hamster [7] behave in the same way. The man-rat [30] hybrid loses the rat chromosomes. It is everybody's experience that the chromosome loss in these hybrids goes on to almost a complete elimination of the "weaker" genome, so to speak. However, when the "stronger" genome is deficient in a function that is essential for survival in a selective medium, the reduced hybrid will necessarily retain the chromosome of the weaker genome bearing the structural gene for that function My colleagues and I have shown [17] that complementation may occur in certain types of man-mouse hybrids even when only the relevant chromosome fragment, with the gene necessary for survival, is retained. This finding proved, at the same time, the occurrence of mitotic separation of X-linked genes in somatic cell hybrids and laid the ground for the mapping of the human X-chromosome via somatic cell genetics [32].

This morning, Dr. *Ruddle* [26], has provided cytological evidence for the occurrence of chromosomal rearrangements in the man-mouse hybrids by showing us that in a case where the retention of human thymidine kinase was the condition for survival, only part of the human 17-chromosome – on which this gene is located – was retained as translocated fragment on a mouse chromosome.

We have unpublished evidence of the translocation of the human genes for HGPRT and for PGK on the mouse genome in somatic hybrids of the same type. All this tells us that interspecific somatic cell hybrids not only undergo chromosomal segregation but also chromosomal exchanges. We do not know, however, whether exchanges between human chromosomes also occur in these hybrids, in particular between homologous chromosomes.

Extreme examples of unstable somatic cell hybrids are those reported by *Harris* and colleagues [9] between HeLa or mouse cells with hen erythrocytes. Soon after fusion, these cells behave as "heterokaryotes" rather than "true hybrids" since their nuclei remain individually distinct for a few weeks. Then the hen's genome is pulverized and is probably completely eliminated. However when the other parental line is deficient in HGPRT, the continuous propagation of the heterokaryons in HAT-selective medium leads to the retention of the hen HGPRT though no hen chromosomes or other hen genetic markers can be detected [5].

This situation seems to be the nearest to that of bacterial "heterogenotes" where "partial heterozygosis" at a given locus arises as a consequence of bacterial conjugation or phage-transduction [18].

It is obvious that these types of extremely reduced hybrids offer interesting potentials for the correction of genetic defects at the level of the diseased tissue cultured in vitro since they end up having the abnormal function corrected by complementation though they retain, more or less, the original karyotype. In principle, these heavily reduced hybrids having lost also the transplantation antigens of one parental cell, should be retained without difficulty if transplanted back into the animal donor of the other parental cell.

The man-rat hybrid is the most suitable type of unstable interspecific hybrid for attempting the correction of genetic defects in human cells cultred in vitro, since this hybrid is the only one that seems to retain the entire human genome while losing very rapidly the chromosomes of the other parental species. For instance, if a human lymphoblastic strain deficient in HGPRT, from a patient with the so-called Lesch and Nyham syndrome, is fused with rat-fibroblasts normal for this enzyme, a prolonged selection in HAT should eventually lead to the retention of only the rat HGPRT locus. If this were the case, the surviving "heterogenotes" could be returned to the patient without risk of rejection. Whether this procedure would be sufficient to restore HGPRT in his peripheral blood and, in this case, to cure the disease, remains a legitimate wonder.

Directed chromosome loss

Pontecorvo [23] has recently been able to induce directed chromosome loss in interspecific hybrids by treating with U.V. or 5-BUDR, before fusion, the parental cell type whose chromosomes he wanted eliminated eventually from the hybrid genome. Under these conditions the damaged genome, when fused with the nontreated cell, is fragmented and readily eliminated except for those chromosome fragments that may be required for the survival of the hybrid in a selective medium.

This is a quick way for the production of reduced hybrids from whatever type of somatic cell hybrid, including the intraspecific ones which – as we have seen – are stable under normal conditions.

Needless to say, such a procedure represents a significant step forward for studying genetics in somatic cell hybrids since it creates the conditions of obligatory chromosomal segregation and mitotic recombination. Its relevance to the "therapeutical illusions", discussed in the previous paragraph, is also obvious since this technique may allow a much faster production of heavily reduced hybrids (not necessarily interspecific) which end up being heterogenotes only at the locus bearing the defective gene which one wants to "correct".

Attempts to cell transformation with nonviral DNA

Though the uptake of nonviral DNA by mammalian cells has been clearly demonstrated by several authors [1], DNA-induced transformation of mammalian cells is far from being unequivocally proven [1]. The great majority of the experiments performed so far with this intent involved attempts to the transfer of drug resistance.

The common drawback of all the experiments of this type lies in the fact that in none of them was the characterization of the "transformed phenotype" performed at the level of the primary gene product. Moreover, it was usually uncertain whether both the "marker" to be transformed and that to be "selected for" were actually the expression of chromosomal genes.

Among these experiments, those of *Szybalski* and *Szybalska* [34], *Bradley, Roosa* and *Law* [4] and *Roosa* and *Bailey* [24] describe the correction of cell behaviour toward 8-azaguanine.

The first is an example of DNA-mediated correction from resistance to nonresistance to this drug. The last two are examples of the opposite type of correction from sensitivity to nonsensitivity. Since drug resistance to 8-azaguanine depends on the absence of HGPRT and vice-versa, the implication of these experiments is that HGPRT-negative cells may be corrected by addition of DNA derived from HGPRT-positive cells and vice-versa.

In all these experiments, both the DNA-donor cells and the DNA-recipient ones belonged to the same species so that a direct characterization of the gene product in the alleged transformants could not be attempted. Thus, it remains possible that the clones recovered from the selective media actually might have been revertants of some kind even though their frequency among the DNA-treated cells were well above that of the controls.

After the demonstration given by *Shin* et al. [28] that the electrophoretic mobility of human HGPRT derived from diploid fibroblastic cells is different from that of mouse cells and of human HeLa cells, my colleagues and I decided to repeat the experiments of *Szybalski* and *Szybalska* in a fashion that would allow the direct characterization of the alleged transformants in case these were to occur again.

The plan of the experiment included the following steps:

1. treatment of mouse HGPRT-negative cells (the so-called A9 and RAG lines which have been isolated through 8-azaguanine selection [14, 12]) with undegraded DNA derived from HGPRT-positive HeLa or mouse L-cells;

2. treatment of human HGPRT-negative male diploid cells (obtained from a skin explant derived from a patient with the *Lesch* and *Nyham* syndrome who are known to carry a structural mutation at this X-linked locus [25]) with DNA derived from HGPRT-positive HeLa or mouse L-cells;

Fig. 1. Recovery of HAT-resistant clones from HGPRT-deficient mouse lines.

Recipient line	Total no. cells scrutinized (x 10^6)	Donor DNA (100 $\gamma/10^7$ cells)	Sendai virus	Average no. of clones rescued (per 10^8 cells)
A9	40	–	–	2.5
A9	60	–	+	none
A9	20	A9	–	5
A9	20	A9	+	5
A9	80	L	–	none
A9	80	L	+	2.5
A9	100	HeLa	–	4 to 9
A9	100	HeLa	+	none
Rag	10	–	+	10
Rag	10	L	–	10
Rag	20	L	+	5
Rag	10	HeLa	–	10 to 30
Rag	20	HeLa	+	10

Recipient cells, *A9* and *RAG,* are two well known HGPRT-deficient mouse lines resistant to 8-azaguanine and unable to grow in HAT.

Cells chosen as a donor of DNA were the same *A9* and the *HGPRT-positive mouse L-cells* or *human HeLa cells.* For each experiment, 100 γ of donor DNA was added to 10^7 recipient cells just before performing a Sendai-virus mediated fusion within the recipient cells. Control experiments were performed in exactly the same way with the omission of DNA or Sendai-virus treatment or both.

Though the highest incidences of HAT-resistance clones were found in the experiments with added DNA, there seems to be no clear correlation between the two phenomena nor between the incidence of HAT-resistance clones and the Sendai-virus treatment.

3. addition of 100 γ DNA/10^8 cells (approximately one genome per recipient cell) of the undegraded-DNA via Sendai virus mediated cell fusion performed within the recipient strain in the hope of increasing the uptake of DNA and minimizing the risk of digestion by cell nucleases;

4. transfer of the treated cells and of the different controls (DNA-treated and untreated cells with and without the use of Sendai virus) into HAT-selective medium for the isolation of HGPRT-positive clones;

5. characterization of HGPRT of the HAT-resistant clones for discriminating between revertants and true transformants;

6. characterization of the same clones for G6PD and phosphoglyceratekinase, PGK, (two additional X-linked markers in mammals) to screen for the possible, though extremely unlikely, event of cotransformation.

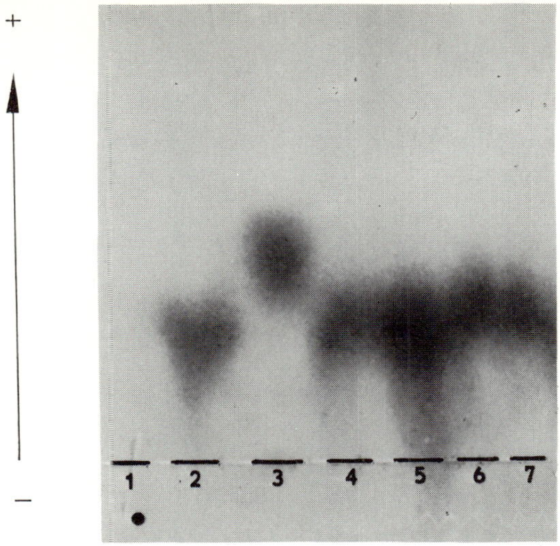

1. A9-cells
2. L-cells
3. HeLa-cells
4. A9 + HeLa DNA
5. A9 + L DNA + inactivated Sendai virus
6. A9 + A9 DNA + inactivated Sendai virus
7. A9 + A9 DNA

Fig. 2. Electrophoretic characterization of the HGPRT produced by some of the HAT-resistant clones rescued from the experiments listed in Fig. 1.

The electrophoretic mobility of HGPRT is clearly different in lysates from mouse (channel 2) and human cells (channel 3). The *A9* recipient cells are entirely deprived of enzyme activity (channel 1). The electrophoretic mobility of all the HAT-resistant clones listed in Fig. 1 was always found to be of "mousetype", independently from having the cells received or not received DNA, of whatever type, and/or treatment with inactivated Sendai virus. Channels 4 to 7 show the HGPRT pattern of some of these clones. Electrophoresis performed on cellogel according to Shin et al. [28].

Up to now only part of these experiments have been completed. Technical details about DNA-preparation, fusion procedure, selection and characterization of HGPRT-positive clones will be reported in full elsewhere.

These preliminary results are summarized in Fig. 1 where it can be seen that HAT-resistant clones have been found both among DNA-treated and untreated A9 and RAG cells with frequencies ranging from 2×10^{-8} to 30×10^{-8}.

Though the highest number of HAT-resistant clones were found in the group of experiments where RAG cells had been treated with HeLa DNA in the presence of Sendai virus, there is obviously no definite correlation between the yield of surviving clones and the DNA-treatment.

The electrophoretic characterization of the HGPRT produced by some of the HAT-resistant clones can be seen in Fig. 2 where its murine nature appears very clear.

This was found to be the case for all HAT-resistant clones which appeared in the control group as well as in the DNA-treated group and irrespectively from the kind of DNA added.

It is not yet decided whether these HGPRT-positive clones have to be considered instances of back-mutation or of derepression of the murine repressed HGPRT-gene.

We believe that the latter possibility is more likely to be true since the behaviour of these "revertants" in 8-azaguanine is strikingly different from that of the L-cells wherefrom they were originally derived. The experiments aimed at clarifying this point are still in progress, and I prefer to leave their discussion to the final report of these data.

Even at this stage, however, the experiments summarized above cast some doubts on the earlier evidences of DNA-transformation involving the same cell drug-resistance marker.

Nevertheless, an encouraging conclusion emerges from all these studies: the possibility of obtaining the correction of "genetic defect" by simple growth of the diseased cells in vitro in the appropriate selective medium. Whenever successful, this procedure would represent the best strategy for possible applications to medical practice since spontaneous revertants would obviously be the most suitable for a therapeutical transplantation of "corrected cells" back into their patient donors.

Addition of isolated DNA-fragments: a better approach?

One of the main obstacles to induce genetic transformation of mammalian cells with DNA extracted from similar sources may lie in the enormous amount of non-relevant material present in the undegraded preparations that are currently used for such purpose.

To put it in simple language, our own efforts to transform a single gene function with the addition of a whole mammalian genome per cell, with all of its innumerable genes and redundant attributes, may well be like searching for a needle in a haystack. Even the application of powerful selection procedures cannot be that much of a help if the chance for the event to occur is beyond detectability, given the size of the populations of recipient cells per experiment. On the other hand, the size of these experimental populations could not, for technical reasons, be adequately increased nor could the ratio of the added DNA per cell without lethal effects on the recipient cells.

It is basically because of these difficulties that most experimenters prefer to attempt genetic transformation of mammalian cells with viral DNA of different source, though, to the best of my knowledge, a comparative study of the uptake of viral versus nonviral DNA in these cells has not been performed so far, under the same experimental conditions. However, the above mentioned attitude is fully justified, as we all know, by the relative easiness with which "malignant transformation" of mammalian cells can be induced by certain DNA and RNA tumor viruses such as the Symian virus SV40, the Polyoma virus, the Freund virus, etc.

The association of the viral genomes with the chromosomal DNA of these types of "transformed cells" is supported by three lines of evidence: (1) the successful hybridization of cellular DNA with in vitro RNA complementary to viral RNA [27]; (2) the recovery of live virus when SV40-transformed but virus-free cells are fused with other cells which are susceptible to the virus infection [13, 35]; (3) the gradual loss of SV40-induced antigens alongside the loss of human chromosomes from man-mouse somatic cell hybrids carrying a human parental genome transformed with SV40 prior to fusion [36].

At a close inspection, even these evidences do not permit the establishment of whether the relationship between viral and host DNA is episomal-like or involves a true integration. However, they justify the general feeling that the "directed" transformation of a given genetic defect in mammalian cells is more likely to be achieved with the use of DNA-fragments such as the ones incapsidated by pseudovirions from Polyoma or SV40-viruses [22] or included into transducing phages [16] and intact viruses [19].

The successful use of a lambda-transducing phage (containing the complete galactose operon) to correct human galactosemic cells known to be deficient in UPD-transferase and of the Herpes simplex virus (HSV), positive in TK activity, to correct TK-deficient mouse cells, are striking examples of the potentialities of the latter two strategies[1]). I do not need to elaborate on these experiments after the detailed accounts given by the authors and the exhaustive discussions that followed their presentations.

I may be excused, however, if I come back to the usual question: are these corrected cells a true example of genetic transformation? Again we don't know for certain whether or not the corrected cells synthesize the missing product as coded by the donor DNA, nor do we know whether other "unwanted" exogenous gene products are synthesized "as well" or "instead of".

---

[1]) Of course the best strategy along these lines would be that of trying to induce the correction of a gene defect in mammalian cells with the addition of the isolated homologous normal DNA-gene, which can apparently now be synthetized in vitro with the help of a reverse transcriptase, provided that the purified messenger RNA from the normal gene is available (cfr. [9a]).

In the favourable case that the translation of the specific piece of viral DNA is eventually proved (e.g., through the biochemical characterization of the relevant gene product in the corrected cells), we still would be left to wonder whether and how the added DNA-fragment is associated with the host chromosome bearing the locus for that specific function.

These questions are somehow very similar to those that microbial geneticists asked themselves over a decade ago: Were chromosomal locations of prophages on the bacterial chromosome strictly specific? Were episomic elements part of its linear structure? Could episomes in the integrated state undergo genetic recombination with neighbouring regions of the bacterial chromosome?

We all know that these questions were answered only when sufficient data on the fine genetic structure of the bacterial chromosome became available. Thus, a great deal of ground work is necessary before we may be able to ask, rather than answer, similar questions in mammalian cells. Yet, the prospects for gathering the missing information are no longer unrealistic now that we can use in mammalian cells many of the "paraphernalia" that proved to be so helpful in microorganisms.

For instance, somatic cell hybridization could help straight off to decide whether or not the correction of TK-deficiency, reported by Dr. *Munyon* in virus-treated mouse cells, involves the chromosomal integration of the relevant viral DNA fragment.

For this purpose, it would be better to repeat the experiments on a human TK-deficient strain. If the result were the same, the alleged transformants should be fused with TK-deficient mouse cells, and the resulting TK-positive somatic hybrids should be cloned respectively in the HAT and 5-BUDR media. Since the human TK-gene is known to be located on chromosome E-17, it follows that the HAT clones should retain it regularly and produce the TK of viral-type only in the case that the viral TK gene is present in an integrated form on this chromosome. For the same reason, the 5-BUDR clones should regularly lose both the chromosome and the enzymatic function.

The same strategy could be utilized to check the integration of the exogenous DNA in HGPRT-deficient human cells should anyone also succeed to cure these cells in vitro with the help of viral DNA. However, this could not be done for the alleged transformants of Dr. *Merril* involving the human UDP transferase gene since the location of the relevant locus in the human genome is yet unknown.

In conclusion, more than anything else, we need data on the genetic mapping of the mammalian genomes and of man in particular. Dr. *Ruddle* and Dr. *Bodmer* have already stressed the potentials of somatic cell hybrids for this purpose. I would like now to complement their presentations by telling you briefly about our own work on the genetical and cytological mapping of the human X-chromosome, performed with the help of the same tools.

Mapping of the human X-chromosome: results, strategies and perspectives

As soon as it became clear that the man-mouse somatic cell hybrid was a powerful experimental tool for studies on human linkage [3, 20, 37], my colleagues and I wondered how far this type of unstable hybrid and others behaving the same way could be utilized to build up a statistical and cytological map of the human X-chromosome [32].

We focused our attention on this particular chromosome for several reasons:

1. this is the only human chromosome for which numerous markers are known (over a hundred) with reasonable estimates of meiotic recombination between some of them;
2. at least four of these markers are expressed in cultured cells (HGPRT, G6PD, PGK, and alpha-galactosidase) and can be distinguished from the corresponding markers of mouse and/or hamster cells;
3. in view of the X-linkage of the human HGPRT locus, the HAT-selection of somatic hybrids between human cells normal for this marker and HGPRT-deficient mouse or hamster cells was expected to lead to the obligatory retention of the human X-chromosome or at least of the portion bearing the HGPRT locus; the opposite situation was expected from the 8-azaguanine selection;
4. several reciprocal X-autosomal translocations are known in man involving the transfer of an active genetic segment of the X-chromosome on one of the autosomes and vice-versa: the hybridization of these aberrant human strains with HGPRT-deficient mouse or hamster cells represented an obvious challenge for the cytological mapping of the available X-linked markers and possibly of several autosomal ones;
5. last but not least, the use of a group of human markers with known linkage relationships offered a way to test the reliability of the somatic cell hybrids of the kind referred to for studies on human linkage and to compare the mapping data based on mitotic segregation with those already available from meiotic recombination.

I am now going to summarize very briefly the main conclusions of our studies with little reference to their sequential development. Full details can be found elsewhere [7, 15, 17, 32, 33].

First of all, it became clear that a prolonged HAT-selection of the somatic hybrid of the kind HGPRT-deficient mouse x normal human cells leads to the accumulation of reduced hybrids with a split X-chromosome [17, 30] which may have lost one or more of the other X-linked markers.

Fig. 3. Mitotic separation of 3 X-linked markers in a series of man-mouse somatic cell hybrids.

Hybrid cell line	HGPRT	G6PD	PGK
1. A9/HRBC(2)	+	+	+
2. A9/HRBC(2) HIH	+	+	+
3. A9/CF	+	+	+
4. A9/ADC	+	+	+
5. A9/4XY	+	+	–
6. A9/HRBC(1)	+	–	–
7. A9/JW	+	–	–

The above describes the mitotic separation of three human X-linked markers in somatic cell hybrids between HGPRT-deficient mouse cells (*A9*) and human cells of different type with normal HGPRT activity. All the seven hybrids listed in this table (which are the same as those reported in [17]) were expected to retain the human X-chromosome (where the HGPRT locus is located) with all other X-linked markers (G6PD and PGK). However, two of these hybrids lost both human G6PD and PGK (rows 6 and 7) and one lost only PGK (row 5). From cloning studies, it appeared clear that, in general, whenever G6PD was lost, PGK was also lost, but PGK could be lost alone. This situation could be best explained assuming the linear sequence of the three loci as reported in Fig. 4.
(A full account on these data will be reported elsewhere, Siniscalco et al [33].)

This interesting phenomenon offered us the possibility to guess the linear sequence of HGPRT, G6PD, and PGK along the X-chromosome (Figs. 3 and 4) but makes the detection of new X-linked markers impossible unless special measures are taken such as stopping the HAT-selection soon after the isolation of the hybrid (Fig. 5).

Better types of somatic hybrids for the purpose of X-linkage detection are the ones between TK-deficient mouse or hamster cells and normal or HGPRT-deficient human cells. In these hybrids, the condition for survival in the HAT medium is the retention of human-TK while the human X-chromosome can be lost at random like every other one. The breakage of the X-chromosome can, of course, still occur in rare cells, but these are not accumulated by selection. Thus, the screening of new X-linked markers can be easily fulfilled by correlating the retention or loss of the "test markers" with that of one or more of the already known X-linked markers ("main markers"). The different kinds of strategies to be utilized for detection of X-linkage and X-chromosome mapping are reported in Fig. 5.

The potentials of somatic hybrids between HGPRT-deficient mouse or hamster cells with human cells carrying X-autosomal translocations have been recently confirmed by a study which led to the cytological localization of the PGK locus on the long arm of the human X-chromosome and of the loci for HGPRT and G6PD on its short arm [7]. From these and other unpublished observations, involving a human strain carrying a different type of translocation, it appears that the

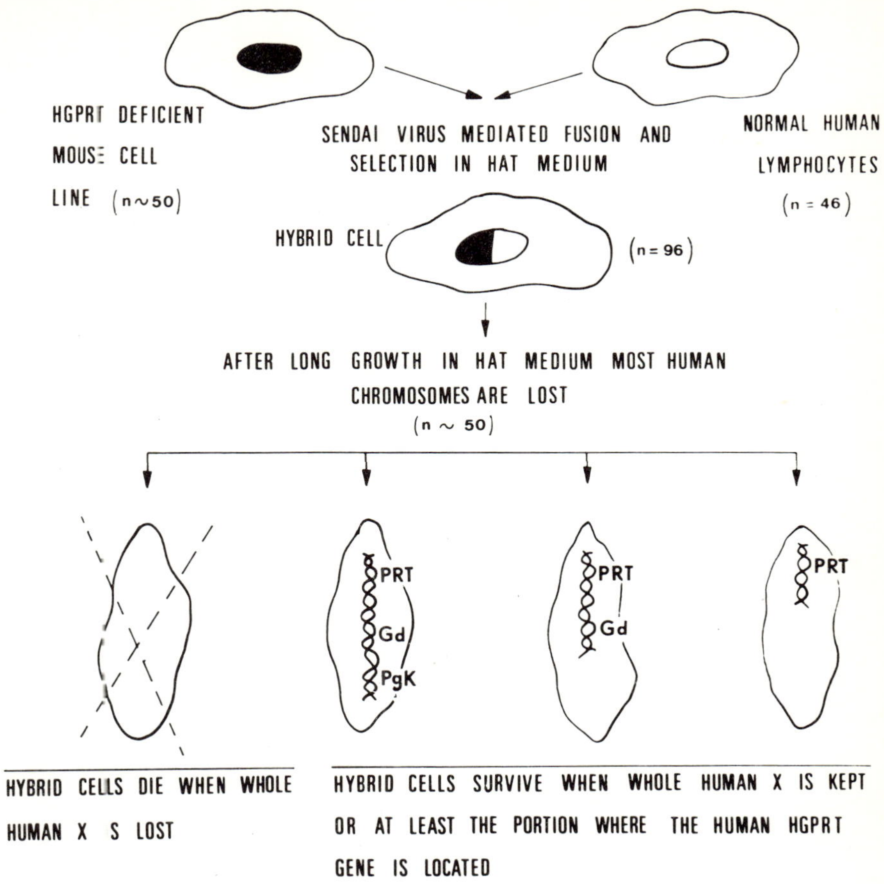

Fig. 4. Segregation of three human X-linked loci in the man-mouse somatic cell hybrid. Mapping of HGPRT, G6PD, and PGK along the human X-chromosome, based on the mitotic separation reported in Fig. 3.

linear sequence of these three markers along the X-chromosome is HGPRT-G6PD-PGK as already suggested by the series of man-mouse hybrids which underwent X-chromosome breakages in culture.

We have good reasons to believe that the G6PD locus is the nearest to the centromere, but we cannot say whether on the short or on the long arm side of it.

Adopting the strategies for X-linkage detection, we have been able to confirm the X-linkage of human PGK [15] and α-gal [8] in somatic cell hybrids of the kind referred to above. The cytological localization of the latter locus is on the way.

Fig. 5. Strategies for X-linkage detection and X-chromosome mapping in man using somatic cell hybrids.

Type of hybrid				Procedure	expectation and analysis
A1 A2	Human HGPRT-deficient cells or Normal human lymphocytes	x	Tk-deficient mouse or hamster cells (Chinese or Syrian)	Isolation and cloning of hybrid cells in HAT	Clones *with* or *without* human-X: screening for absolute correlation or loss of known human X-linked markers and of human-X with that of the "test markers" will permit identification of "new" X-linked markers in man.
A3	Human normal cells (lymphocytes)	x	HGPRT-deficient mouse or hamster cells	Isolation of hybrid in HAT followed by prompt cloning in HAT, 8-aza-guanine and MEM-TH	HAT-clones retain regularly human-X, 8-azaguanine clones lose it regularly, MEM-TH clones retain or lose it, randomly. Screening of new X-linked markers as above.
B1	Human normal cells (lymphocytes)	x	HGPRT-deficient mouse or hamster cells	Isolation of hybrid in HAT followed by long propagation in the same medium (at least six months) and, eventually, cloning in HAT, 8-azaguanine and MEM-TH	Occurrence of clones bearing a split X-chromosome. HAT clones will necessarily retain human HGPRT, 8-azaguanine clones lose it. Other X-linked markers may or not be retained depending on their location. Screening for mitotic separation of already known X-linked markers followed by cytological studies will permit "genetic" and "cytological" mapping.
B2	Human normal lymphocytes irradiated before fusion	x	HGPRT-deficient mouse or hamster cells	Isolation of hybrids in HAT followed by prompt cloning in HAT, 8-aza-guanine and MEM-TH	As above
C	Human lymphocytes from donors carrying an X-autosomal translocation	x	HGPRT-deficient mouse or hamster cells	Isolation of hybrids in HAT followed by prompt cloning in HAT, 8-aza-guanine and MEM-TH	Obligatory retention of X-short arm (where human HGPRT is located). Random loss or retention of other known X-linked markers depending on their location along the X. Screening for segregation of already known X-linked markers followed by cytological studies. Screening also for a large number of autosomal markers to spot *de novo* linkage associations that may have arisen as result of the translocation.

Strategies *A1, A2, A3* are suitable for X-linkage detection; *B1* and *B2*, for X-chromosomal mapping following X-chromosome breaks in culture; *C*, for cytological mapping of the X-chromosome as well as of the "translocated autosome" when it happens that suitable biochemical markers are located on the piece of autosome involved in the translocation.

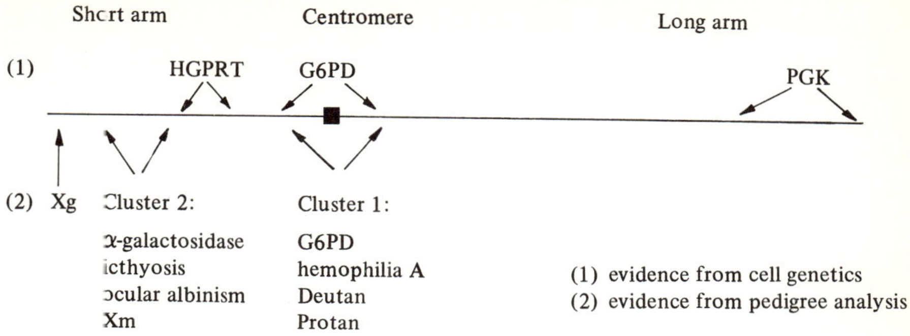

Fig. 6. A tentative map of the human X-chromosome using information from cell genetics and pedigree analysis.

The evidence supporting the linear sequence and cytological location of HGPRT, G6PD, and PGK comes from somatic cell genetics and is reported in full in [7, 17, 33].

Unpublished observations of our group have confirmed, using the same approach, the X-linkage of the structural locus for α-galactosidase; its cytological location is on the way.

The evidence from pedigree analysis, supporting the location of Xg at the end of the short arm as well as the existence of the two "clusters" of closely linked loci is "circumstantial" in the sense that direct linkage comparisons have been done only for some of these loci. The gene sequence within each cluster is not known since the fiducial limits of the recombination estimates are very broad.

A full account on the relevant data can be found in the proceedings of a recent symposium on chromosome mapping in man (see [39]).

A tentative map of the human X-chromosome is reported in Fig. 6 using information from both somatic cell genetics and pedigree analysis.

Even though this map is still very rudimental, there can be little doubt that it can now grow fast in a matter of a few years. When this will have happened, but only then, can we ask ourselves a number of interesting questions.

Is there a direction in the transcription of linked genes along the human X-chromosome? How does the occurrence of a translocation (in vivo or in vitro) affect the transcription and translation of genetic messages pertaining to X-linked genes?

Can exogenous DNA, coding for instance for normal HGPRT, be integrated at the appropriate chromosomal site in HGPRT-deficient cells and, if so, with what effect on the neighbouring genes?

Epilogue

I would like to close this quick outlook on the correction of genetic defects in cultured mammalian cells with another personal opinion. It seems to me unquestionable that studies on the transfer of genetic information in mammalian cells will turn out to be more and more revealing to understand the basic problems of cellular and molecular biology, namely cell regulation and differentiation. However, we should honestly recognize that the time required for these studies to be of help in practical medicine is not likely to be shorter than that necessary for everyone of us to be able to walk on the moon now that we vaguely know how to get there.

References

[1] *Bhargava, P. M.*, and *Shanmugan, G.* 1971. *In* Progress in nucleic acid research and molecular biology, eds. *J. N. Davidson* and *W. E. Cohn.* Vol. 11, pp. 103–192. New York: Academic Press.

[2] *Bodmer, W.* This volume, p. 371 ff.

[3] *Boone, C. M.*, and *Ruddle, F. H.* 1969. Biochem. Genet. **3,** 119.

[4] *Bradley, T. R.*, *Roosa, R. A.*, and *Law, L. W.* 1962. J. Cell. Comp. Physiol. **60,** 127.

[5] *Cook, P. R.* 1970. J. Cell. Sci. **7,** 1.

[6] *Engel, E.*, *McGee, B. J.*, and *Harris, H.* 1969. Nature, Lond. **223,** 152.

[7] *Grzeschik, K. H.*, *Allerdice, P. W.*, *Grzeschik, A.*, *Opitz, J. M.*, *Miller, O. J.*, and *Siniscalco, M.* 1972. Proc. Nat. Acad. Sic., US **69,** 69.

[8] *Grzeschik* et al. (in preparation).

[9] *Harris, H.*, *Watkins, J. F.*, *Ford, C. E.*, and *Schoefl, G. I.* 1966. J. Cell Sci. **1,** 1.

[9a] *Inder* and *Kacian,* 1972. Nature New Biology **235,** 163–1969.

[10] *Kao, F. T.*, *Johnson, R. T.*, and *Puck, T. T.* 1969. Science **164,** 312.

[11] *Kao, F. T.*, and *Puck, T. T.* 1970. Nature, Lond. **228,** 329.

[12] *Klebe, R. J.*, *Chen, T. R.*, and *Ruddle, F. H.* 1970. J. Cell Biol. **45,** 74.

[13] *Koprowski, H.*, *Jensen, F. C.*, and *Steplewski, Z.* 1967. Proc. Nat. Acad. Sci., US **58,** 127.

[14] *Littlefield, J. W.* 1963. Proc. Nat. Acad. Sci., US **50,** 568.

[15] *Meera Khan, P.*, *Westerveld, A.*, *Grzeschik, K. H.*, *Deys, B. F.*, *Garson, O. M.*, and *Siniscalco, M.* 1971. Am. J. Hum. Gen. **23,** 614.

[16] *Merril, C. R.* This volume, p. 329 ff.

[17] *Miller, O. J.*, *Cook, P. R.*, *Meera Khan, P.*, *Shin, S.*, and *Siniscalco, M.* 1971. Proc. Nat. Acad. Sci. US **68, 116.**

[18] *Morse, M. L.*, *Lederberg, E. M.*, and *Lederberg, J.* 1956. Genetics **41,** 758.

[19] *Munyon, W.* This volume, p. 343 ff.

[20] *Nabholz, M.*, *Miggiano, V.*, and *Bodmer, W.* 1969. Nature, Lond. **223,** 358.

[21] *Nadler, H. L.*, *Chacko, C. M.*, and *Rachmeler, M.* 1970. Proc. Nat. Acad. Sci., US **67,** 976.

[22] *Osterman, J. V.*, *Waddell, A.*, and *Aposhian, H. V.* 1970. Proc. Nat. Acad. Sci., US **67,** 37.

[23] *Pontecorvo, G.* 1972. Nature **230,** 367.

[24] *Roosa, R. A.*, and *Bailey, E.* 1970. Cell Physiol. **75,** 137.

[25] *Rubin, C. S., Dancis, J., Yip, L. C., Nowinski, R. C.*, and *Balis, M. E.* 1971. Proc. Nat. Acad. Sci.,US **68, 1461**.

[26] *Ruddle, F. H.* This volume, p. 299 ff.

[27] *Sambrook, J., Westphal, H., Shrinivasan, P. R., Dulbecco, R.* 1969. Proc. Nat. Acad. Sci., US **60,** 1288.

[28] *Shin, S., Meera Khan, P.*, and *Cook, P. R.* 1971. Biochem. Genet. **5,** 91.

[29] *Silagi, S., Darlington, G.*, and *Bruce, S. A.* 1969. Proc. Nat. Acad. Sci., US **62,** 1085.

[30] *Siniscalco, M.* 1970. Atti Ass. Genet. Ital. **XV,** 1.

[31] *Siniscalco, M., Klinger, H. P., Eagle, H., Koprowski, H., Fujimoto, W. Y.*, and *Seegmiller, J. E.* 1969. Proc. Nat. Acad. Sci., US **62,** 793.

[32] *Siniscalco, M.* 1970. *In* Control mechanisms in expression of cellular phenotypes, ed. M. A. Padykula, pp. 205–231. New York: Academic Press.

[33] *Siniscalco, M., Shin, S., Kooy-Veltkamp, C., Meera Khan, P.*, and *Miller, O. J.* submitted to Cytogenetics.

[34] *Szybalska, E. H.*, and *Szybalski, W.* 1962. Proc. Nat. Acad. Sci., US **48,** 3026.

[35] *Watkins, J. F.*, and *Dulbecco, R.* 1967. Proc. Nat. Acad. Sci., US **58,** 1396.

[36] *Weiss, M. C.* 1970. Proc. Nat. Acad. Sci., US **66,** 79.

[37] *Weiss, M. C.*, and *Green, H.* 1967. Proc. Nat. Acad. Sci., US **58,** 1104.

[38] *Westerveld, A., Visser, R. P. L. S., Meera Khan, P.*, and *Bootsma, D.* 1971. Nature New Biol. **234,** 20.

[39] Proceedings, IV international congress of human genetics, Paris 1971. In press.

Discussion

Ringertz: With respect to combinations which involve interspecific hybrids involving human chromosomes, which ones are really known to retain full sets of human chromosomes, eliminating the other species?

Siniscalco: Well, I know only of the man-rat. I think that the man-monkey should be tried out since in the case of the mouse-monkey hybrid, the monkey's genome is the one eliminated. It remains to be seen who is the winner between man and monkey.

Milgrom: Just a short comment concerning the question of possible acceptance of interspecies hybrids which lose cell-surface antigens. I think this is entirely possible. In studies which I described yesterday, one clone of a man-mouse hybrid lost its "manness" completely as far as surface antigens are concerned, but still retained some human chromosomes. So, it appears that there are some chromosomes which carry no information for surface antigens. Therefore, it seems feasible that a man-rat hybrid may retain some rat chromosomes without being recognized as foreign by the human host.

Krone: One of the few examples of deletion mapping in man is the tentative localization of the gene for 6-PGD on the chromosome which we now call 22, by Dr. *Fialkow*. If 6-PGD is linked to the PGM1 locus, one might wonder whether patients with chronic myelogenous leukemia have been investigated for their PGM1 genotype.

Ruddle: I am not familar with all the data on deletion mapping, but there are obvious pitfalls. The deletion may be only apparent, there being a cryptic translocation to another chromosome. The deletion may be associated with a depletion of enzyme activity, but the enzyme's structural gene need not be located in the deleted segment. Genome imbalance could result in enzyme activity depression.[1)]

Merril: *Schwatz, Cook,* and *Harris,* 1971 (Nature, **230,** 5–8) have reported that following fusion of chick erythrocytes to mouse fibroblast cells, there is complete disappearance of the chick chromosomes, and yet the cells still retain some chick gene function. What effect might such a phenomena have on the mapping experiments that are being performed?

Siniscalco: May be Dr. *Croce* could comment on that because, as I said already, he has made the same type of hybrid utilizing the chick fibroblasts instead of the chick erythrocytes. He has seen, in this case, the retention of some of the chick chromosomes, apparently very small ones. In the case of the mouse-chick erythrocyte hybrid, it may be, as Dr. *Harris* and his colleagues have suggested, that the piece of genetic information which is essential for the survival of the hybrid line has been translocated onto a mouse chromosome. However, other interpretations are still not firmly excluded, such as a back-mutation at the mouse locus for HGPRT with a gene product that minics the chick HGPRT. If you remember, the electrophoretic mobility of the HGPRT in this type of hybrid has been described as being "chicklike" and not exactly "chick". Another puzzling property of these hybrids is that they yield in 8-azaguanine, a number of revertants which is much greater than the number of resistant clones that can be obtained from the original mouse line. We have observed the same phenomenon in the series of A9-revertants that I described before. I wonder what does all this mean. Perhaps these deficient cell lines selected in 8-azaguanine are not "structural mutants" after all.

Deák: *Harris,* group in Oxford has shown that in hybrid cells obtained from chick erythrocytes fused with mouse A9 ($HGPRT^-$) cells, no chick chromosomes can be seen. These hybrids nevertheless synthesize chick HGPRTase. The frequency at which such hybrids were obtained is approximately 10^{-4}. The frequency of back-reversion of these hybrid cells to 8-azaguanine resistant cells is about 10^{-2}. The chick gene for HGPRTase is, therefore, not stably incorporated into the mouse genome. These hybrids are of particular interest in that no chick-specific antigens are detectable on the surface of these cells. This fact might be of considerable importance in the correction of metabolic defects in vivo as the corrected cell would not be recognized by the recipient animal as foreign.

Fischer: I have a question directed to all fusionists, the answer to which would be helpful for immunologists and embryologists. There is one cell-type in a foetus which does not express foreignness on its surface: the trophoblast of the foetal placenta. Without the trophablast no pregnancy – that means no nine-months graft – would be possible. Does anybody have experience with fusion of cells where one partner is a trophoblast cell? I raise this question because of *George Klein*'s fascinating finding that Ehrlich ascites tumor cells, when fused to other tumor cells, will suppress the tumor antigens of the latter. My question is: Will the trophoblast act in a similar fashion?

Milgrom: Dr. *Fischer*'s remarks are correct as far as expression of transplantation antigen on trophoblasts or Ehrlich's carcinoma cells are concerned. On the other hand, expression of species-specific surface antigens is not impaired on these cells. Therefore, the experiments on any suppressor mechanisms could pertain only to intraspecific but not interspecific hybrids.

---

1) (Note added in proof stage:) Recent studies carried out by *Ruddle* show that PGM1 can be assigned to chromosome number 1 in man. (See *Ruddle* et al. 1972. Science **176,** 1429.)

Hotchkiss: It is still a rather "heroic era" in the experimentation on interspecies hybrids when one does not yet predict whether the hybrid cell will be viable nor which chromosomes will be retained. These outcomes remind me strongly of more classical studies of chromosome balance within the species, in which certain sets, trisomic or otherwise, survive and others do not. The milieu in which this is worked out is the intracellular cytoplasm or nucleoplasm. I suggest that one might expect to be able to tip the balance right at the beginning by choosing different nutritional environments for the postfusion cell. By contriving to encourage the loss of one or a few mouse chromosomes, the cell might become increasingly unfavorable to other mouse chromosomes, so that quite an other result could be obtained with major retention of human chromosomes. This would recommend some degree of return to experimentation with some major variants of the culture medium – nutrients, carbon dioxide, and pH, etc. I imagine the culture of hybrid cells tends to be routinized and left for the assistants to carry out, so that the resources of this other formerly heroic experimentation are largely set aside and pushed into the past[1]).

Ruddle: I think this is a very good comment. We just don't know enough yet about conditions and how they affect the segregation of the chromosomes, and I suspect that the segregation of the chromosomes is, to some extent, under the influence of selection pressures. This was a point that was discussed yesterday following Dr. *Klines* talk. And I think your remark about the autocatalytic loss of chromosomes is also quite a useful one, especially if we make the assumption than an intact genome of either one of the parents is necessary for the survival of a cell, especially in the interspecific combinations, and if a genome becomes nullosomic for a particular chromosome (which may actually be the result of selection pressure, say, on the part of nutrients), then this may predispose the loss of remainder of that parental genome. A very good comment.

1) Shortly after the meeting a paper appeared (*J. Jami* and *S. Grandchamp*, Dec. 1971. Proc. Nat. Acad. Sci., U.S. **68**, 3097–3101) suggesting that chromosome loss is the result of interaction between the two cells, in part a function of the replication state of the two genomes.

Bacterial Gene Expression in Mammalian Cells

Carl R. Merril, Mark R. Geier, and John C. Petricciani*)

Laboratory of General and Comparative Biochemistry, National Institute of Mental Health, Bethesda, MD. 20014, USA

* Laboratory of Pathology, Division of Biologics Standards, National Institutes of Health, Bethesda, Maryland 20014, USA

Summary: Bacteriophage, lambda and its transducing derivatives were used to infect human fibroblasts. Transcription of the bacteriophage genome was detected by hybridization with lambda DNA. Lambda-specific RNA (0.2 %) reached a maximum four days after infection. Translation was detected by the use of lambda carrying the galactose operon of *E. coli.* Human fibroblasts from a galactosemic patient, lacking the ability to make active α-D-galactose-1-phosphate uridyl transferase, produce the transferase enzyme following infection with the transducing phage.

Problem

Direct manipulation of purified genetic material was first reported in 1944 by *Avery, MacLeod*, and *McCarty* [1]. They found that the physical properties of the pneumonia bacteria could be specifically altered by the addition of purified genetic material, DNA, from another type of pneumococcus.

Since these initial experiments in the mid 1940's, intensive investigations have been made utilizing genetic material, primarily in bacteria and the viruses which readily infect these bacteria. There has been a great deal of speculation as to whether such genetic manipulations could be conducted in man and other higher organisms. *Ledoux* [17] extensively reviewed a large number of experiments up to 1965 which show that eukaryotic cells can take up exogenous DNA, and, in 1969, *Ledoux* and *Huart* [13] reported that DNA from bacteria (Micrococcus lysodeikticus), which is

Manuscript received: 10 December 1971

taken up by the root cells of barley seedlings, appears to be integrated with the nuclear DNA. This was done with heteropycnic labeled DNA (DNA of different densities). A similar experiment was performed by *Hill* and *Hillova* [11], in which exogenous mouse DNA was taken up by chicken cells and integrated into the chicken genetic material. These experiments utilizing density and radioactively labeled DNA are complemented by the demonstration (*Schwartz* et al. [27]) that, following fusion of chick erythrocytes and mouse fibroblasts, all the chicken chromosomes are lost by "premature chromosome condensation", however, some of the cells retained and replicated a functional chicken gene for inosinic pyrophosphorylase.

Demonstrations of the functional activity of exogenous DNA administered to eukaryotic cells have also been reported (see review by *Ledoux* [17], including the appearance of inosinate pyrophosphorylase in human cells deficient in this enzyme when treated with DNA from normal cells [30].

A recent example of DNA mediated transformation is that described by *Ottolenghi-Nightingale* [24] in which skin cells from an albino mouse embryo synthesized melenin after exposure to a nucleoprotein fraction of cells from pigmented mice.

Fox and *Yoon* [4] reported DNA induced transformation in the fruit fly, Drosophila. They have proposed an exosome model to explain this data in which DNA segments enter the cells of treated individuals, but never integrate into the linear structure of the chromosome.

It has been known for some time that foreign DNA can be carried into eukaryotic cells by viruses. Oncogenic viruses which can transform normal cells into cancer cells have been studied intensively in this regard. Recently, *Qasba* and *Aposhian* [25] have shown that mouse embryo DNA, coated with the coat proteins of polyoma virus, is taken up by human embryo cells and that 7 % is found in the nuclei 24 hours after infection. *Rogers* [26] has shown that viruses which induce cell destruction and pathology in one species may express only certain genes in another species. The Shope papalloma virus (SPV) which induces warts on rabbits has no such effect in man, mice, rats, dogs, or monkeys. This virus does induce arginase, however, and rabbits infected with SPV have low blood arginine levels. The blood arginine levels of mice, rats, and monkeys innoculated with the SPV also have lowered arginine levels as do a large number of the people who have studied this virus ("low enough so as not to be confused with low normal range of fasted controls").

Paul Munyon and his associates working with L cells which lack the enzyme thymidine kinase has shown that Herpes simplex irradiated with UV light (to decrease its ability to kill cells) is capable of transforming some of the cells so that they are capable of producing thymidine kinase [20].

Zinder and *Lederberg* [32], demonstrated that bacterial viruses are capable of carrying genes from one bacterium to another (in this case it was the genes necessary for synthesis of certain amino acids). This phenomena is termed transduction and has been investigated in detail for the specialized transducing phage lambda [10]. This virus is normally associated with *E. coli* and is capable of transducing the galactose operon of *E. coli.* If the genetic code is truly universal, it might be possible to effect genetic changes in eukaryotic cells using bacteriophages as specific gene carriers. As a model system to test for this possibility, lambda transducing virus carrying the galactose operon from *E. coli* was used to infect human fibrolasts derived from a patient with galactosemia. Galactosemia is an inborn error of metabolism in which little or no α-D-galactose-1-phosphate uridyl transferase enzyme activity is present [13, 31].

In order for lambda virus to function in a eukaryotic cell, three major events must occur: first, the virus must be capable of getting into the cell; second, its DNA must be exposed and transcribed into RNA, and, finally, the RNA must be translated into protein. The uptake of lambda virus by the human cells was studied by electron microscopy. Transcription was studied by DNA-RNA hybridization techniques which demonstrated the appearance of lambda-specific RNA following infection with the phage [16]. Translational events were demonstrated by the occurrence of transferase activity in the galactosemic human fibroblasts, following infection with the lambda galactose transducing phage [5, 19].

Materials and methods

Electron microscopy of human cells was performed on the human diploid fibrolast cell line CCL-72, obtained from the American Type Culture Collection, Rockville, Maryland, grown in monolayer culture with medium MAB 87/3 [9]. The cells were examined before and after exposure to lambda (1 hour at 37 °C). Staining was with lead citrate and uranyl acetate (Fig. 1). A more positive identification of viruslike particles was made by the use of immunoelectron microscopy in which an antibody preparation to a given antigen is conjugated with peroxidase enzyme. When the peroxidase-antibody complex is incubated with cells, it can be identified within the cell, thus localizing a specific antigen. Antibody to $\lambda_{CI}857S_7$ [7, 8] was produced in a sheep [12], and the peroxidase-antibody conjugate was prepared and used as described previously [21, 22, 28]. After treatment with the peroxidase-labeled lambda-antibody, incubation in substrate (3–3′ diaminobenzidine), and staining with lead citrate and uranyl acetate, the intranuclear particles appeared dense and granular (Fig. 1), presumably to specific virion staining associated with peroxidase reaction product.

Transcription of lambda virus was studied by using DNA-RNA hybridization techniques developed by *Gillespie* and *Spiegelman* [6] and *Das* and *Spiegelman* [3]. These techniques employed the use of single-stranded lambda DNA fixed to cellulose nitrate filters to detect uridine-5-^{3}H labeled lambda-specific RNA [5, 6]. The experimental design is outlined in Fig. 2.

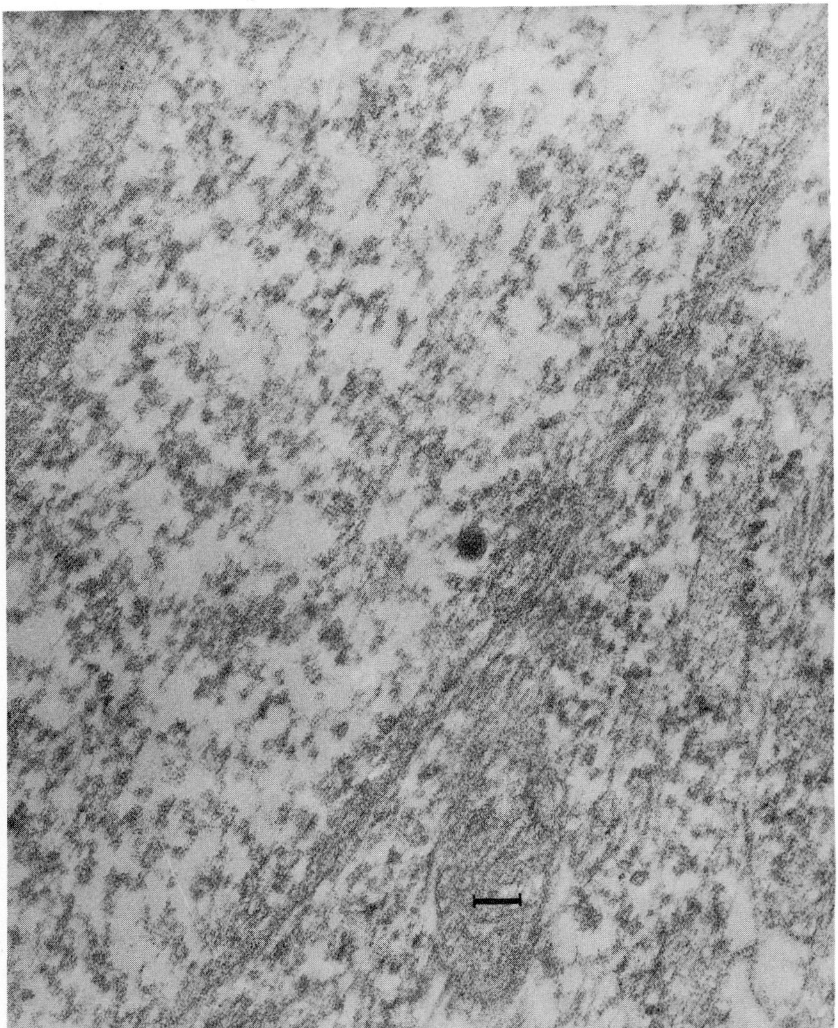

Fig. 1. Electron micrograph of CCL-72. Cell first exposed to lambda for 1 hour, then to peroxidase-labeled antibody and substrate, and stained with lead citrate and uranyl acetate. One densely stained particle is present within the nucleus near the nuclear membrane (scale represents 100 nm).

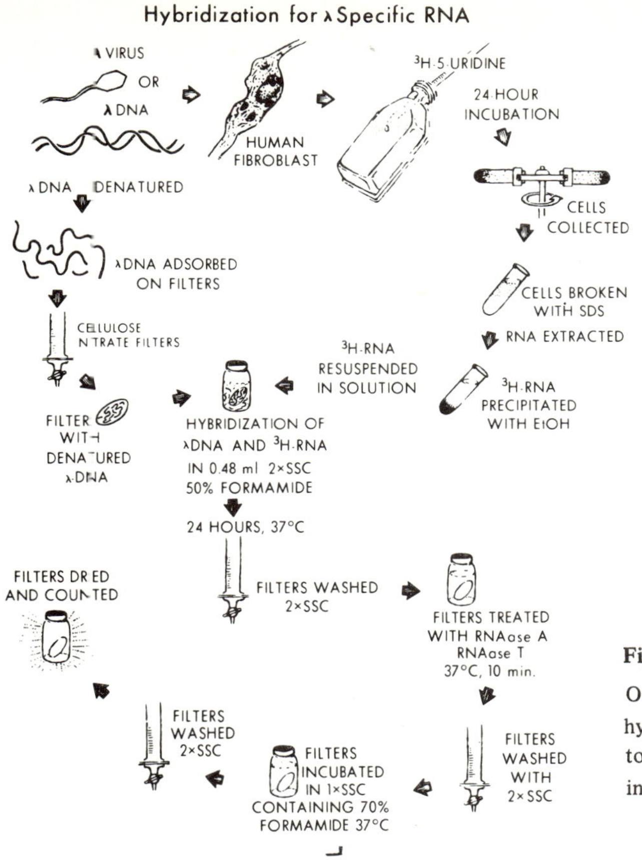

Fig. 2.

Outline of DNA-RNA hybridization technique used to detect lambda-specific RNA in human fibroblasts.

Virus was purified by precipitation in polyethylene glycol and banding in CsCl [23]. The purified virus was dialyzed against .015 M Tris-HC1, pH 7.4, containing .01 M $MgSO_4$. These virus stocks were diluted into Earle's balanced salt medium without bicarbonate (Earle's BSS) prior to use. Lambda DNA was prepared by phenol extractions and dialysis in .01 × SSC is .15 M NaC1 and .015 M Na citrate, pH 7.0). Lambda DNA filters were prepared by passing alkali denatured DNA through cellulose nitrate filters, followed by extensive washing with 2 × SSC. The human fibroblasts, CCL-72, were grown in Eagles Minimal Essential Medium (MEM) with penicillin, streptomycin, and 10 % fetal calf serum. These cells were inoculated with lambda virus in 2 ml of Earle's BSS for 45 minutes. Then, 20 ml of Eagles MEM was added and incubation continued at 37 °C. Twenty-four hours before termination of incubation, the cells were labeled with uridine-5-^{3}H.

The cellular RNA was extracted by a modified version of the method of *Biswal* and *Benyesh-Melnick* [5, 2]. Hybridization was carried out in 2 × SSC and 50 % formamide, followed by extensive washing with 2 × SSC, treatment with RNase A and T_1, and incubation in 1 × SSC containing 70 % formamide for 10 minutes at 37 °C. This was followed by extensive 2 × SSC washes after which the filters were dried and counted in a toluene scintillation system [5].

Translation was studied by utilizing human fibroblasts which do not make active α-D-galactose-1-phosphate uridyl transferase (CCL-72) [19]. These cells were infected with $\lambda pgal_8 C_I 857 (K^+ T^+ E^+)$ which contains the complete galactose operon. Cells were lysed and assayed for enzyme activity before and after infection with this virus (Fig. 3). As a control, some cell cultures were infected with $\lambda pgal_8 C_I 857$ $(K^+ T^- E^+)$, which has a mutation in the transferase gene (generously made for us by *M. E. Gottesman*). The cells were grown and infected as described above. Following infection, the cells were lysed by freezethawing and then centrifuged at 12,500 × G for 2 minutes. The supernatant was assayed for transferase activity as outlined in Fig. 3. The reaction mixture contained 5 μl of D-galactose-^{14}C-1-phosphate (4.5 × 10^{-4}M, New England Nuclear, 227 mc/mmol), 5 μl of H_2O, 5 μl of glycine buffer (1 M, pH 8.7), 10 μl of UDP-glucose (4 × 10^{-4}M), and 10 μl of crude cell lysate. The assay was performed at 37 °C and 5 μl aliquots were removed and applied to the thin layer chromatograph plates PEI (Brinkman). Unlabeled UDP-Gal (2 μl, 10 μmol/ml) was spotted at the origin as a marker. The chromatogram was run for 4 cm in IN formic acid and 17 cm in IN sodium formate, pH 3.6. The position of the UDP-Gal could be determined by the UV absorption of the cold marker (Fig. 4).

Autoradiography of the chromatogram was performed with X-ray film (Kodak NS-2T) (Fig. 4) and the UDP-Gal spots were cut out and placed in a toluene-TLA (Beckman) scintillation system for quantitative counting. All spent medium was tested for bacterial contamination and the cells were periodically monitored for the presence of mycoplasma. All such tests were negative in the experiments reported here.

Results

The experiments performed to illustrate an effect of lambda virus in eukaryotic cells have yielded supportive evidence: when human cells in vitro were examined before and after exposure to lambda virus, only those which were exposed revealed lambda-like particles. These viruslike particles were found in approximately one-third of the cell nuclei. Similar intracellular particles were not found in uninfected cells. By the use of peroxidase-lambda antibody complex followed by staining, these intranuclear particles appeared dense and granular. Intracytoplasmic particles were occasionally noted which were similar to those found in the nucleus.

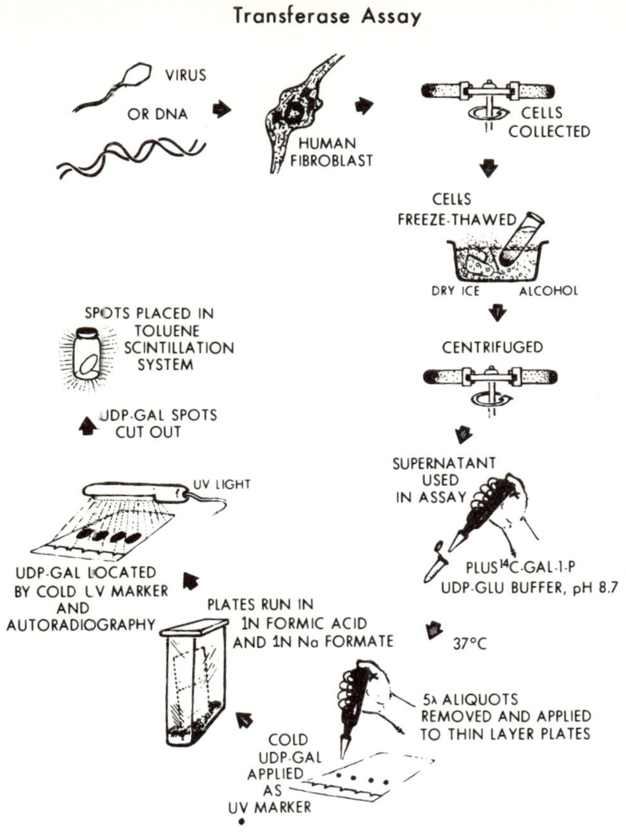

Fig. 3. Outline of assay for α-D-galactose-1-phosphate uridyl transferase activity in human fibroblasts.

Transcription of the virus DNA in the human cells was supported by the discovery of cullular RNA which hybridized with lambda DNA following infection of the cells with lambda virus or its DNA [5]. Within 4.5 days after infection, 0.2 % of the cellular RNA was lambda-specific, while there was no lambda-specific RNA detectable (less than 0.005 %) in the uninfected control flasks. Cultures continued for 41 days after infection continued to produce lambda-specific RNA at about the same level. As a further test for contamination with homology to *E. coli* DNA, the labeled RNA from noninfected and lambda infected cells were hybridized with *E. coli* DNA filters. No significant radioactivity could be detected in either case on these filters. As an additional control, a lambda lysogen (*E. coli* M5107) was added to a tissue culture flask (10^6 bacteria, a higher titre would have made the media cloudy which

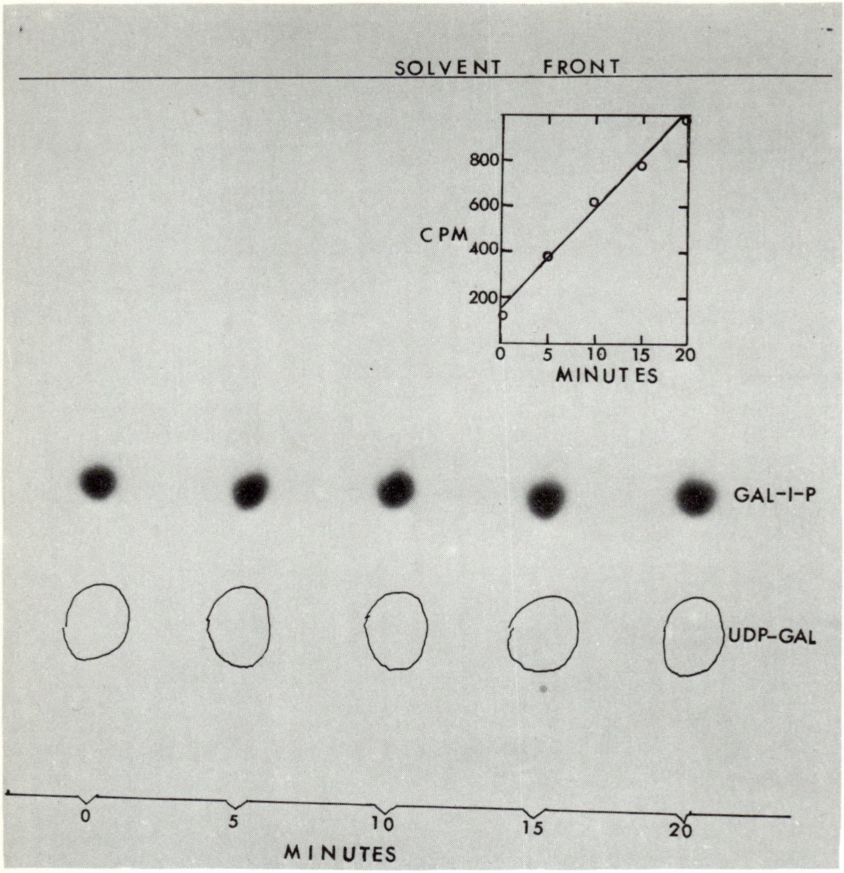

Fig. 4. Autoradiography of chromatogram used in transferase assay. Aliquots 5 μ 1 were applied at 5 minutes intervals. The reactant ^{14}C-Gal-1-P is visible in the autoradiogram while the product UDP-Gal becomes more visible as its concentration increases. The circles represent UDP-Gal markers (nonradioactive) which were detected by UV absorption. The cells used in this assay were normal rat fibroblasts. Insert demonstrates the linearity of the assay as a function of time when the UDP-Gal are cut out and counted in a liquid scintillation system.

would have been visually observed) containing Eagle's MEM without antibiotics and a monolayer of human cells CCL-72. The cells were then labeled for 24 hours with ^{3}H-uridine and treated in the normal manner. No lambda-specific RNA could be found during hybridization of the RNA produced. Presumably the PBS rinses, normally employed prior to harvesting the cells, were sufficient to wash out the contaminating lysogens.

Experiments with $\lambda C_I 857N$ am 7, am 53 which makes almost no lambda-specific RNA in *E. coli* [14, 15, 16, 29] can be substituted for a normal lambda without significantly affecting the hybridizable RNA.

Translation of the λpgal transducing virus was detected by the appearance of transferase enzyme in CCL-72 fibroblasts following infection with either virus or its DNA. Uninfected galactosemic cells gave an activity $\leqslant 0.2$ units (mμmol. of UDP-GAL/60 min/mg of protein). No difference in enzyme activity could be detected between uninfected galactosemic fibroblasts and fibroblasts infected with lambda virus or λpgal ($K^+T^-E^+$). The λpgal ($K^+T^+E^+$) DNA infected cells gave enzyme levels (between 6.0 to 15.5 units) which were generally greater than the λpgal T^+ whole virus infected cells (2.0 to 10.1 units). Infected cell cultures have been monitored through repeated subcultures up to 41 days, and enzyme activity has remained relatively constant at the level achieved four days after infection. Cells were grown in the presence of streptomycin (50 μg/ml) and penicillin (100 units/ml). Mycostatin (50 units/ml) was tested in cultures with no effect on enzyme production or lambda-specific RNA levels. Cyclohexamide (1 mg/ml) which blocks protein synthesis in mammalian cells [12] but not in bacteria [21] prevented enzyme production in cultures infected with λpgal.

Experiments were undertaken to determine whether λpgal T^+ enhances intracellular galactose metabolism. Cells infected with λpgal have survived from 1 to 3 weeks longer than control cells when both were grown in media containing 100 mg percent galactose and dialyzed fetal calf serum.

Discussion

The identification of intracellular lambda viruslike particles in human fibroblasts following infection suggests that these cells can incorporate viral particles of nonmammalian origin, while the detection of lambda-specific RNA and transferase activity are evidence of transcriptional and translational events. There is some indication that the lambda genome is not being segregated out under these conditions since both the lambda-specific RNA and transferase enzyme activity shows no decrease in levels up to 41 days after infection. It has not yet been determined whether the functional viral genome is residing in the nucleus, mitochondria, or cytoplasma.

Conclusions

The results of our experiments suggest that it might be possible to introduce a selected bacteria gene into eukaryotic cells and alter specific metabolic pathways. The development of techniques to transfer genes in cells should prove as powerful a tool toward understanding cellular processes, genetic organization, and control in eukaryotics as it has been in the prokaryotes. How rapidly these techniques will be expanded to treat human genetic abnormalities is completely open to question. However, work on biochemical effects of viruses and foreign DNA should stimulate a more intense investigation of viruses currently used in our vaccination programs.

We generally speak of genetic engineering as a possible new human tool; however, our civilization is in a large part based on the unrecorded, forgotten selective breeding experiments performed by our ancestors. Their emperical and intuitive genetic manipulations have had a revolutionary effect on some of the life forms which exist on the major land masses of this planet. By working directly with the genetic material, the rapidity and extent of change may be increased by several orders of magnitude. Perhaps such techniques will prove useful in combating one of the largest medical problems, starvation. An approach to this problem might involve the use of bacterial genes to allow plants to fix nitrogen directly instead of relying on the symbiosis between a few plants and the nitrogen-fixing bacteria.

Acknowledgements: We are grateful to Dr. *Ruth L. Kirschstein*, Dr. *Giulio L. Cantoni*, and Dr. *Sol. Spiegelman* for their support and encouragement during the course of these experiments. Dr. *Max Gottesman* provided phage and bacteria; his criticisms and helpful advice were substantial contributions to this study. Dr. *Robert S. Krooth* made valuable comments and suggestions for which we are especially thankful.

References

[1] *Avery, O. T., MacLeod, C. M., and McCarty, M.* 1944. Studies on the chemical nature of the substance inducing transformation of pneumococcal types. J. Exp. Med. **79**: 137–158.

[2] *Biswal, N., and Benyesh-Melnick, M.* 1969. Murine sarcoma-leukemia-virus complex in transformed cells. Proc. Nat. Acad. Sci. U. S. **64**: 1372–1379.

[3] *Das, M., and Spiegelman, S.* Personal communication.

[4] *Fox, A. S., and Yoon, S. B.* 1970. DNA-induced transformation in Drosophila: Locus specificity and the establishment of transformed stocks. Proc. Nat. Acad. Sci. U. S. **67**: 1608–15.

[5] *Geier, M. R., and Merril, C. R.* 1972. Lambda phage transcription in human fibroblasts. Virology. **47**: 638–643.

[6] *Gillespie, D., and Spiegelman, S.* 1965. A quantitative assay for DNA-RNA hybrids with DNA immobilized on a membrane. J. Mol. Biol. **12**: 829–42.

[7] *Goldberg, A. R., and Howe, M.* 1969. New mutations in the S cistron of bacteriophage lambda affecting host cell lysis. Virology **38**: 200–202.

[8] *Goldberg, A. R., and Howe, M.* 1969. Erratum. Virology **39**: 152.

[9] *Gorham, L. W., and Waymouth, C.* 1965. Differentiation in vitro of embryonic cartilage and bone in a chemically defined medium. Proc. Soc. Exp. Biol. Med. **119**: 287–90.

[10] *Hershey, A. D., and Dove, W.* 1971. Introduction to lambda In The bacteriophage lambda, ed. *A. D. Hershey.* New York: Cold Spring Harbor Laboratory.

[11] *Hill, M., and Hillova, J.* 1971. Recombinational events between exogenous mouse DNA and newly synthesized DNA strands of chicken cells in culture. Nature **231**: 261–65.

[12] *Horwitz, M. S., and Scharff, M. D.* 1969. The production of antiserum against viral antigens. *In* Fundamental Techniques in Virology, eds. *Habel* and *Salzman*, pp. 253–62. New York: Academic Press.

[13] *Kalckar, H. M., Anderson, E. P., and Isselbacher, K. J.* 1956. Galactosemia, a congenital defect in a nucleotide transferase. Biochim. Biophys. Acta **20**: 262–68.

[14] *Konrad, M.* 1968. Dependence of early lambda bacteriophage RNA synthesis on bacteriophage-directed protein synthesis. Proc. Nat. Acad. Sci. U. S. **59**: 171–78.

[15] *Konrad, M.* 1970. Gene expression in bacteriophage lambda. II. Regulation of RNA synthesis in vivo by the N protein. J. Mol. Biol. **53**: 389–400.

[16] *Kourilsky, P., Marcund, L., Sheldrick, P., Luzzati, D., and Gros, F.* 1968. Studies on the messenger RNA of bacteriophage lambda. I. Various species synthesize early after induction of the prophage. Proc. Nat. Acad. Sci. U. S. **61**: 1013–20.

[17] *Ledoux, L.* 1965. Uptake of DNA by living cells. *In* Progress in Nucleic Acid Research and Molecular Biology, eds. *Davidson* and *Cohn*, Vol. 4, pp. 231–67. New York: Academic Press.

[18] *Ledoux, L. and Huart, R.* 1969. Fate of exogenous bacterial deoxyribonucleic acids in barley seedlings. J. Mol. Biol. **43**: 243–62.

[19] *Merril, C. R., Geier, M. R., and Petricianni, J. C.* 1972. Bacterial virus gene expression in human cells. Nature **233**: 398–400.

[20] *Munyon, W., Kraiselbrud, E., Davis, D., and Mann, J.* 1971. Transfer of thymidine kinase to thymidine kinaseless L cells by infection with ultraviolet-irradiated herpes simplex virus. J. Virol. **7**: 813–20.

[21] *Nakane, P. K., and Pierce, G. B.* 1966. Enzyme labeled antibodies. Preparation and application for localization of antigens. J. Histochem. Cytochem. **14**: 929–31.

[22] *Nakane, P. K., and Pierce, G. B.* 1967. Enzyme labeled antibodies for the light and electron microscopic localization of tissue antigens. J. Cell Biol. **33**: 307–18.

[23] *Nash, H. A., and Merril, C. R.* 1969. DNA of biotin-transducing lambda bacteriophage. J. Mol. Biol. **43**: 357–59.

[24] *Ottolenghi-Nightingale, E.* 1969. Induction of melanin synthesis in albino mouse skin by DNA from pigmented mice. Proc. Nat. Acad. Sci. U. S. **64**: 184–89.

[25] *Qasba, P. K., and Aposhian, V. H.* 1972. DNA and gene therapy: transfer of mouse DNA to human and mouse embryonic cells by polyoma pseudovirions. Proc. Nat. Acad. Sci. U. S. **68**: 2345–49.

[26] *Rogers, S.* 1970. Genetic engineering – a potentially invaluable aid to medicine and mankind. New Sci. **45**: 194–206.

[27] *Schwartz, A. G., Cook, P. R., and Harris, H.* 1971. Correction of a genetic defect in a mammalian cell. Nature **230**: 5–7.

[28] *Shabo, A. L., Petricianni, J. C., and Kirschstein, R. L.* In preparation. Immunologic localization of herpes zoster and semian virus 40 in cell culture with peroxidase labeled antibody.

[29] *Skalka, A., Butler, B., and Echols, H.* 1967. Genetic control of transcription during development of phage lambda. Proc. Nat. Acad. Sci. U. S. **58**: 576–83.

[30] *Szybalska, E. H., and Szybalski, W.* 1962. Genetics of human cell lines, IV. DNA-mediated heritable transformation of a biochemical trait. Proc. Nat. Acad. Sci. U. S. **48**: 2026–34.

[31] *Tedesco, T. A., and Mellman, W. J.* 1971. Galactosemia: evidence for a structural gene mutation. Science **172**: 727–28.

[32] *Zinder, N. D., and Lederberg, J.* 1952. Genetic exchange in salmonella. J. Bacteriology **64**: 679–699.

Discussion

Konings: Did you check your phages for possible contamination with transferase enzyme?

Merril: The experiments were performed with both DNA and whole virus. The DNA was treated with pronase and multiple phenol extractions to assure that no contaminating enzyme was present. We also assayed all phages used in these experiments for transferase activity. Occasionally a preparation of phage was contaminated with enzyme, but most of the activity could be eliminated by rebanding in cesium chloride.

Konings: Did you check on synthesis of phage-specific proteins?

Merril: We have just begun to check for these proteins using the electron microscope and antibody staining methods. We have also begun some complementation in vitro by looking for virus tail-specific proteins.

Migeon: Could you tell us something about the transferase activity in normal human skin fibroblasts, and how the activity you were able to induce in mutant cells compares to the wild-type?

Merril: Normal fibroblasts which we have investigated produced between 11 to 36 mμ mol of UDP-Gal/60 min/mg of protein. The number of units we obtained in the various experiments on galactosemic fibroblasts varied from 6 to 15 for λ pGal DNA infection and 2 to 10 units for whole virus infection.

Klingmüller: Obviously your phages get into the cells, and you have shown electron micrographs documenting phages even in the cell nucleus. How do you think they get there? Are they engulfed by some kind of phagocytosis or what?

Merril: The electron microscope studies which were reported were performed one hour after infection. A micropinocytotic mechanism for the entry of virions into cells has been shown for vaccinia virus, adenovirus, and others as well as for structures as large as chloroplasts. A similar process would seem to be the most likely mechanism by which λ enters human cells. As to how these particles arrived in the nucleus, we have no idea at present.

Klingmüller: Is the λ DNA integrated into the recipient cell genome? And if you don't know yet, would you think that by cooperation with Dr. *Ruddle* you could perhaps answer this question, using his nice system of man-mouse cell hybridization?

Merril: I would hope so, but this was also the reason our group was anxious to extend the work to the fruit fly. In very preliminary experiments, it appears that the fly does not make the transferase enzyme. However, one difficulty is that unless the flies are grown under sterile conditions, bacterial contamination will give some enzyme activity. In one experiment, we infected the fruit fly eggs, using the techniques developed by *Fox,* with λ pGal DNA and obtained enzyme activity in the adult flies with no evidence of contamination. If we can reproduce this experiment, it might be possible to utilize the genetic techniques developed for the fruit fly to determine whether the virus can integrate into a chromsome.

Bodmer: I think it is very important in these transfer experiments to show that the gene product you believe you are making corresponds to the DNA or other genetic information that has been added. We did hear evidence from Dr. *Munyon* that the gene product was from the herpes virus. I wondered, Dr. *Merril,* what your evidence on that was because I gather one can do electrophoreisis for the transferase enzyme. The other question, which I think is important in your work, is the quantitation. You say you add 6 x 10^4 to 10^5 PFU per cell. What proportion of that really gets in? I suppose one can calculate that, if most of it gets in, you have got about a cell's worth of DNA inside, which is really quite a lot.

Merril: I really don't have complete answers for either of those questions. We are working on the enzyme now. We have been trying to establish an SV-40 transformed line of galactosemic cells, so that we can produce more cell material for this type of study. As far as how much DNA gets into the cells, I am not really sure.

Hofschneider: Do you assume that not the λ DNA itself, but an induction of a cellular gene causes the effect, or what else is the background of your question?

Bodmer: Well, we have heard from Dr. *Siniscalco* that there is a question of possible induction of strange types of revertants, and I think one has to keep one's options open as to what the nature of the new expression is.

Merril: One control that may help clarify this to some degree was the use of λpGal with a point mutation in the transferase gene. This control would tend to rule out a general stimulation or reversion by DNA or virus.

Hennig: What is the evidence that this DNA is integrated into chromosomes? Especially in the light of *Gurdon*'s work, one could consider that it stays in the cytoplasm.

Merril: There is no evidence at this point that it is integrated into the chromosome. In fact, I don't really even know where in the cell the virus is functioning. The fact that you have electron-microscope pictures of the virus in the nucleus does not indicate that that is its site of action at all.

Glick: Dr. *Siniscalco,* the evidence for increased reversions induced by DNA was obtained with Hela DNA, which was positive for the enzyme. Of course, you showed that there was a reversion to mouse enzyme, not human. My question is, have you ever used other kinds of DNA, perhaps from phage or some other virus, to demonstrate that the effect of DNA really is nonspecific?

Siniscalco: Well, I could not go very slowly through the slides. One of them showed that in some experiments the added DNA had been derived from the same recipient line. As for the quantitative effect of DNA on the frequency of "revertants", though the highest values were found in the experiments with added DNA (of any kind), this was by no means a clear-cut and constant effect.

May I ask a question of Dr. *Merril,* now? I was particularly puzzled by the fact that in the experiment where he added phage-DNA, the activity of uridylic transferase found in the alleged "transformants" is almost twice as high as that of the control fibroblastic line WI 38. Does this mean that every cell of the transformed line produces the enzyme, may be in greater amounts than the wild-type human cell?

I have been wondering whether this remarkable result may not be the consequence of a kind of massive hybridization between man and phage. If I am correct, the added genetic material was a lot: about 6 x 10^4 viral genomes per mammalian cell. There has been a report of hybridization between man and mosquitoes, why not between man and phages?

Merril: Enzyme activity is not necessarily the same for both the mammalian and bacterial enzymes, one or the other may have a much greater efficiency. It might just be that a few cells were transformed but, because of the nature of the enzyme, more substrate is converted per minute.

Yarmolinsky: Dr. *Merril* very briefly mentioned that λ NN (a mutant in which λ replication is largely inhibited by two amber mutations) was capable of giving as great an effect in an animal cell as the wild-type lambda. This result might or might not be interpreted as being due to suppression of an amber mutation. I know that Dr. *Merril* is aware of the possible significance of this, but perhaps others here have not caught the hint that one might use amber mutants in transducing λ to find out whether animal cell lines exist which have or do not have suppressors. If a cell line is found that does not have a suppressor, then there exists a phage that carries a suppressor, so that one can be introduced into an animal cell. Thereafter it should be possible to look for amber mutants in animal viruses, for example, which would be very valuable to have.

Konings: Another very important conclusion which can be drawn from Dr. *Merril*'s work is that bacterial messengers can be translated in eukaryotic systems. There are also some indications that the reverse might be the case. However, we have learned from Dr. *Woodland*'s paper that there is no indication that bacterial messengers (phage f2 RNA) can be translated in the oocytes. I would like to ask Dr. *Merril* to comment on these differences in observations.

Merril: I don't really know why no protein was produced in this system. Perhaps the RNA phages which are single-stranded RNA cannot be translated into protein in their system. I really don't know the answer.

Hofschneider: As we have learned now, phage DNA as well as complete chromosomes can be used to genetically alter cells. Now a method has been described which uses pulverized chromosomes. Is there anybody here in the audience who could comment on this method and what has been obtained by it?

Johnson: As far as pulverization of chromosomes is concerned, I think it's a conceivable method for integrating small quantities of DNA into new genomes. So far, the only example I can think of in the literature is that of *Schwatz, Cook,* and *Harris,* 1971 (Nature, New Biol. **230,** 5) where a little piece of chick DNA was presumably stably integrated into the A9 line. I think it may be possible to direct chromosome damage, particularly of the S-phase-type, where you have fragments, and possibly to incorporate small pieces in a stable way. But the problem is actually finding those pieces once you've put them in, proving that they are not episomal and that they are integrated into the chromosomes of the host. Now, to Dr. *Munyon,* and I think it might be quite important. In most tissue-culture cell lines, the thymidine kinase gene appears to be most active just after the start of DNA synthesis. The question thus arises of whether the thymidine kinase gene, which has been inserted into the cell, is expressed at one particular time in the cycle? If this were so, it would not only indicate that the gene was incorporated, but that it was integrated in a stable way, in a normal way, into the metabolism of the cell.

Munyon: That would be an extremely interesting series of studies to do. We have not got into them yet, but just exactly how the thymidine kinase is able to functionally integrate into the metabolism of the cell is a mystery.

Transfer of Thymidine Kinase to Cells Lacking This Enzyme by Infection with Ultraviolet Irradiated Herpes Simplex Virus

William Munyon, Judith Mann, Daniel Davis, and Edmundo Kraiselburd

Roswell Park Memorial Institute Buffalo, New York 14203

Summary: L cells which lack the ability to synthesize thymidine kinase (Ltk minus cells) are able to acquire this function due to infection with ultraviolet irradiated Herpes Simplex Virus (HSV). The thymidine kinase (TK) made by these cells is similar to the TK activity induced by a lytic HSV infection of Ltk minus cells and different from the TK activity in Ltk plus cells. These data support the hypothesis that HSV is able to transfer its TK gene to Ltk minus cells (*Munyon* et al. [4]).

Introduction

The purpose of our experiments was to see if we could transfer a particular gene from an animal virus to an animal cell. We selected the gene for thymidine kinase (TK) for these studies. The recipient cell line we used was an L-cell which was lacking TK activity (*Kit* et al. [2]). This cell line will be called Ltk minus. The phenotypic description of this type of cell is that it does not take up thymidine, it is resistant to BUDR, and it dies in HAT medium (*Littlefield* [3]). The virus we used is Herpes Simplex Virus (HSV) which is a double-stranded DNA-containing virus that synthesizes its DNA in the nucleus of the infected animal cell. In lytic infections of cells, HSV induces a new TK activity. Evidence from other workers indicates that the Herpes virus probably contains a structural gene for thymidine kinase (*Hamada* et al. [1]).

Materials and methods

The experiments we carried out were as follows. We took sparse monolayers of Ltk minus cells and infected them with ultraviolet irradiated HSV. We used UV irradiated virus because HSV in tissue culture is a strongly cytopathic virus, although in animals

Manuscript received: 24 January 1972

there is a great deal of reason to believe that it can form nonlytic relationships with cells. The cells were exposed to high multiplicities of this virus, (10^4 virus particles per cell) for one hour. After this the virus inoculum was removed. Less than 1 % of the the inoculated virus is adsorbed to the cells under these conditions.

After the virus particles were removed, a nonselective medium, Eagles medium containing 5 % calf serum, was placed on the cells for approximately 24 hours. At the end of this time, this medium was removed and replaced by a selective medium (TK plus selective medium). This medium has the same selective effect as HAT medium on Ltk minus cells (*Littlefield* [3]). It contains methotrexate (6.0×10^{-7} M), thymidine (1.6×10^{-5} M), adenosine and guanosine (each at 5×10^{-5} M), and glycine (10^{-4} M). These cultures were then incubated for 17 days so that colonies could grow to visable size.

Results

In the absence of infection, no colonies are observed to grow. The frequency of spontaneous reversion of Ltk minus cells to an Ltk plus phenotype is less than 2×10^{-8}.

If cells are infected with the UV-irradiated Herpes virus, colonies do form. For convenience, we divide them into two classes although they show a continuous size distribution.

There are large colonies which have a diameter of between 2 and 4 mm; they contain usually 300 to 1000 cells. These clones, if picked, will grow continuously in the TK plus selective medium. Smaller colonies may continue to grow continuously in culture, but very often they die in TK plus selective medium. These cells we believe, abortively acquire TK activity.

The frequency of total colony formation in this system is as high as 10^{-3}, that is, of the cells planted out, 10^{-3} may form colonies, Of colonies which are capable of permanently acquiring a new thymidine kinase gene, the frequency is 10^{-5} to 10^{-6}.

Factors effecting the number of colonies

The number of colonies is a function of the amount of UV irradiation given to the virus particles. No colonies were observed using unirradiated virus, even if diluted to the point where no cytopathic effect was observed. The greatest number of colonies was observed when virus preparations were irradiated with 23 ergs per second per mm^2 for 9 minutes. This amount of irradiation usually inactivated all virus infectivity. Greater amounts of ultraviolet light decreased the number of colonies.

Another factor which influences the number of colonies is the relative amount of virus applied to the cells. The number of colonies is proportional to the concentration of virus in the inoculum.

A third factor influencing the number of colonies is the duration of the exposure of the infected cells to nonselective medium before this is replaced with TK plus selective medium (see Fig. 1).

Hour of incubation in nonselective medium*)	Average number of colonies per plate	
	Colonies of all sizes	Colonies of 2 mm dia. or greater
0	0	0
2	0	0
4	0	0
8	15	0
10	13	0
13	12	0
22	10	. 20
31	14	. 17

*) nonselective medium is Eagles MEM with 5 % calf serum.

Fig. 1. Effect of the time of incubation in nonselective medium on the number of HSV transformed colonies

If TK plus selective medium is added anytime during the first 4 hours after infection, no colonies formed. The optimum interval for incubation in nonselective medium is from 22 to 31 hours. This data suggests that in order to permit cell transformation, the TK gene must be expressed before the TK plus selective medium is added.

Characterization of cell lines derived from colonies

Cell lines derived from HSV transformed clones differ from the parental Ltk minus cells in several ways. They are able to grow in TK plus selective medium. They are inhibited by BUDR, they take up thymidine and incorporate it into cell DNA, and they contain from 7 to 12 times as much TK activity (depending on the clone being examined) as Ltk minus cells. Cells derived from these clones have therefore been phenotypically changed to Ltk plus cells.

The cells from these colonies do not spontaneously produce virus. They are superinfectable by Herpes virus, and have an efficiency of reproduction of the virus which is approximately the same as the parental Ltk minus cell. The virus yield is approximately 2 to 4 plaque forming units per cell. This sounds like a small yield, but the plaquing system we have for Herpes has an efficiency of 1000 particles per plaque forming unit.

The frequency of reversion of the cells which acquire the TK activity due to virus infection is 10^{-4}. It has not been possible to carry out these experiments with greater numbers of cells because BUDR will not suppress the growth of Ltk plus cells when they are planted out at concentrations of greater than 10^{-4} cells per petri dish.

We have studied the electrophoretic migration of TK activity in polyacrylamide gels from HSV transformed Ltk plus cloned lines, Ltk minus cells, HSV infected Ltk minus cells, and L cells that contain the mouse TK activity (see Fig. 2) (*Munyon* et al. [5]).

Source of TK Activity	Rf of TK*) Activity
L tk minus cells	–
L tk plus cells	. 27
L tk minus cells infected with HSV (8 hours after infection)	. 41
HSV transformed cells**)	. 41

Fig. 2. Migration of thymidine kinase activity in polyacrylamide gel

*) All cell lines also contain a small amount of TK activity that has an Rf of 0.70 to 0.80.

**) HSV transformed cells are cells derived from colonies which form due to infection with ultraviolet irradiated HSV.

The major peak of TK activity in Ltk plus cells has an Rf of o.27. This peak is absent in Ltk minus cells. The major peak of TK activity in HSV infected Ltk minus cells had an Rf of 0.41. The major peak of TK activity in HSV transformed cells was also at 0.41. This data suggested that the TK activity in the HSV transformed cells is derived from a gene carried by HSV. The TK activity in the HSV transformed cells is definitely different from the TK activity in unselected Ltk plus cells.

Acknowledgements: The work reported in this paper was carried out with the aid of grants from the John A. Hartford Foundation and from the United States Public Health Service grant # CA-07745-08.

References

[1] *Hamada, C., Kamiga, T., and Kaplan, A.* 1966. Serological analysis of some enzymes present in pseudorabies virus-infected and noninfected cells. Virology **28:** 271–82.

[2] *Kit, S., Dubbs, D., Piekarski, L., and Hsu, T.* 1963. Deletion of thymidine kinase activity from L cells resistant to bromodeoxyuridine. Expl. Cell. Res. **31:** 297–312.

[3] *Littlefield, J.* 1966. The periodic synthesis of thymidine kinase in mouse fibroblasts. Biochem. Biophys. Acta **114:** 398–403.

[4] *Munyon, W., Kraiselburd, E., Davis, D.,* and *Mann, J.* 1971. Transfer of thymidine kinase to thymidine kinaseless L cells by infection with ultraviolet-irradiated herpes simplex virus. J. Virology **7:** 813–820.

[5] *Munyon, W., Buchsbaum, R., Paoletti, E., Mann, J., Kraiselburd, E.* and *Davis, D.* 1972. Electrophoresis of thymidine kinase activity synthesized by cells transformed by herpes simplex virus. Virology (in press).

Fate of Exogenous DNA in Mammals and Plants

Lucien Ledoux, John Brown*, Pol Charles, Raoul Huart, Michel Jacobs**, Jacques Remy and Claude Watters***

* Present address: Victorian Wheat Research Institute, Horsham, Vic 3400, Australia

** Laboratorium voor Plant Genetica, Vrije Universiteit te Brussels, Belgium

*** On a NATO fellowship
Cell Biochemistry, Radiobiologie, Centre d'Etudes de l'Energie nucléaire CEN-SCK, 2400 Mol, Belgium

DNA in Mammals

Many different living systems were used to study the fate of exogenous DNA [5, 6, 8, 10, 14, 18, 19, 21, 22, 23, 27]. In mammals, a constant finding is the partial or complete destruction of the exogenous DNA. In many cases, however, it was possible to show that a significant amount of DNA escaped the complete destruction and penetrated the living cells. This was evidenced by analyses in CsCl gradients [19] where both the density and the molecular weight of the labelled macromolecules are determined.

It was shown elsewhere that after an intravenous [2, 3, 15, 16, 17, 20] or intraperitoneal [4] injection of 60–100 μg labelled heteropycnic DNA in a mouse, radioactive molecules with the density of the donor DNA were recovered from different organs and tissues of the treated animals.

Essentially, three types of phenomenons were observed: in certain tissues (liver, kidney, muscle, uterus, lung and intestine), the DNA was merely destroyed and reutilized for the *de novo* endogenous DNA synthesis. In others (vagina, and embryos), the exogenous DNA was found as such in the organs, where it resisted hydrolysis for at least 4–5 days. Finally in other organs (ovaries, and spontaneous mammary carcinoma), the recovered labelled DNA molecules banded at a density intermediate between that of the recipient and that of the donor DNA.

Manuscript received: 8 February 1972

In these cases, however, it was not possible to further study the nature of such intermediate molecules (by sonication nor denaturation) because of their low radioactivity. Indeed, in CsCl gradient analysis, it is absolutely necessary to keep the DNA content under a level where artifactual results could be obtained.

In order to improve these factors, we turned toward an infusion technique developed by *C. Watters* and *P. M. Gullin* [29] for the rat (Fig. 1). Quantities of 100–150 μg of DNA per rat were infused at a rate of a few μg/hour. In that case, results such as those presented in Fig. 2 were obtained in a fixed angle rotor following the technique of *Flamm* et al. [7].

It can be seen that label recovery is satisfactory and that, in the absence of a reference DNA, a bimodal distribution can be obtained with the DNA extracted from different organs. A single peak is observable with the kidney.

Figure 3 shows how such DNAs can be analysed. Figure 3a indicates that the heavy component of DNA extracted from bone marrow bands at the density of the heavy *E. coli* DNA used (DNA $^2H^3H$: d = 1.730 g/cm^3). "Cold" micrococcus lysodeikticus (density = 1.731 g/cm^3) was added as a density reference. This is further confirmed by the recentrifugation in a swinging bucket rotor of the heavy component alone (Fig. 3c).

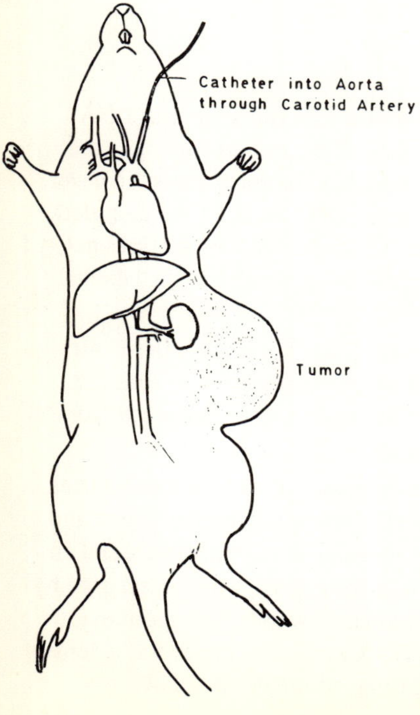

Fig. 1
Carotid cannulation

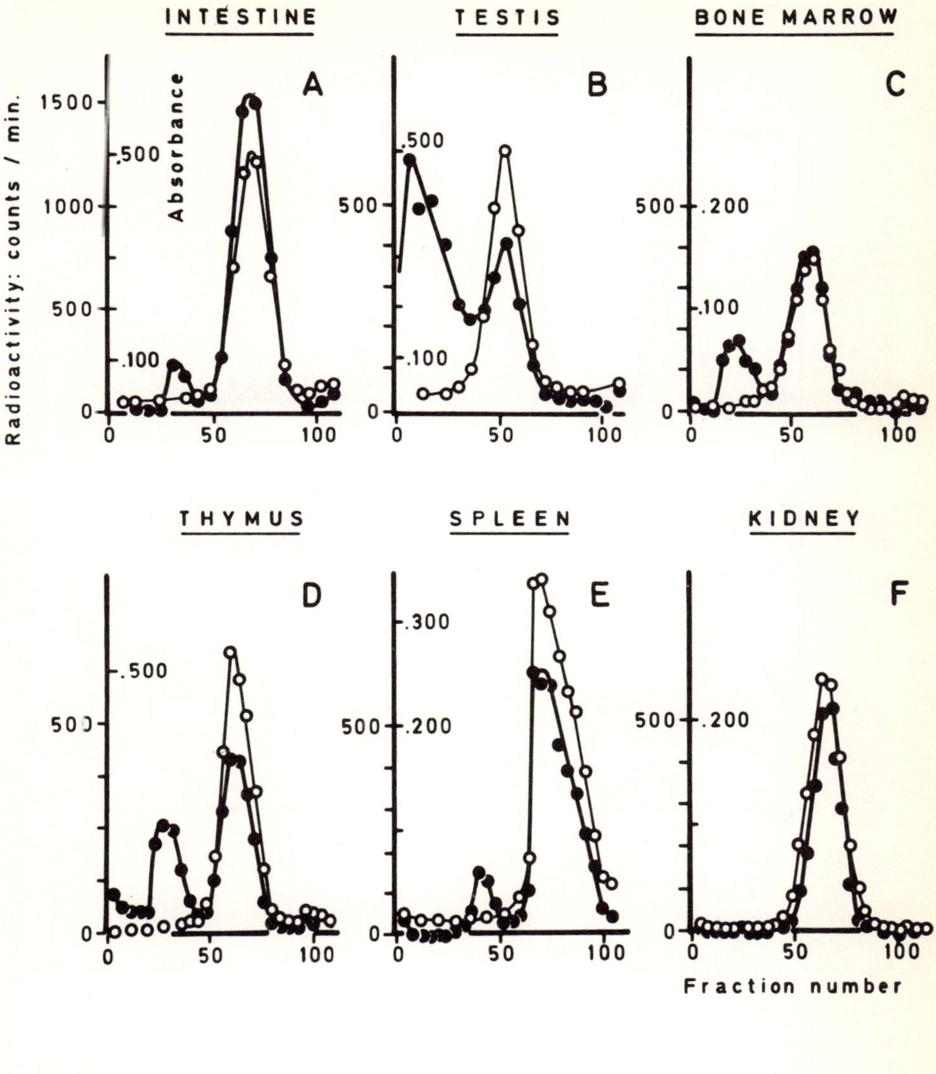

Fig. 2. CsCl density gradient centrifugation (fixed angle rotor) of DNAs prepared from different organs of rats infused with [$^2H^3H$]*E. coli* DNA.

154 μg of [$^2H^3H$]*E. coli* DNA were infused at a rate of about 4 μg/hour during 36 hours, and the animals were killed 48 hours after the arrest of the infusion.

○—○—○: U.V. absorption (= endogenous DNA)

●—●—●: radioactivity

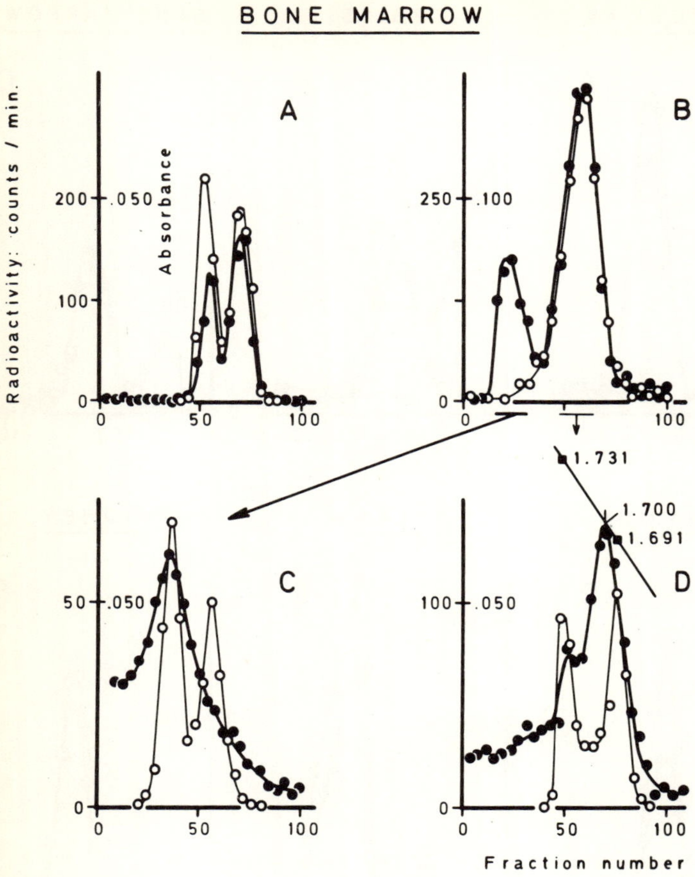

Fig. 3. CsCl density gradient centrifugation of the bone marrow DNA of animals infused with *E. coli* [$^2H^3H$] DNA.

A: in a swinging bucket rotor, with DNA from *M. lysodeikticus* as density marker (d = 1.731 g/cm^3

B: in a fixed angle rotor without density marker

C: reband of the heavy radioactive peak from B in a swinging bucket rotor, with two density markers:

- DNA from *M. lysodeikticus* d = 1.731 g/cm^3
- DNA from ascites tumor cells d = 1.700 g/cm^3

D: reband in a swinging bucket rotor of the light endogenous DNA peak after sonication, with two density markers:

- DNA from *M. lysodeikticus* d = 1.731 g/cm^3
- DNA from *Clostridium perfringens* d = 1.691 g/cm^3

○—○—○ : U.V. absorption

●—●—● : radioactivity

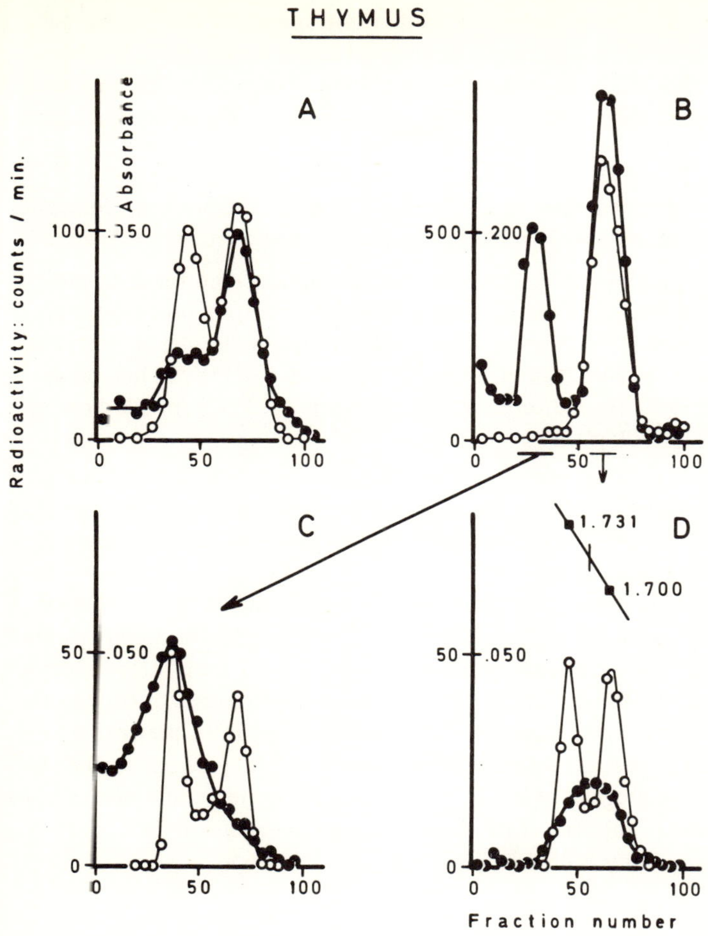

Fig. 4. CsCl density gradient centrifugation of the thymus DNA of animals infused with *E. coli* [^{2}H^{3}H] DNA.

A: in a swinging bucket rotor, with DNA from *M. lysodeikticus* as density marker (d = 1.731 g/cm^3)

B: in a fixed angle rotor without density marker

C: reband of the heavy radioactive peak from B in a swinging bucket rotor, with two density markers:

- DNA from *M. lysodeikticus* d = 1.731 g/cm^3
- DNA from ascites tumor cells d = 1.700 g/cm^3

D: reband in a swinging bucket rotor of the light endogenous DNA peak after sonication, with two density markers:

- DNA from *M. lysodeikticus* d = 1.731 g/cm^3
- DNA from ascites tumor cells d = 1.700 g/cm^3

o—o—o: U.V. absorption

●—●—●: radioactivity

The main peak found in 3b was sonicated (Mullar instrument, 50 kHz at optimal frequency, 60 sec. at 0 °C) and recentrifuged in a swinging bucket rotor (in presence of reference DNA). It presents a significant heterogeneity (Fig. 3d) indicating that, upon depolymerisation, the apparently "endogenous" DNA exhibits heavier components.

Similar results were obtained with thymus (Fig. 4). In that case, however, sonication of the "endogenous" peak leads to a broad distribution, heavier than 1.700 g/cm^3.

Indeed, one or the other type of distribution can be obtained even for the same organ. They seem to depend on the actual result of the sonication process, irreproducible at the molecular level. Anyway, in all cases, heavier molecules are obtained which are never found when the animals are perfused with [3H] thymidine or, as in Fig. 5, with [3H] DNA from *B. subtilis,* whose density (1.703 g/cm^3) is close to that of the rat DNA (1.700 g/cm^3).

These results and others (to be published elsewhere) show that the density of the labelled molecules, recovered after infusion of various radioactive and heteropycnic DNAs, does directly depend on the density of the labelled DNA used which always represents the upper density limit.

These results show integration of foreign DNA to the rat genome. But what is the later fate of this integrated DNA? To answer this question, a rat was first perfused with unlabelled heteropycnic M lysodeikticus DNA (d = 1.731 g/cm^3) for six days. Then it was perfused for 48 hours with [3H]-thymidine. In this case, of course, we were following the newly synthesized molecules produced by DNA-pretreated organs. In the different organs studied, a heterogeneity in CsCl gradient can be observed (Fig. 6). This heterogeneity is real. Let us take, for example, the case of bone marrow. When different portions of the distribution are rebanded in CsCl with reference, the heterogeneity is obvious (Fig. 7).

Fig. 5. CsCl density gradient centrifugation in a fixed angle rotor of the organs or tissues DNAs of animals infused with *B. subtilis* [^{3}H] DNA.
1.12 mg of *B. subtilis* [^{3}H] DNA were infused at a rate of about 8 μg/hour during 140 hours, and the animals were killed 12 hours after the arrest of the infusion.
In I reband in a swinging bucket rotor of the left-hand part of the distribution of the radioactive molecules after ultracentrifugation in CsCl density gradient (fixed angle rotor) of the bone marrow DNA from treated animals (F), with two density markers:
- DNA from *M. lysodeikticus* d = 1.731 g/cm^3
- DNA from ascites tumor cells d = 1.700 g/cm^3

○—○—○ : U.V. absorption

●—●—● : radioactivity

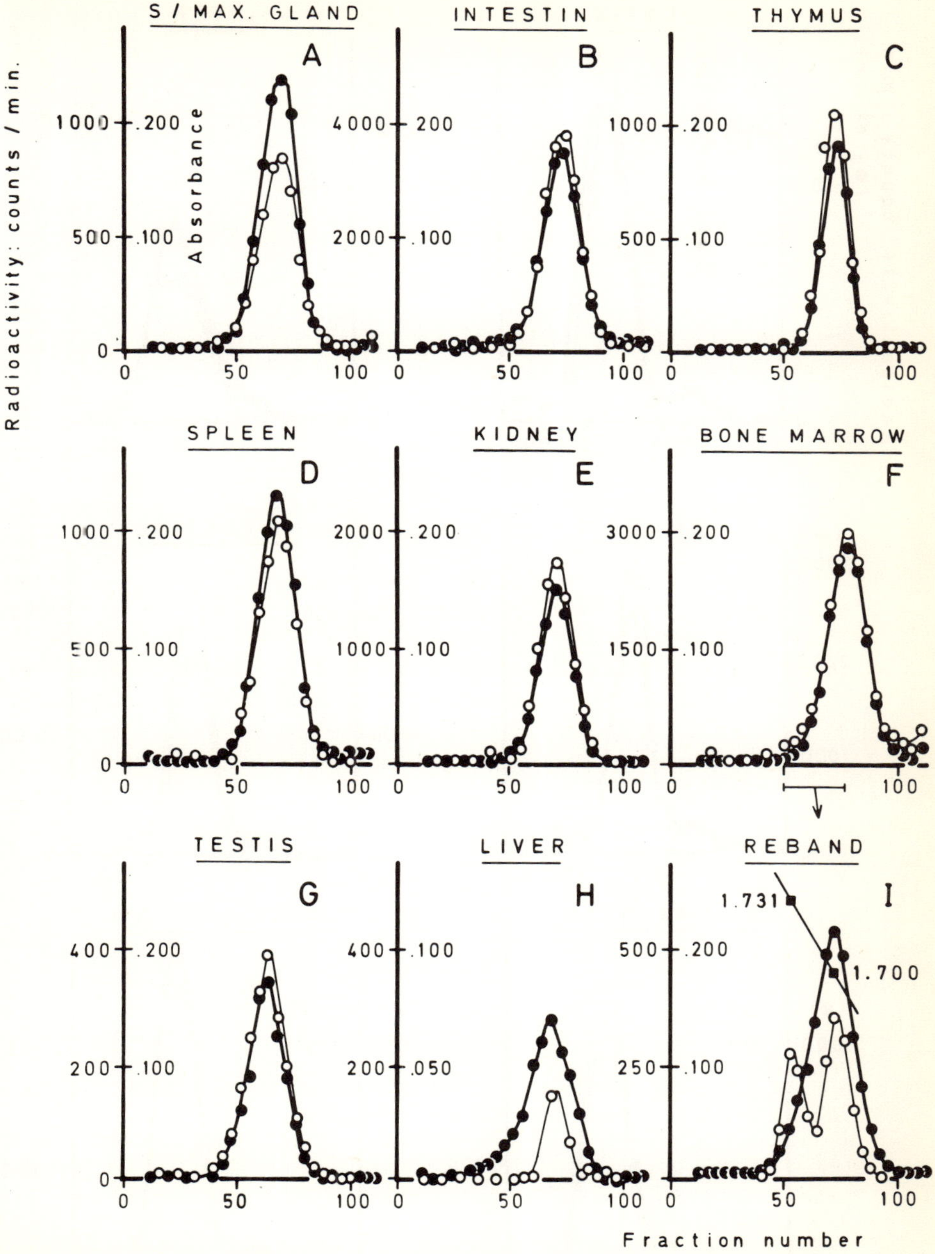
S / MAX. GLAND
INTESTIN
THYMUS
SPLEEN
KIDNEY
BONE MARROW
TESTIS
LIVER
REBAND
A
B
C
D
E
F
G
H
I
Radioactivity: counts / min.
Absorbance
1.731
1.700
Fraction number

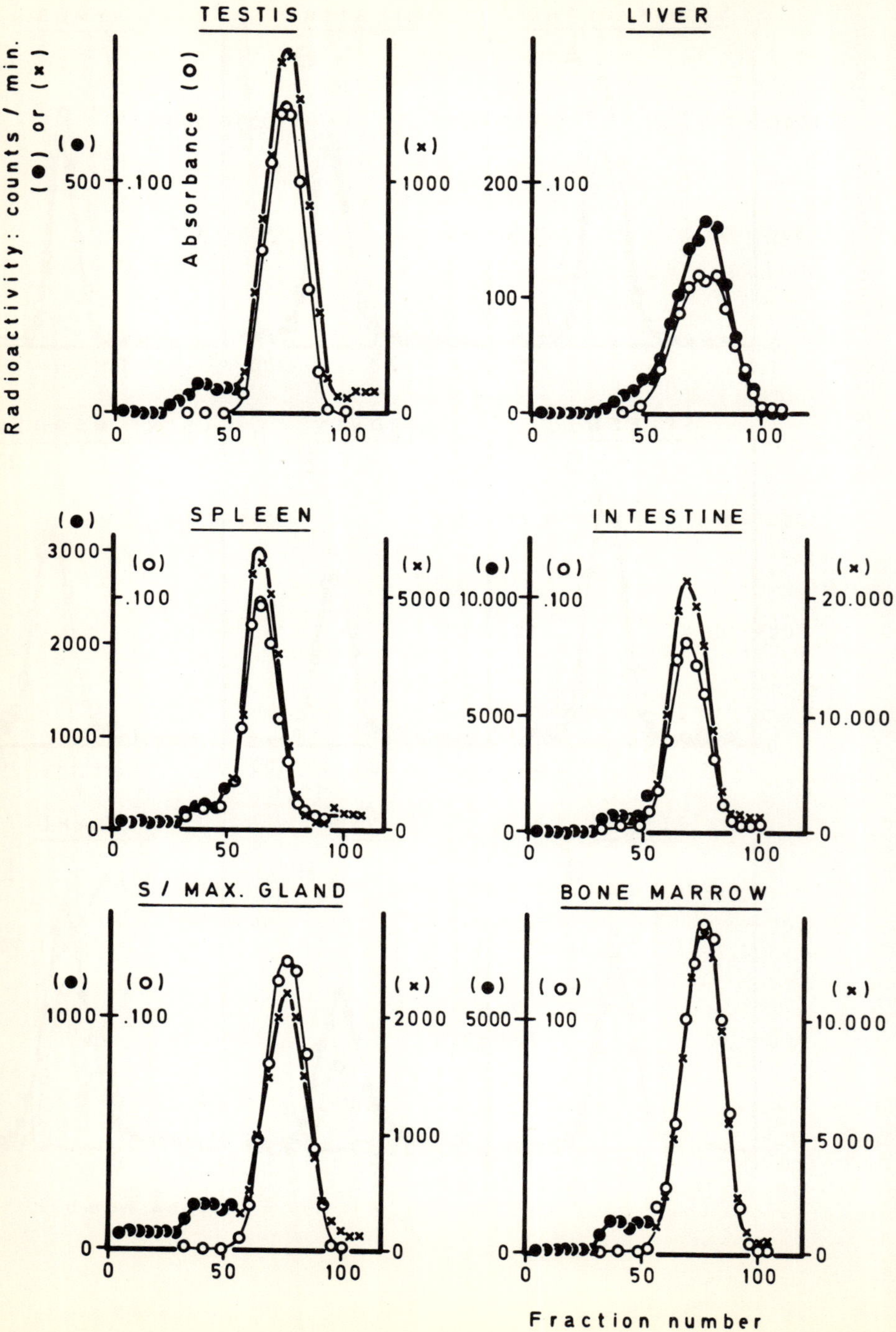
TESTIS
LIVER
SPLEEN
INTESTINE
S / MAX. GLAND
BONE MARROW
Radioactivity: counts / min. (●) or (x)
Absorbance (O)
Fraction number

At this point, a conclusion can be proposed: a DNA slowly infused in the aorta is found, at long time intervals, in different organs. It is partly free (eventually as a consequence of our preparation procedure which leads to fragmentation of the actual endogenous DNA molecules) and partly bound to the endogenous DNA.

Note: That it is bound by covalent linkages and not by artifacts is shown by the fact that molecules, 60 mg heavier, are found in alkaline CsCl. Such denatured DNAs present a pattern corresponding to that found at neutral pH.

On the other hand, the integrated foreign DNA can apparently be subsequently replicated in different organs.

DNA in plants

Recent papers [23, 24] have given some of the biochemical results we obtained with Arabidopsis thaliana, and we shall only briefly comment on them here.

Arabidopsis thaliana is a small crucifer with a life cycle of about 30 days (with continuous illumination). They grow in aseptic conditions on perlite or agar, and in a sterile mineral medium [12, 13]. When seeds are treated with [3H] DNA, such as Streptomyces coelicolor DNA (d = 1.730 g/cm^3), and thereafter cultured, the radioactive molecules found 30 days later in the different organs of the plant are those shown in Fig. 8 (last row); labelled molecules with the density of endogenous and exogenous DNAs plus intermediary molecules can be found. The main surprise being the observation of radioactivity in the flower's stem. This is never observed in plants growing from [3H] thymidine-treated seeds. We obviously are faced here with a problem of redistribution of DNA molecules stored at some site in the plant during early plant growth.

Figure 8 shows that excising the cotyledons at an early stage produced a clear-cut lowering of the radioactivity found in the flowers. This indicates that part of the flower's radioactivity is stored in the cotyledons and redistributed when these cotyledons senesce (at flowering time).

Fig. 6. CsCl density gradient ultracentrifugation (in a fixed angle rotor) of the DNAs extracted from some organs of animals treated

1. with nonlabelled DNA from *M. lysodeikticus* d = 1.731 g/cm^3 during 6 days (solution: 300 μg/ml at a rate of 200 μl/hour – 8.2 mg were infused)
2. and then with tritiated thymidine (10 Ci/mM) during 48 hours (2 mCi were infused).

○—○—○: U.V. absorption

●—●—●: radioactivity

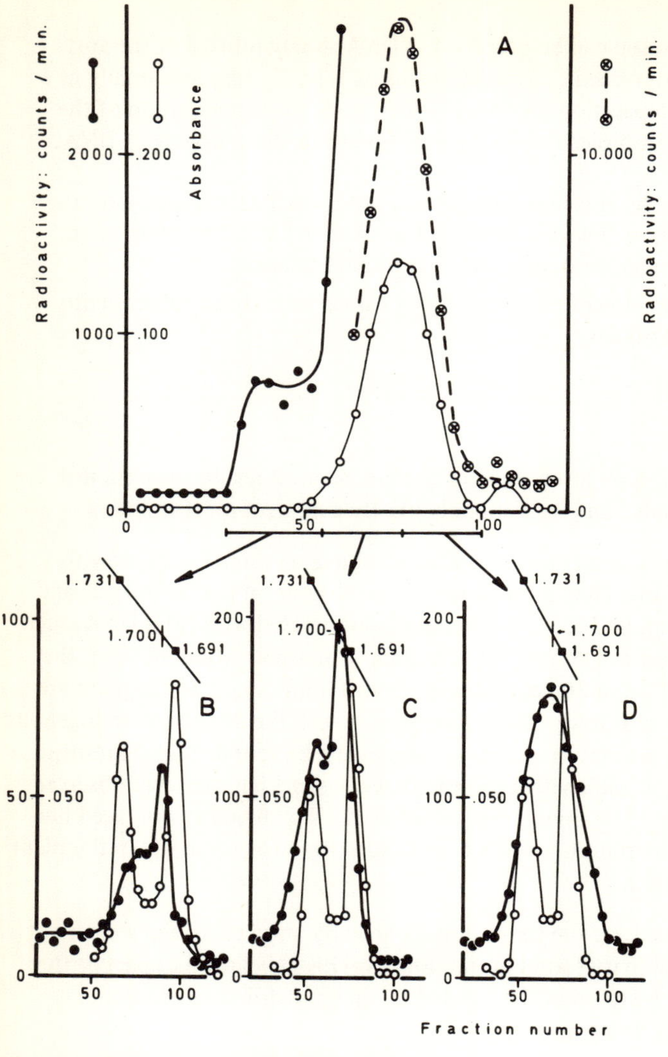

Fig. 7. CsCl density gradient centrifugation

A. in a fixed angle rotor of the bone marrow DNA of animals treated with nonlabelled DNA of *M. lysodeikticus* followed by tritiated thymidine in a swinging bucket rotor

B. of the heavy radioactive peak, with two density markers:
- DNA from *M. lysodeikticus* d = 1.731 g/cm^3
- DNA from *Clostridium perfringens* d = 1.691 g/cm^3

C. of the left hand part of the U.V. distribution, with two density markers:
- DNA from *M. lysodeikticus* d = 1.731 g/cm^3
- DNA from *Clostridium perfringens* d = 1.691 g/cm^3

D. of the right hand part of the U.V. distribution, with two density markers:
- DNA from *M. lysodeikticus* d = 1.731 g/cm^3
- DNA from *Clostridium perfringens* d = 1.691 g/cm^3

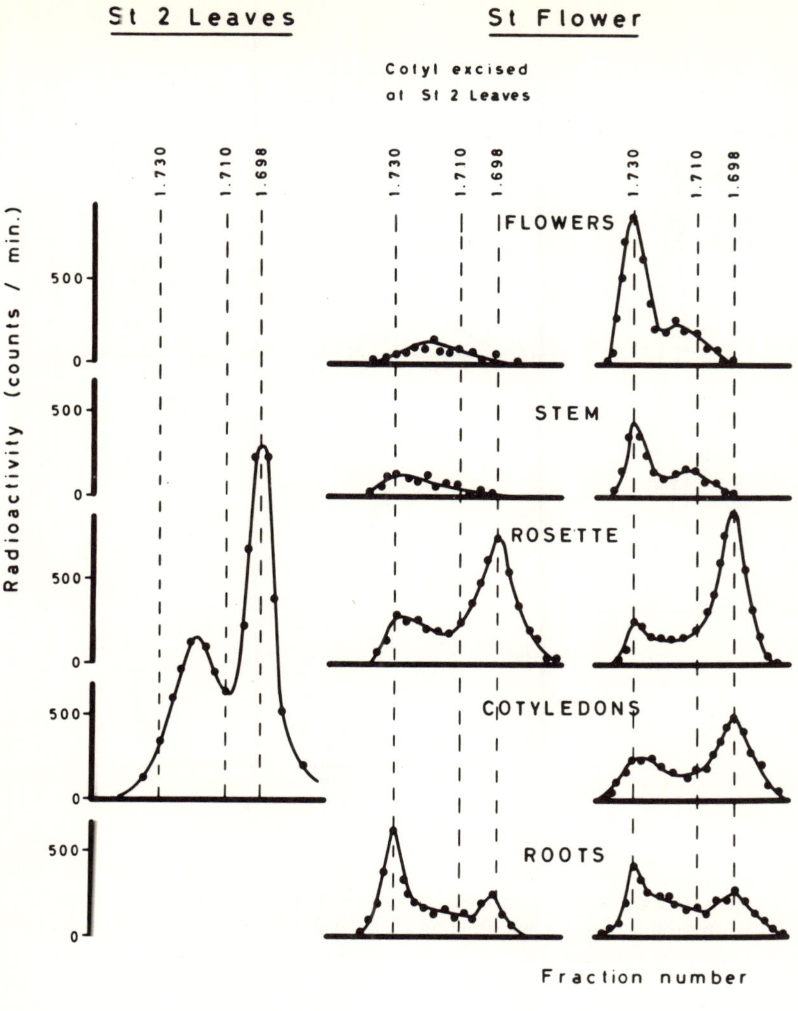

Fig. 8. CsCl gradient centrifugations of DNA isolated from different organs of groups of 80 plants grown from seeds treated for 4 days with S. coelicolor [^{3}H] DNA, density 1.730 g/cm^3.

1. Plants cotyledon excised at the two-leaf stage;
2. no excision. The dotted lines represent the densities of (a) S. coelicolor; (b) *E. coli,* (c) Arabidopsis DNA (1.730, 1.710, and 1.698 g/cm^3, respectively).

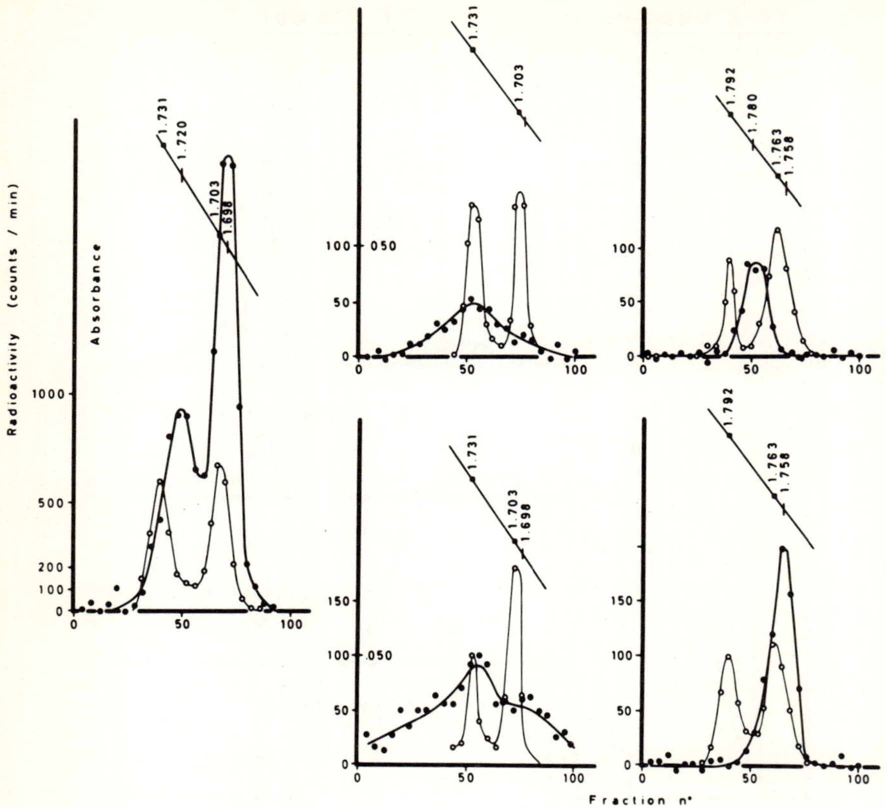

Fig. 9. CsCl gradient centrifugation of DNA isolated from 80 cotyledons excised from plants at the two-leaf stage, grown for seeds treated for 4 days with S. coelicolor DNA, density 1.730 g/cm^3. (1) Starting material from which heavy and light fractions were pooled separately, (2) heavy fraction analysed after sonication, (3) light fraction analysed after sonication, (4) heavy fraction analysed in an alkaline CsCl gradient, (5) light fraction analysed in an alkaline CsCl gradient. Density markers were: (1) *M. lysodeikticus* DNA, 1.731 g/cm^3; (2) *B. subtilis* DNA, 1.703 g/cm^3. Upon denaturation, the density of the markers was (3) 1.787 g/cm^3, and (4) 1.763 g/cm^3. Other densities are shown in g/cm^3.

o : ultraviolet absorption

• : radioactivity

The analysis of the DNA found in the cotyledons excised at the two-leaf stage is shown in Fig. 9.

The heavy and light components are separately pooled and analysed after sonication or OH-denaturation. Results show that depolymerisation of the long double-stranded DNA into small double-stranded pieces (by sonication) allows the expression of the constitutive heterogeneity of the long pieces. Let us note here that sonication brings the M. W. from 10^7 to about 10^5, and that a piece of heavy DNA of 10^5 daltons in a 10^7 daltons fragment would not appreciably change its density and would escape our analysis.

On the contrary, a heavy 10^5 dalton piece among light 10^5 dalton ones would show up in a CsCl gradient.

Figure 10 shows the surprising finding that the foreign DNA also appears to be detectable in the offspring.

Presently, the recovered amounts of radioactive molecules are insufficient to proceed to their analysis with sonication and denaturation. We could only say that labelled molecules with, apparently, intermediate densities depending on that of the DNA used are found in the progeny, while no radioactivity is found in the progeny of [3H] thymidine-treated plants.

It is possible, however, to analyse the DNA found in the offspring of Arabidopsis plants treated with a heavy nonradioactive DNA, such as the M. lysodeikticus DNA ($d = 1.731\ g/cm^3$), by incubating seedlings with [3H] thymidine.

In that case, as shown elsewhere [13, 23], a density heterogeneity is found in the population of the labelled DNA molecules obtained from the F1. Upon treatment of successive generations of plants, with the same M. lysodeikticus DNA, an increased heterogeneity can be observed together with a shift of density of the heavier molecules toward the density of the foreign DNA used.

Figure 11 shows the results of an analysis of the F4 progeny of plants successively treated in F0, F1, F2, and F3, for four days, each time with M. lysodeikticus DNA.

It can be noted that after four such successive treatments, the heavy component of the DNA population has reached a level where it can be easily detected both in absorbance and in radioactivity.

The analysis of the DNA population shows that both the heavy and the light fractions recovered are heterogeneous upon sonication but homogeneous upon denaturation.

Here again, a model where the foreign DNA becomes bound as double-stranded pieces joined by covalent linkages to endogenous double-stranded pieces can be proposed.

With Arabidopsis thaliana, we actually have shifted from the biochemical to the biological aspects of DNA uptake, and to a study of its expression in the recipient organisms and in their progeny.

Different mutants of Arabidopsis thaliana, presenting a mutation in one of the major steps of the biosynthetic pathway of thiamine [28] were used. Let us note that, in Arabidopsis, four loci are responsible for all the encountered mutations.

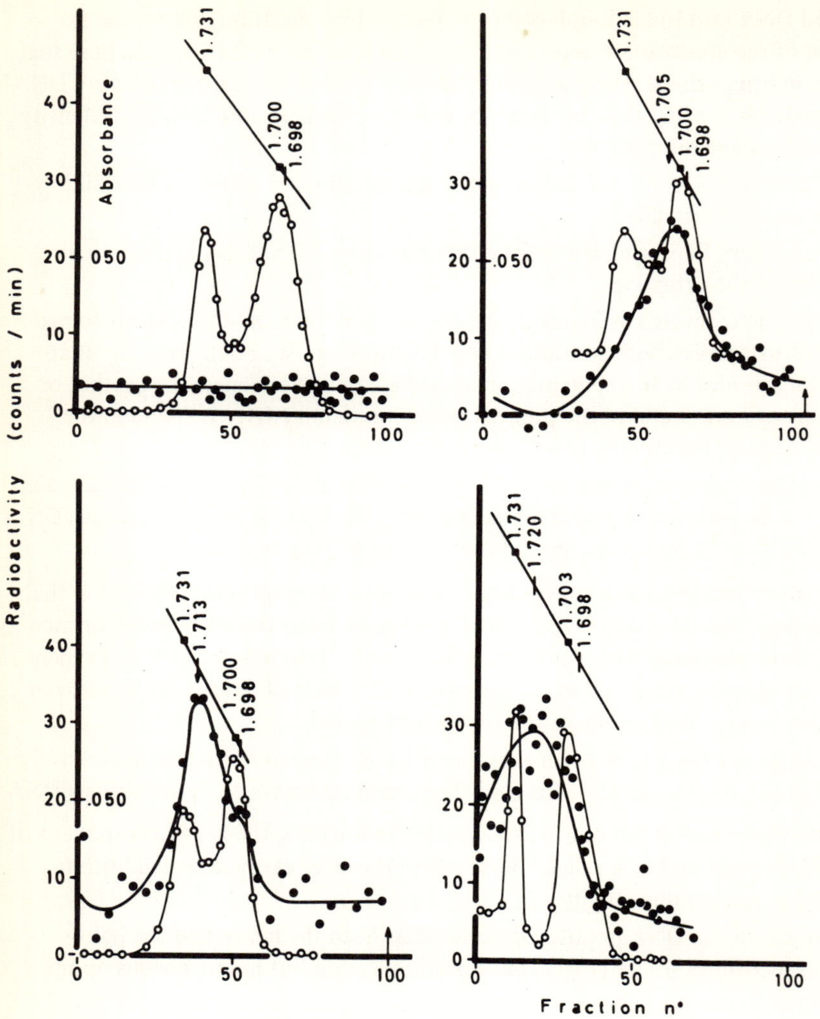

Fig. 10. CsCl gradient centrifugation of DNA isolated from the fruits of plants grown from seeds incubated for 4 days with different radioactive solutions. (1) [^{3}H] thymidine; (2) *E. coli* [^{3}H] DNA, density 1.710 g/cm^3; (3) *E. coli* [^{3}H^{3}H] DNA, density 1.730 g/cm^3; (4) S. coelicolor [^{3}H] DNA, density 1.730 g/cm^3. Density markers were: (1) *M. lysodeikticus* DNA, 1.731 g/cm^3; (2) calf thymus DNA, 1.700 g/cm^3; (3) *B. subtilis* DNA, 1.703 g/cm^3. Other densities are shown in g/cm^3.

○ : ultraviolet absorption

● : radioactivity

The mutants on which we shall comment here are conditional lethal mutants; they die a few days after the development of their cotyledons but survive and develop normally in the presence of the requested metabolite or in the presence of thiamine ($10^{-6} - 10^{-5}$ M). The frequency of the mutagen-induced mutation is, following Redei, about $2.96 \cdot 10^{-4}$ per genome ($0.74 \cdot 10^{-4}$ per locus). The rate of spontaneous mutation for the thiamine deficiency appears to be lower than $5 \cdot 10^{-6}$.

When seeds of the different mutant types were treated with different bacterial DNA, corrections could be obtained with a surprisingly high efficiency (several percent in most cases). The plants grew almost normally and produced seeds.

The progeny of such corrected plants was analysed by self-fertilisation or crossing with either the wild or the original mutant type.

Results, to be described in details elsewhere, can be exemplified with the case of a thiamine mutant (th-), corrected with *E. coli* wild-type DNA.

1. self fertilisation
   No segregation can be obtained, up to the F3 generation (at least). Such a segregation could be expected on the basis of a transformation for one locus of only one of the somatic gametogenic cells or of its chromosome pairs.

   The appearance of a suppressor or of a back mutation could therefore hardly explain these results.
2. Crossing with wild type Arabidopsis
   The first offspring showed no segregation, while the other ones (X2, X3, X4) did segregate (Fig. 12) after self-fertilisation. This is a clear-cut indication that the treatment with the bacterial DNA has produced a correction which can be transmitted to the progeny and which can be removed upon crossing the corrected plant (or its progeny) with a wild-type plant.

Let us note here that the analysis of reciprocal crosses (to be published elsewhere) shows that cytoplasmic heredity is not likely to be involved in the observed correction.

It is of course very tempting to associate our biochemical to our biological data and to say that a DNA-mediated transformation can be obtained through the integration in the host genome followed by replication of pieces of foreign DNA. To our mind, however, the existence of a gene product of the donor type in the corrected plant has to be ascertained before a final conclusion can be reached. A similar restriction also would hold for any study of the biological effects of foreign DNA in mammals.

Acknowledgements: The excellent technical aid of Mrs. *Swinnen-Vrancks* and Mr. *Geldorf* is gratefully acknowledged.

Part of this work was made under the contract D.A.J.A. from the United States Army.

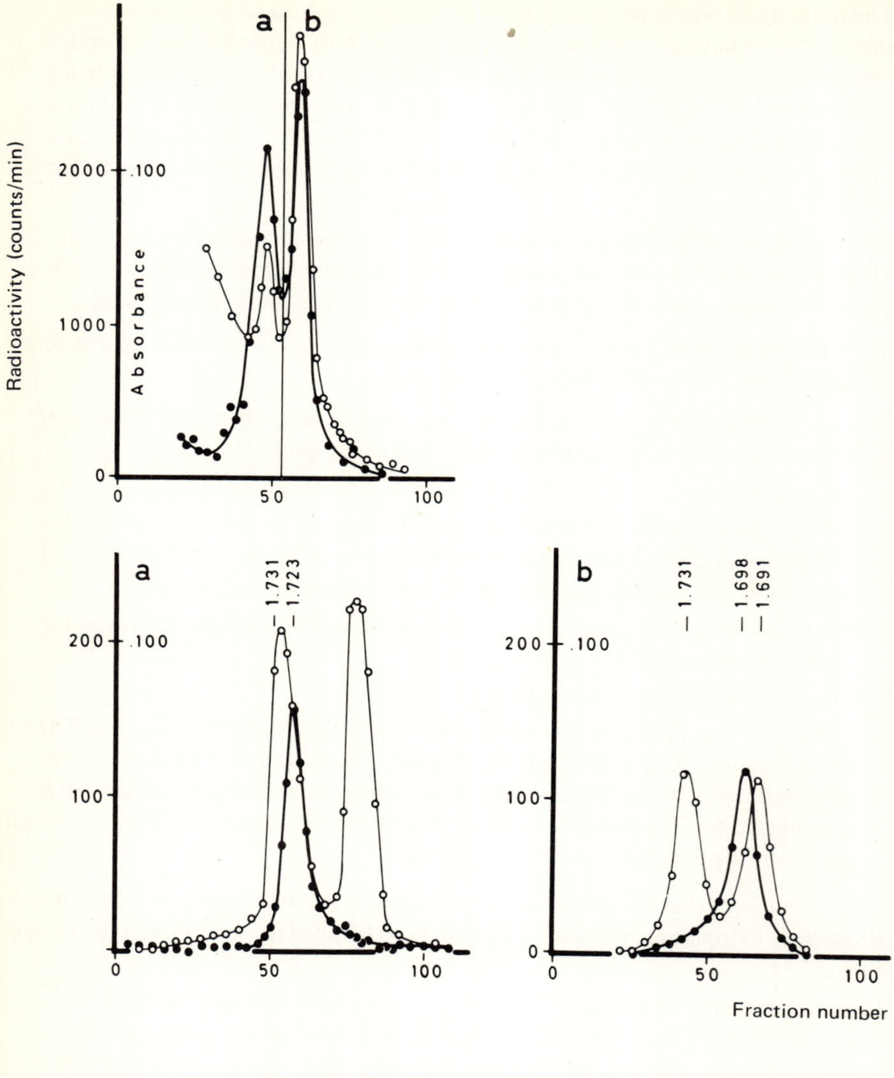

Fig. 11. CsCl gradient centrifugation of DNA isolated from F4ML plants treated with [^{3}H] thymidine.

a, b: pooled fractions, recentrifuged

+ U.S.: neutral CsCl gradients, after sonication

OH^-: alkaline CsCl gradients

Density markers: S. coelicolor DNA: 1.730 g/cm^3, C. perfringens 1.691 g/cm^3.

○ : ultraviolet absorption

● : radioactivity

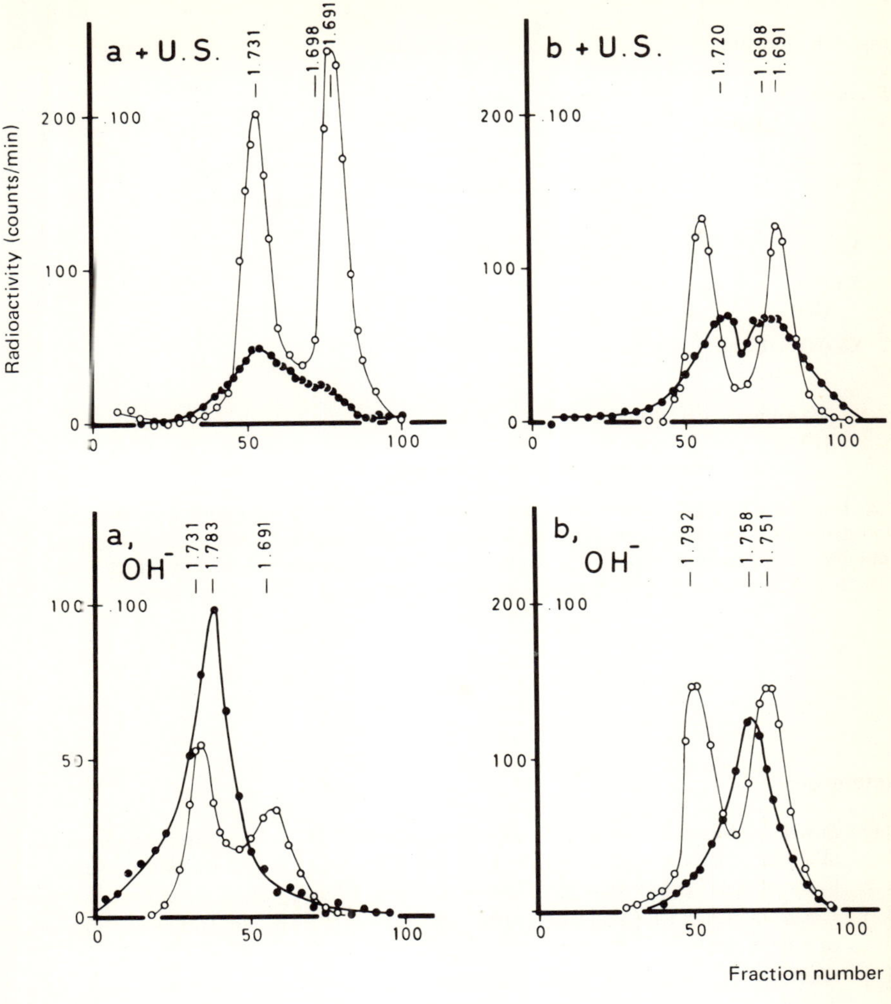

(Fig. 11, 2nd part)

$th_C^- \times th_C^-$ (self-fertilisation) $\xrightarrow{F1, F2, F3}$ no segregation

$th_C^- \times th^+$ (crossing) $\xrightarrow{X1}$ no segregation

Example: $th_C^- \times th^+$

	Phenotype		
	Normal (N)	"Viable mutant" (V)	Lethal mutant
X2	86	14	
X3 (N)	86	7	7
(V)	3	69	28
X3 (N) + thiam. 10^{-5}M	100		
X3 (V) + thiam. 10^{-5}M	100 100		

Fig. 12. The X2 progeny contained normal plants (N) and plants with limited growth (V) and patches of discoloration at the basis of the limbs. Characteristics of the viable thiamine mutants were already described. Those plants produced low amounts of seeds, which were tested (in X3) separately from the seeds obtained from the normal-looking plants.

References

[1] *Brown, J., Huart, R., Ledoux, L., Swinnen-Vrancks, J.* 1971. High specific radioactivity of a heavy fraction in progency of "Arabidopsis Thaliana" treated by "Micrococcus Lysodeikticus" DNA. Arch. Int. Physiol. Biochem. **79**, 820–21.

[2] *Charles, P.* 1967. Etude du sort du DNA injecté chez la souris. Ph. D. Thesis, University of Lille, France.

[3] *Charles, P.* and *Ledoux, L.* 1966. Localisation et sort (structures primaires et secondaires) de DNA exogènes absorbés par les organes génitaux de souris injectées par voie intraveineuse. Arch. Int. Physiol. Bioch. **74,** 912–14.

[4] *Charles, P., Remy, J.* and *Ledoux, L.* 1968. Etude des carcinomes de souris injectées de DNA 3H bacteriens. 1. Influence de la méthode d'injection. Arch. Int. Physiol. Bioch. **77,** 146–47.

[5] *Charles, P., Remy, J.* and *Ledoux, L.* 1971. Fate of hexogenous bacterial DNA in ascites tumor cells. Informative molecules in biological systems, NATO symposium, ed. L. Ledoux, p. 88, Amsterdam: North-Holland.

[6] *Coumanne, C. Jacqmard, A., Kinet, J. M., Bodson, M., L. Ledoux* and *R. Huart* 1971. Translocation and distribution of bacterial DNA. *In* "Sinapis Alba". Arch. Int. Physiol. Bioch. **79,** 823–24.

[7] *Flamm, W. G., Birnstiel, M. L.* and *Walker, P. M. B.* 1969. Preparation fraction, and isolation of single strands of DNA by isopycinic ultracentrifugation in fixed angle rotors. *In* Subcellular components: preparation and fractionation, eds. G. D. Birnie and S. M. Fox, pp. 125–55. London: Butterworths.

[8] *Fox, M.* and *Ayad, S. R.* 1971. Replication of repaired and integrated DNA in P388 lymphoma cella. *In* Informative molecules in biological systems. NATO symposium, ed. L. Ledoux, p. 99. Amsterdam: North-Holland.

[9] *Hess, D.* 1970. Versuche zur Transformation an höheren Pflanzen. Z. Pflanzenphysiol. **63,** 31.

[10] *Hill M.* and *Hillova, J.* 1971. Recombinational events between exogenous mouse DNA and newly synthesized DNA strands in chicken cells in culture. Nature New Biology **231,** 261–65.

[11] *Huart, R., Swinnen-Vrancks, J., Jacobs, M.* and *Ledoux, L.* 1970. Etude des descendants de plantes d'«Arabidopsis Thaliana» traitées par des DNA bacteriens. I, Presence et replication du DNA exogène. Arch. Int. Physiol. Bioch. **78,** 591–93.

[12] *Jacobs, M.* 1964. Nutrient recipes for selection and growing of biochemical mutants, Arab., Inf. Serv. Göttingen, **1,** 36.

[13] *Jacobs, M.* 1967. Mutagénèse expérimentale et analyse des mutants induits chez «Arabidopsis Thaliana» (L) Heynh, Ph. D Thesis, University of Brussels.

[14] *Laval, J., Malaise, E.* and *Laval, F.* 1971. Uptake and fate of heterologus DNA in mice fibroblast, in vitro. *In* Informative molecules in biological systems, NATO symposium, ed. L. Ledoux, p. 124. Amsterdam: North-Holland.

[15] *Ledoux, L.* 1967. La pénétration et le sort des acides desoxy-ribonucléiques dans les cellules vivantes. Agregation Thesis, University of Brussels.

[16] *Ledoux, L.* and *Charles, P.* 1966. Uptake of exogenous DNA by mouse embryos. Exp. Cell. Res. **45,** 498–501.

[17] *Ledoux, L.* and *Charles, P.* 1967. Présence et effets du DNA exogène dans les carcinomes mammaires spontanés de souris injectées de DNA par voie intraveineuse. Arch. Int. Physiol. Bioch. **75,** 363–65.

[18] *Ledoux, L.* and *Charles, P.* In Press. Fate of exogenoux DNA in mammals. *In* Uptake of informative molecules by living cells, NATO course, ed. L. Ledoux, Amsterdam: North-Holland.

[19] *Ledoux, L.* and *Charles, P.* In press. On the use of preparative CsCl gradients. *In* Uptake of informative molecules by living cells, NATO course, ed. L. Ledoux. Amsterdam: North-Holland.

[20] *Ledoux, L., Charles, P.* and *Srogel, M.* 1967. Fate of exogenous DNA in blood and hyperplastic genital tract of female mice. Nature **214,** 1241–43.

[21] *Ledoux, L.* and *Huart, R.* 1969. Fate of exogenous bacterial deoxyribonucleic acids in barley seedlings. J. Mol. Biol. **43,** 243–62.

[22] *Ledoux, L.* and *Huart, R.* In press. Fate of exogenous DNA in plants. *In* Uptake of informative molecules by living cells, NATO course, ed. L. Ledoux. Amsterdam: North-Holland.

[23] *Ledoux, L., Huart, R.* and *Jacobs, M.* 1971. Fate of exogenous DNA in Arabidopsis Thaliana. II. Evidence of replication and preliminary result at the biological level. *In* Informative molecules in biological systems, NATO symposium, ed. L. Ledoux. p. 159. Amsterdam: North-Holland.

[24] *Ledoux, L. Huart, R.* and *Jacobs, M.* 1971. Fate of exogenous DNA in Arabidopsis Thaliana. Eur. J. Biochem. **23,** 96–108.

[25] *Ledoux, L.* and *Jacobs, M.* 1969. Redistribution, lors de la floraison, des DNA exogénes absorbés par des graines d'«arabidopsis thaliana» Arch. Int. Physiol. Biochem. **77,** 568–69.

[26] *Li, S. L.* and *Redei, G. P.* 1969. Genelocus specificity of the glucose effect in the thiamine pathway of the angiosperme Arabidopsis. Plant Physiol. **44,** 225.

[27] *Lurquin, P., Balluet, H.* and *Ledoux, L.* 1971. Uptake and fate of bacterial DNAs in the growing chicken embryos. *In* Informative molecules in biological systems, NATO symposium, ed. L. Ledoux, p. 334. Amsterdam: North-Holland.

[28] *Redei, G. P.* (1965). Genetic blocks of thiamine synthesis in the angiosperme "Arabidopsis". Amer. J. Bot. **84,** 834.

[29] *Watters, C.* and *Gullino, P. M.* 1971. Translocation of DNA from the vascular into the nuclear compartment of solid mammary tumors. Cancer Res. **31,** 1231–43.

Discussion

Glick: I have obtained data similar to those of Dr. *Ledoux*, which suggest integration of isologous donor DNA with host DNA in cultured L1210 mouse leukemia cells. The donor DNA is labelled with tritiated bromodeoxyuridine, so its density is greater than that of the host DNA. When nuclear DNA from DNA-treated cells is centrifuged on a CsCl density gradient, it contains a radioactive peak intermediate between the peaks corresponding to donor and host DNA. When this intermediate peak is sonicated, both a heavier radioactive peak and a lighter UV peak result with densities similar to control donor and host sonicated DNA peaks. Heat denaturation of the intermediate peak only results in a single peak located between the control donor and host denatured DNA peaks. These data indicate that both strands of the donor DNA become covalently bound to the host DNA in an end-to-end fashion, as Dr. *Ledoux* has shown. However, much more DNA is bound than could be due to the donor DNA simply attaching to the ends of host DNA molecules *in situ.* Since several other laboratories have also obtained similar evidence for integration of donor DNA in eukaryote cells, we have proposed a model to explain such evidence (*J. L. Glick,* and *A. Majumdar,* in press , *J. Theoret. Biol.*).

Figure 1 summarizes a description of this model. A loop is formed in the host genome, followed by doubled-stranded insertion of the donor DNA but with only single-stranded linages at opposite ends of the donor DNA (Fig. D1/A). DNA synthesis then occurs on both sides of the loop, completing the formation of a double-stranded loop (Fig. D1/B). Upon extraction of such integrated donor DNA, one might expect the ends of the loops to be preferentially susceptible to breakage because of the single-stranded linkages. Thus, hybrid molecules would be isolated containing donor DNA linked end-to-end to host DNA.

This model is also consistent with a number of biologic studies, including perhaps the one just referred to by Dr. *Ledoux* concerning genetic transformation in plants. For example, *Fox* and his associates (*A. S. Fox, S. B. Yoon* and *W. M. Gelbart,* 1971, Proc. Natl. Acad. Sci., U.S. **68,** 342) have shown that drosophila can be transformed by allogeneic DNA. The inserted donor allele maps at the same site as the analogous host allele without excision of the host allele, such that both alleles are capable of genetic expression. The loop model obviously permits both alleles to be present parallel to one another.

Figures D1/C and D1/D simply demonstrate how the loop model permits semiconservative replication of the integrated segments of donor DNA during the normal replication cycle of host DNA.

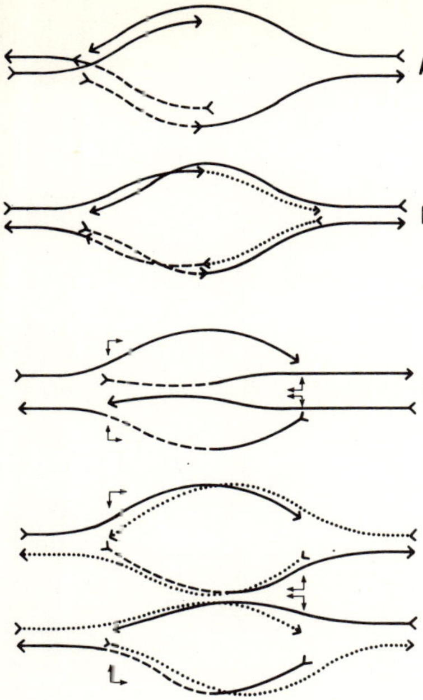

Fig. D1. A model for integration of donor DNA in eukaryote cells. Solid, dashed, and dotted lines represent host, donor, and newly synthesized DNA strands, respectively. Polarity of each strand is also shown. Arrows at right angles to the host DNA designate sites on the template strands where synthesis of DNA is initiated. Arrows parallel to the host DNA indicate the direction in which each new strand grows.

Introduction to Session III

P. H. Hofschneider

Max-Planck-Institut für Biochemie, 8 München 2, Goethestr. 31

The title of our last session is, "Present and Future Application to Medical Practice", and I would like to explain briefly our purpose in naming the subject this way. It is absolutely clear that no one here in the audience would want to breed a better human race by genetic engineering and, therefore, we can well leave out a science-fiction-like discussion of this subject. I would like to interpret this topic in the context of exactly to what basic biological research can practically contribute to the welfare of our society. In order to be able to speak about this, it is inevitable to know exactly where we stand scientifically. I have, therefore, asked Dr. *Bodmer* to take over the difficult task of summarizing the results of the last day and to draw some conclusions.

Some Reflections on Mechanisms and Prospects of Genetic Exchange*)

Walter F. Bodmer

Genetics Laboratory, Department of Biochemistry, University of Oxford, England

I feel very hesitant at taking on the difficult task of summing up what has been said so far as we come to the beginning of the end of our meeting. It would clearly be very difficult and also uninteresting to attempt any complete or systematic summary of everything that has been said. I have therefore assumed that my task is to try, if possible, to weave some sort of a common thread through what we have heard and to try and relate what has been discussed so far to what may be discussed under applications later on in the afternoon.

I think many of us would agree that one, if not the major, common thread to all of biology is evolution. With this in mind, I should like first to discuss a few evolutionary considerations relating to the question of genetic exchange and recombination. If any process exists in nature, one assumes that it is there for a reason; and by reason one means that it has been selected for, namely has had, some advantage throughout evolution. We are therefore naturally led to ask the question about mechanisms of genetic exchange and recombination: why are they there? What is their advantage in terms of biological fitness? These are questions to which population geneticists and evolutionists have addressed themselves for some time. The most commonly quoted argument is one that goes back to *H. J. Muller* and *R. A. Fisher* in the 1930s. They pointed out that the existence of recombination in its broadest sense allows new mutations that have arisen in different individuals to be put together in a single individual with a much higher efficiency than is possible only by mutation in the absence of any form of sexual mechanism. This ability to accumulate different sources of genetic information and so, presumably,

*) This is an adaptation of the transcript of the actual talk given at the meeting. No specific references are included since it was assumed that relevant references could be found in the other published papers of the workshop.

Manuscript received: 27 January 1972

new advantageous combinations, is of great advantage for accelerating the evolutionary rate of the population. Those populations which have sex should be able to evolve much faster. This is the explanation for the existence of sexual mechanisms and recombination that is now widely accepted. There is one feature of this as an evolutionary mechanism which makes it very different from most of the mechanisms that one usually thinks about: it is an advantage not to the individual, but to the population. Some of us may think sex is an advantage to the individual; from the evolutionary point of view, it appears that it is the population which has the sexual mechanism that can evolve faster. The selection is between whole populations and not between the individuals within a population. There has actually been much argument as to the efficiency of mechanisms which work at the population rather than the individual level, and some people think that sex and genetic recombination may be almost the only traits which have evolved through population selection.

The total recombination mechanism is very complex and has many different components. There must, therefore, be very many genes that are needed for successful sexual recombination, whether it be in phages, bacteria, or in higher organisms. The population advantage of recombination must, however, depend on the proper functioning of the complete system and so the advantage cannot really exist until the total system has evolved and is functional. This poses a paradox for the evolution of recombination which is similar to that posed by the evolution of any complex organ. I think that the only way out of this dilemma is that the individual components of the system must have evolved for reasons other than their contribution to the end product. We have many examples of this in what we assume to be the basic recombination mechanisms at the DNA level, which include a number of enzymes whose primary function could be in DNA metabolism.

It is very important to realize that sexual mechanisms and genetic exchange comprise two very different and quite separate components. On the one hand there is the mechanism by which different individuals or organisms get together and somehow juxtapose their genetic information so that it can be reassorted or exchanged. In classical genetic terms, this includes all those aspects of the sexual mechanism which lead up to the process of synapsis, which is the pairing of chromosomes that is seen in cytological preparations of cells in meiosis. The second aspect of recombination, which is much more specific, is the actual exchange of genetic information at the level of coherent genetic units, commonly called chromosomes. Genetic recombination and reassortment can occur independently of the existence of recombination in this second more restricted sense. The independent segregation of whole chromosomes during meiosis is a process of genetic exchange and recombination from the biological and population point of view. This segregation contributes to the assortment that underlies the arguments for the advantage of sex, without there being any actual recombination of genetic material in the sense in which it was discussed during the first day of this meeting. It is important to realize, when thinking about the evolutionary advantages of genetic mechanisms, that there are many mechanisms of exchange which do not even go as far as synapsis, but involve only whole genetic units or chromosomes. In bacteria, for example, the transfer of drug resistance factors and even bacterial conjugation itself may occur without any recombination in the more restricted sense, but these are certainly important mechanisms of exchange. Another mechanism, which we have heard a great deal about yesterday and today, for higher organisms is cell fusion and the reassortment of whole chromosomes which may follow such fusion. This is a mechanism of genetic assortment that involves no recombination between the genetic units, but which is now of great practical imprtance. It seems possible this was the precursor of fertilization and meiosis in higher organisms as we see them now. The mechanisms by which one arrives at the alignment of the units of genetic information so that they can exchange their content is a topic that has, on the whole, been studiously ignored during most of this meeting. I think that this is because these mechanisms are still so very poorly understood.

The mechanisms which lead up to synapsis are, on the whole, extremely diverse throughout evolution. On the other hand, the second level of recombination mechanisms which lead to the actual exchange of genetic information within chromosomes seem to have, at least at the DNA level, many features in common, as was discussed at some length during the first day. We believe that we can draw sensible pictures of recombination in terms of DNA strands; and that we know something about the nuclease enzymes that are involved in the breakage of these strands, which must precede their joining with each other. We believe that the basis for homology and its precision is connected with base pairing and exchange of base pairing partners, and we believe that we understand something about the mechanisms by which DNA strands are broken and rejoined to make physically new combinations. I think it should again be emphasized that all the enzymes that are assumed to be involved in this breakage and rejoining process are quite conceivably enzymes that are necessary for DNA metabolism and so may have evolved for reasons that were initially quite unconnected with the process of recombination.

There is at least one very important, perhaps all important, exception to the idea that recombination has evolved only because of the population advantage and that the enzymes needed for it evolved as byproducts of other functions. This exception is the phage lambda type integration system (see the paper by *Yarmolinsky*), which is important for the survival of phage lambda as we know it. The enzymes that are needed for the integration process are, therefore, advantageous for the individual phage. Thus, here is an example of a recombination mechanism which has evolved because of the advantage to the individual organism as a part of its life cycle. The properties of phage μl and its ability to go in anywhere on the *E. Coli* chromosome were discussed by *Zipser.* This phage is indeed a very clever individual that has evolved a system of recombination which has relaxed the requirement of homology between it and its entering host. Here again one can assume that this evolution took place because of the individual advantage to the organism. One is naturally led to inquire whether there are other examples in higher organisms where there may be an individual advantage to recombination type mechanisms. There are, of course, the transforming animal viruses whose genome may be inserted into that of the recipient cell by a mechanism similar to that of phage integration in bacteria (see the papers by *Hirt* and *Mizell*). Another example which has been widely discussed, is the joining together of the constant and variable chain genes of the immunoglobulins during the differentiation of the cells of the immune system. More generally, there may be a case for having recombination mechanisms which function during the differentiation of cells in order to transpose pieces of DNA from one part of the genome to another for control purposes at various stages of development. Another example may be if information such as that contained in the DNA of mitochondria and chloroplasts is inserted or excised in and out of the genome, then this may be done using a recombination mechanism that is analogous to those which have evolved for the phage specific integration systems.

Let us briefly look at the history of recombination and the way that people have thought about it. It started with observations of exchange of parental genetic markers in suitable crosses, reflected in unequal proportions of parental and non-parental or recombinant genetic types. In those days one drew pictures with just two lines crossing-over, one for each chromosome and said that there was some sort of recombination between these two single entities. Then at the next stage, it was realized that one had to think in terms of duplicated chromosomes, or chromatids, and so four lines were drawn, one for each chromatid and now the lines which crossed over referred to chromatids. Finally, we now know that chromosomes are made of DNA duplexes, and on the whole we think that there is just one double helix that extends along the length of the chromosome. So now we draw pictures in which the crossover lines refer to exchange at the level of single DNA strands. The line drawings have not changed very much, but their interpretation has. The models that have been suggested for the ways in which exchange at the level of the DNA strands could occur came from two different sources. First, they arose as attempts to explain anomalous recombination events observed in fungi, and second from studies on the physical fate of recombinant molecules in phage systems, such as lambda, and in bacterial transformation systems, such as Pneumococcus and B. Subtilis. Many models have been proposed for ways in which DNA strands can exchange, be broken, and be repaired, and there is endless opportunity for constructing more models, especially in the absence of decisive data. I think it is fair to say that the next step is actually to find those enzymes which are involved in these processes, to study in detail how they work, and to find out what are the restrictions, if any, on the types of models that one can draw up in terms of DNA strands. The ultimate goal is, presumably, the reconstruction in vitro of the actual details of the recombination process at the level of the DNA. The first paper of the meeting, by *Radding,* provided an example of the progress that has been made recently in this direction. It is interesting to see some of the differences that arise in the models according to the way that the initial breaks occur in the strands of two recombining duplexes. The differences appear mainly to be due to the sorts of enzymes which are postulated to do the breakage. Clearly, as we were reminded on the first day, one should not assume that there is a single universal mechanism. Rather, there may be a variety of different ways, all of them perhaps quite similar, in which one can solve the problem of putting together different pieces of DNA to form a single piece, or of inserting segments of DNA into one or other strand of a longer molecule.

The significance of DNA repair processes in recombination has often been emphasized. There are, however, two very different sorts of repair. One of them is the sort which *Trautner* and his colleagues have, by implication, been studying, where there is excision of one or more DNA bases at positions where two associating DNA strands have noncomplementary bases because of a mutational difference. This excision may be associated with more extensive removal of bases and subsequent repair around the primary nonhomologous base pair positions. The second, quite different reason for having some sort of repair arises if in the process of exchanging base pairing partners between two DNA duplexes there is a gap left in one or another strand of either duplex – that is, if breakage and rejoining are not precise at the level of the individual base pair. Such gaps must, presumably, be filled in by repair DNA synthesis. What are the restrictions on the exchange process, if any? Can exchange actually occur between adjacent base pairs? At least until recently there have been few attempts at detailed molecular model building aimed at determining what the constraints on the process of DNA exchange might be. There is, however, apparently work by *Siegel* and *Alberts,* in press (JMB), describing the building of models that suggest that precise exchange, leaving no gaps, is possible. This may, of course remove the need for the second sort of repair postulated for mending gaps left by untidy exchange of base-pairing strands. However, clearly not all exchanges are necessarily precise, so that some form of untidy exchange of strands followed by repair may still be a major feature of recombination mechanisms at the DNA level.

A recurring theme in this meeting, to which I will return later, has been the question of nonhomologous associations. At this point I would just like to emphasize the implications of two observations. The first is the evidence that μl phage does not depend on homology for insertion into the bacterial chromosome. The second comes from *Khorana's* data on the chemical synthesis of tRNA. He pointed out, if I understand the data correctly, that the *E. Coli* ligase could, under certain circumstances, mediate end-to-end joining of DNA molecules without any single-stranded overlaps and so without any need for base-pairing homology. I think that this is an extremely important observation because, as I mentioned earlier, it has nearly always been assumed that a basic requirement for exchange at the DNA level is that there exists some base-pairing homology. Even the case of the μl phage could be explained by a model in which the phage creates its own homology. This could be, for example, by adding a few base-pairs to a free end of one DNA strand in the molecule that the phage wants to get into, which base-pairs are homologous to a corresponding sequence occurring somewhere in its own genome. This would then be an apparent exception to the need for homology, allowing entry of the phage into apparently nonhomologous regions but still by a process of base-pairing alignment. End-to-end joining, if it occurs at all frequently, is something that is completely different and new and that may be very important for understanding some of the nonhomologous associations which are now being described more and more frequently. I think there is, incidentally, an interesting analogy between *Khorana*'s studies on the complete in vitro synthesis of double-stranded DNA molecules and *Radding* and coworkers attempts, using larger naturally occurring molecules, to construct in vitro recombinants.

I have already commented on how very little has been said about the ways in which DNA molecules may get together and align themselves in preparation for exchange. *Yarmolinksy* mentioned the possible involvement of regions of DNA which are being transcribed in recombination events. In bacterial DNA transformation, it has been suggested that a region of the DNA of the recipient may be membrane associated, and that this is the way in which the incoming donor DNA can find its right place in the recipient. Such membrane associated DNA stretches may, indeed, be regions that are being transcribed.

A critical test of the biological activity of *Khorana*'s in vitro synthesized DNA molecules might be to demonstrate their activity in a DNA transformation system. The problem in doing this, at least with most of the transformation systems I am familiar with, is that small molecules, even if they have the right homology, are apparently physically unable to take part in the process which leads up to the alignment of the donor and recipient DNA molecules. This inability is presumably not connected with the details of exchange at the molecular level, but has something to do with the mechanisms leading up to synapsis. One solution might be to insert the artificial tRNA gene into a long, nonspecific, synthetic DNA duplex and hope that this can solve the problem of synapsis, allowing the homologous regions to find each other for subsequent recombinational insertion.

I should like now to turn to the main subject of this morning and yesterday, namely to some of the ways in which exchange in higher organisms can be studied. One of the main contrasts with the earlier part of the meeting is that the mechanisms of exchange have not, on the whole, depended on recombination in the more restricted sense, but have been based on those features of the recombination process which precede synapsis. One of the major themes in this second part of the meeting has been what might loosely be called "heterology." Heterologous systems have dominated the discussion of how one can do genetic manipulation of higher organism cells in culture.

These heterologous systems can, very roughly, be divided into three categories. The first category, about which we have heard a great deal, is fusion. The fusion is mostly between somatic cells. However, nuclear transplantation is also a form of fusion, namely of nuclei with cells, as is also the work on acetabularia. Under heterologous systems, one should perhaps include not only species differences, but also differences between cell types within any given species.

Cell fusion has, so far, been one of the major tools for the genetic manipulation of higher organism cells, and will probably continue to be so for some time. *Ruddle* and *Siniscalco* have described very fully some of the ways in which it can be used to do more or less conventional genetic analysis. *Siniscalco,* for example, described how cell fusion, without chromosome segregation, can be used as a test both for dominance and for complementation. In this context, the distinction between intra- and interspecific crosses must be kept in mind. The interspecific cross, notably between mouse and human cells, is very useful for genetic studies because of the chromosomal segregation that follows from the loss of human chromosomes. The intraspecific cross is chromosomally relatively stable. Thus, in most crosses between cells from the same species, the resulting hybrid retains most of the chromosomes of both parental components, though there may subsequently sometimes be some segregation. It is this relative stability of the intraspecific cross that makes it the analog of forming a diploid, and so allows one to test for dominance and complemtation without the complication of genetic segregation. This is the key to understanding some of the complicated and elegant experiments described by *Klein,* as the first step in this work is to look for dominance in crosses of malignant by nonmalignant cells.

The phenotype of a population of cells with respect to malignancy is established by injecting cells into a mouse, and asking whether or not the mouse survives. If, in a series of injections using a given cell line, the proportion of cases where the mouse dies is above some threshold, then the cell line is said to be malignant, but not otherwise. The threshold is chosen for optimum discrimination of the malignant and nonmalignant (or sometimes high and low malignant) phenotypes. Consider now, for example, the cross between nonmalignant A9 cells (A9 is a mouse L-cell derivative) and the highly malignant mouse Erlich ascites tumour cells. The initial hybrid products from this intraspecific cross are relatively stable, as already discussed, and appear to have a nonmalignant phenotype suggesting that malignancy is recessive. The complication in these experiments, however, is that a very large number of cells is being injected, among which there must be some with mutations and other genetic changes including chromosome losses. If some such changes lead to a reversion of the nonmalignant hybrid cells to malignancy, the reverted cells will be picked out by selection and will grow and kill the animal. This accounts for the fact that even though the hybrid cell population as a whole has the nonmalignant phenotype, there is a proportion of animals in which tumours do grow out. This is quite analogous to plating out a streptomycin sensitive bacterial culture on streptomycin and finding a small number of resistant colonies. The cells from tumours derived from the nonmalignant hybrids themselves have the malignant phenotype, and tend to have fewer chromosomes than the original hybrid cells. These malignant products have presumably arisen through genetic segregation in what was initially a relatively stable hybrid. Imposing genetic segregation on the initial test for dominance thus gives a result which is consistent with malignancy being recessive. It also provides the opportunity for studying the joint segregation of malignancy with other penotypes, in particular transplantation and tumour antigens, as further illustrated by the work discussed by *Milgrom.* The study on malignancy is just one example of the way in which techniques for genetic manipulation based on cell fusion can be used to study a given phenomenon.

Another type of cell fusion which has great potential interest, namely that between sperm and somatic diploid cells, was discussed by *Croce.* Crosses between sperm and eggs are hardly novel when they are done naturally, but they are novel if, as described by *Ericsson,* they are done in the test tube, especially when they involve combinations of species that do not normally produce fertile offspring. The types of mixtures of early embryonic cells which have been described by *Graham* and *Tarkowski* are another example of fusion, though here, for example in the formation of tetraparental mice, the cells themselves do not necessarily fuse. The work on acetabularia, (described by *Schweiger*) is, as already mentioned, another example of fusion. It is perhaps surprising that all the functions studied, even those that one thought might be unique to the chloroplast, seem to be determined by the nucleus. The ultimate chloroplast specific function, which was not mentioned, is the chloroplast DNA itself. Is this also controlled by the nucleus? The classical experiments of cytoplasmic inheritance of chloroplast differences do, of course, clearly indicate that, in some cases at least, there must be some functions that are determined by the chloroplast genome. The fate of mitochondria in hybrid cells poses an almost exactly analogous problem which we are now working on in Oxford.

The study of heterokaryons, which is another aspect of cell fusion, is concerned not so much with genetic analysis itself, but with understanding the control processes involved in mixing information of very different kinds. This approach is exemplified by the studies described by *Ringertz* and *Deak* on the significance of the ribosomes for the control of protein synthesis, and by *Johnson* on the factors which control mitosis. An understanding of such control processes may be essential for the interpretation of the results of many of the studies with heterologous systems.

The second of the three major categories of heterologous systems is characterized by the work described by *Woodland* in which the effects of injecting foreign substances into a cell are studied. For example, are injected foreign RNA messages translated? The work described by *Merril* is analogous except that it involves DNA rather than RNA, and indirect injection into a human fibroblast, rather than into a frog egg or oocyte. If I understand the work correctly, there is so far no evidence for actual genetic exchange between the inserted DNA and the recipient DNA, so that the main question concerns the expression of the information encoded by the donor DNA.

The last of the three categories of heterologous systems involve actual DNA exchange. As in bacteria, there are two main classes of DNA exchange. The first is transduction, in which DNA is carried into cells by viruses. The attempt to get transduction with pseudovirions, as described by *Hirt* and *Axelrod,* is really an exact parallel to generalized transduction with phages in bacteria. *Munyon*'s studies on the transduction of the thymidine kinase gene with Herpes virus may, on the other hand, represent a more specialized form of transduction, still very different, however, from that mediated by phage lambda. The second form of DNA exchange involves purified DNA and is analogous to DNA mediated bacterial transformation. Extensive results using purified DNA in extraordinarily heterologous situations have been described by *Ledoux.*

There are, I believe, two major levels at which all this "heterology" is expressed. The first is at the level of the control of expression of genetic information. Some of the results of genetic experiments with human-mouse and chicken-mouse hybrids, as has been pointed out, imply that the mouse cell can translate human and chicken messages. Taking us even further afield is the evidence that the frog can translate mammalian messages, and that perhaps the mammalian cell can even transcribe and translate bacterial DNA sequences. One can look at these extensions of homology at the molecular level as extensions of the fact that the genetic code itself is universal. I always feel that the universality of the code is perhaps the best evidence for evolution because it is impossible to believe that exactly the same code could have evolved independently more than once. Thus, the common evolutionary origin of all organisms can be equated with the origin of the genetic code as we see it now. The compatibility of transcriptional and translational mechanisms across wide differences between species is just an indication that these mechanisms may have become fixed at an earlier stage of evolution than has been assumed so far.

The second level of heterology is expressed in genetic exchange between widely disparate organisms. The evolutionary reasons why this is surprising are (1) that one believes that the nature of an organism is determined by its genetic information, (2) that evolution leads to divergence of organisms and their genetic information, and (3) that homology of the genetic information is needed for genetic recombination to be successful. This leads to the expectation that the more divergent the organisms, the smaller will be the chance of successful recombination between them. It is becoming clear that the weak link in the chain is the need for homology at the genetic level. This is an example of the way that nature is turning out to be less tidy than we like to think it ought to be. Another major example of nature's untidiness that we have heard a great deal about here (see the contributions of *Hurwitz* and *Duesberg*) and elsewhere is the reverse transcriptase.

What might be the mechanisms by which heterologous exchange of genetic information takes place? I have already mentioned two possibilities, namely end-to-end connection of DNA molecules mediated by ligase, (described by *Khorana*) and synthesis of a homologous sequence. A third possibility is that there exist homologous sequences, which may be quite short, that are widespread throughout the animal and plant kingdom and that are used for heterologous insertion. It is this last possibility that I should like to discuss a little further. Highly repetitive chromosomal DNA sequences are revealed by the recently developed Giemsa banding and quinacrine fluorescence techniques, as discussed by Ruddle. Understanding the nature of this repetitive DNA may be of great importance for the molecular biology of higher organisms. It now appears that there are very extensive DNA sequences in the chromosomes of higher organisms which are extremely simple, which are highly repetitive, and which probably do not code for information that the organism makes use of in any direct way during its differentiation. One of the first and most extreme examples of such sequences was the mouse satellite DNA which is found near the centromeres of most of the mouse chromosomes. There are suggestions that the other bands that are found in higher organism chromosomes using the Giemsa and quinacrine techniques, may just be lesser expressions of the same phenomenon of repetition of simple, meaningless, DNA sequences. These bands may reflect similar types of material to the satellites, but with sequences which are not quite as repetitive. Though it is known that satellite DNAs differ substantially even between closely related species, it nevertheless seems possible that, since the repetitive sequences are so very simple, they might indeed be suitable sites for heterologous insertion. Complex heterologous DNAs may quite probably contain sequences that are homologous to some of the repetitive sequences, allowing insertion into repetitive rather than informative regions of the chromosome. This could, and should, be looked for in, for example, the very striking experiments that have been reported by Ledoux. It should not be difficult to obtain evidence for physical insertion of donor DNA into repetitive recipient sequences. Heterologous insertion by end to end addition, if it occurs, may also take place preferentially in repetitive DNA regions. The building up of repetitive sequences may also to some extent occur by end to end addition. Perhaps repetitive sequences represent a form of metabolic DNA, a concept which was until quite recently very alien to most molecular biologists. I think one should also mention, when discussing non-homology, that there is no evidence that the classical phenomena of chromosomal translocation and insertion involve any form of homologous exchange. In fact, if anything, these have been thought of as examples of nonhomologous exchanges. This possibility is further emphasized by *Ruddle*'s evidence for translocations involving both human and mouse chromosomes. Perhaps such chromosomal exchanges also occur preferentially in repetitive DNA regions. Certainly, chromosomal breakage in mammalian systems often occurs in the centromere regions, which are the very regions which tend to have the most highly repetitive material.

Why is so much effort devoted to the search for, and the development of, techniques of genetic exchange and genetic manipulation? I think that the obvious answer for most of us who are involved in essentially basic biological research is that genetic manipulation has been in the past, and will continue to be in the future, a major tool for understanding cellular functions, especially at the molecular level. This has been the case most clearly with microbial genetics, and the obvious hope is the successful extension of this approach to higher organisms. Genetic manipulation, as emphasized by *Merril,* is nothing new. Selective breeding of one sort or another, which is certainly a form of genetic manipulation, has been practiced by mankind, consciously and unconsciously, since the beginnings of agriculture several thousand years ago. Conscious planned genetic maipulation dates back to Mendel so that we should perhaps reserve for him the true title of "first genetic engineer!" Nevertheless, the mechanisms discussed at this workshop of heterologous and other forms of exchange in mammalian systems in the test tube provide an opportunity for genetic manipulation which is much wider than we have had before.

I now should like to discuss two features of these new techniques for genetic manipulation. The first is the importance of selective techniques. This has been seen many times in the development of genetics but still needs to be emphasized. Selective techniques date back, I believe, to *H. J. Müller* and his elegant techniques in Drosophila for detecting mutations and studying mutation rates. In microbial genetics perhaps their most striking success came with *Lederberg* and *Tatum*'s demonstration of sexual recombination in bacteria. Now again in the development of mammalian somatic cell genetics, we see the importance of selective techniques. Selection is used, of course, to make mutants, to produce hybrids, and to obtain the malignant products from a relatively stable nonmalignant hybrid. Selective techniques may also, for example, help in picking up translocated chromosomes and in so doing extend the approaches discussed by *Siniscalco* for genetic mapping. ing a little chromosomal loss may lead to an autocatalytic process of chromosome loss giving abundant segregation. This would lead to a mechanism of genetic analysis which parallels very closely *Pontecorvo*'s and coworker's well known studies with Aspergillus. They made stable diploids, which were forced to segregate chromosomes using suitable selective techniques. As a result, the diploids progressively lost one set of chromosomes, apparently at random, and so reverted to haploidy. Selective techniques may also, for example,help in picking up translocated chromosomes in so doing use the approaches discussed by *Siniscalco* for genetic mapping.

The other major feature of present techniques of somatic cell genetics is the use of heterologous systems. It is, however, very important to emphasize that heterologous systems are not by any means a necessary part of genetic manipulation at the cellular level in higher organisms. They may, in fact, have serious disadvantages, when studying control mechanisms. Mixing different species could well mix up their control mechanisms so that it may be much better to use mutations and intraspecific crosses. One of the overriding advantages of heterologous systems is that they provide a natural source of genetic markers, namely the expected evolutionary differences between species. I think that this is common to almost all the examples of heterologous systems that have been discussed. Of course, in the case of cell fusion and its application to genetic analysis, one is also making use of the very important phenomenon of chromosome segregation in human-mouse and other interspecific crosses. Another possible interest in heterologous systems is illustrated by further applications of the sort described by *Merril,* where advantage is being taken of the accumulated, extensive, and detailed knowledge of bacterial and phage systems. Thus, although heterologous systems are not a necessary part of genetic manipulation in higher organisms, including especially man, they may be extremely useful.

I think it is important to say again that one of the major reasons why the basic scientist studies these phenomena is because he believes that genetic manipulation is an important tool in learning to understand basic cellular processes. Thus, I would argue that at the present time practical applications of these techniques are a fallout from basic research on genetic manipulation. I think it is particularly with respect to applications that one should emphasize that heterologous systems by no means necessarily provide anything new. Though it may seem to be more sensational, there is no reason why it should be better to put a phage transferase gene into human fibroblasts than to put in another human transferase gene. Perhaps quite on the contrary. Systems involving, for example, some form of transduction coupled with appropriate selective techniques may well be developed for the latter possibility. Another approach would be to use reverse mutation, as pointed out by *Siniscalco.*

One other point that I should like to emphasize in relation to what is often called genetic engineering is that this can be interpreted in two very different contexts. On the one hand there are changes at the somatic level. Engineering of somatic cells manipulates the phenotype and so contributes to euphenics, as *Lederberg* has called it. On the other hand, there is the more specific and much more difficult task of manipulating the germ cells as a prelude to eugenics. This difference must be kept very clearly in mind. The engineering of germ cells is a great deal more difficult and so is still a much more remote possibility. It must be remembered that the techniques that we are talking about are, at the present time, hard enough, to do on a limited scale in our own laboratories under carefully controlled conditions. Their application on a wide scale may clearly be extremely difficult.

Our main responsibility as scientists is surely not to frighten people about what might or might not be done in the future with the spending of enormous amounts of money for which there may be many other better uses. But we should, at least, begin a major attempt to educate people, especially those who are not scientists, as to what can be done now and what might really be possible in the future. Scientific knowledge must be promulgated to a sufficient extent so that these issues can be more readily discussed by all people, scientists and nonscientists alike.

Discussion

v. Ehrenstein: I would like to mention a different view on one of the topics you presented. It concerns the repetitive sequences in DNA. Rather than – as you have put it – they, being common or very similar among different species, I would submit that they may represent one of the major differences between species. I remind you of the work of *Brown, Wensink,* and *Jordan* who found that the ribosomal DNAs of the closely related species *Xenopus laevis* and *Xenopus mulleri* differ in their nucleotide sequence. The nucleotide sequences which code for the 28S and 18S rRNA were indistinguishable. The difference between the two ribosomal DNAs of the two species lies exclusively in the repetitive spacer regions. The chromosomes of two different species are easily distinguished from one another by their banding pattern. These heterochromatic regions in the chromosomes are most likely due to repetitive DNA sequences. Of course, one expects also differences in the unique DNA sequences. We know, however, that the amino acid sequences of a variety of proteins isolated from two closely related species are not very different. The hemoglobin of a chimpanzee is different by only one amino acid from that of human hemoglobin. One could entertain the notion that divergence of the repeated DNA sequences accompanies very early or may even be one of the causes for the divergence of species.

Bodmer: I am aware that there is a great deal of species specificity in the repetitive sequences. On the other hand, the primary repetition sequences may be very short, and it may be very easy to find, in any given added DNA, a sequence which is homologous to this. The question of whether the repetitive sequences are the source of differences between organisms must be considered in the context of all the arguments as to how and why species become separated from each other. Why do they become reproductively isolated? What are the isolating mechanisms? One also has to puzzle over the fact that the satellite sequences do seem to evolve very quickly. I am afraid I would disagree with your suggestion that these satellite sequences are the primary source of differentiation between species. I think there may be some unknown force causing divergence of repetitive DNA sequences which is not connected with the cause of major divergence between species. The fact that the hemoglobins of the primates differ by only one amino acid is true, but there may, after all, be a few hundred thousand informative sequences with respect to which this can be the case.

v. Ehrenstein: The detection of amino acid sequence differences between proteins of two species rather reflects on the sophistication of the analytical techniques available. I cannot help but be impressed with the fact that with relatively coarse methods like hybridization or Quinacrine-mustard staining, one can detect major differences.

Bodmer: But there are major differences within the species that don't lead to reproductive divergence. The varieties of domestic dogs look extremely different, but they are all part of the same species, and they are not reproductively isolated. So I think one has to be careful in extrapolating from a phenotype, such as these banding patterns on the chromosomes, to the function of that phenotype in leading to reproductive isolation.

Stern: Since your request is unexpected, I do not have the slides to present specific data. I might perhaps make one general comment about the discussion on the incorporation of heterologous DNA, and then a few equally general remarks about recombination. I am very hesitant to accept as fact all the reports we have heard on heterologous DNA incorporation into cells of higher organisms. I would much rather stay away from the subject because the current mood is to welcome unorthodox findings. In our laboratory, my colleague, Dr. *Hotta,* made many unsuccessful attempts to repeat the published works of Dr. *Ledoux* who presented new and intriguing findings to the conference. Indeed, Dr. *Hotta* visited Dr. *Ledoux'* laboratory but was still unable to repeat the results after the visit. He may yet attempt another one. Thus far he has obtained two major results concerning the incorporation of heterologous DNA from bacteria into higher plants: (1) In homogenates of tissues which had previously absorbed bacterial DNA, there is a residue of such DNA which is undegraded and which remains attached to the nuclear membrane. (2) Covalent integration of this DNA into host plant DNA has been found to occur, to a small degree, only under conditions of radiation damage. Under such conditions, one gets a messy physiological situation in which the cells are probably inviable. I therefore remain skeptical of experimental claims and give priority to securing experimental confirmation rather than to theoretical debate. I much prefer discussing mechanisms of recombination in higher organisms. It is obvious that meiosis is a complex system in which one principal and major aim is to intermix genes between homologous chromosomes. To complement the earlier discussions on recombination in microorganisms, it should be of interest to generalize from the results obtained in biochemical studies of higher organisms. At the molecular level, the mechanisms we have thus far identified are very similar to those found in microorganisms. We have thus come to expect that if a novel mechanism is discovered in microorganisms, a similar one will be found sooner or later in higher systems. Most recently, Dr. *Hotta* and I have published a set of initial studies on a protein which is similar in many respects to the gene 32-protein found in T-4 infected bacterial cells. Mutants at that locus can neither recombine nor replicate. Our studies reveal that, in all meiotic cells thus far examined, this protein is formed during the prophase stages of meiosis in association with lipid, probably as part of the nucelar membrane. The prophase of meiosis is generally regarded as the interval during which genetic exchanges occur. Thus, nuclei in meiotic prophase from species as distant as lily and man contain a protein which is similar to gene 32-protein. The capacity of this protein to catalyze the renaturation of single-stranded DNA has been demonstrated physicochemically and by electron microscopy. Biochemical studies such as these together with the numerous cytogenetic analyses point to an impressive number of mechanisms in meiosis which assure homologous exchange and exclude heterologous ones. It is perhaps this preoccupation with meiosis that makes the apparently facile acceptance of reports on conserved heterologous DNA integration somewhat perplexing. I am of course not referring to the viral systems which have been so intensively analyzed.

Hofschneider: I have one question concerning the heterologous systems. It is well known that for example in *E. coli* restiction enzymes can degrade foreign DNA. Is anything known on restriction mechanisms in eukaryotes? Dr. *Merril* has to use a very high multiplicity of infection to get transformation of cells by λ. Is perhaps most of the λ DNA degraded by restriction enzymes?

Bodmer: Well, I am no expert on the DNA insertion experiments in higher organisms, but the one comment I should like to make is that, having worked on bacterial transformation and it's complications, I was naturally skeptical, as I see Dr. *Stern* is, about similar experiments in higher organisms. I think that one's worries were at two levels. The first is that the DNA would get degraded when it enters the cell and that it couldn't possibly stay around long enough to find where it should go. I think it is fair to say that there is a great deal of evidence now from published works, which is against this prejudice. One of the remarkable features of the DNA uptake studies is that there seems to be no major degradation of DNA that gets into cells. I gather this is also true of the systems with which *John Gurdon* and his colleagues work. So I can't answer the question on restriction in detail, but it would seem that there is no close parallel to this in higher organisms. The other problem which one worries about still exists, namely, how does the DNA get to the right position on the chromosome? Whatever one's view on the observations that have been made, it's nice to think of ways in which they could be explained.

Therapy of Genetic Diseases in Man and the Possible Place of Genetic Engineering

Walter Fuhrmann

Institut für Humangenetik der Justus Liebig-Universität, Giessen, Germany

Summary: Conventional therapy of genetic diseases has been very successful in selected disorders. It can be applied on a large scale and a great number of patients may benefit. Treatment, however, in most cases has to be continued for life and may be costly and strenuous. Moreover, at best it succeeds in correcting the phenotype but it does not alter the genotype. In many other cases even correction of the phenotype cannot be achieved. The search for early and lasting correction of the primary defect, therefore, is urgent but has been unsuccessful so far. Attempts and speculations on practical application of genetic engineering in human disease states are reviewed and difficulties pointed out. It is concluded that for some time to come research on the exact definition of genetic diseases and elucidation of the primary defect must have priority in medical genetics. This will constitute the basis for more effective palliative therapy as well as for possible correction of the primary defect in the future. Genetic exchange is a valuable research tool in this respect as well as in mapping of the human genome; its direct application as therapeutic principle, when at all possible, will be restricted to a small number of selected cases.

When I was asked to review the present scene of clinical genetics and therapy and its outlook for the future in regards to genetic engineering, I could only speculate on what would be said by the preceding speakers involved in the experimental work. I, therefore, thought it appropriate to add a word of caution to those in the audience who might be too ready to assume that therapy of genetic diseases will be revolutionized in the near future by direct manipulation of the genetic material, an impression that could arise easily if one were not familiar with the achievements and problems of medical genetics. With great satisfaction I then noted the sober and cautious view taken by almost all speakers. To set the scale right, one should not simply list the difficulties still unsolved, but also try to balance the possible benefits of genetic engineering against the achievements of conventional therapy.

Manuscript received: 13 January 1972

Conventional therapy of genetic disease

All presently available treatment of inherited disease in man is symptomatic. It aims at a change of the phenotype, without alteration of the genetic information. The place of action is somewhere between the altered primary gene product and the phenotypic manifestation. Such treatment could frequently be instituted without exact knowledge of the primary genetic defect and even before inheritance was recognized. Methods of intervention include surgical repair, prostheses, and drug therapy. More elegantly, the environment has been adapted to the patient's needs, for example by diet therapy which had dramatic results in certain inborn errors of metabolism, as for example in phenylketonuria, galactosemia, and fructose intolerance.

A somewhat more convincing approach is substitution of a missing or functionally defective gene product. With few exceptions this has been successful only in cases of humoral substances acting when circulating in the blood stream.

Closely related in principle are procedures which aim at an influence on the regulation of hormone or enzyme synthesis and action. A well established case is the cortisol treatment of patients with the adrenogenital syndrome. These patients have a defective cortisol synthesis, but the major symptoms derive from the resulting dysregulation of adrenocortical hormones, some of them with virilizing effect. Cortisol substitution restores the regulatory cycle.

Such measures can be applied on a large-scale basis; a great number of patients may benefit, but continuous treatment is required. Such treatment may be costly and strenuous, and results usually are far from perfect. Nevertheless, the number of patients is increasing who are enabled to lead an almost normal life and to propagate their defective genes. For the majority of genetic diseases, however, no treatment is available as yet. Here any possibility of betterment will be appreciated.

Attempts toward causal therapy

What is wanted ideally is a therapeutic approach to correction of the primary defect, which could be instituted early enough, to prevent irreversible damage and, if possible, do away with the need for continuous treatment. What has been achieved along this line? An early idea was the transplantation of glands in order to give a patient a permanent internal supply of a hormone. This usually failed due to lack of specificity and, among other reasons, due to immunologic intolerance and degeneration of the transplant. As a logical extension one may view the attempts recently undertaken with variable success to transplant bone marrow cells from healthy donors to patients with specific defects, or to transplant thymus to patients with congenital absence of the organ. Success of bone marrow transplantation has been reported particularly in immunodeficiency diseases and will be discussed later by Dr. *Good* [7]. It could offer a chance also in certain hemoglobinopathies, if one succeeded to establish sufficiently large cell clones able to produce normal hemoglobin, or in the mucopolysaccharidoses, if such clones could be shown to supply to the recipient's cells the correction factors found by *Neufeld* and her group [5, 15].

In hemophilia A, it has been shown by transplantation experiments in dogs and by organ perfusion that the antihemophilic factor, factor VIII, or its precursor is produced mainly in the liver and is activated especially by the spleen. From these experiments a therapy appears possible by either transplantation of liver and spleen to a patient or perhaps by the infusion of cell suspensions from these organs [3, 10, 11, 14, 29].

Use of in vitro mutation or gene transfer?

Major problems in all such therapeutic attempts remain the graft versus host reactions and the immunosuppression often required to facilitate the acceptance of the graft. This could be avoided if the old suggestion of *Tatum* [23, 24] could be realized. This was to culture cells of the patient himself and, by induced mutation and selective cloning, produce a clone of reverted cells which could be reimplanted into the patient to supply the missing (or defective) gene product.

As Professor *Henry Harris* and his colleagues at Oxford have shown, genetic repair could also be achieved in cultured mammalian cells by hybridization procedures without necessarily introducing or retaining genes specifying new antigens, which would cause homograft rejection on reimplantation or reinjection. Such cells, therefore, should be accepted and be able to propagate. A draw back at present is that reverted cells produced by the latter method don't have sufficient genetic stability [21]. They tend to lose the newly introduced genetic material. If these obstacles could be overcome, this then would constitute the use of genetic engineering and exchange in practical therapy but only for a small fraction of cells. The obvious difficulties of such an experiment have prevented its realization until now. Besides, this method could be applied only to diseases, correctable by factors secreted by these cells into their environment or perhaps again in certain hemoglobinopathies.

Specific mutagenesis not available

An old dream in correction of genetic defects is specific mutation in vivo. Although particular bases may be attacked preferentially by certain mutagenic substances, and particular regions within genes may be altered in preference by different mutagens, no specificity of mutagenesis is known in the sense needed for therapeutic use in vivo or even in vitro. Even a refined technique, which may serve to bring a chosen mutagen in close contact with a specified region of the genetic material, would certainly not give the desired effect, but would result in various mutations at this site and be accompanied by so much "noise" elsewhere as to do more harm than good. No one has suggested such methods during this meeting.

Gene transfer in vivo as a therapeutic principle

Much more promising for in vivo treatment and more easily accessible was the attempt to use viruses to introduce a new information into host cells. In this way one could hope to infect a great number of cells and supply them with useful additional genetic information. A possibility to realize this concept was suggested by three observations: In 1959, *Rogers* [18] found that rabbit cells infected with Shope rabbit papilloma virus contain an enzyme that degrades arginine. This arginase induced by the virus can be distinguished by several properties from the enzyme naturally occurring in organs of healthy rabbits. Further studies of the same author in 1966 [19] showed that persons who were in close contact with the virus and the infected rabbits had a diminished arginine concentration in blood. The decrease reached up to about 50 % of normal values and persisted in some individuals for as long as 30 years. The persons studied had never shown any ill effects due to the apparent infection. Finally, two sibs were observed who suffered from a hyperargininemia due to a lack of arginase [26, 27].

The occurrence in sibs suggested autosomal-recessive inheritance of the defect in the urea cycle at the point of degradation of arginine to ornithine and urea. Clinical features included mental and motor retardation, seizures, electro-encephalographic anomalies, spastic paresis, periodic spells of vomiting, and liver enlargement. Use of a diet low in proteins was unsuccessful. In view of the experience with people accidentally infected and in the absence of other effective therapy, it was tempting to try the effect of an infection with the Shope virus. This has been undertaken by *Terheggen* in 1970 at Cologne, Germany, in cooperation with Drs. *Lowenthal*, Antwerp, Belgium; *Colombo*, Berne, Switzerland; and *Rogers*, Oak Ridge, USA. (I quote this unpublished work with permission of Dr. *Terheggen* [25].)

A virus suspension was injected intravenously to the younger child, but no change could be observed in the serum arginine concentration. A higher dose of virus given to the elder sib likewise remained without notable result. Meanwhile, another child had been born to this family in whom the diagnosis of the same metabolic error could be established from cord blood. Since experimental infection of rabbits, rats, and donkeys consistently led to decreased arginine levels [20], this youngest child again was inoculated with the virus, now using an again higher dose of virus corresponding to the dose being used in the animal experiments. Results of this attempt cannot be evaluated as yet.

Certainly one could have had little hope for clinical improvement in the older children since therapy of metabolically caused mental retardation is usually successful only if instituted very early in life. Chances for the youngest child could be much better. It must be noted, though, that only about one-half to one-third of all persons exposed to the virus and studied by *Rogers* had a diminished serum arginine concentration, and then the decrease amounted to only about 50 %, whereas serum arginine concentration in the affected children was increased 7 to 13 times.

It is unlikely that we will find many naturally occurring viruses, which specify enzymes, able to correct genetic defects in man, and that also satisfy the requirement of doing no harm to the cell – particularly, if one wants to be sure about late effects. Most often the viral enzyme will also be different from the missing enzyme in certain of its properties and kinetics.

The prospects of *de novo* synthesis of genes opened by Professor *Khorana's* work [1, 8] immediately led to speculation that eventually it will become possible to produce pieces of DNA exactly resembling the missing information and to introduce them into the cells of patients. To many popular writers this problem seems close to solution, but I don't believe it is. There is very little hope that injected naked DNA could effectively reach its proper place of action within the cell. I will not discuss here the problems concerning the possible means of transferring such artificial or elsewhere derived genes to the cell by the use of viruses or pseudovirions, because this has been done by others much more competent than I. I may only point to the work of *Osterman* and associates [17] who have shown that pseudovirions used for transport of DNA fragments indeed can be uncoated by animal cells, and to the work of *Merril* and coworkers [12, 13] who, as you have just heard, succeeded in transducing human fibroblasts in culture from a patient with galactosemia by infection with Lambda phages carrying the *E. coli* galactose operon including the missing enzyme alpha-D-galactose-l-phosphate uridyl (G. P. U.) transferase. They could demonstrate that the phage DNA under these conditions was transcribed in human cells.

It remains an open question, of course, whether a phage could also be used to transport selected bacterial genes or, by some technical modification, artificially synthesized DNA stretches into human cells in vivo. A special problem is raised by the fact that in order to produce pseudovirions in the usual fashion, the artifical DNA first would have to be incorporated into the nuclei of the host cells of the virus. *Strauss* [22], and with some variation *Klingmüller* [9], therefore, recently pictured the possibility of going one step further and producing artificial pseudovirions by coating the synthesized DNA pieces with proteins also derived synthetically or isolated from virus particles, thus facilitating their entrance into living cells by simulating the steps known in viral infection.

No less important an obstacle is that, with a few exceptions, we do not know the primary gene defect in most human inherited diseases. In some we can pinpoint at least a certain enzyme activity that seems missing, in most of the over 1500 monogenic human diseases known not even this can be done. A prominent example of popular misquotation is diabetes mellitus. Since insulin is used successfully in therapy and since the amino-acid sequence of insulin has been known for some time, it was concluded that within the near future it will become possible to assemble a DNA sequence coding for this enzyme. Such an artificial gene could then be introduced into the cell. Notwithstanding the difficulties for such transfer, already discussed, this speculation also does not account for the fact that the insulin molecule consisting of 51 amino acids is derived from a much larger proinsulin with 84 amino acids, and that from a given amino-acid sequence it is not possible to unequivocally deduce the natural sequence of DNA triplets. Moreover, from the fact that externally supplied insulin is effective in therapy, it does not follow that the primary defect in diabetes is a defect of the structural gene for insulin. Many studies indeed have disproved this assumption. Much rather diabetes mellitus includes several genetically distinct entities with different causes on various levels of regulation, synthesis, and transport [30].

On the long road from the transduction of some fibroblasts in culture to therapeutically useful transduction in an intact mammalian organism or in man, more profound difficulties will be encountered. It will be necessary to introduce the new DNA parts into the genome of the cell so that successive cell generations will preserve the corrected information. Such preservation apparently took place in the experiments of *Merril* and coworkers, but it could not be said whether this was due to integration of the new information into the genome, due to plasmid like existence in the cytoplasm, to interaction with mitochondria, or some other as yet unknown mechanisms. Cells supplied with foreign genes by the technique of *Harris* and coworkers [21] tend to shed the introduced genetic material.

Clinical success most likely may come first in a case where the correction of the defect in a considerable number of cells could suffice perhaps to supply a factor to the blood stream. It will be much harder to correct a defect in all or nearly all cells of the body in similar fashion.

Problems of gene regulation and control

Will it be necessary and will it be possible to place the newly introduced gene under the genetic control at the correct locus? Little more than nothing is known about the exact position of the probably not less than 40 000 and possibly more human structural genes. On the other hand, if, as *Ohno* assumed [16], most of the human DNA is noninformational, would the exact position of a foreign gene matter? Will the defective DNA unit, retained in the genome, interfere with effective control of the new gene? A defective gene product coded by the original DNS may also compete with the correct protein for its binding sites.

It is impossible to answer these questions at present. The answers, moreover, will most probably be different for various loci and genes. However, we know very well of cases in which part of the normal genome is present in excess, and this usually leads to severe developmental disturbance and often death of the organism. Such examples are supplied by the trisomies or partial trisomies – best known of the group is mongolism. It may well be that the actual defect here derives mainly from disturbed regulation; but could we be sure that transduction or other means of gene transfer would not also alter regulation? Admittedly the amount of genetic material present in excess in mongolism is much greater than the one probably used for therapeutic gene transfer, but much smaller additional parts of a chromosome than a chromosome 21 have been associated with human maldevelopment. On the other hand, some marker chromosomes with little excess of material obviously are tolerated without apparent ill effect.

If the newly introduced gene does not come under proper regulatory control, could it not exhaust the protein synthesizing capacity of a cell at least to the expense of other functions?

There is a real danger in genetic engineering with the help of viruses that infection may turn out to be oncogenic because many viruses used in these experiments can transform cells to potentially neoplastic cells and cause neoplasms in animals [4].

Very little is known about gene regulation in higher organisms. Most researchers favored the view that the vast majority of DNA in higher organisms and man is used for control purposes and that control may be extremely complicated (an opinion that recently has been put forward again by *Crick* [2]). Only a few weeks ago *Ohno* [16] suggested that mammalian regulatory systems will turn out to be much less complicated than generally anticipated. According to him, they should be as simple or even simpler than the lac-operon system of *E. coli*. In the latter case, some hope may be justified that one could introduce not only defined genes but also the appropriate control regions attached to them.

At present very little is known about possibilities to influence gene expression directly and specifically by influencing gene control. One may, of course, view hormone therapy as a first step on this road. Further progress will have to await the analysis of genetic regulatory systems in mammals.

Conclusions

We must conclude that there is little hope that genetic exchange or other forms of genetic engineering can be applied effectively in the therapy of human inherited disease in the near future. The achievement of genetic repair involving not only a limited number of body cells but all cells including those of the gonads appears highly improbable indeed. This ultimate goal alone could be named "radical therapy." It is questionable whether it ever can be reached – certainly not in our generation's lifetime, and equally certain not on a large-scale basis for a much longer time.

Considering the economy of means and man power, if therapy of inherited disease is the matter in question, at the present time priority should be given in medical genetics to research on the exact definition of entities of genetic disease and elucidation of the basic defects and mechanisms involved. For this, the techniques of gene transfer and exchange may contribute very important information. This analysis will constitute the basis for the second step: search for effective, though palliative therapy. It also is the prerequisite for any future therapy aimed at the correction of the primary defect.

Genetic exchange has proved to be an excellent research tool; it will help to map the human genome, and in the future it also may have an application as a therapeutic principle in selected cases. It, therefore, should be explored thoroughly. However, it would be unwise, even unethical, to seek publicity with precocious promises and speculations regarding clinical therapy which will not be fulfilled and again may bring genetic research into disrepute. For some time to come, prevention will remain the paramount task in human genetics. Genetic counseling will be more important than genetic engineering.

References

[1] *Agarwal, K. L., Büchi, H., Caruthers, M. H., Gupta, N., Khorana, H. G., Kleppe, K., Kumar, A., Ohtsuka, E., Rajbhandara, U. L., Van de Sande, J. H., Sgaramella, V., Weber, H., and Yamada, T.* 1970. Total synthesis of the gene for an alanine transfer ribonucleic acid from yeast. Nature **227**: 27.

[2] *Crick, F.* 1971. General model for the chromosomes of higher organisms. Nature **234**: 25.

[3] *Dodds, W. J.* 1969. Hepatic influence on splenic synthesis and release of coagulation activities. Science **166**: 882.

[4] *Dulbecco, R.* 1969. Cell transformation by viruses. Science **166**: 962.

[5] *Fratantoni, J. C., Hall, C. W., and Neufeld, E. F.* 1969. The defect in hurler und hunter syndromes. II. Deficiency of specific factors involved in mucopolysaccharide degradation. Proc. Nat. Acad. Sci., USA **64**: 360.

[6] *Gatt, R. A., Meuwissen, H. J., Allen, H. D., Hong, R., and Good, R. A.* 1968. Immunological reconstitution of sex-linked lymphopenic immunological deficiency. Lancet **2**: 1366.

[7] *Good, R. A.* 1972. Cellular engineering – an approach to treatment of genetically determined disease. This volume, p. 411 ff.

[8] *Khorana, H. G.* 1972. The synthesis of transfer RNA genes. This volume, p. 89 ff.

[9] *Klingmüller, W.* 1971. Heilung von Erbkrankheiten durch gezielte Eingriffe in das Erbgut. Biologie in unserer Zeit **1**: 87.

[10] *Marchioro, T. L., Hougie, C., Radge, H., Epstein, R. B., and Thomas, E. D.* 1969. Hemophilia: Role of organ homografts. Science **163**: 188.

[11] *Marchioro, T. L., Hougie, C., Radge, H., Epstein, R. B., and Thomas, E. D.* 1969. Organ homografts for hemophilia. Transplant. Proc. **1**: 316.

[12] *Merrill, C. R.* et al. 1972. Bacterial gene expression in mammalian cells. This volume, p. 329 ff.

[13] *Merrill, C. R., Geier, M. R., and Petricianni, J. C.* 1971. Bacterial virus gene expression in human cells. Nature **234:** 398.

[14] *Müller-Berghaus, G.* 1971. Zur Pathophysiologie der Hämophilia A. Dtsche Med. Wschr. **96**: 1723.

[15] *Neufeld, E. F., and Fratantoni, J. C.* 1970. Inborn errors of mucopolysaccharide metabolism. Science **169**: 141.

[16] *Ohno, S.* 1971. Simplicity of Mammalian Regulatory Systems inferred by Single Gene Determination of Sex Phenotypes. Nature **234**: 134.

[17] *Osterman, J. V., Waddell, A., and Aposhian, H. V.* 1970. DNA end gene therapy: Uncoating of polyoma pseudovirus in mouse embryo cells. Proc. Nat. Acad. Sci. USA **67**: 37.

[18] *Rogers, S.* 1959. Induction of arginase in rabbit epithelium by the shope rabbit papilloma virus. Nature **183**: 1815.

[19] *Rogers, S.* 1966. Shope papilloma virus: A passenger in man and its significance to the potential control of the host genome. Nature **212**: 1220.

[20] *Rogers, S.* 1971. Genetic engineering. 4th International Congress of Human Genetics, Paris 1971. Paper presented and abstract published in: International Congress Series No. 233, p. 9. by Amsterdam, Excerpta Medica.

[21] *Schwartz, A. G., Cook, P. R., and Harris, H.* 1972. Correction of a genetic defect in a mammalian cell. Nature New Biology **230**: 5.

[22] *Strauss, B. S.* 1971. Clinical application of genetic engineering. Prospects for the Future. *In* Lipid storage diseases, eds. *Bernsohn* and *Grossman*. New York–London. Academic Press.

[23] *Tatum, E. L.* 1965. The nature of the revolutionary new biology. *In* The control of human heredity and evolution, ed. *Sonneborn*, New York. Macmillan.

[24] *Tatum, E. L.* 1966. The possibility of manipulating genetic change. *In* Genetics and the Future of Man, ed. *Roslansky*. Amsterdam: North-Holland Publishing Comp.

[25] *Terheggen, H. G.* 1971. Personal communication.

[26] *Terheggen, H. G., Schwenk, A., Lowenthal, A., von Sande, M., and Colombo, J. P.* 1969. Argininaemia with arginase deficiency. Lancet **2**: 748.

[27] *Terheggen, H. G., Schwenk, A., Lowenthal, A., van Sande, M., and Colombo, J. P.* 1970. Hyperargininämie mit Arginasedefekt. Eine neue familiäre Stoffwechselstörung. Z. Kinderheilkunde **107**: 298, und **107**: 312.

[28] *Warkany, J.* 1959. Congenital Malformations in the Past. J. Chron. Dis. St. Louis **10**: 83.

[29] *Webster, W. P., Zukoski, C. F., Hutchin, P., and Penick, G. D.* 1970. Hepatic and extrahepatic synthesis of factor VIII (abstract). XIIIth Int. Congr. Hematol., München 1970, p. 222.

[30] Editorial: Diabetes mellitus: Disease or syndrome. 1972. Lancet **1**: 583.

Discussion

Passarge: When we talk about possibilities of genetic engineering, we must ask about its impact at the clinical or medical level. There exists an applicability gap. Over the past eight to nine years, I have been involved in problems of medical genetics and have examined well over a thousand patients and families with congenital diseases. It is perhaps important to point out to this audience that of those problems present at birth, only a rather small fraction is amenable to genetic engineering in the sense we have defined so far. To date, the vast majority of birth defects remains unexplained in their etiology; human geneticists continue to define many diseases in terms of their etiology,but a relatively small minority follows a simple Mendelian genetic mode of inheritance with a genetic defect that could be defined at the molecular level in the near future. As to the vast majority, we either know nothing or little about the etiologic factors, and their multifactorial nature is unlikely to be completely understood for some time to come. Therefore, we are here concerned about a very small sector only, and I think it is important to realize this, as important as our efforts are towards specific therapy of patients with a single gene defect. Furthermore, those who are at risk for such a disorder which could be treated genetically in a sophisticated manner, it will not be possible to realize a genetic problem until an affected child is already born. Although it would be desirable to determine the existence of a genetic risk in advance and initiate the earliest possible therapy (and I think Dr. *Vogel* might perhaps come back to some of the population aspects), i.e., before a child is born, but due to mutations and a number of other factors, a large fraction of patients will continue to be born with congential disorders which can neither be prevented nor expected in advance and by which effective genetic therapy is impossible and symptomatic treatment is the only alternative.

With your permission, I would like to add a small comment to Dr. *Bodmer's* talk. Dr. *Bodmer*, you suggested that perhaps *Mendel* was the first medical genetic engineer, but I would like to submit that Laban and Jacob were the first. I don't have the text with me – I wonder whether anyone here has the Bible along – but, if I recall the story from Genesis correctly, Jacob had to labor for seven years to be able to marry Rachel. Then the next morning after the wedding, he discovered Laban had given him Rachel's older sister Leah, which was perhaps kind of a genetic engineering attempt on the part of Laban. Now, as Jacob insisted that he wanted to marry Rachel, he had to labor for another seven years watching Laban's sheep or goats. Part of the bargain was that when Jacob left he could take all black animals along, most of which were white at the time. The official version is, of course, that Jacob, during the mating of the sheep, let them look at black and white sticks. After fourteen years he managed to have black sheep mostly, which he took with him.

Bodmer: I am quite prepared to stand corrected. It is very hard to decide who was the first genetic engineer – perhaps it was Adam and Eve between them. Its a question of consciousness, I suppose.

Fuhrmann: If I may answer Dr. *Passarge's* last question, I think the story from the Bible could, at best, be called an example of "euphenics", that is a change of phenotype in the desired direction without altering the genotype. It has been quoted as an example of the "photographic theory" by *Warkany*, 1959 see ref. [28] p. 87 from the book Genesis, Chapter 30. Jacob tried to obtain a large number of speckled offspring from Laban's flock from which all the black and speckled sheeps and goats had been removed. He succeeded by peeling branches so that white streaks appeared on the rods and then placed them in the watering troughs of the flocks. As the flocks conceived before the troughs, they produced speckled and spotted offspring.

Eugenic Aspects of Genetic Engineering

Friedrich Vogel

Institut für Anthropologie und Humangenetik der Universität, D 6900 Heidelberg, Germany

Summary: It is very unlikely that the human society, when left to itself, would deteriorate genetically in the future. Different trends can be observed at present which will probably continue, some of which are favourable while others are unfavourable. Genetic engineering will not become a major tool for improvement of the human race, but, in the context of other genetic measures, it will help to alleviate genetic diseases in single persons and families.

Some scientists have expressed the opinion that manipulation of the genetic material through genetic engineering is necessary as the human race would otherwise degenerate or, at least, would not be capable of tackling the increasingly severe problems which future generations will have to face [10, 11, 17].

Are they right? This question has to be answered by the human population geneticist.

In order to do this in a reasonably sound way, he will evaluate present trends, extrapolating them into the future. Doing so, he shares the fate of all futurologists: Unexpected scientific discoveries may change future development in all directions, falsifying our prognosis.

With this necessary qualification, it seems feasible to answer two questions:

1. *When left to itself, would the human race indeed be destroyed by an overwhelming genetic burden? Is it desirable or even necessary to improve the genetic quality of human populations by genetic engineering?*
2. If we were to arrive at the conclusion that genetic improvement is necessary, *can it be achieved by engineering techniques?* How do the prospects compare to those of more conventional eugenic measures?

Manuscript received: 22 December 1971

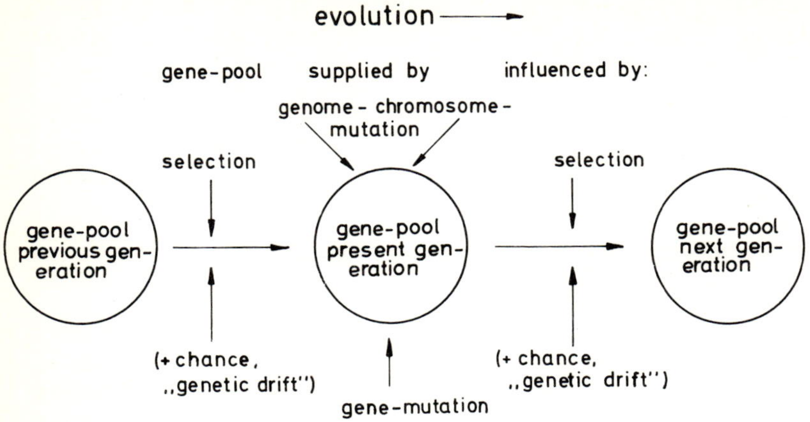

Fig. 1

In the following, I shall mainly try to examine the first question: Which future trends can be predicted when there would be no conscious manipulation? Wherever feasible, however, I shall also discuss the second question.

Three main factors influence the genetic composition of a population: *Mutations* continuously create new genetic variability; *natural selection* chooses those genes or combinations of genes which are best adapted to the enviromental conditions for propagation into future generations; *chance deviations* acting in finite populations do not only cause shifts in gene frequencies, but can lead to random fixation or random extinction of alleles. This decay of variability occurs in complete absence of natural selection. (For references cf. [3, 13].

Which are the presently visible trends for these three factors? The third factor, *genetic drift,* can be discussed very briefly: The world population increases, and marriage restrictions, which have been so important in the past for a more or less complete isolation of population groups, will gradually disappear. Both trends will have the effect that chance fluctuations, contrary to their indisputable importance for human evolution in the past, will become quite insignificant in the future. Therefore, we are left with the two factors causing systematic changes in gene frequencies – mutation and selection.

Let us start with *mutations*! They may conveniently be subdivided into three groups: numerical and structural chromosome aberrations, and gene mutations.

In accordance with the opinion shared by most geneticists, we can assume that practically all chromosome aberrations and the great majority of gene mutations are unfavourable for the individual as well as for the population; most of the chromosome aberrations kill the zygote during embryonal development. A strong minority survives up to birth or still longer, but the patients suffer from severe anomalies or malformations. About 0.5 % of all newborns show such chromosome aberrations due to new mutation.

Gene mutations will frequently lead to inherited disease with a simple Mendelian mode of inheritance, and up to now we have come to know more than 2,000 of those conditions [16]. Other gene mutations lead to defects in multifactorial genetic systems.

The combined evidence from different sources justifies the conclusion: *Influences enhancing the overall mutation rate will be unfavourable.*

That modern civilization has presented us with such influences is so well known that the mere fact is almost not worthwhile mentioning; so much has been discussed about ionizing radiation and chemical mutagens. We do not know enough, on the other hand, about the extent to which this additional load will enhance the natural, "spontaneous", mutation rate.

For *ionizing radiation*, a very crude estimate is possible on the basis of data from mammalian radiation genetics. I shall give the estimated figures without supplying the underlying rationale:

A doubling of the radiation from natural sources (3–4 R within one generation, if given chronically with a very small dose rate) would enhance the spontaneous mutation rate by about 2 % (see [24]). With irradiation with a high dose rate, the increase would be about four times higher.

This result is certainly no reason for us to neglect the problem, as every percent is expressed in thousands of instances of individual suffering; however, it is a very far shot away from imposing on humanity the danger of a general breakdown. The obvious remedy is careful radiation protection, probably supplemented in future times by special protective drugs which are now under investigation. Admittedly, these are very conventional measures.

Much more complex and difficult to evaluate is the problem of a possible additional genetic load due to *chemical mutagens.* In the present state of affairs, quantitative predictions are completely impossible; too little is known about which chemicals are mutagenic at all, which is their pattern of action in human germ cells, and which part of the breeding population is actually exposed to mutagens [5, 9, 25].

Research in this field is one of the most urgent problems in human genetics.

In conclusion, increased exposition to ionizing radiation and chemical mutagens will enhance the mutation rate; presumably this is an unfavourable tendency.

However, as has been appreciated only rarely, there are also tendencies towards a reduction of mutations.

Most numerical chromosome aberrations observed in humans are trisomies; there is one chromosome too many. The most frequent cause of trisomy is meiotic non-disjunction, which mainly occurs in the female germ cell. The probability for this event increases very steeply with the age of the mother, especially in women over 35. However due to the general acceptance of family planning in Western societies, reproduction in women who have already passed this critical age is decreasing, and more and more children are being born by women in the more favourable age groups, i.e., between 18 and 30. In West Germany, this effect may have reduced the relative incidence of Down's syndrome among newborns between 1950 and 1969 by about 25 % ([23] and Fig. 2). Corresponding figures exist for Japan [15] and for Great Britain [21]. So much about trisomies. Some gene mutations, however, show an association with parental age as well. Here, the father's age is the critical factor [18, 25]. The association is not quite as pronounced as in trisomies; a definite decrease of the mutation rate, however, can safely be anticipated.

Taking into account all trends for mutation rates, we arrive at the conclusion that there are *tendencies in both directions* – increases as well as decreases, unfavourable as well as favourable. There is no way to predict which of these two tendencies will dominate in the long run. However, a monotonous and quick deterioration due to increased mutation pressure is very unlikely, to say the least.

Even without any change in the mutation rate, such a genetic deterioration could be possible, if *selection pressure* against these mutations would decrease, i. e., if their chance to be transmitted to later generations would increase.

For patients with numerical chromosome aberrations, this is presently not so, and it must not be expected for the future: Most bearers of x-chromosomal aberrations are sterile, and patients with autosomal trisomies are so severely malformed and mentally retarded that reproduction is very rare indeed. If I am not very much

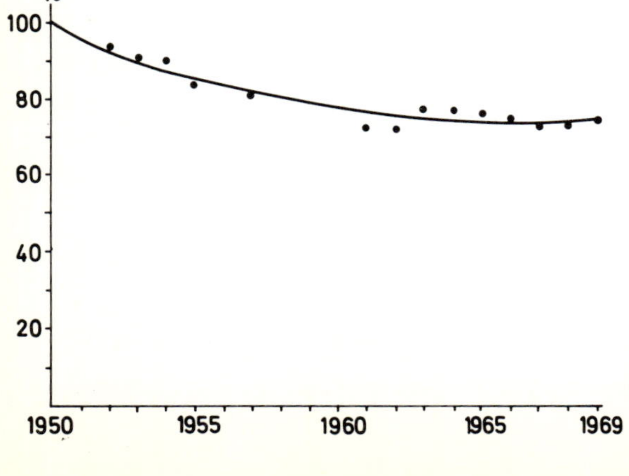

Fig. 2

Decrease of the relative incidence of Down's syndrome in West Germany from 1950–1969.

mistaken, this will not change in the future. Many patients whose genetic defects are due to point mutations, on the other hand, and who would have died young in earlier times, are now able to reproduce, as medical treatment helps them to overcome their defects. Dr. *Fuhrmann* has discussed the problem; let me add only two examples – hemophilia A and B, and retinoblastoma, the malignant eye tumour in little children.

No special gift of prophecy is required in order to predict that these inherited diseases will become more frequent when selection pressure decreases. However, this increase will not be unlimited, but a new genetic equilibrium between mutation and selection will appear; we shall have to learn to live with a higher number of such patients. Fig. 3 shows the calculated increase for retinoblastoma [23]. This burden, however, will not be too severe as medical treatment will help increasingly more patients to have a more or less normal life.

Besides, also for this group of diseases, we can hopefully expect one favourable tendency With the increasing acceptance of family planning in general, people will become more aware of the fact that birth of children with hereditary diseases can frequently be prevented when genetic risks are known and taken into account. Some statistical data, for example in Huntington's Chorea, show that even today birth control is being practiced in order to avoid affected progeny [20]. For everybody with personal experience in genetic counseling, this result will come as no surprise. In these situations, genetic counseling will be of increasing importance. Besides, the methods of prediction will be refined especially due to the possibilities of *prenatal diagnostics.* Even today it is possible to diagnose some hereditary diseases in

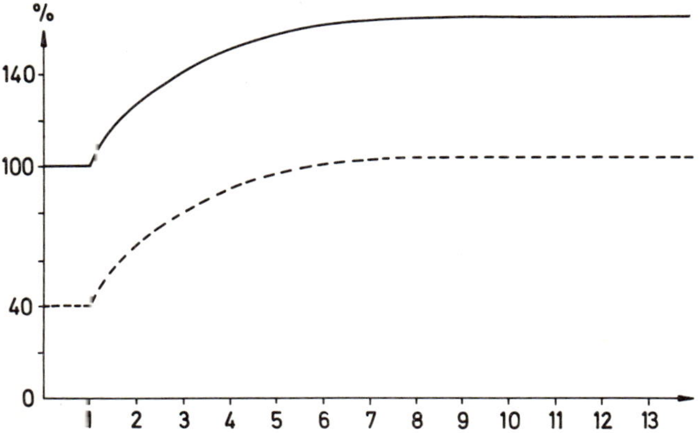

Fig. 3. Increase of the frequency of retinoblastoma after selection relaxation due to improvement of medical therapy.

embryonic cells taken out of the mother's uterus, and to interrupt pregnancy wherever required. Taking into account the efforts made all over the world to refine these methods, it is obvious that many more inherited defects will be found by prenatal diagnostics in the future [4].

In this group of inherited diseases with simple modes of inheritance, *genetic engineering* has its best chances whether artificial infections with benign viruses, transformation of liver cells with DNA of normal people possibly by use of pseudovirions, cell fusions, or artificial back mutations with special chemicals are concerned.

However, there seems to be one border which will be very difficult to cross: A good number of mutations lead to degenerative diseases of the nervous or muscular system. Some examples are Huntington's Chorea, progressive neural muscular atrophy, and the different types of muscular dystrophy [2]. Apparently all these mutations affect genes which are either specifically human, or those which humans have in common with other mammals. The biochemical basis is unknown so far in almost all of them. I feel unable to imagine any virus containing homologous genes that could be transferred into humans. Unfortunately, however, our conventional methods of therapy are also quite inefficient in this group of diseases. Will it always be impossible for us to cross this border, or will the transfer of DNA from other human cells be helpful?

Whatever our answer to this question may be, for natural selection these inherited diseases with simple modes of inheritance will be of minor importance even in the future. Why is this so?

Also in the past, they have caused only a very minor share of mortality before and during the age of reproduction. That only 200 years ago more than 50 % of the population died before reaching adulthood was due to very different reasons: Most persons died from *infectious diseases,* while others died from malnutrition. As *Haldane* has stressed repeatedly, the genetic composition of the human race at present should be determined mainly by adaptation to different and changing infections.

In recent times, this selection has decreased so much that in Western societies more than 95 % of all newborns will reach adult age. What are the genetic consequences?

Probably some will be unfavourable, while others will be favourable. Again, let us first mention the unfavourable trends.

Since the first chordata have developed, a complicated defense system against foreign antigens from external as well as from internal sources has been built up – presumably under strong selective pressure. From the genetic nature of this system [8]; – for genetic defects cf. [6]; – it can be concluded that it can only be maintained by continuously ongoing selection. Selection relaxation will lead to a gradual decline – comparable to the rudimentary eyes of animals living in dark caves for many generations. If our present opinion about the role of the immune system is not completely wrong, this decline will lead to an increased susceptibility to infections, which could force our descendants to live in an artificial environment protected by antibiotics. Besides, the number of malignant tumours will also increase. However, we have good reasons to hope that this process will go on very slowly.

Much quicker, on the other hand, some favourable consequences of selection relaxation due to disappearing infections will become visible. In tropical countries, we know some anomalies and diseases of the erythrocyte. They attained a high population frequency because the heterozygotes have a selective advantage against the malaria infections which have been ubiquitous in these regions up to very recently. The sickle cell gene is a classical textbook example (cf. [1]); hemoglobin C in Western Africa, hemoglobin D mainly in parts of India, hemoglobin E in Thailand, the thalassaenias, and the defects of the enzyme glucose-6-phosphate dehydrogenase are other cases in point (see [4]). Malaria has been efficiently eradicated, and unless it reappears in the future for one reason or the other, anomalies mentioned will become less frequent fairly quickly.

Here, nature has undergone a kind of *compromise*: For the advantage of an increased resistance against malaria, the price of many cases with inherited diseases had to be paid. As soon as malaria has been eradicated, the compromise is not needed any more and will gradually disappear.

These diseases of the erythrocyte are the best known examples; it is very likely that a comparable mechanism is also existing for the polymorphism of the AB0 blood groups, which is responsible for many cases of mother-child-incompatibility (see [26]).

To sum up, also in infectious diseases, certain drawbacks of selection relaxation are counteracted by other advantages. Which of the two will be more important cannot be decided.

For malnutrition, similar genetic compromises are likely, the disappearing of which could lead to genetic improvement. On the other hand, selection relaxation similar to the anticipated deterioration of the immune system could also impair other vital functions. In fact, this tendency is obvious even today when civilized populations are compared, for example, as to colour vision, visual and acustic acuity, and many other functions with hunters and gatherers virtually living under stone age conditions (see [19]).

We have mentioned repeatedly that conscious planning of reproduction is one of the main factors to be taken into account in every prognosis. This factor, however, may not only have favourable but also unfavourable consequences. In our population, one subgroup exists for which the reduced ability to make plans for their own future is characteristic – the mildly feebleminded. And, indeed, their average reproduction exceeds the population average [22]. Now, for those genetic factors which influence intelligence, there is strong positive assortative mating. Therefore, we have not to expect that the population average of intelligence will decrease, or that brilliant persons will become less frequent. However, the number of mildly feebleminded will increase – all the more tiresome as modern society provides progressively fewer positions for such people.

I do not know how this tendency could be avoided; most likely, possibilities have to be found to induce this group to some family planning as well.

Genetic engineering, however, will have no chances. First, the genes concerned are again specifically human. Secondly, the genetic component of "intelligence" is exceedingly complex and highly polygenic. Very many different genes cooperate, and a delicate equilibrium of regulations is required for the development of a human brain. Besides, most cases of mild oligophrenia are not due to defects of single genes but to inadequate functioning of many genes [27]. The complexity of this system becomes apparent when we remember that almost all numerical and structural chromosome aberrations, autosomal and – up to a certain degree – also x-chromosomal will cause oligophrenia [7, 12]. This effect is independent of the special chromosome concerned and can be seen independently of whether the

a) Unfavourable trends:	**Probable significance:**
Mutation rate increase due to ionizing radiation	Not very significant
Mutation rate increase due to chemical mutagens	Unknown
Higher reproduction of patients with inherited diseases	Probably not very significant
Deterioration of normal abilities due to "selection relaxation"	Significant in the long run
Higher reproduction of feebleminded persons	Depending on social factors; possibly significant
b) Favourable trends:	
Elimination of genetic adaptations with otherwise unfavourable effects, c.e infections or malnutrition	Unknown; possibly significant
Decrease of mutation rates for chromosome aberrations and point mutations due to decreasing parental age	Significant within short time
Voluntarily decreased reproduction in families with genetic diseases	Unknown; will become more important in the future

Fig. 4. Favourable and unfavourable trends (without genetic engineering)

amount of genetic material is increased in trisomies or decreased in deletions. Hence, the integrity of the whole chromosome set seems to be required for normal brain development.

To me, at least, it is obvious that the problem to manipulate such a complex system of regulations by genetic engineering will not be solved for a long time to come. The speculation that the average intelligence of whole populations could be enhanced artificially are, in my opinion, not warranted. This will be impossible unless we are confronted with completely unexpected developments.

In conclusion, let me summarize the trends visible to date (Fig. 4).

Which of the two tendencies, favourable or unfavourable, will be stronger in the long run cannot be predicted to date. It will largely depend on the behaviour of future generations. Will our descendants find a way to protect themselves against ionizing radiation and chemical mutagens? Will they proceed in the reasonable regulation of reproduction, taking into account genetic risk factors? Or, to put the question in a more general way: Will they manage to create an environment which damages their biological inheritance as little as possible? This opens to them a wide field of reasonable "eugenic" measures. Within the variety of reasonable precautions, methods of genetic engineering will have a relatively small place. They will not change the fate of whole populations, but, in a good number of single cases, they will hopefully help the doctors to manage some problems in a somewhat better way.

References

[1] *Allison, A. C.* 1964. Polymorphism and natural selection in human populations. Cold Spring Harbor Symposium on Quantitative Biology, vol. 29, Human Genetics, New York. p. 137.

[2] *Becker, P. E.*, (ed.) 1966. Infantile progressive pinale Muskelatrophie. *In:* Werdning-Hoffmann, Handbuch für Humangenetik, vol. 5, 1. Stuttgart: Thieme, p. 336.

[3] *Crow, J. F.*, and *Kimura, M.* An Introduction to Population Genetics Theory. 1970. New York, Evaston and London: Harper & Row.

[4] *Emery, A. E. H.* 1970. Antenatal diagnosis of genetic disease. *In:* Modern trends in human genetics, ed. Emergy, London; Butterworths.

[5] *Fischbein, L. W.* et al. 1970 Chemical mutagens in man's environment, Academic Press, New York.

[6] *Fudenberg, H.* 1971. Primary immunodeficiences. Pediatrics **47**, 5.

[7] *Hienz, H. A.* 1971. Chromosomenfibel. Thieme, Stuttgart.

[8] *Hilschmann, N.* et al. 1969. Structure and Formation of Antibodies. Bayer-Symposium I "Current Problems in Immunology". Berlin, Heidelberg, New York: Springer Verlag.

[9] *Hollaender, A.* ed. 1971. Chemical mutagens, Vols. 1 and 2, New York and London: Plenum Press.

[10] *Huxley, J.* 1963. The future of man – evolutionary aspects. *In:* Man and his future, Ciba Symposium 1962. London: Churchill Ltd.

[11] *Lederberg, J.* 1963. Biological future of man. *In:* Man and his future, Ciba Symposium 1962, London: Churchill Ltd.

[12] *Lenz, W.* 1967. Anomalien der Autosomen unter besonderer Berücksichtigung des Schwachsinns. *In:* Handbuch für Humangenetik, vol. 5,2. Stuttgart: Thieme, p. 340.

[13] *Li, C. C.* 1955. Population genetics. Chicago: The University of Chicago Press.

[14] *Livingstone, F. B.* 1967. Abnormal Hemoglobins in Human Populations. Aldine Publishing Company: Chicago.

[15] *Matsunaga, E.* 1965. Measures affecting population trends and possible genetic consequences. Unit. Nat. World Populations Conference, Belgrad, Aug.–Sept.

[16] *McKusick, V. A.* 1968^2. Mendelian Inheritance in Man. Catalogs of Autosomal Dominant, Autosomal Recessive, and X-Linked Phenotypes. William Heinemann Medical Books Ltd.: London.

[17] *Muller, H. J.* 1963. Genetic progress by voluntarily conducted germinal choice. *In* Man and his future, Ciba Symposium 1962, London: Churchill Ltd.

[18] *Penrose, L. S.* 1955. Parental age and mutation. Lancet **2**: 3/2.

[19] *Post, R.* 1971. Possible cases of relaxed selection in civilized populations. Humangenetik **13**: 253–84.

[20] *Reed, T. E.*, and *Neel, J. V.* 1959. Huntington's chorea in Michigan. Am. Journ. of Human Genetics **11**:2.

[21] *Richards, B. W.* Mongolism: The Effect of Trend in Age at Childbirth on Incidence and Chromosomal Type. J. of Mental Subnormality, **13**, 3 (1967).

[22] *Vogel, F.* 1961. Lehrbuch der allgemeinen Humangenetik. Berlin–Göttingen–Heidelberg Springer.

[23] *Vogel, F.* 1967. Wie stark ist die theoretische Häufigkeit von Trisomie-Syndromen durch Verschiebungen im Altersaufbau der Mütter zurückgegangen? Zool. Beiträge, N.F., **13**: 451.

[24] *Vogel, F.* et al. 1969. Strahlengenetik der Säuger. *In:* Fortschritte der allgemeinen und klinischen Humangenetik, ed. P. E. Becker et al., Vol. 1. Thieme, Stuttgart.

[25] *Vogel, F.*, and *Röhrborn, G.* eds. 1970. Chemical mutagenesis in mammals and man. Springer, Berlin–Heidelberg–New York.

[26] *Vogel, F.*, and *Helmbold, H. W.* 1972. Blutgruppen: Populationsgenetik und Statistik. *In* Handbuch für Humangenetik, vol. 1,4. Stuttgart: Thieme, p. 129.

[27] *Zerbin-Rüdin, E.* 1967. Idiopathischer Schwachsinn. *In* Handbuch für Humangenetik, vol. 5,2. Stuttgart: Thieme, p. 158.

Discussion

McClearn: I appreciated very much Dr. *Vogel's* balanced view that things are perhaps not going downhill at an accelerating pace. With respect to one of his items concerning an unfavorable trend, that is the higher reproductive rate of feebleminded individuals, it might be pointed out that there are some recent demographic data in the United States by *Bajema* and by *Higgins, Reed,* and *Reed* that suggest that some of the earlier analyses that had led to predictions of IQ decline had omitted a very salient point. A number of low-IQ individuals have zero reproduction, and when these zero-sized families are included in the calculations, no net reproductive disadvantage to individuals with higher IQs is evident. It is not a linear effect in any case, but at the present time the best projection one could make is neither a net increase nor a net decrease in IQ based on present reproductive patterns.

Bodmer: The question of the reduction in the incidence of mongolism with a reduction in the average age of reproduction is, of course, an interesting one that has been discussed quite a lot. I think it is worth pointing out, however, that these changes must be relatively transitory because primitive populations presumably have an age of reproduction which is much lower than that in the developed countries, at least until recently. I was a little puzzled by your statement about the effects of voluntary curtailment of reproduction in families with genetic disease. Since so much genetic disease is recessive, I would have thought one would make the assumption that curtailment is only in matings between heterozygotes. This leads effectively to selection against recessives, which tends to be very inefficient and would have little effect. I think we all agree that it may not be wise to put large amounts of money into genetic engineering when, as you point out, the genetic complement of the human race may not be deteriorating, though maybe it will if we don't have *environmental* engineering. One should however, distinguish individual from population effects. Thus one should distinguish the advantage to the individual who is helped, from what might be the effect on the frequency of any particular trait in the population. I very much agree with what you say about the effects on the population, but we shouldn't forget the benefits to the individual of being able to help with respect to genetic diseases.

Vogel: With respect to the last point, I completely agree with you, and I am not opposed to any genetic engineering attempt. Even if we only find some therapeutical tools which can help a very small minority of patients, then we still have to follow this up. You are also right when you say that many of the genetic defects are recessive, but even in the case of recessive defects, if those parents who have one child would refrain from having the second or have a prenatal diagnosis made, then there would already be an appreciable population effect.

Bodmer: In terms of changing gene frequency, I don't think it would be very appreciable, because it is, effectively, selection against the recessive. Since most of the genes are in the heterozygote, the frequency of the disease is a function of the gene frequency; therefore the effect on the frequency of the phenotype would still, I think, be fairly marginal.

Vogel: This is completely correct, of course. However, we are not so much interested in the effect on the gene frequency; we are interested in the effect on the phenotype frequency, the frequency of hereditary disease, and this could be reduced in this way. I think we agree on this point.

Bresch: Let me underline one word, which was mentioned, I think, for the first time in this meeting, the word *amniocentesis.* Dr. *Vogel* mentioned it. If our motive for future research will be that of helping mankind, then, in my view, amniocentesis is much more promising than all genetic engineering. In the future, we will be able to recognize even heterozygotes in biochemical tests, then abortion could follow. Thus, when a PKU-child is detected, one could trace that pedigree of both parents' families to make sure that all couples get only healthy children, rather than children with a genetic disease whom we try to cure of their dieseases afterward. Furthermore, I think it will be much easier scientifically to develop methods for prenatal diagnosis of many diseases than to find a fitting virus to change the germ line which would have to be the final aim of genetic engineering.

Vogel: I agree with Dr. *Bresch* that for future developments amniocentesis and prenatal diagnosis will be more useful than genetic engineering. However, I also wish to put some water into this wine. There are, after all, only a very limited number of patients with whom you can make prenatal diagnosis because you have to diagnose the defect of the patient in the amnion cell or amnion fluid, and this is only possible in a minority of cases. I think that from the numerical point of view, classical genetic counselling is much more important on the basis of Mendelian laws. This may sound very conservative, but I think this will be the most useful way for a long time to come.

Glick: I would like to take exception to what Dr. *Bresch* just said, but that would be political dynamite in the United States. Let me give you an example. There are over two million Negroes in the United States who are heterozygotes for the hemoglobin S trait and over 50 thousand Negroes who have sickle cell anemia. Now, to tell these people – over 7 % of the Negroes in the Unites States – that they would be targets for abortion, would aggravate racial problems which already exist and lead to charges of genocide. I don't think that is the answer, and I don't think medical history has ever gone in the direction of "let's cut it out and not try to cure it."

Bodmer: Sickle cell anemia is perhaps one of the largest single public health problems arising from a genetic disease, and one that is being discussed a great deal. It has been suggested by my colleague *Cavalli-Sforza* among others, that one way one might solve a great deal of that problems is to discourage matings between heterozygotes. This is a process that is a little analogous to voluntary curtailment of reproduction once you know you've got a recessive child. The suggestion that he made, which, in fact, has been taken up with some interest by the WHO, is that simply education to understand the problems (since the frequency of mating between heterozygotes is not that high) may be a great help in actually reducing the incidence of individuals with anemia, which is a severe problem. Here it is a question not of genetic counselling after marriage but of premarital counselling, which really might be of great help.

Hausmann: I would like to comment on your remark about the "political dynamite" amniocentesis would represent in the United States. As far as I remember, the only large-scale amniocenthesis program which has ever been carried out was done in the Unites States. I think the Jewish community in New York has developed such a program regarding Tay-Sachs disease. Of course, this is the main point one has to make: it must be on a voluntary basis. I think that nobody has suggested that amniocentesis would be a compulsory check to be done in any kind of suspicious case.

Migeon: Perhaps I might say a word about the Tay-Sachs program since there seems to be some confusion about its intent. The program was established in Maryland, primarily to determine the feasibility of mass population screening for identification of carriers, as a means of preventing genetic disease. Since heterozygotes for Tay-Sachs disease could be readily identified, and since an affected child could be detected prenatally by amniocentesis, it seemed possible to avoid the birth of even the firstborn child affected with the disease.

Hofschneider: Could you do me a favor and explain to the nonmedical people what amniocentesis is? Perhaps there are many who don't know it.

Migeon: Amniocentesis is the process by which a small amount of fluid is removed from the amniotic sac in which the fetus is swimming. This is carried out sometime between the 12th and 16th weeks of gestation when the relative size of the fetus and sac is appropriate for removal of about 10 to 20 cc of fluid. This fluid contains cells that have been shed from amnion and the fetus itself and are, therefore, fetal in origin. Both cells and fluid may be used for a variety of diagnostic purposes. The cells can be examined directly for some of their biochemical characteristics, or they may be cultivated to provide proliferative populations of cells for karyotyping and enzyme studies. The cell-free fluid also may be used for chemical analysis. Unfortunately, the number of genetic diseases which can be detected by amniocentesis is limited to the relatively few which have a well-defined enzyme defect or some marker characteristic of the disease, and then only if the marker is found in amniotic fluid cells. For example, a disease like phenylketonuria, attributable to the absence of an enzyme in liver cells, cannot be diagnosed from studies of amniotic fluid cells since cells of nonhepatic origin in normal individuals also lack the pertinent enzyme.

The point I wish to make is that the Maryland Tay-Sachs program, a very extensive one, is involved in trying to detect *adult carriers* of the disease before their first baby is born. The carrier state is being identified, not in fetuses by amniocentesis, but in consenting adults by blood analysis. Amniocentesis is performed only when both the mother and the father of the fetus have been shown to be carriers, and the pregnancy terminated only if the fetus is affected, and only then on parental demand. In fact, one of the important prerequisites to amniocentesis is that an affected fetus be distinguishable from a carrier, so that (as in the case of recessive diseases) you are not in the position of terminating three-fourths of the pregancies any one person might have. I think we must consider the psychological repercussions of our actions. The other point I wish to make is that since each of us carries an estimated three to five lethal genes, selection against heterozygotes might be detrimental to all of us.

Siniscalco: I waited for the last minute to say the following because I did not want to discourage the discussion on the items raised in this session. My feeling is that for the time being and for a long time to come, it is the genetically abnormal patient who is helping the molecular biologist and not the other way round. Yet, even if medically useless, the kind of experimentation that we came here to discuss should be encouraged and emphasized since it may obviously lead to a better understanding of the genetic mechanisms controlling the normal cell differentiation and its errors.

Regarding the much publicized plans about the massive screening of genetic defects before birth, I would like to remind the enthusiastic supporters of such a colossal enterprise that we still live in a world where about one-quarter of the individuals are irreversibly damaged in their biological fitness because of malnutrition during fetal life or soon after birth. Thus, given the limitations of the funds available to preventive medicine, it would make more sense to make every effort to prevent the wastage of good genetic makeups before embarking oneself in the impossible task of eradicating the "bad genes" from mankind.

Last but not least, how do we decide about "bad genes"? The fact that certain genes, which are lethal in homozygous condition such as the sickle cell gene and the gene for cystic fibrosis, are maintained at a very high equilibrium frequency, leads to the conclusion that they may be beneficial for the species even though deleterious to the individuals under given circumstances. To pull these genes out from our genetic pool may turn out to be an irreversible disaster for the species. Thus, it is our task as geneticists to convince the public and the governments that the birth of genetically handicapped persons is the price that the species has to pay for its survival and consequently the care for such patients and for their families must be everybody's concern. I hope it is clear, however, that what I said so far is not meant to underestimate the need and the importance of prenatal screening in the so called "risk families", nor to discredit the encouraging efforts that are being made for the "phenic" (rather than "genic") therapy of genetic diseases.

Vogel: I completely agree with Dr. *Siniscalco.* What I wanted to give you was a simplified view, but Dr. *Siniscalco* has told you about some of the discussions which are now going on between human geneticists and human population geneticists. Take diabetes, for example. It is a very difficult question whether it would be useful to reduce the genes for diabetes in the populations; it might even be very useful to maintain and cultivate the diabetes genes in the population because, as Neel has said, maybe the diabetic has a genotype that can cope better with malnutrition. So if mankind faces a situation where the intake of calories will be short, maybe diabetic phenotypes will help us to survive as a species. This is possible.

Migeon: I think there's some misunderstanding about the word "lethal". I was referring to genetic lethals – detrimental genes that, in double dose, may result in the birth of an abnormal fetus, who because of his disease is subsequently unable to reproduce.

Sinsheimer: Well, I suppose how one looks at genetic engineering depends on whether one is a pessimist or an optimist, and I suspect that is genetic in itself. So maybe the first thing is to engineer everybody to be optimists, then we won't have to discuss it any more. But it did seem to me that the discussion that was going on previously was too dour.

It is easy enough to outline all the difficulties, and we are professionally always well aware of them, for what we can do now is, indeed, very limited. Nonetheless, it seems to me that the rate of progress in science is such that if you had held a conference on genetic exchange say twenty years ago in 1951, anybody who had prognosticated that we would know as much about it as we do now would have been considered a grievous optimist. So it doesn't seem to me unlikely that if Schering holds a repeat of this conference in 1991 – some of us may be around to attend – I would guess there would really be some therapies of some relatively simple genetic diseases available by that time. It doesn't seem that unreasonable to me.

While, of course, there are other very important things to do, I would refer to the point that Dr. *Bodmer* made that, *to the individual,* these are very serious problems. And *collectively,* of course, it is a very serious problem. The statement has been made, and I have no reason to doubt it, that 25 % of the hospital beds in the United States are occupied by people who are suffering from genetic disease – that's a major problem. We could put it another way, if you estimate the life-years that are lost per year, since genetic disease frequently starts relatively early, the total for genetic disease is several times what it is for heart disease or for cancer, which, of course, strike much later.

So the only point I would make is that I think one could be more sanguine about it than some of the previous comments have been. For two reasons: one, I anticipate continued progress and, secondly, it is clear that there are many phenomena in genetics and biology that we do not understand yet; when we do understand them, this knowledge will afford new possibilities beyond those that we can presently envision for genetic therapy, and I think these will become evident. The voice of the optimist.

Cellular Engineering – An Approach to Treatment of Genetically Determined Disease

Robert A. Good

Departments of Pathology, Pediatrics, and Microbiology, University of Minnesota Hospitals, Minneapolis, Minnesota 55455, USA

Summary: Immunodeficiency diseases are infrequent but not rare diseases which present major problems for diagnosis, treatment and prevention. Since Bruton originally described agammaglobulinemia we in Minneapolis have discovered and studied more than 500 cases of immunodeficiency disease. Most of the patients with these diseases appear to be suffering from disorders of development of the lymphoid apparatus which are genetically determined.

Correction of some of these immunodeficiency diseases has been possible. The immunodeficiency associated congenital absence of thymus often associated with failure of development of parathyroid glands and abnormalities of great vessels and outflow tracts of the heart (DiGeorge syndrome) has been corrected by transplantation of allogeneic thymus. This procedure now has a strong experimental basis from work carried out by Biggar et al. in our laboratories. Severe dual system immunodeficiency (Swiss type agammaglobulinemia) of both autosomal recessive and x-linked recessive types has repeatedly been completely corrected by bone marrow transplantation using matched sibling donors. Efforts to overcome the HL-A barrier in marrow transplantation have regularly been frustrated by the production, with this therapeutic approach, of fatal graft-versus-host reactions. Efforts to overcome this immunogenetic barrier will be described. The successful marrow transplants that overcome weaker immunogenetic barriers are associated with the development of blocking antibodies capable of inhibiting a killer effect of circulating sensitized lymphocytes on cells of the recipient host.

The approach to establishing lymphoid cell lines from such patients is discussed.

We have come a long way over the past three days and covered some rough roads. Although I am perhaps over my depth in considering genetic engineering with this august body, I will attempt in my dissertation to present a hopeful perspective. I would not be pleased if we were to leave this meeting giving support to *Piet Hein* who wrote in one of the "Grooks": "Experts have their expert fun, ex cathedra telling one, just how nothing can be done." Further, in listening to the biblical focus this afternoon, I was provoked to view my own work in a biblical context.

Manuscript received: 20 March 1972

Perhaps the first euphenics was practiced when Eve was created from man's rib. In the product, there can be no question the Lord really did a job this time out. Perhaps this should be our ideal. When we get really good at cellular engineering utilizing marrow transplantation, we should have much more to offer than is now the case. We will, for the moment, continue to accept the first euphenics as a model.

I have been concerned over the past 30 years with using experiments of nature to help unravel the relations of structure to function in the lymphoid system [25]. Often, the patients who represent these experiments of nature have unusual diseases. Occasionally very rare diseases. Such patients, as mentioned earlier in this symposium by *Siniscalco,* have often taught us much about fundamental issues in biology. Among such patients are those who, having perturbations of the mechanisms of the bodily defense, have taught us much about how the healthy remain healthy. In reality, these patients often represent genetically based model systems. These must be studied thoroughly, defined as precisely as possible, and, of course, whenever possible treated effectively to prevent or delay the lethality of the process. We have no real rights as physicians to study such patients if we are not at the same time bound to use all the resources of modern medicine to treat them in the most effective fashion available. To date, our contributions to the welfare of most of those unique experiments of nature that relate to issues of bodily defense have not been great. Antibiotic treatment of their bacterial infections, gamma globulin therapy providing a degree of passive immunization, supportive therapy for exudative enteropathies, buddy system plasma therapy, treatment of their autoimmunities and malignancies, and supportive pulmonary hygiene have been all that is available. Often it is not enough.

I would, in the few minutes available to me this afternoon, like to describe some developments that have taken place in my clinics and laboratories which I believe bring us to a position where we can reconstruct the lymphoid apparatus and, indeed, help some of these patients in a dramatic way by a process I have referred to as cellular engineering. Somewhat of a Pandora's box was opened when Col. *Ogden Bruton* discovered a child who was very susceptible to infection and who had agammaglobulinemia [5]. Prior to the antibiotic era, these patients did not exist as such, because they died early of their infections [17]. Among the agammaglobulinemic patients are those who cannot produce antibodies. They fail to develop immunoglobulins. On immunoelectrophoretic analysis, a pattern is developed which is grossly deficient in all major immunoglobulin fractions. These patients lack IgM, IgA, IgD, and IgE and have only the tiniest amount of IgG. These patients suffer from a genetically-determined disease which is transmitted as an x-linked recessive trait, and stimulation with the most powerful antigens we have available fails to bring about demonstrable antibody synthesis [12].

These patients have a particular defect in their lymphoid tissue, and this is regular in this rare disease. If you look at the lymph nodes you will find quite a well-developed population of lymphocytes in the deep cortical areas of the lymph nodes. In the far cortical areas, by contrast, you can see that they have very few cells, and they do not develop germinal centers. Plasma cells are also lacking. Life for these patients is a succession of life-threatening bacterial infections. The organisms causing their major problems are the high-grade encapsulated pyogenic pathogens. So these patients have one attack of infection with pneumococcus, streptococcus or Hemophilus influenza, meningococcus or Pseudomonas aeruginosa. Life for them in the sea of microorganisms that we have chosen as our ecological niche is but a succession of such infections [18].

These patients fail to form antibodies because they lack the plasma cell and antibody secreting population of lymphoid cells. Large basophilic lymphocytes and plasma cells that appear in the lymphoid tissue and reticular tissue following antigenic stimulation of normal persons just are not present in the lymphoid tissues of such children. And you can look very exhaustively in their tissues, and you will not find such cells even though you may have repeatedly stimulated the lymphoid tissue with potent bacterial, viral, or fungus antigens. But such patients do not lack immunity. Indeed, they have the capacity to develop what we call cellular immunity. They will develop a positive tuberculin reaction. They will develop inflammation based on cellular immunity following appropriate antigenic stimulation to a wide variety of antigens. Further, you can take their circulating lymphocytes,transfer this type of specific reactivity to an immunologically normal individual who is not sensitive, and that person promptly becomes sensitized [19]. This sensitivity is higly specific for the antigen to which the agammaglobulinemic patient has been sensizited. Thus, patients with what I have called the Bruton-type agammaglobulinemia failed to form immunoglobulins and antibodies and plasma cells on the one hand, but can develop delayed allergic responses very well on the other. They have a normal thymus, and they have a normal thymus-dependent T-cell population, according to our new terminology for the lymphoid functions just pointed out for you by Dr. *Rajewsky.*

These patients are patients for whom we now have developed an experimental model. One can create essentially the same problem in the chicken by going into the egg and taking out the strange and peculiar lymphoepitheilial organ located at the posterior end of the gastrointestinal tract that is called the bursa of Fabricius. If you get that organ out completely at an early stage of embryonation, e.g., 15–16 days, these chickens will be just like the agammaglobulinemic humans. They will have a good population of T-cells that differentiate in the thymus, will have a perfectly adequate cell-mediated immune response, but will lack all the known chicken immunoglobulins. The discovery that the bursa is essential to the development of antibodies was made by *Glick* and *Chang,* graduate students at Ohio State University

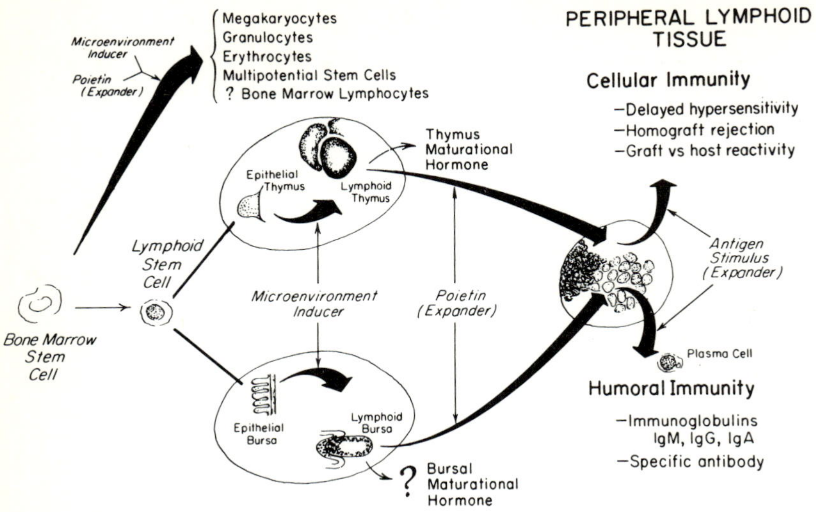

Fig. 1. Scheme depicting development of the lymphoid system along two separate lines.

in the mid-1950's [14, 15]. *Glick*'s experiments coupled with our study of patients with thymona and immunodeficiencies led to our analysis of the influence of the thymus in rabbits, mice, rats, and hamsters and the demonstration that the thymus, like the bursa of Fabricius, is essential to the development of immunity. After much work and considerable backing and filling, we and others have established that the thymus subserves development of a population of lymphocytes that subserve the cellular immunities. It is now clear (presented in Fig. 1) that there actually exist, in the marrow of the bone and in fetal liver and yolk sac, stem cells which can be differentiated into a variety of hematopoietic elements. This differentiation depends upon the microchemical environment in which these cells find themselves. One line of differentiation can be to a population of cells that subserve the functions of cellular immunity, and another line of differentiation – that in birds is directed by the bursa of Fabricius – gives rise to the cells that engage in antibody production. Now this scheme, which we constructed in 1967, has actually been formally tested making use of a variety of markers to follow the cells through cell development [28–33]. One such marker is a T-6 chromosome marker by which one can, in syngenic mice, follow the cells from place to place. In addition, we use the surface antigenic markers on the lymphocytes. For example, the theta-isoantigen is a fine marker for T-cells if one uses properly absorbed antiserum [3]. Also one can use functional markers. One such marker is the response of the cells of peripheral blood or lymphoid tissue to phytohemagglutinin. This response we have now formally

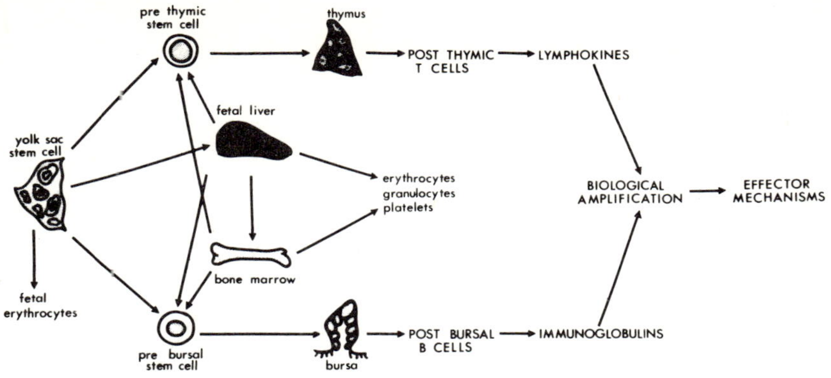

Fig. 2. Traffic patterns and lineage of T and B cell systems as analyzed from studies of cell movements, surface markers, chromosomal markers, and functional markers.

shown to be exclusively a response of T-cells – the cells that have differentiated under the thymic influence. B-cells do not respond to PHA stimulation under usual conditions. The functions of cellular immunity include capacity to initiate graft-versus-host reactions, a major component of solid tissue allograft rejection, major elements of defense against certain virus, fungi, and facultative intracellular pyogenic pathogens. Even the process of immune surveillance belongs to the functions of cells of this class. Thus we can now actually trace the odyssey illustrated in Fig. 2 in a formal way from yolk sac, where cells will differentiate only to a terminal population of fetal erythrocytes, to fetal liver, through prethymic cells, to postthymic cells which can be turned on by an immunological event, to produce substances that can engage a complicated biological amplification system to activate the effector mechanisms. The latter involve very primitive processes like phagocytosis, vascular reactivity, inflammation, and coagulation. This same stem cell population can follow a second pathway that gets to the fetal liver and ultimately to the bone marrow. In birds, this population develops into prebursal cells and, under the influence of the bursa, to postbursal B-cells that become the immunoglobulin-producing cells. The latter also operate through a different set of amplification systems to engage the major effector mechanisms listed above.

Studies of the cells of the two systems reveals a heterogeneity that is getting more and more complex by the day. Cell lines attributable to the thymic influence include a prethymic stem cell that is theta negative. This cell comes under the influence of the thymus and thus differentiates to postthymic stem cell. The latter can be further differentiated by the indirect influence of the thymus to a T-1-cell that can be pressed by antigen into another compartment, which is immunocompetent T-cell. Some of these immunocompetent T-cells are long lived and some are short lived. They can be turned on to a lymphokine-producing cell that engages the amplification systems or to a memory cell. There is some evidence that there is even greater heterogeneity in this T-cell system than is indicated above and that perhaps parallel as well as longitudinal compartments exist. I find this complexity not at all surprising. We see in the B-cell system an extraordinary heterogeneity of bursal-derived or thymus-independent populations. In this line of development, a prebursal cell comes under bursal influence and becomes a postbursal stem cell or B-cell. This element located at first only in the bursa and later also in bone marrow and spleen. One can completetly reconstitute a bursectomized agammaglobulinemic chicken, with syngeneic cells of this postbursal class, whether they be in the bone marrow after bursal involution or in the bursa itself prior to its involution or even if they are present in other parts of the lymphoid tissue [37]. The bursal line of cells seems to have to go through an IgM-producing step to achieve capacity to produce IgG. They ultimately become specifically Ig-secreting cells by a further step of differentiation. Thus B-producing cells and B-secreting cells can now be sharply distinguished in this cell line. We are not certain whether the IgA-, IgD- and IgE-producing cells have to go through the IgM step because it has not yet been possible to analyze this question in the same way that we analyze the IgM-IgG sequential relationship. The important thing to remember is that there is great heterogeneity in this B-cell system, and that like the T-cell system the B-cell system operates through complex amplification and effector mechanisms. Now let's look in this perspective at the diseases of immune deficiency. Almost all of these are genetically determined. A few, like in utero rubella infection which produces hypergammaglobulinemia IgM and hypogammaglobulinemia IgG are congenital and not genetic, but the great majority are genetically determined (Figs. 3 and 4).

One type transmitted as an autosomal recessive seems to have a stem cell population that fails to develop into normal adult hemoglobin-producing cells, granulocytes, platelets, or into either of the two separate lymphoid lines. Of course these patients can live for only a few days in our sea of microorganisms. There are two forms, one an autosomal recessive trait and the other an x-linked recessive, where failure to develop either the T- or the B-cell system occurs. The other hematopoietic systems seem to develop quite well, although I am not quite certain if they always develop absolutely normally. There are patients who are born without a thymus. Such patients generally lack parathyroid glands as well and probably also the calcitonin-producing clear cells derived from third and fourth pharyngeal pouch. In the immunological perspective, these patients are lacking the T-cell system. They may also fail to form some antibodies because of the role of T-cell-B-cell interactions in antibody production mentioned early by *Rajewsky.* But other patients, like the ones I mentioned as the first experiment of nature in this series, are born essentially without the B-cell system, but their T-cell system is intact. And then there is a congeries of diseases – we have now studied more than 500 cases of primary immunodeficiency

Disease	Genetic basis
1. Bruton-infantile immunodeficiency	X-linked recessive
2. Wiskott-Aldrich syndrome	X-linked recessive
3. Ataxia-telangiectasia	Autosomal recessive
4. Severe dual system immunodeficiency (true Swiss)	Autosomal recessive
5. Severe dual system immunodeficiency (Gitlin-Boston type)	X-linked recessive
6. Immunodeficiency and short-limbed dwarfism	Autosomal recessive
7. Cartilage-hair syndrome	Autosomal recessive
8. Fatal granulomatous disease with abnormality of G-6-PD stability	X-linked recessive
9. Chronic granulomatous disease with deficiency of glutathione peroxidase	Autosomal recessive
10. Lipochrome histiocytosis with granulocyte malfunction	Autosomal recessive
11. Myeloperoxidase deficiency associated with recurrent candidal sepsis	Autosomal recessive
12. Common variable immunodeficiency with circulating B cells	Autosomal dominant
13. Common variable immunodeficiency with circulating B cells	Automsomal recessive
14. Transient hypogammaglobulinemia of infancy	Autosomal recessive in some cases

Fig. 3

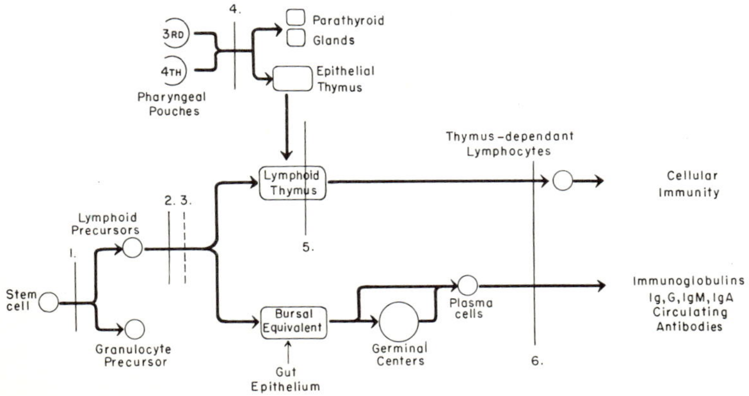

Fig. 4. Proposed blocks in differentiation and development of the lymphoid system which could account for the major immunodeficiency diseases.

disease in Minneapolis – and most of them are diseases in which there is one degree or another of involvement of both the T- and B-cell system. These include patients such as those with Louis-Barr's ataxia-telangiectasia associated with immunodeficiency, the Wiskott-Aldrich syndrome, and what I have called the common variable form of immunodeficiency.

Let's look at the patients with the so-called DiGeorge syndrome and consider lymphoid tissue of such patients. These are patients born without a thymus. They lack T-cells, but these patients have absolutely normal numbers of plasma cells in their lymphoid tissues. They have well-developed far cortical areas in their lymph nodes, they develop germinal centers well following antigenic stimulation, but they lack cells in the deep cortical regions of the node. This disease is regularly a fatal disease, and the patients generally die of atypical acid-fast organisms, one or another of the low-grade facultative intracellular bacterial pathogens, or they may get a fungus infection and die of that. They may be vaccinated and die of generalized or progressive vaccinia infection. The correction of this disease can be accomplished, or apparently has been accomplished by transplantation of fetal thymus [1, 7]. This treatment apparently achieves full correction of the structure of the lymph nodes and corrects at least some of the immunological and functional deficits of the DiGeorge syndrome. There remain some questions about correction of the DiGeorge syndrome because sometimes, just as in the experimental model of this disease which occurs as an autosomal recessive in nude mice, the thymus is not completely absent. Thus late development of cellular immunological capacity may occur [11]. Another form of disease is the one we call the Swiss-type agammaglobulinemia because it was originally described in Switzerland and has been analyzed primarily by *Hitzig* and *Willi* [20], *Barandun* [2], and *Tobler* and *Cottier* [36]. In this disease, the patients suffer from a condition inherited as an autosomal recessive. They fail to develop the T- or B-cell systems normally. Very often when a physician tries to help these children by vaccinating them, he destroys them with generalized vaccinia or vaccinia gangrenosa. Alternatively, they experience fungus infection or virus infection and die. They may be killed by vaccination. If they encounter the virus of measles, they develop overwhelming Hecht's giant cell pneumonia. And these patients just cannot grow up. Figure 5 is a tonsil from a normal child of two years of age, and you see the well-developed germinal center structure, the tremendous lymphoid accumulation in this region.

By contrast, in patients with the severe dual system immunodeficiency disease, a failure to develop any of this lymphoid structure is characteristic (Fig. 6). Here you see a characteristic tonsilar crypt from such a patient – no lymphoid elements are to be found. The normal lymph node, after antigenic stimulation, shows the germinal center development, and great accumulations of lymphocytes in the deep cortical areas, and in the medullary cords one encounters magnificent accumulations of plasma cells (Fig. 7). But the patients with the dual system immunodeficiency

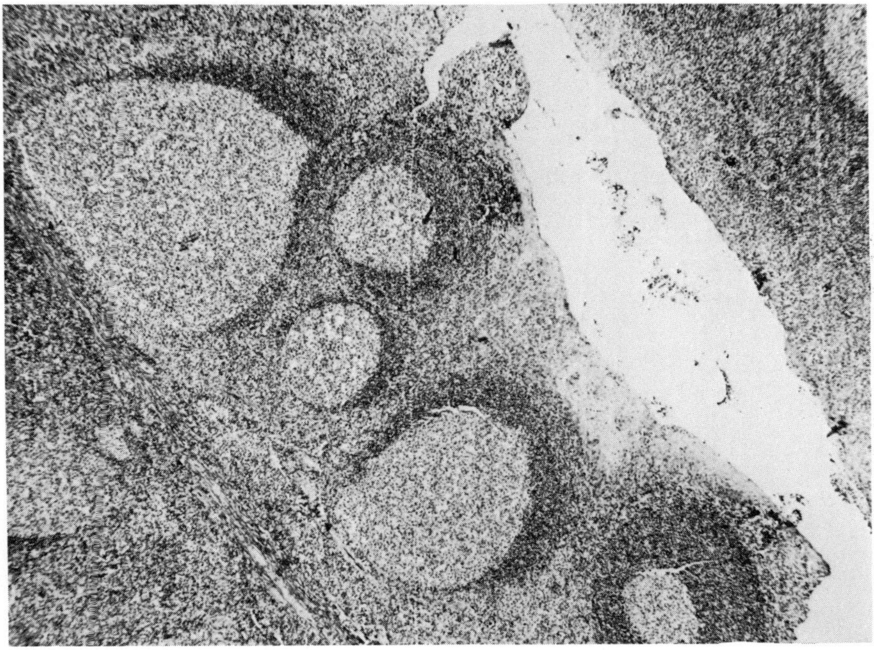

Fig. 5. Normal tonsil. Note huge germinal centers and extensive lymphocytic development – many plasma cells are also present.

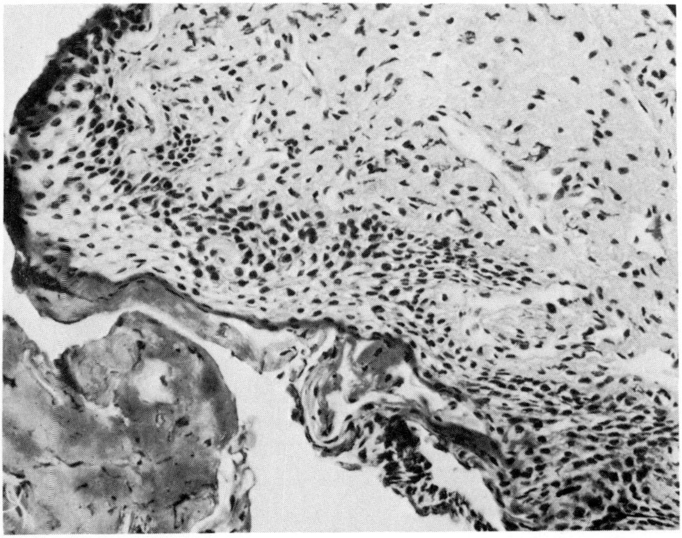

Fig. 6. Tonsil from patient with combined system immunodeficiency. Epithelial crypt is present, but there is no lymphoid cell accumulation.

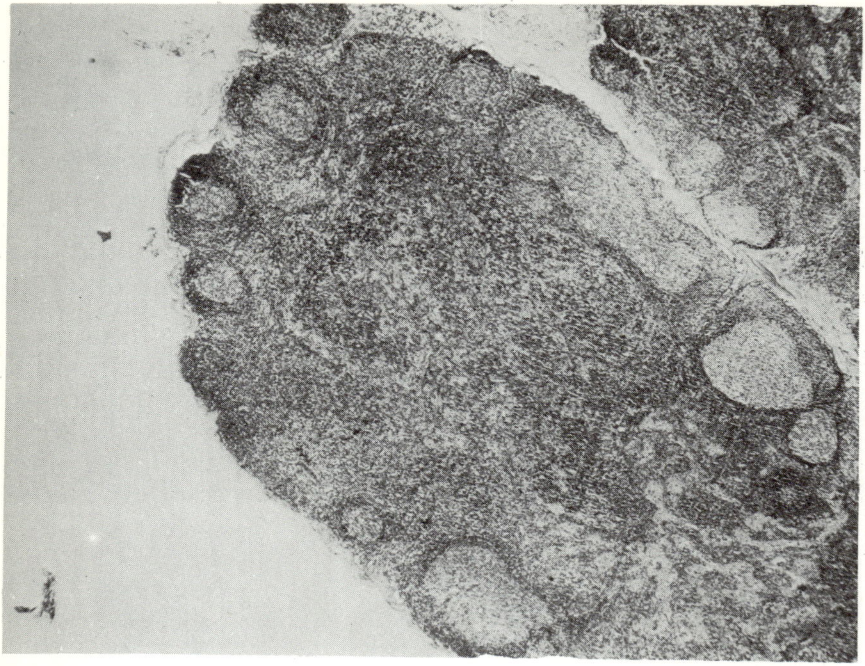

Fig. 7. Normal lymph node following secondary antigenic stimulation. Note germinal center development, deep cortical thymic dependent lymphoid tissue development, and medullary cords.

have a lymph node that contains almost no lymphoid cells. There are no germinal centers, no deep lymphocytes in thymic-dependent deep cortical areas, and no plasma cells. Because it is not obscured by these lymphoid elements, the stromal reticulum of the nodes can be clearly visualized (Fig. 8). The thymus of the normal child has cortex, medulla, Hassall's corpuscles, lymphoid, and stromal epithelium as the basis of its structure. By contrast, the thymus of a patient with a severe dual system immunodeficiency disease, although always being present, looks like the thymus of a fetus in a very early stage of development. The lobes and lobules are present but lack lymphoid structure. No Hassall's corpuscles are to be seen. Here then is the disease we figured from analysis of the scheme I just outlined for you where the patient has a developmental abnormality located at the point where the two systems of lymphoid cells have not yet separated from one another. This form of immunodeficiency can be transmitted either as an autosomal recessive or an x-linked recessive condition. We reasoned that if this were indeed the case, these patients should have a disease that might be correctable by marrow transplantation.

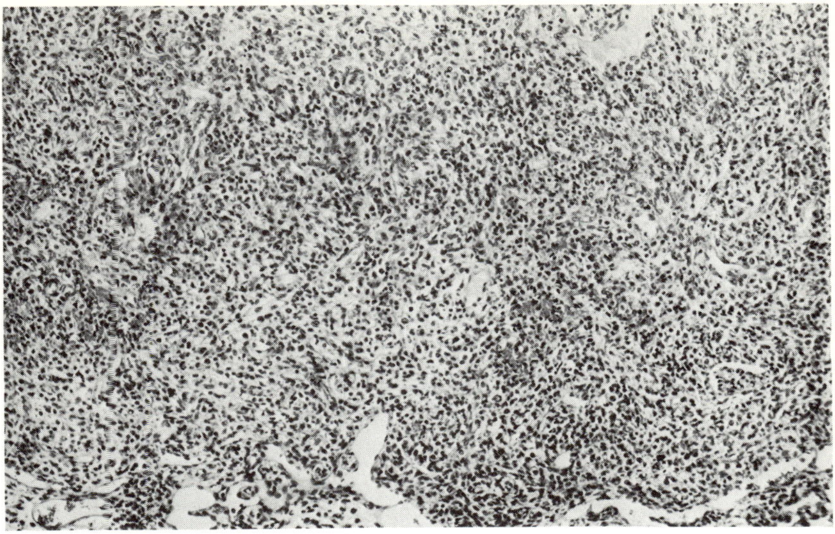

Fig. 8. Lymph node from child with autosomal recessive form of combined system immunodeficiency.

This we considered likely because in the bone marrow of the normal person there should be stem cells which could be differentiated to both T- and B-cells in the patient. We gave one of our little children with dual system immunodeficiency – a certainly fatal disease – a transplant of fetal liver cells. The patient was reconstituted immunologically with respect to both B- and T-cell functions. He became able to reject a skin allograft that had been in place for some two months, and he developed antibody-producing cells and began to produce antibodies. Unfortunately with this immunological reconstitution he developed a fever, rash, and liver involvement and then all manifestations of a graft-versus-host reaction. The latter was lethal. In this instance the liver had come from a fetus [21]. But many different tissues including bone marrow, peripheral blood, and fetal thymus have been repeatedly transplanted to these patients [27]. They either will not correct the immunodeficiency, or correcting the immunodeficiency they destroy the children with a graft-versus-host reaction. This was the situation in which we found ourselves when we realized that we had long been working experimentally with a very relevant system in Minneapolis. We had been working with what we called for many years the Minnesota system for studying transplantation immunity. In this approach you must cheat just a little bit. Instead of transplanting across major histocompatibility determinants, you match the mice up with respect to the major histocompatibility determinants. In mice these are governed by the H_2 histocompatibility locus. The mice are then mismatched only at minor H_1 and H_3 loci. And when you do that, we have found that you can make all sorts of fantastic transplantation experiments. For example, with this system we found it easy to produce so-called immunologic tolerance. We could transplant on hypophysis into a conditioned animal and the hypophysis would be fully functional and exercise all of the normal pituitary functions. Dr. *J. B. Aust* and his students even did experiments in which they castrated female C3H mice, transplanted orthotopically Ce ovaries, and did reproduction experiments in which C3H mice bearing Ce ovaries produced pure bred Ce strain offspring. This for mice was a dramatic form of organ and cellular engineering, and it was based on matching according to major histocompatibility determinants.

One thing we learned using this Minnesota system is that we could not kill animals with a graft-versus-host reaction if the recipient Fl animal in which we produced a graft-versus-host reaction was matched with the animal donating the lymphoid cells with respect to the antigens controlled at the H2 major histocompatibility locus [16]. Differences at non-H2 were, of course, present in these experiments. Nonetheless, lethality attributable to GVH did not occur in this system whereas death was regular when we dealt with a histocompatibility antigen mismatch at the major locus. One of the great advances in transplantation immunology in recent years has been beginning identification of the genetic determinants responsible for the major histocompatbility systems subserving transplantation rejection in man, dogs, monkeys, rats, and chickens as well as in mice. In transplantation biology, what seems to be true of the mouse is true of man and all the other species. There exists basically a region on one of the somatic chromosomes at which multiple alleles operate. The latter seem to control major aspects of transplantation immunity. I believe *Bodmer* was the first to suggest this very interesting relationship. This meant to us that within families the game might be very very different if reconstitution of immunologic capacity were attempted across only a minor rather than major histocompatibility barrier. While this was in mind [22], another patient was presented to us from the family depicted in Fig. 9. This family had been thoroughly analyzed, and two cases of dual system immunodeficiency had been defined in Boston. Both patients had been shown to have severe dual system disease. In this instance, the severe dual system immunodeficiency was transmitted as an x-linked recessive trait. Our little patient also had the lethal form of severe immunodeficiency. He had four sisters and one of these little sisters, an 8-year-old, matched with the patient by mixed leukocyte culture. I won't go into details, but the HL-A determinants of the patient and his sister were not perfectly matched. This sib pair of our family represented the first of the now popular recombinant families that have established for us the special meaningfulness of the MLC locus [38]. The sister matched well, however, with our little patient in the MLC test, and we transplanted her bone marrow to our little boy. The little boy developed a graft-versus-host reaction but just as was to be expected from the mouse experiments, the reaction was mild. Our little patient was completely corrected immunologically following this transplant.

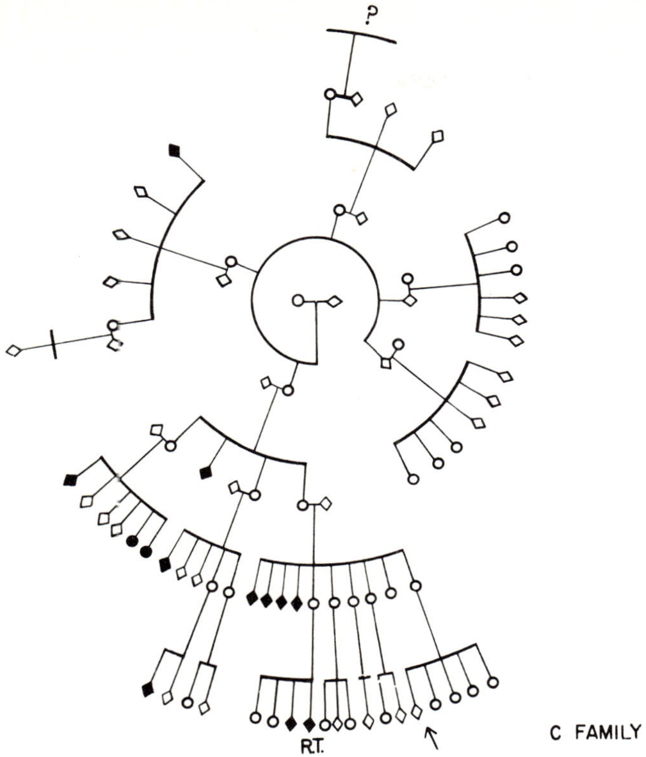

Fig. 9. Tree of family of child whose dual-system immunodeficiency was corrected by transplantation of bone marrow.

He developed capacity for PHA response. He became able to form cellular immunity to 2–4 DNFB. He developed small lymphocytes that he previously had lacked. He began to be able to produce isohemagglutinins, and he produced all of the immunoglobulins that had been lacking. His blood levels of immunoglobulins went progressively from very low to normal levels. He began to be able to form typhoid antibodies, he developed plasma cells, and he even developed a thymus shadow on the roentgenogram. We had succeeded by bone marrow transplantation to completely reconstitute the immunological system in a child with severe dual system immunodeficiency of the x-linked type [13]. In our enthusiasm for doing this transplant of the marrow with this matched sibling donor, we had accepted another mismatch. This little girl was of blood group 0 and the little boy was of blood group A. He was completely corrected immunologically, but after the correction he developed first a hemolytic anemia and then an aregenerative pancytopenia, so now he had an immunologically-induced pancytopenia. We had a recourse however. We still had his little sister and her MLC-matched bone marrow. We transplanted once again the bone marrow of the matched sibling donor into our little patient and completely corrected his hematological disease. We switched his blood type A to 0 and completely reconstituted his entire hematopoietic apparatus. The initial transplant had established some 25 % of the bone marrow cells that were clearly of the female karyotype. After the second transplant, the mitotic figures in the marrow by direct squash preparation jumped to almost 100 % female karyotype. This child remains well today some four years after the first transplant. He has full immunologic vigor. He is a perfectly normal and healthy boy in every respect. He deals with the usual infections and antigens in his environment. He still is a complete chimera. By that I mean he is of blood group 0. Interestingly, his blood group 0 cells possess at their surface the Lewis antigens attributable to his blood group A genetic heredity [34] and they also possess at their surface a determinant that *Flossie Cohen* has called a "weak blood group A antigen". Both of these, we think, are put on the cell surface after it has been differentiated as an 0 cell with the genetic determinant attributable to the sister's genotype. All the cells in the marrow are of female karyotype, all the responding cells in the peripheral blood are of female karyotype. This has to be taken as a heady experience. If this observation had just stood by itself, it would have taught us a great deal about immunological issues and provided much encouragement for ultimately dealing with these patients. But the experience doesn't stand alone, and it gives us regularly the possibility of reconstructing the entire immunologic apparatus when a matched sibling donor is available. Then to be able to switch permanently the blood type, as we did in this instance, is even more challenging. That is what we mean by cellular engineering, and, in this case, the results of this engineering enterprise were most dramatic to say the least.

Using these same basic manipulations and the same line of reasoning, we have corrected several additional cases among the immunodeficient children, which we have been studying in Minneapolis. We have since reconstituted fully four additional children in Minneapolis.

Thus far there have been, to our knowledge, 15 such cases around the world. They have been cured of clinical and laboratory evidence of disease by treatment with marrow from matched sibling donors. These children all have been fully reconstituted as far as the immunological system is concerned. The first case has thus far been the only one where it was necessary to switch the blood type. Many obstacles still remain, of course. No one has yet, to my satisfication, bridged the major histocompatibility barrier by marrow transplantation with complete reconstitution of a child with severe dual system immunodeficiency.

How are these children surviving with the foreign marrow in their bodies? Are they tolerant – they are surely chimeric. Dr. *David Jose* working with us [24] showed that these chimeric children have lymphocytes that will destroy their own fibroblasts. But they do not injure the body of the recipient child. The reason they do not appears to be that the patients possess blocking antibody, clearly an IgG antibody that can be removed from circulation on solid immune adsorbant columns. These antibodies can be eluted from that column in a form capable of showing the blocking activity. We believe it is this blocking or enhancing antibody that is interfering with this type of cellular assault. I call this adaptation immunodeviation, and I believe it is similar to the enhancement phenomena of tumor immunity long ago described by *Kaliss* [26].

Rebecca Buckley [6] has claimed and has presented some evidence for having bridged a major histocompatibility barrier using a blocking antibody from the blood of a maternal grand multipara to permit transplantation of the father's marrow cells. But Buckley did not have good markers, and she surely did not achieve full immunologic reconstitution by her HL-A mismatched marrow transplant.

Drs. *Park, Biggar,* and I have recently fully reconstituted a child by crossing the major histocompatibility barrier using very small marrow transplants. Indeed, we have reconstituted one patient quite completely beginning with only one million marrow cells. *DeKoning* et al. [9] have isolated stem cell fraction that is not responsive to PHA or to allogeneic cells and have, with this apparent stem cell fraction, reconstituted patients with dual system immunodeficiency. Reconstitutions have subsequently been reported from Los Angeles, Madison, Boston, Leiden, and Bern, as well as from Minneapolis. These efforts surely represent a most interesting and provocative experience. We are still not satisfied about the major barrier, and we are going after the solution of this problem in a number of ways.

From the first efforts at correction of severe dual system immunodeficiency, it is clear that cellular engineering can be used as an extraordinary approach to the correction of genetically determined disease. We have already corrected two different genetically determined abnormalities, the x-linked and autosomal recessive forms of dual system immunodeficiency. We have corrected fully an immunologically induced pancytopenia, and we have switched the blood type. Subsequently, it appears that correction of aregenerative pancytopenia has been achieved in basically the same way [35], namely by using matched sibling donors along with intense immunosuppression.

One of the things that has come from all of these studies is that we know a lot more (because we have been able to test our schemes in a new way) about the organization and development of the immunological system. We already have been developing approaches to overcome the major transplantation barrier, and these approaches are beginning to work in the laboratory and in the clinic. Perhaps they can soon be applied regularly to man.

We now also have been able to put a number of these immunodeficiency diseases in the test tube, so to speak, using tissue culture techniques. Dr. *B. Park* of my laboratory has developed a method where, in normal children, using only 20 cc of blood he can get a permanent diploid lymphoid cell line started in tissue culture with 50 % regularity. This method doesn't involve adding any virus – we of course don't know whether virus is present. Indeed the EB virus may very well be part of the selection process. No virus, however, is intentionally added and no other cell components or supernatants are added to these cultures. With such cell lines, we can literally culture kilogram quantities of these diploid cells from patients with genetically determined diseases. Such cell lines may make it more directly and more efficiently possible for people like Dr. *Siniscalco, Bodmer,* and *Harris* to study these diseases. Some of the cell lines have genetic and enzymatic markers that can be quantified and the manipulations of which could be analyzed. For example, we already have permanent cell lines from patients with the fatal granulomatous disease I originally defined [4, 21]. Three different cell lines from three different forms of the latter disease are now available in tissue culture. These are diploid cell lines. Further, we have cell lines from five or six different patients with what I have called the common variable form of immunodeficiency, and we are struggling to get the other forms of immunodeficiency into culture.

Let me emphasize that we look upon these patients with immunodeficiencies as model systems. They suffer from relatively rare diseases. They are, however, asking us important questions, and they provide a powerful approach to investigations in immunobiology because they pose critical questions which we then can take to the laboratory to get definitive answers. But we do feel, as you all must, about the clinical problems you face – the need to do something about their disease besides trying to destroy the heterozygote state. The latter is surely a futile effort and must be deplored. The diseases however are occurring and being presented to us. The patients are with us. We as physicians would welcome any kind of genetic engineering that would permit an even better approach to the treatment of this problem than our cellular engineering. Our cellular engineering, however, does provide an approach that works. If you allow your imagination to run a little bit, you can see that being able to switch the blood types, if we would do it reproducibly and across all histocompatibility barriers, might give us an approach to many diseases in addition to rare primary immunodeficiencies. Further, Dr. *E. Yunis* and I have been able to define one component of the aging process – the immunodeficiency of aging. Further, we have been able to manipulate separately both B-cell immunity and T-cell immunity by chemical methods and nutritional restrictions. Knowing that there are two and not one immunity system predicts this will be so.

I am also certain we will rapidly develop other means of treating, especially with transplantation methodology, a number of other genetically determined diseases.

Already Drs. *Desnick, Krivit, Najarian* [10], and their associates are correcting by renal transplantation inborn errors of metabolism like Fabry's x-linked glycospingolipidosis and Gauchet's disease. Others will come, and we think these approaches are only the beginning. Hopefully if progress occurs normally, we can expect, as in the biblical study alluded to above, to create quite a few beautiful ladies and maybe some from the bone marrow of man if not from his entire rib.

Acknowledgements: Aided by grants from The National Foundation-March of Dimes, American Heart Association, American Cancer Society, and U.S. Public Health Service (AI-08677, NS-02042, and HE-06314) and John A. Hartford Fdn.

References

[1] *August, C. S., Rosen, F. S., Filler, R. M., Janeway, C. A., Markowski, B.,* and *Kay, H. E. M.* 1968. Implantation of a foetal thymus, restoring immunological competence in a patient with thymic aplasia (DiGeorge's syndrome) Lancet **2,** 1210.

[2] *Barandun, S.* 1959. Das Antikorpermangelsyndrom. Basel/Stuttgart: Benno Schwabe & Co.

[3] *Boyse, E. A., Old, L. J.,* and *Stockert, E.* 1968. An approach to the mapping of antigens on the cell surface. Proc. Natl. Acad. Sci. **60,** 886–93.

[4] *Bridges, R. A., Berendes, H.,* and *Good, R. A.* 1959. A fatal granulomatosis of childhood. The clinical, pathological and laboratory features of a new syndrome. A.M.A. J. Dis. Child. **97,** 387–408.

[5] *Bruton, O. C.* 1952. Agammaglobulinemia. Pediatrics **9,** 722–28.

[6] *Buckley, R. H., Amos, D. B., Kremer, W. B.,* and *Stickel, D. L.* 1972. Incompatible bone marrow transplantation: Use of immunological enhancement. New Engl. J. Med. **285,** 1035–42.

[7] *Cleveland, W. W., Fogel, B. J., Brown, W. T.,* and *Kay, H. E. M.* 1968. Foetal thymic transplant in a case of DiGeorge's syndrome. Lancet **2,** 1211–14.

[8] *Cooper, M. D., Peterson, R. D. A.,* and *Good, R. A.* 1965. Delineation of the thymic and bursal lymphoid systems in the chicken. Nature **205,** 143–46.

[9] *DeKoning, J., Dooren, L. J., van Bekkum, D. W., van Rood, J. J., Dicke, K. A.,* and *Radl, J.* 1969. Transplantation of bone marrow cells and foetal thymus in an infant with lymphopenic immunological deficiency. Lancet **1,** 1223–27.

[10] *Desnick, R. J., Allen, K. Y., Simmons, R. L.,* and *Najarian, J. S.* 1971. Treatment of Fabry's disease. Correction of enzymatic deficiency by renal transplantation (abstract). Central Soc. for Clin. Res. J. Lab. Clin. Med. **78,** 789–90.

[11] *Fudenberg, H., Good, R. A., Goodman, H. C., Hitzig, W., Kunkel, H. G., Roitt, I. M., Rosen, F. S., Rowe, D. S., Seligmann, M.,* and *Soothill, J. R.* 1971. Primary Immunodeficiencies. Report of a World Health Organization Committee. Pediatrics **47,** 927–46.

[12] *Gatti, R. A.,* and *Good, R. A.* 1970. The immunological deficiency diseases. *In* Medical Clinics of North America, ed. M. G. Wohl, **54,** 281–307. Philadelphia: W. B. Saunders.

[13] *Gatti, R. A., Meuwissen, H. J., Allen, H. D., Hong, R.,* and *Good, R. A.* 1968. Immunologic reconstitution of sex-linked lymphopenic immunologic deficiency. Lancet **2,** 1366–69.

[14] *Glick, B.* 1970. The bursa of Fabricius: A central issue. Bio Sciences **20,** 602–4.

[15] *Glick, B., Chang, T. S.,* and *Jaap, R. G.* 1956. The bursa of Fabricius and antibody production. Poultry Sci. **35,** 224–25.

[16] *Good, R. A.* 1970. Progress toward a cellular engineering. J.A.M.A. **214,** 1289–1300.

[17] *Good, R. A.,* and *Bergsma, D.* 1968. Immunologic deficiency diseases in man. Birth Defects Original Article Series. New York: National Foundation Press.

[18] *Good, R. A., Finstad, J.,* and *Gatti, R. A.* 1970. Bulwarks of the bodily defense. *In* Infections agents and host reactions, ed. S. Mudd, pp. 76–114. Philadelphia: W. B. Saunders.

[19] *Good, R. A., Zak, S. J., Jensen, D. R.,* and *Pappenheimer, A. M.,* Jr. 1967. Delayed allergy and agammaglobulinemia (abstract). J. Clin. Invest. **36,** 894.

[20] *Hitzig, W. H.,* and *Willi, H.* 1961. Hereditare lymphoplasmacytare dysgenesis ("Alymphocytose mit agammaglobulinamie") Schweiz. med Wchnschr. **91,** 1625.

[21] *Holmes, B., Quie, P. G., Windhorst, D. B.,* and *Good, R. A.* 1966. Fatal granulomatous disease of childhood: An inborn abnormality of phagocytic function. Lancet **1,** 1225–28.

[22] *Hong, R., Gatti, R. A.,* and *Good, R. A.* 1968. Hazards and potential benefits of blood transfusion in immunological deficiency. Lancet **2,** 388–89.

[23] *Hong, R., Kay, H. E. M., Cooper, M. D., Meuwissen, H., Allen, M. J. G.,* and *Good, R. A.* 1968. Immunological restitution in lymphopenic immunologic deficiency syndrome. Lancet **1,** 503–6.

[24] *Jose, D. G., Kersey, J. H., Choi, Y. S., Biggar, W. D., Gatti, R. A.,* and *Good, R. A.* 1971. Humoral antagonism of cellular immunity in children with immune deficiency reconstituted by bone marrow transplantation. Lancet **2,** 841–43.

[25] *Kagan, B. M.,* and *Stiehm, E. R.,* eds. 1971. Immunologic incompetence. Chicago: Yearbook Medical Publishers.

[26] *Kaliss, N.* 1958. Immunological enhancement of tumor homografts in mice: A review. Cancer Res. **18,** 992–1003.

[27] *Miller, M. E.* 1968. Graft-versus-host reaction in man with special reference to thymic dysplasia. *In* Immunologic deficiency diseases in man, eds. R. A. Good, and D. Bergsma, pp. 257–63. New York: National Foundation Press.

[28] *Stutman, O.* 1970. Hemopoietic origin of cells responding to phytochemagglutinin in mouse lymph nodes. *In* Fifth leukocyte culture conference, ed. J. Harris, pp. 671–81. New York: Academic Press.

[29] *Stutman, O.,* and *Good, R. A.* 1971. Immunocompetence of embryonis hemopoietic cells after traffic to thymus. Transpl. Proc. **3,** 923–25.

[30] *Stutman, O., Yunis, E. J.,* and *Good, R. A.* 1969. Carcinogen-induced tumors of the thymus. IV. Humoral influences of normal thymus and functional thymomas and influence of post-thymectomy period on restoration. J. Exp. Med. **130,** 809–19.

[31] *Stutman, O., Yunis, E. J.,* and *Good, R. A.* 1969. Thymus: An essential factor in lymphoid repopulation. Trans. Proc. **1,** 614–15.

[32] *Stutman, O., Yunis, E. J.,* and *Good, R. A.* 1970. Studies of thymus function. II. Cooperative effect of newborn and embryonic hemopoietic liver cells with thymus function. J. Exp. Med. **132,** 601–12.

[33] *Stutman, O., Yunis, E. J.,* and *Good, R. A.* 1970. Studies of thymus function. I. Cooperative effect of thymic function and lympho-hemopoietic cells in restoration of neonatally thymectomized mice. J. Exp. Med. **132,** 583–600.

[34] *Swanson, J., Crookston, M. C., Yunis, E. J., Azar, M., Gatti, R. A.,* and *Good, R. A.* 1971. Lewis substances in a human marrow transplantation chimera (Letter to the Editor). Lancet **1,** 396.

[35] *Thomas, E. D., Buckner, C. D., Storb, R., Neiman, P. E., Fefer, A., Clift, R. A., Slichter, S. J., Funk, D. D., Bryant, J. I.,* and *Lerner, K. E.* In Press. Aplastic anemia treated by marrow transplantation. Lancet.

[36] *Tobler, R.,* and *Cottier, H.* 1958. Familiare lymphopenic mit agammaglobulinemie und schwerer monilliasis. Helv. paediat. Acta **13,** 1333.

[37] *Toivanen, P.,* and *Linna, T. J.* 1971. Reconstitution of cyclophosphamide bursectomized chickens. Fed. Proc. **30,** 529.

[38] *Yunis, E. J.,* and *Amos, B.* 1971. Three closely linked genetic systems relevant to transplantation. Proc. Natl. Acad. Sci. **68,** 3031–35.

Discussion

Kluge: Is the presence of the thymus during the whole life span necessary for differentiation of the T-cell system? Or, let me ask, what happens after thymectomy, does the T-cell system disappear?

Good: Well, I think that the experimental data in the mouse is very clear on this. If you want to completely illiminate the T-cell system, you have got to do some quite nifty juggling. What you do is to take out the thymus in the neonatal mouse, give the mouse fetal irradiation, then reconstitute him with cells from the fetal liver that have been treated with antitheta antibody and compliment. Such a mouse is a T-less mouse. We call him Minnesota variety, a super-duper doubleduty T-less mouse. These animals don't have a T-cell system, and they cannot survive for any appreciable time in a conventional environment. If you simply take out the thymus in the neonatal period in some strains of mice, there is a period where the mice may function quite well immunologically. Especially if they are in a germfree environment, they function quite normally. However, they gradually show a loss of the small T-cell compartment. If you take out the thymus as late as 60 days of age, after about a year the long-lived population of lymphocytes begins to disappear. This occurs in anticipation of the normal process of loss of the T-population that takes place in many strains with aging. Giving a thymus back, after having taken it out, will completely reconstitute the animal; we have completely reconstituted these T-less mice by putting a little wet membrane of thymus anlagen under the renal capsule. Such a transplant is derived from an 11- or 12-day-old mouse embryo. You put the anlagen of thymus under the renal capsule and it becomes a beautiful little thymus and fully reconstitutes the immunologic function. I think that it is very clear that the T-population can be maintained and expanded if enough of the T-population is present. We look on the thymus as being essential all through life but experiencing a programed involution with aging in many strains of animals, including man. The thymus serves two separate functions. It is essential to the differentiation of T-cells and is essential to the expansion of this population through an indirect humoral function.

Droege: One of the central points of this meeting was supposedly the question about the clinical prospects of genetic exchange. At the end of this meeting, I think everyone appreciated learning about one successful example where a genetic disease was cured by transfer of healthy genetic material in the form of cells. Now, I would like to ask one specific question on the blocking factor. Is it possible that the blocking antibody may only crossreact serologically with IgG, or do you have further evidence for the identity of the blocking factor?

Good: I really don't know what this material is, and I don't think anybody does. All I can say is it has determinants on its heavy chain that are reacting with antisera produced against an isolated gamma chain of the human immunoglobulin. That doesn't at all mean that we have it identified; it's about a 7 S, maybe a little bit larger, and we are doing studies to see whether or not we can dissociate from it a possible antigenic fragment. It is thought that it might be an antigen antibody complex. We also think it important to define it in terms of the subclass of immunoglobulin. It is interesting that in each of the children that have been corrected where the studies have been carried out, this blocking material has appeared in the circulation about the time of disappearance of the graft vs. host reactions. So we don't know what it is. It could be another class or subclass of immunoglobulin. I am very much interested in your work, which suggests that within the thymus there may be a set of B-cells which secrete just such an immunoglobulin. I think you should pursure.

Siniscalco: I have a short question for Dr. *Good* and a suggestion from somatic cell genetics as an additional tool to investigate the genetic nature of X-linked agammaglobulinemia.

The question: Did it ever happen that you transplanted bone marrow from an Xg-negative to an Xg-positive individual? If so, did such individuals show a mosaic of Xg(+) and Xg(–) red cells? My question is, of course, related to the hypothesis that the Xg-substance, like the Lewis substance, may be synthesized elsewhere and then adsorbed on the red cells. If such hypothesis could be proved true, it might explain why, unlike every other X-linked marker thus far studied, the female heterozygous for Xg fails to show a mosaic of positive and negative red cells as expected from the Lyon phenomenon.

However in view of a recent report by *Ceppellini* and his colleagues of an XX/XY chimera with Xg-positive and negative red cells, this hypothesis has lost credit. You might be able to say the last word on the matter if the transplantation referred to above has ever been done.

Good: Well, unfortunately we haven't carried out such studies – we have just not had the appropriate combination. I think they are very relevant and extremely important studies to carry out. As far as the relationship of the two different forms of X-linked immunodeficiency goes, the family studies are very commanding. None of them would rule out an X-linked dominant. I think, further that it would be very interesting to analyze this with one of the newer methods and to try to link the genetic determinant precisely to the X-chromosome. We haven't attempted to do this yet, but would welcome an opportunity to study this question with you.

Siniscalco: I would like now to illustrate a design of an experiment that may help to understand the nature of the genetic defect responsible for the so-called X-linked agammaglobulinemia. As it is well known, the structural loci for the human gammaglobulins are autosomal. So the suggestion has been made that the X-linked agammaglobulinemia may be due to a kind of "regulatory" phenomenon governed by an X-linked structural gene. This hypothesis could perhaps be tested with the use of somatic cell hybrids. As it has been repeatedly illustrated, the interspecific somatic cell hybrids – the man-mouse and man-hamster kind – progressively lose the human chromosomes.

When the animal parental cell is deficient in HGPRT, these types of hybrids tend to retain the entire human X-chromosome when grown in the HAT-selective medium, and to lose it when grown in the 8-azaguanine medium (See *Ruddle,* pp. 299 ff. and *Siniscalco,* pp. 307 ff. of this symposium). Thus the hybridization of a HGPRT deficient mouse or hamster line, able to produce immunoglobulins, with human cells of different types may permit one to find out whether and how the production of immunoglobulins in the hybrid cells is affected by the presence of the human X-chromosome. The human cells should always be normal for HGPRT but derived from individuals with normal gammaglobulin production or affected by X-linked agammaglobulinemia. For a complete picture, the human cells utilized in this type of experiment should include fibroblastic as well as lymphoblastic diploid cells.

I hope Dr. *Good* finds this experimental plan worthwhile to the extent of giving us some of his precious cell lines.

Good: That's a very elegant experiment, and we would like very much to cooperate with you. Let me tell you one of the problems.

Hofschneider: Is this a principal problem or . . .

Good: Directly to this point. We have been unable, as yet, to establish a cell line of the T-cells from humans who lack B-cells. (We believe the lymphoid cell lines that have been established are all of a B-cell nature.) Dr. *Park* and I are working on this, and as soon as we get such a cell line, we'll let you have it to study.

Prematurity and Uniqueness in Scientific Discovery

Gunther S. Stent

Department of Molecular Biology, University of California, Berkeley, Cal. 94720, USA

One of the depressing by-products of the fantastically rapid progress that was made in molecular genetics in the past twenty-five years is that now merely middle aged participants in its early development are obliged to look back upon their early work from a depth of historical perspective that, in the case of biological specialties that came into flower in earlier times, had opened up only after all the witnesses of the first blossoming were long dead. I have been trying to make virtue out of necessity and actually exploit this singular position for fathoming the evolution of a scientific field. Thus, in looking back on the history of molecular genetics from the viewpoint of my own experiences, I have found that two of its most famous incidents, *Oswald Avery*'s identification of DNA as the active principle in bacterial transformation and, hence, as genetic material [2], and *Watson* and *Crick*'s discovery of the DNA double helix [25], illuminate two general problems of cultural history. The case of *Avery* brings, I think, insights into the question of whether it is meaningful, or merely tautological, to allege that a discovery is "ahead of its time", or *premature.* And the case of *Watson* and *Crick* may be used, and in fact *has* been used, to consider the question of whether there is anything *unique* in a scientific discovery, in view of the likelihood that if Doctor A had not discovered fact X today, Doctor B would have discovered it tomorrow.

Prematurity

Four years ago I published a brief retrospective essay on molecular biology [21] with particular emphasis on its origins. In this historical account, I mentioned neither

Manuscript, received: 13 December 1971

Avery's name nor DNA-mediated bacterial transformation. My essay brought forth a letter-to-the-editor by *Carl Lamanna* [14], who complained that "it is a sad and surprising omission that...

Stent makes no mention of the definitive proof of DNA as the basic hereditary substance by *O. T. Avery, C. M. MacLeod* and *H. McCarty*. The growth of [molecular genetics] rests upon this experimental proof... I am old enough to remember the excitement and enthusiasm induced by the publication of the paper by *Avery, MacLeod* and *McCarty*. *Avery*, an effective bacteriologist, was a quiet, self-effacing, non-disputatious gentleman. These characteristics of personality should not prevent the general scientific public represented by the audience of *Science* to let his name go unrecognized."

I was taken aback by *Lamanna*'s letter and replied [22] that I agree that I *should* have really mentioned in my essay *Avery*'s proof in 1944 that DNA is the hereditary substance. But, I went on to say, in my opinion, it is not true that the growth of molecular genetics rests upon *Avery*'s proof. For many years that proof had actually made a surprisingly small impact on geneticists, both molecular and classical, and it was only the *Hershey-Chase* experiment of 1952 which caused those people to focus on DNA. The reason for this delay was neither that *Avery*'s work was unknown to or mistrusted by geneticists nor that the Hershey-Chase experiment was technically superior. Instead, *Avery*'s discovery, so I declared, had been merely "premature". And in the last two sentences of my reply to *Lamanna* I sketched out the argument about prematurity that I shall try to develop here in somewhat greater detail.

My *prima facie* reason for considering *Avery*'s discovery premature is that it was not appreciated in its day. But is it, in fact, *true* that *Avery*'s discovery was not appreciated? *Lamanna*, for example, mentions his own excitement and enthusiasm induced by the publication of *Avery*'s paper, and several participants in the 1946 Cold Spring Harbor Symposium on Heredity and Variation in Microorganisms have told me that *Avery*'s discovery formed the subject of intense discussion at that symposium. So how can I say that it was not appreciated? By lack of appreciation I do not mean that *Avery*'s discovery went unnoticed, or even that it was not considered important. What I *do* mean is that no one seemed to be able to do much with it, or build upon it, except for the students of the transformation phenomenon. That is to say, *Avery*'s discovery had virtually no effect on general genetic discourse.

By way of support of this allegation, I invite examination of the 1946 Cold Spring Harbor Symposium volume [27]. It contains a paper by *McCarty, Harriet Taylor*, and *Avery*, whose main concern is not the meaning of the discovery for genetics but the elucidation of the role of serum in the DNA-mediated transformation phenomenon. Although many of the other papers of the volume are followed by discussants' remarks, no discussant of the *McCarty, Taylor*, and *Avery* paper is on record. Only five of the other 26 symposium papers refer to *Avery*'s discovery:

Three phage workers, *T. F. Anderson, A. D. Hershey*, and *S. E. Luria*, venture the opinion that the phenomenon is probably of wide biological importance. *L. Dienes* concludes that since DNA, "a substance without apparent organization", *Avery's* discovery means that "bacteria possess a mechanism for the exchange of hereditary characteristics, [that is] different from the usual sexual processes", and *S. Spiegelman* is under the impression that *Avery* discovered "the induction of a particular enzyme with a nucleoprotein [sic] component". Neither *Max Delbrück* nor *J. Lederberg* and *E. L. Tatum* mention *Avery* at all in their now famous 1946 symposium papers.

An even more convincing demonstration of the lack of appreciation of *Avery's* discovery is provided by the 1950 Golden Jubilee of Genetics symposium "Genetics in the 20th Century" [9]. Here some of the most eminent geneticists of that time presented essays that surveyed the progress of the first 50 years of genetics and assessed its present status. Only one of the 26 essayists saw fit to make more than a passing reference to *Avery's* discovery, then six years in the past, namely *A. E. Mirsky*, who still expressed some doubts that the active transforming principle is really pure DNA. *H. J. Muller's* 1950 symposium essay on the nature of the gene contains no mention of *Avery* or DNA.

So, *why* was *Avery's* discovery not appreciated in its day? Because it was "premature". But is this really an explanation or is it merely an empty tautology? In other words, is there a way of providing a criterion of the prematurity of a discovery other than its failure to make an impact? Yes, there is such a criterion: *A discovery is premature if its implications cannot be connected by a series of simple logical steps to contemporary canonical knowledge*. This criterion is not to be confused with that of an *unexpected* discovery, which *can* be connected with the canonical ideas of its day but might overthrow one or more of them. For instance, the recent finding of a "reverse transcriptase" [3, 24] would fall into the category of unexpected discoveries – provided, of course, that the function attributed to that enzyme of catalyzing the assembly of a DNA replica from an RNA template can eventually be shown to occur in vivo. Although prior to that finding, it had been generally assumed by molecular geneticists that there is no reverse flow of "information" from RNA to DNA, there is no difficulty at all in understanding such a process from the viewpoint of current ideas of polynucleotide synthesis.

Why could *Avery's* discovery not be connected with canonical knowledge? By 1944, DNA had long been suspected of exerting *some* function in hereditary processes, particularly after *R. Feulgen* had shown in 1924 that DNA is a major component of the chromosomes. But the then current view of the molecular nature of DNA made it wellnigh inconceivable that DNA *could* be the carrier of hereditary information. First of all, until well into the 1930s DNA was generally thought to be merely a *tetranucleotide* composed of one residue each of adenylic, guanylic, thymidylic, and cytidylic acid. Secondly, even when it was finally realized by the early 1940s that the molecular weight of DNA is actually much higher than that demanded by the tetranucleotide theory, it was still widely believed that the tetranucleotide is the basic repeating unit of the large DNA polymer in which the four purine and pyrimidine bases recur in regular sequence. DNA was therefore viewed as a monotonously uniform macromolecule which, like other monotonous polymers such as starch or cellulose, is always the same no matter what its biological source. The ubiquitous presence of DNA in the chromosomes was, therefore, generally explained in purely physiological or structural terms. Instead, it was usually to the chromosomal protein that the informational role of the genes had been assigned since the great differences in the specificity of structure that exist between heterologous proteins in the same organism, or between homologous proteins in different organisms, had been appreciated since the beginning of this century. The conceptual difficulty of assigning the genetic role to DNA had by no means escaped *Avery*, for in the conclusion of his paper [2] he states that "if the results of the present study of the transforming principle are confirmed then nucleic acids must be regarded as possessing biological specificity the chemical basis of which is as yet undetermined".

However, by 1950, the tetranucleotide theory had been overthrown, thanks largely to the work of *Erwin Chargaff* [5] who showed that, contrary to the demands of that theory, the four nucleotide bases are not necessarily present in DNA in equal proportions. *Chargaff* found, furthermore, that the exact base composition of DNA differs according to its biological source, suggesting that DNA may not be a monotonous polymer after all. So when, two years later, *Hershey* and *Chase* [10] showed that upon infection of the host bacterium at least 80 % of the phage DNA enters the cell whereas at least 80 % of the phage protein remains outside, it was now possible to connect *their* conclusion that DNA is the genetic material with canonical knowledge. For *Avery*'s "as yet undetermined chemical basis of the biological specificity of nucleic acids" could now be envisaged as the precise sequence of the four nucleotide bases along the polynucleotide chain. The general impact of the *Hershey-Chase* experiment was immediate and dramatic. DNA was suddenly in and protein was out, as far as thinking about the nature of the gene was concerned. Within a few months, there arose the first speculations about the genetic code, and *Watson* and *Crick* were inspired to set out to discover the structure of DNA.

Naturally, the case of *Avery* is only one of many premature discoveries in the history of science. I have presented it here for consideration mainly because of my own failure to appreciate it when I joined *Delbrück*'s phage group and took the Cold Spring Harbor Phage course in 1948. Since then, I have often wondered what my later fate would have been if only I had been intelligent enough to appreciate *Avery*'s discovery and infer from it four years before the *Hershey-Chase* experiment that DNA must also be the genetic material of the phage.

Probably the most famous case of prematurity in the history of biology is that of *Gregor Mendel*, whose discovery of the particulate nature of heredity in 1865 had to await 35 years before it was "rediscovered" at the turn of the century. *Mendel*'s discovery made no immediate impact, so it can be argued, because the concept of discrete hereditary units could not be connected with the (mid 19th century) canonical knowledge of anatomy and physiology. Furthermore, the statistical methodology by means of which *Mendel* interpreted his data was wholly foreign to the way of thinking of his contemporary biologists. By the end of the 19th century, however, chromosomes, mitosis, and meiosis had been discovered, and *Mendel*'s results could now be accounted for in terms of microscopically visible structures and processes. Furthermore, by then the application of statistics to biology had become commonplace. In some respects, however, *Avery*'s case is a more dramatic example of prematurity than *Mendel*'s. Whereas *Mendel*'s discovery seems to have been hardly mentioned by anyone until its rediscovery, *Avery*'s discovery *was* widely discussed, and yet could not be appreciated for eight years.

A striking example of delayed appreciation of a discovery in the physical sciences, as well as an explanation of that delay in terms of the concept to which I refer here as prematurity, has been provided by *Michael Polanyi* [18]. In the years 1914–1916, *Polanyi* published a theory of the adsorption of gases on solids which assumed that the force attracting a gas molecule to a solid surface depends only on the position of that molecule, but not on the presence of other molecules, in the force field. Despite the fact that *Polanyi* was able to provide strong experimental evidence in favor of his theory, it was generally rejected. Not only was the theory rejected, but it was considered so ridiculous by the leading authorities of the time that *Polanyi* believes continued defense of his theory would have ended his professional career had he not managed to publish work on other more palatable ideas. The reason for the general rejection of *Polanyi*'s adsorption theory was that, at the very time he put it forward, the role of electrical forces in the architecture of matter had just been discovered. And hence, there seemed to be no doubt that gaseous adsorption must also involve electrical attraction between gas molecules and solid surfaces. That point of view, however, was irreconcilable with *Polanyi*'s basic assumption of the mutual independence of individual gas molecules in the adsorption process. Instead of *Polanyi*'s theory, the theory of *I. Langmuir*, which did envisage a mutual interaction of the gas molecules of the kind expected from electrical forces, found general acceptance. It was only in the 1930s after *F. London* developed his new theory of cohesive molecular forces based on quantum mechanical resonance rather than electrostatic attraction, that it became conceivable that gas molecules *could* behave in the way in which *Polanyi*'s experiments indicated they are actually behaving. Meanwhile, *Langmuir*'s theory had become so well-established, and *Polanyi*'s had been consigned so authoritatively to the ash can of crackpot ideas, that *Polanyi*'s theory was rediscovered only in the 1950s.

We may now consider whether the notion of prematurity is actually a useful historical concept. First of all, is prematurity the only possible explanation for the lack of contemporary appreciation of a discovery? No, evidently not. For instance, *Lamanna* suggested the "quiet, self-effacing, non-disputative" personality of *Avery* as the cause for the failure of general recognition of his discovery. And *Chargaff* [7] is another believer in the idea that personal modesty and reticence for self-advertisement accounts for lack of contemporary appreciation. For instance, *Chargaff* has attributed the 75-year hiatus between *F. Miescher's* discovery of DNA in 1869 and the appreciation of its importance to *Miescher* being "one of the quiet in the land", who lived when "the giant publicity machines, which today accompany even the smallest move on the chess-board of nature, with enormous fanfares were not yet in place". Indeed, the 35 year hiatus in the appreciation of *Mendel's* discovery is often attributed to *Mendel* having been a modest monk living in an out-of-the-way Moravian monastery. Hence, the notion of prematurity provides an alternative to the – in my opinion, for the cases mentioned here, false – invocation of lack of publicity as an explanation for delayed appreciation.

But, more importantly, does the prematurity concept pertain only to retrospective judgements made with the wisdom of hindsight? No, I think it can be used also to judge the present. For some discoveries have been made recently that are still premature at this very time. One example of here-and-now prematurity is the alleged finding that sensory information received by an animal can be stored in RNA or other macromolecules.

About ten years ago, there began to appear reports by experimental psychologists purporting to have shown that the memory trace, or engram, of a task learned by a trained donor animal can be transferred to a naive recipient animal by injecting or feeding the recipient with an extract made from the tissues of the donor.[1]) At that time, the central message of molecular genetics that nucleic acids and proteins are "informational macromolecules" had just gained wide currency, and the facile equation of sensory information with genetic information soon led to the proposal that macromolecules – DNA, RNA, or protein – store memory. As it happens, the experiments on which the macromolecular theory of memory is based have been very difficult to repeat, and the results claimed for them may indeed not be true at all. But it is significant that few neurophysiologists have even bothered to check these experiments, despite everybody's having heard about them and being aware that the possibility of chemical memory transfer would constitute a fact of capital importance. The lack of interest of neurophysiologists in the macromolecular theory of memory can be accounted for by recognizing that this theory, whether true or false, is clearly premature: there is no chain of reasonable inferences by means of which our present, albeit very imperfect, view of the functional organization of the

[1]) For a critical summary, see *Quarton* [19].

brain can be reconciled with the possibility of its acquisition, storage, and retrieval of experiential information by encoding such information in nucleic acid or protein molecules. Thus for the community of neurobiologists there is no point in devoting its time to checking on experiments whose results, even if they were true as alleged, could not be connected with canonical knowledge.

The concept of here-and-now prematurity can be applied also to the troublesome subject of extrasensory perception, or ESP. During the summer of 1948, while taking the Cold Spring Harbor phage course, I happened to witness a heated argument between two future mandarins of molecular biology, *S. E. Luria* and *R. E. Roberts*. *Roberts* was then interested in ESP and felt it had not been given fair consideration by the scientific community. As far as I remember, he thought one might be able to set up some experiments with molecular beams which could provide more definitive data on the possibility of mind-induced departures from random distributions than *J. B. Rhine*'s [20] then much discussed card-guessing procedures. *Luria* declared that not only was he not interested in *Roberts* proposed experiments, but that in his opinion it was unworthy of anyone claiming to be a scientist even to discuss such rubbish. How could an intelligent fellow like *Roberts* entertain the possibility of phenomena totally irreconcilable with the most elementary physical laws? Moreover, a phenomenon which is manifest only to specially endowed subjects, as claimed by parapsychologists to be the case for ESP, is outside the proper realm of science, which must deal with phenomena accessible to every observer. *Roberts* replied that far from *his* being unscientific, it was *Luria* whose bigoted attitude toward the unknown is unworthy of a true scientist. The fact that not everybody has ESP only means that it is an elusive phenomenon, such as musical genius. And just because a phenomenon cannot be reconciled with what we now know, we need not shut our eyes to it. On the contrary, it is the *duty* of the scientist to try to devise experiments designed to probe its truth or falsity.

It seemed to me then that *both Luria* and *Roberts* were right, and, in the intervening years, I often thought about this puzzling disagreement, unable to resolve it in my own mind. Finally, five years ago I read *C. W. Churchman*'s review [8] of a book on ESP, and I began to see my way toward a resolution. *Churchman* set forth that there are three different possible scientific approaches to ESP. The first of these is that the truth or falsity of ESP, like that of the existence of God or the immortality of the soul, is totally independent of either the methods or findings of empirical science. And hence, an adherent of the tenets of logical positivism would relegate ESP to the class of meaningless propositions. Thus the problem of ESP is defined out of existence. I imagine that this was more or less *Luria*'s position.

Churchman's second approach is to reformulate the ESP phenomenon in terms of currently acceptable scientific notions, such as *unconscious* perception or conscious fraud. This procedure is not as arbitrary as it might seem on first sight, because the "extra" in extrasensory perception is a conceptually fuzzy negative property anyhow. Thus, rather than defining ESP out of existence, it is trivialized. The second approach probably would have been acceptable to *Luria* too, but not to *Roberts.*

Finally, the third approach is to take the proposition of ESP literally and to attempt to examine in all seriousness the evidence for its validity. That was, more or less, *Roberts'* position. But, as *Churchman* points out, this approach is not likely to lead to satisfactory results. Parapsychologists can maintain with some justice that the existence of ESP has *already* been proven to the hilt, since no other set of hypotheses of psychology has received the degree of critical scrutiny that has been given to ESP experiments. And many other phenomena have been accepted on much less statistical evidence than that which has been offered for ESP. The reason that *Chuchman* advances for the futility of a strictly evidential approach to ESP is that, in the absence of a hypothesis of how ESP *could* work, it is not possible to decide whether any set of relevant observations can be accounted for *only* by ESP, to the exclusion of alternative explanations. *Chuchman* thus applies to the problem of ESP the principles of *Karl Popper*'s "hypothetic deductive" theory of scientific discovery, according to which facts gain scientific meaning only within the framework of preconceived hypotheses.

After reading *Churchman*'s review, I realized that *Roberts* would have been ill-advised to proceed with his ESP experiments, not because, as *Luria* claimed, they would not be "science", but because any positive evidence he might have found in favor of ESP would have been, and would still be, premature. That is, until it is possible to connect a phenomenon like telepathy with canonical knowledge of, say, electromagnetic radiations and neurophysiology, no demonstration of its occurrence could be appreciated.

Is the lack of appreciation of premature discoveries merely attributable to the intellectual shortcoming of scientists, who, if they were only more perceptive, would give immediate recognition to any well-documented scientific proposition? *Polanyi* is not of that opinion. Upon reflecting on the cruel fate of his theory half a century after first advancing it, he declares that "this miscarriage of the scientific method could not have been avoided.... There must be at all times a predominantly accepted scientific view of the nature of things, in the light of which research is jointly conducted by members of the community of scientists. A strong presumption that any evidence which contradicts this view is invalid must prevail. Such evidence has to be disregarded, even if it cannot be accounted for, in the hope that it will eventually turn out to be false or irrelevant."

This is a view of the operation of science rather different from that commonly held, under which acceptance of authority is seen as something that must be avoided at all costs. The good scientist is seen as an unprejudiced man with an open mind who is ready to embrace any new idea supported by the facts. As the history of science shows, its practitioners do not appear to act according to that popular view.

Uniqueness

Four years ago I prepared a review [23] of some reviews of "The Double Helix", *James Watson*'s autobiographical account of his and *Francis Crick*'s discovery of the structure of DNA [25]. Among the reviews of *Watson*'s book, which I then analyzed, was a rather unfriendly one by *Chargaff* [6]. *Chargaff* observes that scientific autobiography is "a most awkward literary genre". Must such works, he says, "give the impression of having been written for the remainder tables of bookstores, reaching them almost before they are published". The reasons for this, according to him, are not far to seek: scientists "lead monotonous and uneventful lives and... besides often do not know how to write". In addition to these allegations of fact, however, *Chargaff* adduces also a general principle belonging to the philosophy of cultural history: "There may also be profounder reasons for the general triteness of scientific autobiographies. *Timon of Athens* could not have been written, *Les Desmoiselles d'Avignon* not have been painted, had *Shakespeare* and *Picasso* not existed. But of how many scientific achievements can this be claimed? One could almost say that, with very few exceptions, it is not the men that make science, it is science that makes the men. What A does today, B or C or D could surely do tomorrow."

I was very surprised to find *Chargaff* state this, to me obviously bizarre, principle and wrote in my review of reviews, "Quite apart from the intrinsic impossibility of subjecting this view of artistic and scientific evolution to any test, and hence quite apart from its nugatory historicism, *Chargaff* surely realizes that the foundation of great writing is depth rather than uniqueness of experience. So why does he bother to make that empty point? Aha! *Watson* and *Crick* are not such great fellows anyhow, because if they had not found the structure of DNA in March of 1953, someone else would surely have found it by the following April".

Since writing my review, I have from time to time discussed this proposition with scientific colleagues and asked them whether they, by any chance, think like *Chargaff* that there is an important qualitative difference between the creations of art and of science, namely that the former are *unique* and the latter *commonplace*. And again to my surprise I found that most of my colleagues seem to agree with *Chargaff*. Yes, they say, it is quite true that we would not have had *Timon* or *Les Desmoiselles* had *Shakespeare* and *Picasso* not existed, but had *Watson* and *Crick* not existed, we would have had the DNA double helix anyhow. Contrary to my first impression, therefore, it does not seem to be all that obvious that this proposition has little philosophical or historical merit. Hence, I shall now attempt to develop an argument that shows that the proposition of a profound difference between arts and sciences in regard to uniqueness of their creations is false.

Before discussing the proposition of uniqueness of creation, it is necessary, however, to make an explicit statement of the meaning of "art" and of "science". My understanding of these terms is based on the view that both the arts and the sciences are activities that endeavor to discover and communicate truths about the world. The domain to which the artist addresses himself is the inner, subjective world of emotions. Artistic statements, therefore, pertain mainly to relations between private events of affective significance. The domain of the scientist, in contrast, is the outer, objective world of physical phenomena. Scientific statements, therefore, pertain mainly to relations between or among public events. This, it must be emphasized, is a *traditionalist* view, since it maintains that the transmission of information, and the perception of meaning in that information, is the central content of both arts and sciences. And hence a creative act on the part of either artist or scientist would mean his formulation of a new meaningful statement about the world, an addition to the accumulated capital of what is sometimes called our "cultural heritage". An alternative to the traditionalist attitude is what can be called the *action* view of art and science, which holds that what counts is what the artist or scientist *does*, rather than any possible semantic content of his work. From the action viewpoint, an artist is a person who employs as a medium of self-expression the methods of the plastic, tonal or verbal arts, and a scientist is a person in a white coat whose self is realized by working in a laboratory. The notion of a cumulative cultural heritage is foreign to this view, as is any objective criterion of creativity. *Chargaff's* proposition is obviously false, as well as irrelevant, from the action viewpoint: here every artistic and scientific act is by definition unique, since it has never been done before and will never be done again in exactly the same way. And whether it has or has not been done is of no historical import in any case. But the proposition *could* have merit under the traditionalist viewpoint since it is at least conceivable that *Shakespeare,* and only *Shakespeare,* could have formulated the semantic structures represented by *Timon*, whereas persons other than *Watson* and *Crick* might have made the communication represented by their 1953 *Nature* paper, *A Structure for Deoxyribonucleic Acid* [26].

First of all, everyone will surely agree that the exact word sequence that *Watson* and *Crick* published in *Nature* would not have been written had they not existed, no more than the exact word sequence of *Timon* would have been written without *Shakespeare*, at least not until the famous monkeytypists complete their work at the British Museum.[1]) So both creations are, from that point of view, unique. But, of course, we are not concerned with the exact word sequence. We are concerned

[1]) In the following, I focus only on *Timon* as the paradigm of uniqueness in art since drama, being a verbal artform, is more congruent with language and, hence, easier to discuss than painting. It is my good luck that *Chargaff* did not choose a *Bach* cantata or *Beethoven* sonata as his example, since music is even less congruent with language and, hence, even more difficult to discuss than painting. *Chargaff's* choice of *Timon*, however, is not exactly felicitous for his purpose of illustrating the uniqueness of *Shakespeare's* creation. The quality of this least popular of the Bard's plays is considered to be so uneven that scholars find it unlikely that *Shakespeare* could have been the sole author of the whole piece.

with the *content*. So, we admit that persons other than *Watson* and *Crick* might eventually have described a satisfactory molecular structure for DNA. But then, the character of *Timon* and the story of his trials and tribulations not only might have been written without *Shakespeare*, they *were* written without him. *Shakespeare* merely reworked the story of *Timon* that he had read in *William Paynter's* collection of classic tales, *The Palace of Pleasure*, published 40 years earlier; and *Paynter*, in turn, had used as his sources *Plutarch* and *Lucian*. But, of course, we don't really care about *Timon's* story; what counts are the deep insights into human emotions that *Shakespeare* provides in this play. For he shows us here how a man may make his response to the injuries of life, how he may turn from lighthearted benevolence to passionate hatred toward his fellow men. Can one be sure, however, that *Timon* is unique from this bare-bones standpoint of the work's artistic essence? No, because who is to say that had *Shakespeare* not existed no other dramatist could have provided for us the same insights? Another dramatist would surely have used an entirely different story – as *Shakespeare* himself actually did in his much more successful *King Lear* – to treat this theme, and he *might* have succeeded in pulling it off. The reason why no one has done it since is that *Shakespeare* had already done it in 1607; just as no one discovered the structure of DNA after *Watson* and *Crick* had already discovered it in 1953.

So we are finally reduced to asserting that *Timon* is uniquely *Shakespeare's*, because no other dramatist though he might have brought us more or less the same insights, would have done it quite in the same exquisite way as the Bard. Fair enough. But here we must not short-change *Watson* and *Crick* and take for granted that anyone else who eventually found the structure of DNA would have described in just the same way and produced the same revolutionary effect on contemporary biology as they did. The studies of *Robert Merton* [16] have shown that many scientific discoveries are unique in this only nontrivial sense in which one can apply this term to acts of artistic creation. In examining some of the thousands of cases of so-called "multiple discoveries" in the history of science where the same idea appears to have turned up two or more times independently, *Merton* has found that the discoveries are hardly ever identical. The reason why they are said to be "multiple" is that despite their diversity one can recognize a semantic overlap between them which is transformable into the same set of ideas.

And, so *Merton* has pointed out, to be one of the participants in a multiple discovery case is by no means a sign of run-of-the-mill mediocrity. On the contrary, great scientists are frequently involved in these multiples, their greatness finding reflection in the fact that they have taken part in many more such multiples than each of the other coparticipants. *Merton* found, for instance, that *Lord Kelvin* took part in some 65 multiple discoveries. When one counts the number of his coparticipants, *Kelvin's* stature rises rather than diminishes: he was the equivalent of a corps of lesser scientists. *Kelvin's* contribution was unique, not in that the corps of lesser scientists could not have done what he did, but in that no *single* man reproduced the whole of his work.

To demonstrate that what A did today could have been done very differently by B et al. tomorrow, and hence that men *do* make science, I have invented a plausible scenario of how we *might* have come to know the structure of DNA if *Watson* and *Crick* (the Doctors A) had not existed.

July 1953: Doctor B infers from X-ray diffraction data of DNA that its structure is a helix of two intertwined polynucleotide chains. January 1954: Doctor C infers from molecular model building that the nucleotide bases can be accommodated in the double helix of Doctor B by having a purine base on one chain form hydrogen bonds with a pyrimidine base on the other chain, thus holding the two chains together. Because of the facultative tautomerism of the bases, either purine can pair with either pyrimidine. July 1954: Doctor D calls attention to *Chargaff's* brief report [5] four years earlier in the not very widely read Swiss journal *Experientia* that there is an equivalence relation in the nucleotide base composition DNA: the molar contents of adenine and thymine and of guanine and cytosine are equal. Doctor D proposes his *obligatory dinucleotide neighbor* theory to explain *Chargaff's* data: in the DNA polynucleotide chain an adenine base is always followed or preceded by a thymine base and a guanine base is always followed or preceded by a cytosine base. December 1954: Doctor E has isolated trinucleotides from DNA digests in which a thymine nucleotide is bracketed by guanine nucleotides, thus disproving the obligatory dinucleotide neighbor theory of Doctor D. July 1955: Doctor F studies the tautomerism of purine and pyrimidine bases and discovers that for each base *one* tautomeric form prevails over all others and that, in the most common forms, adenine can make good hydrogen bonds only with thymine, and guanine only with cytosine. Doctor F proposes, therefore, that in the DNA double helix the base pairing is specific, incidentally also providing thereby an explanation for the equivalence relation left unexplained since the demise of the dinucleotide neighbor theory. December 1955: Doctor G publishes a theoretical paper pointing out that if the genetic information is encoded into DNA in the form of a specific nucleotide sequence, then the obligatory base pairing proposed by Doctor F means that *both* polynucleotide chains of the double helix of Doctor B carry the same information, albeit in complementary notation. Doctor G concludes his paper with the remark: "It has not escaped my notice that the specific pairing postulated by Doctor F immediately suggests a possible copying mechanism for the genetic material."

Thus under this scenario, we would have reached a mere 30 months later the insights into DNA provided by *Watson* and *Crick* in April 1953. But it would not have been the same thing at all: B, C, D, E, F, and G all made important contributions to that

development, but none of them could have claimed the final product as *his* structure. As Sir *Peter Medawar* [15] has pointed out in his excellent review of *The Double Helix*, "the great thing about [*Watson*'s and *Crick*'s] discovery was its completeness, its air of finality. If *Watson* and *Crick* had been seen groping toward an answer, if they had published a partly right solution and had been obliged to follow it up with corrections and glosses, some of them made by other people; if the solution had come out piecemeal instead of in a blaze of understanding; then it would still have been a great episode in biological history; but something more in the common run of things; something splendidly well done, but not in the grand romantic manner."
The similarity in uniqueness of individual contributions to both arts and sciences becomes even more apparent when *Watson* and *Crick* are compared, not to *Shakespeare*, but to *Aeschylus, Sophocles,* or *Euripides.* For it was these authors of Greek antiquity who discovered tragedy as a vehicle for communicating deep insights into the most basic truths of the affects, and *Shakespeare*, drawing on many earlier sources, finally developed that Greek discovery to its ultimate heights. We must admit that while on the one hand, *Prometheus, Oedipus Rex*, and *Medea* would not have been written had *Aeschylus, Sophocles*, and *Euripides* not existed, on the other hand, someone else probably would have discovered the dramatic possibilities of tragedy anyway. Furthermore, on the one hand, *Shakespeare* might still have written his plays 21 centuries later, but, on the other hand, who is to say that *Timon* would have been exactly the same play had tragic drama been discovered by someone other than the Big Three?

For the purpose of our venture into cultural historicism, the proposal of the operon theory by *Jacob* and *Monod* in 1961 [12] is a better molecular genetic analogue to *Shakespeare*'s *Timon* than *Watson*'s and *Crick*'s double helix. For *Jacob* and *Monod* used the tremendous possibilities opened up by *Watson*'s and *Crick*'s fundamental yet simple discovery to explain and illuminate a very much more complex problem than that faced by *Watson* and *Crick. Jacob* and *Monod* brought molecular genetic thought to a peak of subtlety, so that the efforts of their successors in the domain of the regulation of gene expression often give the impression of what is called "mannerism" in art. Would the operon theory have been developed had *Watson* and *Crick* not existed? Probably, yes. Had *Jacob* and *Monod* not existed? Yes and no. Yes, because in making their theory *Jacob* and *Monod* used some key ideas previously put forward by others, such as the negatively acting cytoplasmic repressor first proposed by *Bertani* [4] and the coordinate regulation of linked genes discovered by *Ames* and *Garry* [1]. So finally someone would have put it all together, somehow. No, because it is unlikely that anyone else would have produced the same masterful synthesis, which had become a classic before it even appeared in print. Furthermore, we now see that the precise form of the original operon theory was highly idiosyncratic, in that it was strongly influenced by *Jacob*'s and *Monod*'s personal preoccupation with the induction of the lactose fermentation enzymes in *E. coli.* Had they not chosen to neglect the "glucose effect" as a confusing side issue, or had they focussed their attention on the arabinose fermentation enzymes, their theory would have taken quite a different form.

Structuralism

I think that the apparent popularity among scientists of the proposition that artistic creations are and scientific creations are not unique is to be attributed to rather widespread contradictory epistemological attitudes toward the events in the outer and inner world. The outer world, which science tries to fathom, is often viewed from the standpoint of *materialism,* according to which events and relations between them have an existence independent of the human mind. Hence the outer world and its scientific laws are just *there*, and it is the job of the scientist to find them. So going after scientific discoveries is like picking wild strawberries in a public park: the berries A does not find today, B or C or D will surely find tomorrow. At the same time, many scientists view the inner world, which art tries to fathom, from the standpoint of *idealism,* according to which events and relations between them have no reality other than their reflection in human thought. Hence there *is* nothing to be *found* in the inner world, and artistic creations are cut simply from whole cloth. Here B or C or D could not possibly find tomorrow what A found today, because what A found had never *been* there. It is not altogether surprising, of course, to find this split epistemological attitude toward the two worlds, since of the two antithetical traditions in Western philosophical thought, materialism is obviously an unsatisfactory approach to art and idealism an unsatisfactory approach to science.[1])

It is only in the past twenty years or so, more or less contemporaneously with the growth of molecular biology, that a resolution of the age-old epistemological conflict of materialism versus idealism was found in the form of what has come to be known as *structuralism* [17]. This development is another example of *Merton's* multiple discovery concept, since structuralism emerged simultaneously, independently, and in different guises in several diverse fields of study, for example in psychology, linguistics, anthropology, and biology.

Both materialism and idealism take it for granted that all the information gathered by our senses actually reaches our mind; materialism envisages that thanks to this information reality is *mirrored* in the mind whereas idealism envisages that thanks to

---

[1]) This is not the view of Marxist philosophy, however, whose entirely consistent epistemological attitude maintains that materialism is the only valid approach to *both* art and science. According to Marxist historical determinism, there can be no uniqueness of creation in either domain: given the socioeconomic infrastructures of Elizabethan and Post-Second-World-War England. *Timon* and the DNA double helix, or their equivalents, would certainly have been produced had men whose names happened to be *Shakespeare* and *Watson* and *Crick* not existed.

this information reality is *constructed* by the mind. But structuralism has provided the insight that knowledge about the world enters the mind not as raw data but in an already highly abstracted form, namely as *structures.* And in the preconscious process of converting step-by-step the primary data of our experience into structures, information is necessarily lost. For the creation of structures, or the recognition of patterns, is nothing else than the selective destruction of information. So since the mind does not, and cannot, gain access to the full set of data about the world, it can neither mirror nor construct reality. Instead, for the mind reality is a set of structural *transforms* of primary data taken from the world. This transformation process is hierarchical in that "stronger" structures are formed from "weaker" structures through selective destruction of information. And any set of primary data becomes meaningful only after a series of such operations has so transformed it that it has become isomorphic with a stronger structure pre-existing in the mind.

Neurophysiological studies which *Stephen Kuffler* [13] and *David Hubel* and *Torsten Wiesel* [11] have carried out on the process of visual perception in higher mammals have not only shown directly that the brain actually operates according to the tenets of structuralism but also offer an easily understood illustration of those tenets. According to these studies, the primary photoreceptors in the retina report the absolute light intensity that reaches the eye from individual points in the visual field. These primary data are not sent on from the retina to the brain, however. They are first transformed in the retina into information about the light-dark *contrast* existing at individual points in the visual field, the absolute intensity data having been largely destroyed in the abstraction process. Upon first reaching the brain, the light contrast data for individual points are then transformed into the light contrast data for individual *straight edges*, or point sets, in the visual field, the information about contrast at individual points being destroyed in that second abstraction process. And at the next level of processing in the brain, the contrast data for individual straight edges are transformed into the corresponding data for sets of *parallel edges*, or sets of point sets in the visual field, entailing further destruction of information about individual edges. It is not yet clear what transformations take place at the next higher level of processing in the visual pathway, but it is certain that the mind experiences reality without knowing the "real" point-to-point light intensity in its surrounding.

Finally, we may consider the relevance of structuralist philosophy for the two problems in the history of science under discussion here. As far as prematurity of discovery is concerned, structuralism provides us with an understanding of why a discovery cannot be appreciated until it can be connected logically to contemporary canonical knowledge.

In the parlance of structuralism, canonical knowledge is simply the set of pre-existing "strong" structures with which primary scientific data are made isomorphic in the mental abstraction process. Hence, data which cannot be transformed into a structure isomorphic with canonical knowledge are a dead end; in the last analysis, they remain meaningless. They remain meaningless, that is, until a way has been shown how to transform them into a structure that is isomorphic with the canon.

As far as uniqueness of discovery is concerned, structuralism leads to the recognition that every creative act in the arts and sciences is *both* commonplace and unique. On the one hand, it is commonplace in the sense that there is an innate, or genetically determined, correspondence in the transformational operations which different individuals perform on the same primary data. With reference to science, cognitive psychology has taught that different individuals recognize the same "chairness" of a chair because they all make a given set of sense impressions from the outer world isomorphic with the same mental structure, or *Gestalt.* With reference to art, analytical psychology has taught that there is a sameness in the subconcious life of different individuals because an innate human *archetype* causes them to make the same structural transformations of the events of the inner world. And with reference to both art and science, structural linguistics has taught that communication between different individuals is possible only because an innate human *grammar* causes them to transform a given set of semantic signals into the same syntactic structure. On the other hand, every creative act is unique in the sense that no two individuals are quite the same and hence never perform *exactly* the same transformational operations on a given set of primary data. But then, while all creative acts in both arts and sciences are commonplace *and* unique, some may be more unique than others.

Acknowledgements: I made an informal presentation of the ideas covered in this essay at a Conference on the History of Biochemistry and Molecular Biology, held in May 1970 at the American Academy of Arts and Sciences. I am indebted to the dozen or so conference participants whose vigorous discussion accompanied my presentation and who helped me focus my ideas more sharply. I am particularly grateful to *Harriet Zuckerman* for calling my attention to *Polanyi*'s paper on prematurity and to *Robert K. Merton* for his assistance in the analysis of the uniqueness of discovery concept.

References

[1] *Ames, B. N.,* and *Garry, B.* 1959. Proc. Natl. Acad. Sci., Wash. **43**: 1453.

[2] *Avery, O. T., MacLeod, C. M.,* and *McCarty, M.* 1944. Studies on the chemical nature of the substance inducing transformation in the pneumococcus. J. Exp. Med. **79**: 137.

[3] *Baltimore, D.* 1970. Viral RNA-dependent DNA polymerase. Nature **226**: 209–211.

[4] *Bertani, G.* 1953. Cold Spring Harbor Symp. Quant. Biol. **18**: 65.

[5] *Chargaff, E.* 1950. Chemical specificity of nucleic acids and mechanism of their enzymatic degradation. Experientia **6**: 201.

[6] *Chargaff, E.* 1968. A quick climb up Mount Olympus. Science **159**: 1448–1449.

[7] *Chargaff, E.* 1971. Preface to a grammar of biology. Science **172**: 637–642.

[8] *Churchman, C. W.* 1966. Perception and deception. Science **153**: 1088–1090.

[9] *Dunn, L. C.* 1951. Genetics in the 20th century. New York: Macmillan.

[10] *Hershey, A. D.*, and *Chase, M.* 1952. Independent function of viral protein and nucleic acid in growth of bacteriophage. J. Gen. Physiol. **36**: 39–56.

[11] *Hubel, D. T.*, and *Wiesel, T. N.* 1968. Receptive fields and functional architecture of monkey striate cortex. J. Physiol. **195**: 215–243.

[12] *Jacob, F.*, and *Monod, J.* 1961. Genetic regulatory mechanisms in the synthesis of proteins. J. Mol. Biol. **3**: 318–356.

[13] *Kuffler, S. W.* 1953. Discharge patterns and functional organization of the mammalian retina. J. Neurophysiol. **16**: 37–68.

[14] *Lamanna, C.* 1968. Science **160**: 1397.

[15] *Medawar, P. B.* 1968. Lucky Jim. New York Review of Books, March 28.

[16] *Merton, R. K.* 1961. Singletons and multiples in scientific discovery. Proc. Am. Phil. Soc. **105**: 470–486.

[17] *Piaget, J.* 1970. Le Structuralisme. Paris: Presses Universitaires de France.

[18] *Polanyi, M.* 1963. Potential theory of adsorption. Science **141**: 1010–1013.

[19] *Quarton, G. C.* 1967. The enhancement of learning by drugs and the transfer of learning by macromolecules. *In* The Neurosciences, eds. *G. C. Quarton, T. Melnechuk*, and *F. O. Schmitt*. New York: The Rockefeller University Press.

[20] *Rhine, J. B.* 1948. The reach of the mind. London: Faber & Faber.

[21] *Stent, G. S.* 1968. That was the molecular biology that was. Science **160**: 390–395.

[22] *Stent, G. S.* 1968. Science **160**: 1397.

[23] *Stent, G. S.* 1968. What they are saying about honest Jim. Quarterly Rev. Biol. **43**: 179–184.

[24] *Temin, H. M.*, and *Mizutani, S.* 1970. RNA-dependent DNA polymerase in virions of Rous sarcoma virus. Nature **226**: 1211–1213.

[25] *Watson, J. D.* 1968. The double helix. New York: Atheneum.

[26] *Watson, J. D.*, and *Crick, F. H. C.* 1953. A structure for deoxyribonucleic acid. Nature **171**: 737.

[27] Heredity and variation in microorganisms 1946. XI Symposium on quantitative biology. New York: Biological Laboratory, Cold Spring Harbor.

Lecturers

Axelrod, David, M. D.

1935	Born in Great Barrington, Massachusetts, USA, on January 7th Education in Medicine at Harvard College and Harvard Medical School
	Positions held:
1960–1961	Intern, Strong Memorial Hospital, Rochester, New York
1961–1962	Medical Resident, Strong Memorial Hospital
1962–1965	Research Associate, National Institutes of Health, Bethesda, Maryland
1965–1968	Staff Virologist, Laboratory of the Biology of Viruses, National Institute of Allergy and Infectious Diseases, Bethesda, Maryland
since 1968	Director, Infectious Disease Center, Division of Laboratories and Research, New York State Department of Health, Albany, New York

Fields of Interest:

Control mechanisms for virus-induced oncogenesis; virus-mediated genetic exchange in mammalian cells.

Bodmer, Walter, F., Ph. D.

1936	Born in Frankfurt am Main, Germany, on January 10th Education at Clare College, Cambridge University, England

Positions held:

1958–1961	Research Fellow, Clare College, Cambridge University
1961	Official fellow
1960–1961	Demonstrator, Department of Genetics, Cambridge University
1961–1962	Fellow, Visiting Assistant Professor
1962–1966	Assistant Professor
1966–1968	Associate Professor
1968–1970	Professor, Department of Genetics, Stanford University School of Medicine, California, USA
since 1970	Professor of Genetics, University of Oxford

Fields of Interest:

Somatic cell genetics, genetics of human tissue antigens and human population genetics.

Brinton, Charles, Ph. D.

1926	Born in Pittsburgh, Pa., USA, on August 15th Education in Physics, Biophysics, and Microbiology at Carnegie Mellon University and University of Pittsburgh

Positions held:

1953–1955	Fulbright and French-Government Fellow, Institut de Radium, Paris
1955–1956	Research Fellow, Department of Biophysics, University of Geneva
1956–1960	Research Associate, Department of Biophysics, University of Pittsburgh
1960–1965	Assistant Professor to Full Professor of Microbiology, University of Pittsburgh, U.S. Public Health Service Career Development Award
since 1965	Professor of Microbiology, University of Pittsburgh

Fields of Interest:

Surface structure and genetics of cells.

Croce, Carlo M., M. D.

1944	Born in Milan, Italy, on December 17th Education in medicine at the University of Rome, Italy

Positions held:

1966–1968	Intern at the University of Rome with Dr. E. Antonini, Biochemistry
1968–1970	Pupil of Giuseppe A. Andres, Experimental Pathology, University of Rome, Italy
since 1970	The Wistar Institute of Anatomy and Biology Philadelphia, Pennsylvania, USA

Field of Interest:

Genetics.

Duesberg, Peter H., Ph. D.

1936	Born in Münster, Germany, on December 2nd Education in Chemistry in Würzburg, Basel, München and Frankfurt

Positions held:

1963	Postdoctoral fellow at the Max-Planck-Institut für Virusforschung in Tübingen
1964–1967	Postdoctoral fellow at the Virus Laboratory in Berkeley, California
since 1968	Assistant-to-Associate Professor of Molecular Biology, University of California, Berkeley, Department of Molecular Biology and Virus Laboratory

Fields of Interest:

Structure and replication of cytocidal and oncogenic animal viruses.

Ericsson, Ronald J., Ph. D.

1935 Born in Belle Fourche, South Dakota, USA, on July 17th
Education in Biology at Colorado State University and University of Kentucky

Positions held:

1957–1960 Working Cowboy
1960–1963 Research Assistant, University of Kentucky, Lexington
1963–1964 Postdoctoral Fellow in Endocrinology, University of Wisconsin, Madison
1964–1971 Senior Research Scientist, The Upjohn Company, Kalamazoo, Michigan, USA
Adjunct Associate Professor, Western Michigan University, Kalamazoo, USA
since 1971 Special Consultant, Schering AG, Berlin, Germany

Fields of Interest:

Fertility control, sexual behavior, reproductive biology, good science.

Fuhrmann, Walter, Prof. Dr. med.

1924 Born in Berlin, Germany, on September 12th
Education in medicine at the Humboldt Universität Berlin and Freie Universität Berlin

Positions held:

1951–1952 Internship, Elizabeth, New Jersey, USA
1953 Fellowship, The Lahey Clinic, Boston, USA
1954–1961 Assistant, Pediatric Clinic of the Free University Berlin
since 1961 Privatdozent in Pediatrics
1963 Research Fellowship in Genetics and Biochemistry, Children's Memorial Hospital, Northwestern School of Medicine, Chicago, USA
1964–1967 Senior Assistant and Privatdozent, Institute of Human Genetics, University of Heidelberg
since 1967 Professor in Human Genetics and Director, Institute of Human Genetics, University of Giessen

Fields of Interest:

Clinical genetics, human developmental genetics, inborn errors of lipid metabolism.

Gibian, Heinz, Dipl. Chem., Dr. Ing., Honorary Professor, Medical Faculty, Free University of Berlin

1916 Born in Munich, Germany, on March 30th
Education in Chemistry at Technische Hochschule München

Positions held:

1942 Research Chemist, Schering AG Berlin
1946 Head of Department of Biochemistry, Schering AG Berlin
1961 Director of Biological Research Departments, Schering AG Berlin
1964 Director of Chemical and Pharmaceutical Research, Schering AG Berlin
1969 Director of Experimental Drug Research, Schering AG Berlin
since 1970 Member of Executive Committee, Pharmaceutical Division, Schering AG Berlin
Chairman of Management Committee Drug Research, Schering AG Berlin
Managing Director, Chemistry, Schering AG Berlin

Fields of Interest:

Biological chemistry in general
Problems of organizing and planning of research

Good, Robert A., M. D. Ph. D.

1922 Born in Crosby, Minnesota, USA, on May 21st
Education in Medicine at the University of Minnesota Medical School and in Biology at the University of Minnesota Graduate School

Positions held:

1944–1945 Teaching Assistant, Department of Anatomy, University of Minnesota
1945–1947 Intern in Pediatrics
1947–1949 Assistant Resident in Pediatrics, University of Minnesota Hospitals
1949–1950 Visiting Investigator, Rockefeller Institute for Medical Research, New York
Assistant Physician to the Hospital of the Rockefeller Institute for Medical Research, New York
since 1950 Attending Pediatrician, Hennepin County General Hospital
1950–1951 Instructor in Pediatrics, University of Minnesota Medical School
1951–1953 Assistant Professor of Pediatrics, University of Minnesota Medical School
1953–1954 Associate Professor of Pediatrics, University of Minnesota Medical School
since 1954 American Legion Memorial Research Professor of Pediatrics, University of Minnesota
since 1960 Consultant on the Pediatrics and Pediatric Contagion Service Hennepin County General Hospital
since 1962 Professor of Microbiology, University of Minnesota
since 1965 Distinguished Service Professor, University of Minnesota
since 1969 Regents' Professor of Pediatrics and Microbiology
since 1970 Professor and Head, Department of Pathology

Fields of Interest:

Immunobiology; immunodeficiency; cellular and genetic engineering.

Graham, Christopher, F., D. Phil.

1940	Born in Drumelzier, Scotland, on September 23rd
	Education in Zoology, Oxford
	Positions held:
1963–1966	Pupil of Dr. J. B. Gurdon, Zoology, Oxford
1966	Pupil of Prof. J. Paul, Biochemistry, Glasgow
1967	Pupil of Dr. D. R. S. Kirby, Zoology, Oxford
	Junior Beit Memorial Fellowship in Medical Research
1968–1970	Sir William Dunn School of Pathology, Oxford
	Medical Research Council Fellowship: Junior Research Fellowship Brasenose College
since 1970	Tutorial Fellow, St. Catherine's College, Oxford; University Lecturer, Zoology Department, Oxford

Field of Interest:

Embryology.

Hill, Miroslav, M. D., Ph. D.

1929 Born in Ostrava, Czechoslovakia, on December 25th

Education in Medicine at the Faculty of Medicine, University of Brno, and in Radiobiology and Molecular Biology at the Institute of Biophysics, Czechoslovak Academy of Sciences, Brno, Czechoslovakia

Positions held:

1953–1958 Instructor in Histology and Embryology, Faculty of Medicine, University of Brno

1959–1962 Junior research worker, Institute of Biophysics, Brno

1962–1963 Research Fellow of the International Atomic Energy Agency (Vienna) at the Laboratory of Biophysics and Radiobiology, Université Libre de Bruxelles, Belgium

1963–1966 Senior research worker at the Institute of Biophysics, Brno

1966–1968 Head of the Department of Cytogenetics, Institute of Biophysics, Brno, Czechoslovakia

1968 Left Czechoslovakia on December 10th

since 1969 Research worker in C.N.R.S., Laboratory of Molecular Biology, Institute Gustave Roussy, Villejuif, France

Fields of Interest:

Intercellular relationships, interactions of foreign genetic material with animal cell, transformation of somatic cell, molecular biology of oncogenic viruses.

Hirt, Bernhard, Ph. D.

1934	Born in Aarau, Switzerland, on October 6th Education in Physics at the University of Bern and later in Molecular Biology at the University of Geneva
	Positions held:
1963	Research Fellow, California Institute of Technology
1964–1969	Swiss Institute for Experimental Cancer Research
1970	Cold Spring Harbor Laboratory, New York
since 1971	Swiss Institute for Experimental Cancer Research, Professor of Biophysics, University of Lausanne

Field of Interest:

Tumour Viruses.

Hofschneider, Peter Hans, Prof. Dr. med. Dr. phil.

Education in Medicine and Biological Sciences at the Universities of Freiburg, Tübingen (Dr. med.), and Heidelberg (Dr. phil.)

Positions held:

1955–1957	Assistant at the university hospitals of Zürich, Basel and Freiburg
1957–1965	Assistant and Dozent, Max-Planck-Institut für Biochemie, Munich
1958, 1964	Visting Scientist at the Laboratoire de Biophysique, Geneva
1966	Scientific Member of the Max-Planck-Institut für Biochemie, Munich Joined appointment: Professor for Biochemistry at the Medical Faculty, University of Munich
since 1972	Director at the new founded Max-Planck-Institut für Biochemie, Martinsried near Munich

Fields of Interest:

Molecular biology of bacteria and viruses.

Hurwitz, Jerard, Ph. D.

1928 Born in New York City, USA, on November 20th
Education in Chemistry at the Indiana University and Biochemistry at the Western Reserve University

Positions held:

1953–1954 American Cancer Society, Postdoctoral Fellow, National Medical Research Council, Mill Hill, London, England
1955–1956 American Cancer Society, Postdoctoral Fellow, National Institutes of Health, Bethesda, Maryland, USA
1956–1958 Instructor of Microbiology, Washington University, St. Louis, Missouri, USA
1958–1960 Assistant Professor of Microbiology, New York University School of Medicine, New York, N. Y. USA
1960–1963 Associate Professor of Microbiology, New York University School of Medicine, New York, N. Y., USA
1963–1965 Professor of Molecular Biology, Albert Einstein College of Medicine, Bronx, New York, USA
since 1965 Professor of Developmental Biology and Cancer, Albert Einstein College of Medicine, Bronx, New York, USA
1969–1970 Sabbatical year, Pasteur Institut, Paris, France

Fields of Interest:

Nucleic acids.

Johnson, Robert T., D. Phil.

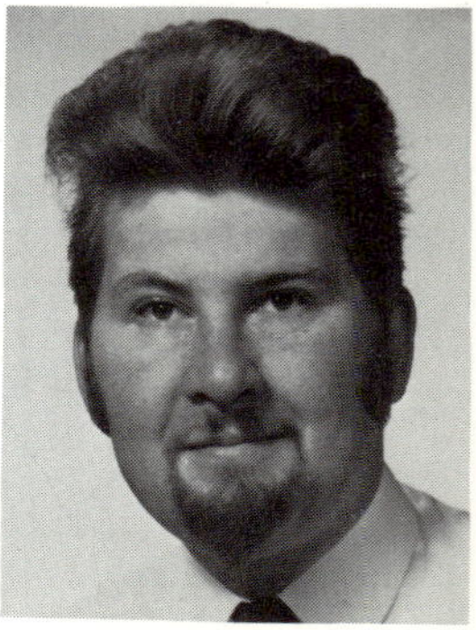

1943 Born in Manchester, Great Britain, on 22 June 1943 Education in Zoology and Cell Biology at the University of Oxford

Positions held:

1968–1970 Research Fellow and Instructor, in Biophysics, University of Colorado Medical Center, Denver, Colorado, U. S. A.
since 1970 Research Assistant, University of Cambridge, Great Britain
since 1971 Research Fellow, Peterhouse, Cambridge

Fields of Interest:

Animal cell fusion and the regulation of cellular reproduction.

Khorana, H. Gobind, Ph. D.

1922	Born in Raipur, India, on January 9th Education (B.Sc. Hons. (I Class) and M.Sc. Hons. (I Class)) at Punjab University, Lahore, India Ph. D. (Organic Chemistry) at University of Liverpool, England, with Prof. A. Robertson, F.R.S.
	Positions held:
1948–1949	Postdoctoral Fellow of Government of India, Federal Institute of Technology, Zürich, Switzerland, with Prof. V. Prelog
1950–1952	Nuffield Fellow, University of Cambridge, England, with Prof. Lord Alexander Todd, F.R.S.
1952–1960	B.C. Research Council, Vancouver, British Columbia, Head – Organic Chemistry Group
1958–1960	Visiting Professor, Rockefeller Institute, New York
1960–1970	Professor and Co-Director, Institute of Enzyme Research, University of Wisconsin, Madison, Wisconsin, USA
1962–1970	Professor, Department of Biochemistry, University of Wisconsin
1964–1970	Conrad A. Elvehjem Professorship in the Life Sciences, University of Wisconsin
1964	Visiting Professor, Stanford University, Palo Alto, California, USA
1966	Visiting Professor, Harvard Medical School, Cambridge, Massachusetts, USA
1967	Honorary Degree, Doctor of Science, University of Chicago
1968	Nobel Prize for Medicine
1969	Honorary Degree, Simon Fraser University, Vancouver, British Columbia
since 1970	Alfred P. Sloan Professor of Biology and Chemistry, Massachusetts Institute of Technology, Cambridge, Massachusetts, USA
1971	Honorary Doctor of Science Degree, University of Liverpool, England

Fields of Interest:

Chemistry and biology of nucleic acids and proteins: chemistry of membranes, phospholipids.

Klein, George, M. D., D. Sc. (Hon.)

1925	Born in Budapest Education in Medicine at Budapest, University and at Karolinska Institutet, Stockholm, Sweden

Positions held:

1945	Instructor in Histology, Budapest University
1946	Instructor in Pathology, Budapest University
1947–1949	Research Fellow, Karolinska Institutet, Stockholm, Sweden
1950	Guest Investigator at the Institute for Cancer Research, Philadelphia, Pa., USA
1950–1957	Assistant Professor of Cell Research
since 1957	Professor of Tumor Biology and Head of the Institute for Tumor Biology, Karolinska Institutet Medical School, Stockholm,Sweden
1961	Visiting Professor, Stanford University, Stanford, Calif., USA
since 1961	Member of the Board of Directors of the Swedish Cancer Society
1966	D. Sc. (Hon.), University of Chicago, Ill., USA

Fields of Interest:

Experimental cell research, cancer research.

Koprowski, Hilary, M. D.

1916	Born in Warsaw, Poland, on December 5th Education in Medicine at the University of Warsaw, Poland

Positions held:

1936–1939	Research Assistant, Department of Experimental Pathology, University of Warsaw, Poland
1940–1944	Research Associate, Yellow Fever Research Service, Rockefeller Foundation and Ministry of Education, Rio de Janeiro, Brazil
1944–1946	Research Associate, Section of Viral and Rickettsial Research, Lederle Laboratories, American Cyanamid Company, Pearl River, New York, USA
1946–1957	Assistant Director, Section of Viral and Rickettsial Research, Lederle Laboratories, American Cyanamid Company, Pearl River, New York, USA
1957 to date	Director, The Wistar Institute of Anatomy and Biology, Philadelphia, Pennsylvania, USA
1957 to date	Wistar Professor of Research Medicine, University of Pennsylvania, Philadelphia, Pennsylvania, USA
1957 to date	Professor of Microbiology, Graduate School of Arts and Sciences, University of Pennsylvania, Philadelphia, Pennsylvania, USA

Fields of Interest:

Biological research with emphasis on cellular biology, virology, tumor research.

Ledoux, Lucien, Ph. D.

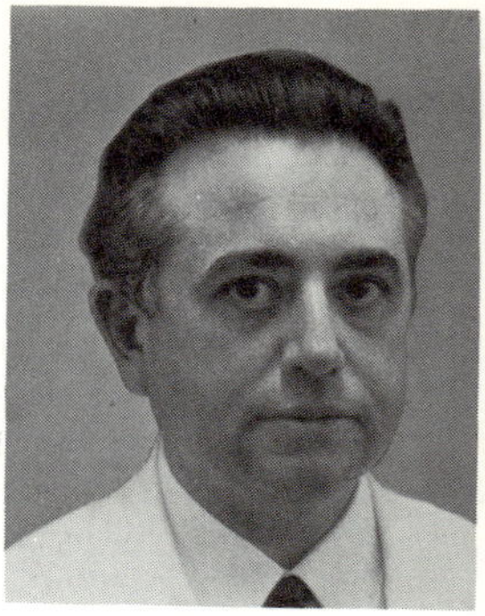

1928 Born in Brussels, Belgium, on October 22th
Education in Chemistry at the University of Brussels
Agregation in Biochemistry

Positions held:

1953–1958 Aspirant, chargé de recherches, chercheur qualifié of the Fonds National de la Recherche Scientifique (lab. Prof. J. Brachet)
since 1958 Head of section (cell biochemistry) at the Centre d'Etudes de l'Energie nucléaire in Mol, Belgium
since 1971 Professor of molecular genetics at the University of Liège, Belgium

Fields of Interest:

Fate and effects of exagenous macromolecules in living systems.

Merril, Carl R., M. D.

1936 Born in New York, N.Y., on December 6th
Education in Chemistry at College of William and Mary, Williamsburg, V.A. (B.S.) and at Georgetown University, School of Medicine, Washington, D.C. (M.D.) USA

Positions held:

1958–1959 Student Fellow, Dept. of Biochemistry, Georgetown University Medical School, Washington, D.C., USA
1960–1962 Student Fellow, Cardio-Pulmonary Physiology Laboratory, Georgetown Univ., School of Medicine, Mt. Alto VA Hospital, Washington, D. C., USA
1962–1963 Intern, U.S. Public Health Service Hospital, Boston, Massachusetts, USA
1963–1969 Research Associate, USPHS, Section on Physical Chemistry, Laboratory of Neurochemistry, NIMH, Bethesda, Md. Research Scientist, Section on Proteins, Laboratory of General and Comparative Biochemistry, Division of Biological and Biochemical Research, NIMH-IRP, Bethesda, Md., USA

Fields of Interest:

Molecular biology: nucleic acid structure and function, subcellular particles, bacteriophages.

Milgrom, Felix, M. D.

1919	Born in Rohatyn, Poland, on October 12th Education in Medicine at the Universities of Lwow, Lublin and Wroclaw, Poland
	Positions held:
1946–1954	Instructor to Professor and Acting Chairman, Department of Microbiology, School of Medicine, University of Wroclaw, Poland
1954–1957	Professor and Chairman of the Department of Microbiology, Silesian University School of Medicine, Zabrze, Poland
1958–1967	Associate Professor to Professor, Department of Bacteriology and Immunology, State University of New York at Buffalo, School of Medicine, Buffalo, New York
since 1967	Professor and Chairman of the Department of Microbiology, State University of New York at Buffalo, School of Medicine, Buffalo, New York

Fields of Interest:

Immunity to infectious diseases, tissue immunology, transplantation immunology.

Radding, Charles Meyer, M. D.

1930 Born in Springfield, Massachusetts, USA on June 18th
Education: A. B. in Biology at Harvard College and M. D. at Harvard Medical School

Positions held:

1956–1957 Intern, Harvard Medical Service, Boston City Hospital
1957–1959 Research Associate, National Heart Institute
1959–1962 Advanced Research Fellow of the American Heart Association, Department of Biochemistry, Stanford University School of Medicine
1962–1965 Assistant Professor of Human Genetics, University of Michigan
1965–1967 Associate Professor of Human Genetics, University of Michigan
1967 Senior Postdoctoral Fellow of the National Science Foundation at the Pasteur Institute, Paris, France
1967–1969 Associate Professor of Medicine, Yale University School of Medicine, New Haven, Connecticut
since 1969 Associate Professor of Medicine; and of Molecular Biophysics and Biochemistry, Yale University School of Medicine

Fields of Interest:

Molecular genetics; genetic recombination, regulation and development.

Ringertz, Nils R., M. D.

1932 Born in Tärnaby, Sweden, on December 16th
Education in Medicine and Cell Biology at Karolinska Institutet, Stockholm

Positions held:

1952–1955 Instructor in Histology at Karolinska Institutet
1956–1960 Research Associate at Institutefor Medical Cell Research and Genetics, Karolinska Institutet
1960–1969 Assistant Professor in Cell Research and Genetics, Karolinska Institutet
1962–1963 Postdoctoral Research Fellow, Dept. of Biochemistry, Tufts University, New England Medical Center, Boston
since 1963 Associate Editor for Experimental Cell Research, Academic Press
since 1969 Associate Professor in Cell Biology, Karolinska Institutet

Fields of Interest:

Regulation of gene expression in eukaryotic cells, quantitative cytochemistry of the cell nucleus.

Ruddle, Frank H., Ph. D.

1929 Born in New Jersey on August 19th
Education at Wayne State University, Detroit, Michigan, USA (B. A., 1953, M. S., 1956), and at the University of California at Berkeley (Ph. D., 1960)

Positions held:

1953–1956 Research Associate Child Research Center of Michigan, Detroit, USA
1958–1960 NIH Fellow, University of California at Berkeley
1960–1961 NIH Postdoctoral Fellow, Dept. of Biochemistry, University of Glasgow, Scotland
1961–1967 Assistant Professor, Yale University, New Haven, Connecticut, USA
1967–1972 Associate Professor, Yale University
1972 Professor, Yale University, New Haven, Connecticut

Fields of Interest:

Cell biology, developmental biology, somatic cell genetics.

Sänger, Heinz Ludwig, Dr. rer. nat.

1928	Born in Grossen Linden, Kreis Giessen/Hessia, Germany, on December 18th Education in Biology at the Justus Liebig-University, Giessen

Positions held:

1960	Research assistant, Institut für Phytopathology, Justus Liebig-University, Giessen
1961–1962	NATO Postdoctoral Research Fellow, Department of Plant Pathology, University of California, Berkeley/Calif., USA
1962–1963	NATO Postdoctoral Research Fellow, Viruslaboratory and Department of Molecular Biology, University of California, Berkeley/Calif., USA
1963–1968	Assistant Professor, Institut für Phytopathology, Justus Liebig-University, Giessen
1969–1971	Privatdozent, Justus Liebig-University, Giessen
since 1971	Professor in Molecular Biology and Virology, Justus Liebig-University, Giessen

Fields of Interest:

Multicomponent plant viruses; defective viruses; infectious nucleic acids.

Schweiger, Hans-Georg, Prof. Dr. med.

1927	Born in Königsberg, on August 21st Education in Medicine and Chemistry at Berlin, Dr. med. (Biochemistry), University of Hamburg

Positions held:

1955–1958	Assistant (Biochemistry), Berlin
1958–1962	Assistant, Max-Planck-Institut für Zellbiologie, Wilhelmshaven
1963–1964	Assistant Professor (Biochemistry), Chicago Medical School
since 1965	Scientific Member, Director, Max-Planck-Institut für Zellbiologie, Wilhelmshaven
1970	Professor of Cell Biology, Berlin (joined appointment)

Fields of Interest:

Biochemical cell biology, nucleocytoplasmic interrelationships, circadian rhythmicity of unicellular organisms.

Siniscalco, Marcello, M. D.

1924	Born in Naples, Italy, on July 31st Education in Medicine at the University of Naples
	Positions held:
1945–1950	Holder of a research-desk at the Marine Biological Station, Naples, Italy
1950–1958	Research Assistant, Department of Genetics, University of Naples, Italy
1952–1953	British Council Scholar, Galton Laboratory, University College London, England
1957–1958	Visiting Lecturer, Galton Laboratory
1958–1962	Assistant Professor of Genetics, University of Naples, Italy
1962–1970	Professor of Human Genetics, Chairman, Department of Human Genetics, University of Leiden, Holland
1968	Visiting Lecturer, Department of Biochemistry, MIT, Cambridge, Massachusetts
1967–1968	Visiting Professor of Genetics, Albert Einstein College of Medicine, New York
1970–1971	Visiting Professor of Genetics, Albert Einstein College of Medicine, New York
since 1970	Professor of Geneticis, Faculty of Science, University of Naples, Italy

Fields of Interest:

Human population genetics, human biochemical genetics, somatic cell genetics.

Sinsheimer, Robert L., Ph. D.

1920	Born in Washington, D. C., USA, on February 5th Education in Quantitative Biology (S. B., 1941) and Biophysics (S. M., 1942; Ph. D. 1948) at Massachusetts Institute of Technology

Positions held:

1942–1946	Staff member, Radiation Laboratory, Massachusetts Institute of Technology
1946–1948	American Cancer Society fellow, MIT
1948–1949	Research Associate, Biology, MIT
1949–1955	Associate Professor of Biophysics, Iowa State College
1955–1957	Professor of Biophysics, Iowa State College
since 1957	Professor of Biophysics, California Institute of Technology
since 1968	Chairman, Division of Biology, California Institute of Technology

Fields of Interest:

Physical and chemical properties of nucleic acids; replication of nucleic acids; bacterial viruses; biological effects of ultraviolet radiation; biological applications of ultraviolet and infrared spectroscopy.

Stent, Gunther S., Ph. D.

1924 Born in Berlin, Germany, on March 28th
Education in Physical Chemistry at the University of Illinois

Positions held:

1948–1950	NRC Merck Fellow, California Institute of Technology
1950–1951	NRC American Cancer Society Fellow, University of Copenhagen
1951–1952	NRC American Cancer Society Fellow, Pasteur Institute, Paris
1952–1959	Assistant Research Biochemist, University of California
1956–1959	Associate Professor of Bacteriology, University of California
since 1959	Professor of Molecular Biology, University of California
1960–1961	NSF Senior Fellow, Universities of Kyoto and of Cambridge
since 1962	External member, Max-Planck-Institute for Molecular Genetics, Berlin
1967–1968	Professor of Arts and Sciences, University of California
1969–1970	Guggenheim Fellow, Harvard Medical School

Fields of Interest:

Molecular genetics, neurobiology.

Trautner, Thomas, Dr. rer. nat.

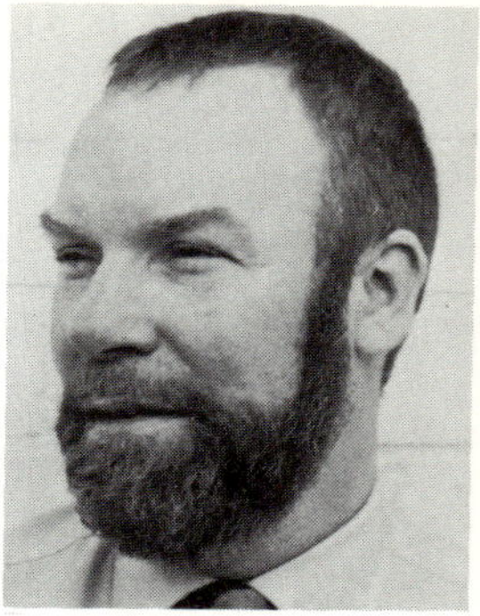

1932	Born in Göttingen, Germany, on April 3rd Education in Biology, Genetics, Biochemistry at Universities in Münster, Göttingen, Urbana (Illinois), Dr. rer. nat. at University of Göttingen
	Positions held:
1957–1964	Postdoctoral Fellow, Assistent, Institut für Genetik, University of Cologne
1960–1961	Postdoctoral Fellow, Department of Biochemistry, Stanford University
1964–1967	Assistant Professor, Dept. of Molecular Biology and Virus Laboratory, University of California, Berkeley, Calif.
1965	Director at Max-Planck-Institut für molekulare Genetik, Berlin
1968	Honorar-Professor, Freie Universität Berlin

Fields of Interest:

Bacteriophage, genetics, genetic recombination, mechanism of transfection and transformation.

Vogel, Friedrich, Prof. Dr. med.

1925	Born in Berlin, Germany, on March 6th Education in Medicine at the Free University of Berlin
	Positions held:
since 1953	Research Associate, Max-Planck-Institut für vergleichende Erbbiologie, Berlin
1957–1962	Also Privat-Dozent in Human Genetics at the Free University of Berlin
since 1962	Full Professor of Human Genetics at the University of Heidelberg

Fields of Interest:

Human genetics, especially mutation, selection in the ABO-system, genetic basis of the normal human EEG.

Woodland, Hugh R., Ph. D.

1943 Born in Surrey, England, on April 11th
Education in Zoology at Oxford University

Positions held:

Doctoral research on Amphibian and Mammalian Development, supervised by Dr. J. B. Gurdon, at Oxford University

since 1969 Continued research on amphibian development, supported by Medical Research Council of Great Britain
Browne Research Fellow, Queen's College, Oxford

Field of Interest:

Molecular biological aspects of development, particularly in Amphibia.

Yarmolinsky, Michael B., Ph. D.

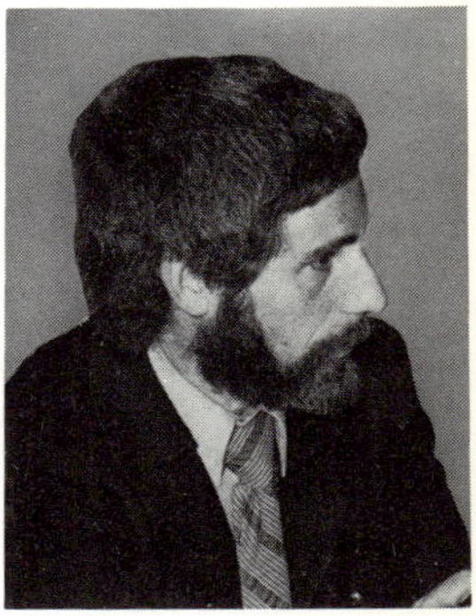

1929 Born in New York City on January 18th
Education in Biology at Harvard College (A. B.) and in Biochemistry at Johns Hopkins University (Ph. D.)

Positions held:

1954–1955 Instructor in Pharmacology, N.Y. University College of Medicine

1955–1958 Officer, U.S. Public Health Service, National Institutes of Health, Bethesda, Maryland

1958–1963 Research Associate-Assistant Professor, Johns Hopkins University, Baltimore, Maryland

1963–1964 National Science Foundation Fellow, Institut Pasteur, Paris

1964–1971 Research Chemist, Section on Microbial Genetics, Laboratory of Molecular Biology, NIAMD, National Institute of Health, Bethesda, Maryland

since 1971 Directeur de Recherche au C.N.R.S., Chief of Microbial Genetics Unit of the Institut de Biologie Moléculaire, Faculté des Sciences, Paris

Fields of Interest:

Molecular biology of temperate bacteriophages.

Zipser, David, Ph. D.

1937 Born in New York City on May 31st
Education in Botany at Cornell University, B. A., and in Biochemistry at Harvard University, Ph. D.

Positions held:

1963–1964 M.R.C. Laboratory, Cambridge, England
1965–1969 Assistant Professor to Associate Professor Columbia University, New York City
since 1970 Senior Staff Investigator, Cold Spring Harbor Laboratory

Fields of Interest:

Molecular biology.

Note: The portraits of Sinsheimer, Khorana, Hurwitz, Duesberg, Klein, Ruddle, and Siniscalco have been printed by the permission of Story Press, Berlin.

Participants

Achtman, M.
c/o Max-Planck-Institut für
Biologie, Abt. Henning
Corrensstr. 38
74 Tübingen, Germany

Adamson, E. D.
Dept. of Zoology
South Parks Road
Oxford, Great Britain

Anders, F.
Genetisches Institut der
Justus-Liebig-Universität
Leihgesterner Weg 112/114
63 Giessen, Germany

Anders, H-P.
Fachbereich Tierexp. Pharmakodynamik
und Chemotherapie der Schering AG

Arber, W.
Biozentrum der Universität Basel
Klingelbergstr. 70
4056 Basel, Switzerland

Axelrod, D.
Div. of Laboratories and Research
New York State Dept. of Health
New Scotland Ave.
Albany, N. Y. 12202, USA

Bauer, H.
Robert-Koch-Institut
Nordufer 20
1 Berlin 65, Germany

Bautz, E. K. F.
Institut für Molekulare Genetik
Berliner Str. 15
69 Heidelberg, Germany

Bautz, F. A.
Institut für Molekulare Genetik
Berliner Str. 15
69 Heidelberg, Germany

Benbow, M.
Division of Biology
California Institute of
Technology
Pasadena, Cal. 91109, USA

Bermek, E.
Max-Planck-Institut für
experimentelle Medizin
Arbeitsgruppe Biochemie
Hermann-Rein-Str. 3
34 Göttingen, Germany

Berndt, M.
Dept. Endokrinpharmakologie
der Schering AG

Bernhard. S.
Scientific Relations der Schering AG

v. Berswordt-Wallrabe, R.
Dept. Endokrinpharmakologie
der Schering AG

Beyersmann, D.
Max-Planck-Institut für
Molekulare Genetik
Ihnestr. 63/73
1 Berlin 33, Germany

Bhargava, A. S.
Fachbereich Tierexp. Pharmakodynamik
und Chemotherapie der Schering AG

Billeter, M.
Universität Zürich
Institut für Molekularbiologie Hönggerberg
8049 Zürich, Switzerland

Bodmer, W. F.
Genetics Laboratory
Dept. of Biochemistry
University of Oxford
South Parks Road
Oxford OX1 3QU, Great Britain

Böhm, R.
Institut für Humangenetik
der Universität Giessen
Ludwigstr. 27
63 Giessen, Germany

Bonhoeffer, F.
Friedrich-Miescher-Laboratorium
der Max-Planck-Gesellschaft
Spemannstr. 37/39
74 Tübingen, Germany

Boon, T.
Rockefeller University
66th St. and York Ave.
New York, N.Y. 10021, USA

Bresch, C.
Institut für Biologie III
der Universität Freiburg
Schänzlestr. 9/11
78 Freiburg/Br., Germany

Brinton, C. C., Jr.
University of Pittsburgh
Dept. of Biophysics
and Microbiology
342 Crawford Hall
Pittsburgh, Pa. 15213, USA

Cassuto, E.
Yale University School of Medicine
Dept. of Internal Medicine
333 Cedar Street
New Haven, Conn. 06510, USA

Chapman, V. M.
Yale University, Dept. of Biology
Kline Biology Tower
New Haven, Conn. 06520, USA

McClearn, G. E.
University of Colorado
Inst. for Behavioral Genetics
Boulder, Col. 80302, USA

Cotae, D.
Max-Planck-Institut für
Molekulare Genetik
Ihnestr. 63/73
1 Berlin 33, Germany

Croce, C. M.
c/o The Wistar Institute
Thirty-Sixth Street at Spruce
Philadelphia, Pa. 19104, USA

Daum, J.
Dept. Mikrobiologische Chemie
der Schering AG

Deák, I. I.
Sir William Dunn School
of Pathology
South Parks Road
Oxford OX1 3RE, Great Britain

Defendi, V.
Imperial Cancer Research Fund
P. O. Box No. 123
Lincoln's Inn Fields
London WC2A 3PX, Great Britain

Dellweg, H.
Inst. f. Gärungstechnologie der
Technischen Universität Berlin
Seestr. 13
1 Berlin 65, Germany

Doerfler, W.
The Wallenberg Laboratory
Uppsala Universitet
Dag Hammarskjöldväg 21
Uppsala, Sweden

Droege, W.
Max-Planck-Institut für Eiweiß-
und Lederforschung, Abt. Hannig
Schillerstr. 46
8 München 2, Germany

Duesberg, P. H.
University of California
Dept. of Molecular Biology
Berkeley, Cal. 94720, USA

Eckardt, F.
Zentralinst. f. Biochemie u. Bio-
physik, Abt. Genetik u. Strahlenbiol.
Habelschwerdter Allee 30
1 Berlin 33, Germany

Ehrenstein, G. von
Max-Planck-Institut f. experimen-
telle Medizin, Abt. Molek. Biologie
Hermann-Rein-Str. 3
34 Göttingen, Germany

El Etreby, M. F. A.
Fachbereich Vorklimische Präparate-
entwicklung der Schering AG
4619 Bergkamen, Germany

Elsner, J.
Dept. Biodynamik der Schering AG

Ericsson, R. J.
Dept. Endokrinpharmakologie
der Schering AG

Feldmann, H.
Institut für Physiol. Chemie
der Universität München
Pettenkoferstr. 14 a
8 München 15, Germany

Fiers, W.
Rijksuniversiteit-Gent
Laboratorium voor
Molekulaire Biologie
K. L. Ledeganckstraat 35
9000 Gent, Belgium

Fischberg, M.
Station de Zoologie Expérimentale
154, route de Malagnou
1224 Chêne-Bougeries (GE)
Switzerland

Fischer, H.
Max-Planck-Institut für
Immunbiologie
Stübeweg 51
78 Freiburg-Zähringen, Germany

Freese, E.
Laboratory of Molecular Biology
National Institute of Neurological
Diseases and Stroke
National Institutes of Health
Building 36, Room 3D02
Bethesda, Md. 20014, USA

Friis, R. R.
University of Southern California
School of Medicine
Dept. of Microbiology
Los Angeles, Cal. 90033, USA

Fuhrmann, W.
Institut für Humangenetik
der Universität Giessen
Ludwigstr. 27
63 Giessen, Germany

Gardner, R. L.
Physiological Laboratory
University of Cambridge
Cambridge CB2 3EG, Great Britain

Gerhards, E.
Forschungsleitung Pharma
Fachbereich Tierexp. Pharmakodynamik
und Chemotheropie
der Schering AG

Gerwin, R.
Max-Planck-Gesellschaft
Residenzstr. 1 a
8 München 1, Germany

Gibian, H.
Forschungsleitung Pharma
Fachbereich Chemie der Schering AG

Gierer, A.
Max-Planck-Institut für Virusforschung
Molekularbiol. Abt.
74 Tübingen, Spemannstr. 35, Germany

Giesbrecht, P.
Robert-Koch-Institut
Nordufer 20
1 Berlin 65, Germany

Glick, J. L.
Associated Biomedic Systems, Inc.
872 Main Street
Buffalo, N. Y. 14202, USA

Goebel, W.
Institut für Mikrobiologie und
Molekularbiologie der
Universität Hohenheim
Kirchnerstr. 30
7 Stuttgart-Hohenheim, Germany

Good, R. A.
University of Minnesota
Research Laboratories of the
Variety Club Heart Hospital
Depts. of Pediatrics and Pathology
Minneapolis, Minn. 55455, USA

Graham, C. F.
Dept. of Zoology
South Parks Road
Oxford OX1 3PS, Great Britain

Günzel, P.
Fachbereich Vorklimische Präparate-
entwicklung der Schering AG
4619 Bergkamen, Germany

Hausen, P.
Max-Planck-Institut für Biologie
Abteilung Beermann
Spemannstr. 34
74 Tübingen, Germany

Hausmann, R.
Institut für Biologie III
der Universität Freiburg
Schänzlestr. 9/11
78 Freiburg/Br., Germany

Heisenberg, M. A.
Max-Planck-Institut für
biologische Kybernetik
Spemannstr. 38
74 Tübingen, Germany

Hennig, W.
Max-Planck-Institut für Biologie
Abteilung Beermann
Spemannstr. 34
74 Tübingen, Germany

Hess, D.
Universität Hohenheim, Lehrstuhl
f. Botan. Entwicklungsphysiologie
Emil-Wolff-Str. 25,
7 Stuttgart 70, Germany

Heumann, W.
Institut für Mikrobiologie der
Universität Erlangen-Nürnberg
Friedrichstr. 33
8520 Erlangen, Germany

Hill, M.
Institut Gustave-Roussy
94 Villejuif, France

Hilz, H.
Inst. für Physiol. Chemie
Martinistr. 52
2 Hamburg 20, Germany

Hirokawa, H.
Max-Planck-Institut für
Molekulare Genetik
Ihnestr. 63/73
1 Berlin 33, Germany

Hirt, B.
Institut Suisse de Recherches
Expérimentales sur le Cancer
Bugnon 21
1011 Lausanne, Switzerland

Hocks, P.
Forschungsleitung Pharma
der Schering AG

Hofschneider, P. H.
Max-Planck-Institut für Biochemie
Goethestr. 31
8 München 2, Germany

Hoppe, G.
Schneebergstr. 69
1 Berlin 46, Germany

Horowski, R.
Dept. Neuropsychopharmakologie
der Schering AG

Hotchkiss, R. D.
Rockefeller University
66th St. and York Ave.
New York, N. Y. 10021, USA

Hurwitz, J.
Albert Einstein College of Medicine
Dept. of Genetics
1300 Morris Park Ave.
New York, N. Y. 10461, USA

Jacob, E.
Max-Planck-Institut für Biochemie
Goethestr. 31
8 München 2, Germany

Jacobson, A.
Max-Planck-Institut für Biochemie
Goethestr. 31
8 München 2, Germany

Jaenisch, R.
Princeton University
Dept. of Biology
Princeton, N. J. 08540, USA

Johnson, R. T.
Dept. of Zoology
University of Cambridge
Cambridge CB2 3EG, Great Britain

Jost, E.
Max-Planck-Institut für
Molekulare Genetik
Ihnestr. 63/73
1 Berlin 33, Germany

Jungblut, P. W.
Max-Planck-Institut für
Zellbiologie
Anton-Dohrn-Weg 59
294 Wilhelmshaven, Germany

Kaerner, H.-Ch.
Max-Planck-Institut für
medizinische Forschung
Jahnstr. 29
69 Heidelberg, Germany

Kano, K.
State University of New York at
Buffalo, School of Medicine
Dept. of Microbiology
335 Sherman Hall
Buffalo, N. Y. 14214, USA

Kaudewitz, F.
Institut für Genetik der
Universität München
Maria-Ward-Str. 1 a
8 München 19, Germany

Keil, T. U.
Max-Planck-Institut
für Biochemie
Goethestr. 31
8 München 2, Germany

Kelly, S.
Dept. of Zoology
South Parks Road
Oxford OX1 3PS, Great Britain

Khorana, H. G.
Massachusetts Institute of Technology
Depts. of Biology and Chemistry
77 Massachusetts Ave.
Cambridge, Mass. 02139, USA

Klein, G.
Dept. of Tumor Biology
Karolinska Institutet
104 01 Stockholm, Sweden

Klenow, H.
Universitetets
Biokemiske Institut B
Juliane Maries Vej 30[2]
2100 København Ø, Denmark

Klingmüller, W.
Institut für Genetik der
Universität München
Maria-Ward-Str. 1 a
8 München 19, Germany

Klotz, G.
Max-Planck-Institut für
Molekulare Genetik
Ihnestr. 63/73
1 Berlin 33, Germany

Kluge, F.
Max-Planck-Institut für Biochemie
Goethestr. 31
8 München 2, Germany

Knippers, R.
Friedrich-Miescher-Laboratorium
der Max-Planck-Gesellschaft
Spemannstr. 37/39
74 Tübingen, Germany

Koch, M. A.
Max-Planck-Institut für
Virusforschung
Spemannstr. 37/39
74 Tübingen, Germany

Konings, R. N. H.
c/o Max-Planck-Institut
für Biochemie, Goethestr. 31
8 München 2, Germany

Koprowski, H.
The Wistar Institute
Thirty-Sixth Street at Spruce
Philadelphia, Pa. 19104, USA

Korte, R.
Fachbereich Vorklinische Präparate-
entwicklung der Schering AG
4619 Bergkamen, Germany

Kroeger, H.
Institut für Genetik der
Universität des Saarlandes
66 Saarbrücken, Germany

Kröger, H.
Robert-Koch-Institut
Nordufer 20
1 Berlin 65, Germany

Krone, W.
Institut für Humangenetik und
Anthropologie der Universität
Albertstr. 11
78 Freiburg/Br., Germany

Kühl, P. W.
Max-Planck-Institut für
Biochemie
Goethestr. 31
8 München 2, Germany

Laudahn, G.
Forschungsleitung Pharma
Fachbereich Klinische Forschung
der Schering AG

Ledoux, L.
Centre d'Etudes de
l'Energie Nucleaire
2400 Mol, Belgium

Lewandowski, L. J.
University of California
Virus Laboratory
Berkeley, Cal. 94720, USA

Lezius, A.
Max-Planck-Institut für
experimentelle Medizin
Chemische Abteilung
Hermann-Rein-Str. 3
34 Göttingen, Germany

Loge, O.
Fachbereich Tierexp. Pharmakodynamik
und Chemotherapie der Schering AG

Losert, W.
Fachbereich Tierexp. Pharmakodynamik
und Chemotherapie der Schering AG

Lutz, D.
Robert-Koch-Institut
Nordufer 20
1 Berlin 65, Germany

Marunouchi, T.
Max-Planck-Institut für
Molekulare Genetik
Ihnestr. 63/73
1 Berlin 33, Germany

Mehring, M.
Dept. Endokrinpharmakologie
der Schering AG

Melchers, F.
Institut für Immunologie Basel
487, Grenzacherstraße
4058 Basel, Switzerland

Merril, C. R.
Laboratory of General and
Comparative Biochemistry
National Institute of Mental
Health, National Institutes
of Health
Building 36, Room 3A15
Bethesda, Md. 20014, USA

Messer, W.
Max-Planck-Institut für
Molekulare Genetik
Ihnestr. 63/73
1 Berlin 33, Germany

Messing, J.
Max-Planck-Institut für Biochemie
Goethestr. 31
8 München 2, Germany

Meyer, A.
Dept. Biodynamik
der Schering AG

Migeon, B. R.
The Johns Hopkins Hospital
Baltimore, Md. 21205, USA

Milgrom, F.
State University of New York at
Buffalo, School of Medicine
Dept. of Microbiology
210 Sherman Hall
Buffalo, N. Y. 14214, USA

Mizell, M.
The Chapman H. Hyams III
Laboratory of Tumor Cell Biology
Tulane University
New Orleans, Louisiana 70113, USA

Müller-Hill, B.
Institut für Genetik der
Universität Köln
Weyertal 121
5 Köln 41, Germany

Munyon, W.
Roswell Park Memorial Institute
666 Elm St.
Buffalo, N. Y. 14203, USA

Munz, E.
Fachbereich Vorklinische Präparate-
entwicklung der Schering AG
4619 Bergkamen, Germany

Neubert, D.
Pharmakolog. Institut der
Freien Universität Berlin
Thielallee 69/73
1 Berlin 33, Germany

Neumann, F.
Dept. Endokrinpharmakologie
der Schering AG

Nishino, Y.
Dept. Endokrinpharmakologie
der Schering AG

Park, B. H.
University of Minnesota
Research Laboratories of the
Variety Club Heart Hospital
Dept. of Pediatrics
Minneapolis, Minn. 55455, USA

Passarge, E.
Institut für Humangenetik
der Universität Hamburg
Abt. Cytogenetik u. Klin. Genetik
Martinistr. 52
2 Hamburg 20, Germany

Pörschke, D.
Max-Planck-Institut für
Biophysikalische Chemie
Am Faßberg
34 Göttingen, Germany

Portman, R.
Max-Planck-Institut
für Biochemie
Goethestr. 31
8 München 2, Germany

Propping, P.
Institut für Anthropologie und
Humangenetik der Universität
Mönchhofstr. 15 A
69 Heidelberg, Germany

Radding, C. M.
Yale University School of Medicine
Dept. of Internal Medicine
333 Cedar Street
New Haven, Conn. 06510, USA

Rajewsky, K.
Institut für Genetik der
Universität Köln
Weyertal 121
5 Köln 41, Germany

Ringertz, N. R.
Karolinska Institutet
Institute for Medical Cell
Research and Genetics
104 01 Stockholm 60, Sweden

Rörsch, A.
Medical Biological Laboratory
P. O. Box 45
Rijswijk 2100, Netherlands

Roland, H.
Dept. Endokrinpharmakologie
der Schering AG

Ruddle, F. H.
Yale University, Dept. of Biology
Kline Biology Tower
New Haven, Conn. 06520, USA

Russo, E.
Institut für Biologie III
der Universität Freiburg
Schänzlestr. 9/11
78 Freiburg/Br., Germany

Saar, U.
Fachbereich Tierexp. Pharmakodynamik
und Chemotherapie der Schering AG

Sänger, H. L.
Arbeitsgruppe Pflanzenvirologie
Justus-Liebig-Universität
63 Gießen, Germany

Sauer, G.
Institut für Virusforschung
Deutsches Krebsforschungszentrum
Thibautstr. 3
69 Heidelberg, Germany

Schaller, H.
Max-Planck-Institut für Virus-
forschung, Biochemische Abteilung
Spemannstr. 35
74 Tübingen, Germany

Schillinger, E.
Fachbereich Tierexp. Pharmakodynamik
und Chemotherapie der Schering AG

Schroeder, T.
Institut für Anthropologie und
Humangenetik der Universität
Mönchhofstr. 15 A
69 Heidelberg, Germany

Schützel, H.
Dept. Biodynamik
der Schering AG

Schuster, H.
Max-Planck-Institut für
Molekulare Genetik
Ihnestr. 63/73
1 Berlin 33, Germany

Schweiger, H. G.
Max-Planck-Institut für Zellbiologie
Anton-Dohrn-Weg 59
294 Wilhelmshaven, Germany

Schweiger, M.
Max-Planck-Institut für
Molekulare Genetik
Ihnestr. 63/73
1 Berlin 33, Germany

Siegert, R.
Max-Planck-Institut für Biochemie
Goethestr. 31
8 München 2, Germany

Siewert, G.
Dept. Mikrobiologische Chemie
der Schering AG

Siniscalco, M.
Albert Einstein College of Medicine
Yeshiva University, Dept. of Genetics
Eastchester Road and
Morris Park Avenue
Bronx, N. Y. 10461, USA

Sinsheimer, R. L.
California Institute of Technology
Division of Biology
Pasadena, Cal. 91109, USA

Spatz, H. Ch.
Max-Planck-Institut für
biologische Kybernetik
c/o Dr. K. G. Götz
Spemannstr. 38
74 Tübingen, Germany

Starlinger, P.
Institut für Genetik
der Universität Köln
Weyertal 121
5 Köln 41, Germany

Staudenbauer, W.
Max-Planck-Institut für
Biochemie
Goethestr. 31
8 München 2, Germany

Steinbeck, H.
Dept. Endokrinpharmakologie
der Schering AG

Stent, G. S.
University of California
Dept. of Molecular Biology
Berkeley, Cal. 94720, USA

Stern, H.
c/o University of Copenhagen
Institute of Genetics
Øster Farimagsgade 2 A
1353 Copenhagen K., Denmark

v. Szczepanski, Ch.
Dept. Arzneimittelchemie
der Schering AG

Tarkowski, A. K.
University of Warsaw
Dept. of Embryology
Krakowskie Przedmieście 26/28
Warsaw 64, Poland

Tiedemann, H.
Physiol.-chem. Institut der
Freien Universität Berlin
Arnimallee 22
1 Berlin 33, Germany

Töpert, M.
Fachbereich Tierexp. Pharmakodynamik
und Chemotherapie der Schering AG

Traub, P.
Max-Planch-Insitut für Zellbiologie
294 Wilhelmshaven, Germany

Trautner, T. A.
Max-Planck-Institut für
Molekulare Genetik
Ihnestr. 63/73
1 Berlin 33, Germany

Vielmetter, W.
Institut für Genetik
der Universität Köln
Weyertal 121
5 Köln 41, Germany

Vogel, F.
Institut für Anthropologie und
Humangenetik der Universität
Mönchhofstr. 15 A
69 Heidelberg, Germany

Vorbrüggen, H.
Dept. Allgem. Synthetische Chemie
der Schering AG

Wecker, E.
Institut für Virologie der
Universität Würzburg
Versbacher Landstr. 7
87 Würzburg, Germany

Wenzel, G.
Max-Planck-Institut für Züchtungs-
forschung (Erwin-Baur-Institut)
5 Köln 30, Germany

Wiener, F.
Dept. of Tumor Biology
Karolinska Institutet
104 01 Stockholm, Sweden

Wittmann, H. G.
Max-Planck-Institut für
Molekulare Genetik
Ihnestr. 63/73
1 Berlin 33, Germany

Woodland, H. R.
Dept. of Zoology
South Parks Road
Oxford OX1 3PS, Great Britain

Yarmolinsky, M.
Institut de Biologie Moléculaire
de la Faculté des Sciences de Paris
9, Quai Saint-Bernard
Paris 5, France

Zabel, I.
Sektion Pharmakol. Charakterisierung
der Schering AG

Zillig, W.
Max-Planck-Institut für Biochemie
Goethestr. 31
8 München 2, Germany

Zipser, D.
Cold Spring Harbor Laboratory
P. O. Box 100
Cold Spring Harbor
N. Y. 11724, USA

Name Index

» Advances in the Biosciences

volume 9

International Conference on Prostaglandins, September 1972

Editor: G. Raspé, Associate Editor: S. Bernhard. This volume contains about 130 papers read at the Schering-Workshop in September 1972, in Vienna. The volume will be published in March 1973

ISBN 3 528 07691 7

volume 8

Schering Workshop on Mechanisms and Prospects of Genetic Exchange December 1973

Editor: G. Raspé, Associate Editor: S. Bernhard. With 155 ill. – New York/Braunschweig: Vieweg 1972, 486 pages.

ISBN 3 528 07690 9

volume 7

Schering Workshop on Steroid Hormone 'Receptors'

Editor: G. Raspé, Associate Editor: S. Bernhard. With 365 ill. – New York/Braunschweig: Vieweg 1971. VIII, 420 pages.

ISBN 3 528 07689 5

volume 6

Schering Workshop on Intrinsic and Extrinsic Factors in Early Mammalian Development, April 1970

Editor: G. Raspé, Associate Editor: S. Bernhard. With 347 ill. – New York/Braunschweig: Vieweg 1971. VIII, 653 pages.

ISBN 3 528 07688 7

volume 5

Schering Workshop on Pharmacokinetics, May 1969

Editor: G. Raspé, Associate Editor: S. Bernhard. With 183 ill. – New York/Braunschweig: Vieweg 1970. VIII, 285 pages.

ISBN 3 528 07687 9

» *Advances in the Biosciences*

volume 4

Schering Symposium on Mechanisms Involved in Conception March 1969

Editor: G. Raspé, Associate Editor: S. Bernhard. With 289 ill. — New York/Braunschweig: Vieweg 1970. VIII, 471 pages.
ISBN 3 528 07686 0

volume 3

Schering Workshop on Steroid Metabolism „In vitro versus in vivo" December 1968

Editor: G. Raspé, Associate Editors: M. Friedrichs, S. Bernhard. With 139 ill. — New York/Braunschweig: Vieweg 1969. VI, 217 pages.
ISBN 3 528 07685 2

volume 2

Schering Symposium on Biodynamics and Mechanism of Action of Steroid Hormones, March 1968

Editor: G. Raspé, Associate Editor: S. Bernhard. With 213 ill. — New York/Braunschweig: Vieweg 1969. VI, 353 pages.
ISBN 3 528 07681 x

volume 1

Schering Symposium on Endocrinology, May 1967

Editor: G. Raspé, Associate Editor: S. Bernhard. With 235 ill. — New York/Braunschweig: Vieweg 1969. VII, 322 pages.
ISBN 3 528 07678 1